AF389083

Artisanal and Small-Scale Gold Mining (ASGM) Related Environmental and Health Problems

Artisanal and Small-Scale Gold Mining (ASGM) Related Environmental and Health Problems

Editors

Masayuki Sakakibara
Win Thiri Kyaw
José Luis Rivera Parra

MDPI • Basel • Beijing • Wuhan • Barcelona • Belgrade • Manchester • Tokyo • Cluj • Tianjin

Editors

Masayuki Sakakibara
Research Institute for
Humanity and Nature
Kyoto, Japan

Win Thiri Kyaw
Research Institute for
Humanity and Nature
Kyoto, Japan

José Luis Rivera Parra
Escuela Politécnica Nacional
Quito, Ecuador

Editorial Office
MDPI
St. Alban-Anlage 66
4052 Basel, Switzerland

This is a reprint of articles from the Special Issue published online in the open access journal *International Journal of Environmental Research and Public Health* (ISSN 1660-4601) (available at: https://www.mdpi.com/journal/ijerph/special_issues/Artisanal_Gold_Mining_Environmental_Health_Problems).

For citation purposes, cite each article independently as indicated on the article page online and as indicated below:

LastName, A.A.; LastName, B.B.; LastName, C.C. Article Title. *Journal Name* **Year**, *Volume Number*, Page Range.

ISBN 978-3-0365-6870-6 (Hbk)
ISBN 978-3-0365-6871-3 (PDF)

Cover image courtesy of Xiaoxu Kuang

Contents

About the Editors

Masayuki Sakakibara

Masayuki Sakakibara (Professor) is an earth scientist with multidisciplinary backgrounds: Geology, Petrology, Astrobiology, Geochemistry, Medical Geology, Geoengineering, and Remediation Engineering. He is currently working as the project leader of the SRIREP Project (co-creation of Sustainable Regional Innovation for Reducing Risk of High-impact Environmental Pollution) at the Research Institute for Humanity and Nature, Kyoto, Japan, and as a professor at the Faculty of Collaborative Regional Innovation and Graduate School of Science and Engineering, Ehime University. His interest in environmental pollution led him to conduct intensive fieldworks and activities to reduce mercury pollution and poverty problems in artisanal and small-scale gold mining (ASGM) areas in Indonesia and Myanmar. He has organized international conferences and seminars, such as Transdisciplinary Research on Environmental Problems in Southeast Asia (TREPSEA) and Transdisciplinary Research and Practice for Reducing Environmental Problems (TRPNEP), which focus on transdisciplinary research and practice, and has developed various regional innovations for reductions in the environmental pollution levels of ASEAN countries.

Win Thiri Kyaw

Win Thiri Kyaw (Dr.) is a medical doctor and researcher in Medical Science, Neurology, and Clinical Pharmacology. She is currently working as the subleader of the SRIREP Project (Co-creation of Sustainable Regional Innovation for Reducing Risk of High-impact Environmental Pollution) in the Research Institute for Humanity and Nature, Kyoto, Japan, conducting environmental and health impact assessments of ASGM activities in Mandalay Region, Myanmar, and Indonesia. Her research interest is in promoting the health of ASGM communities, which drives her to collaborate with researchers from various backgrounds, key stakeholders, and communities through mercury-free society networks, reducing environmental problems and promoting health in Myanmar and Indonesia.

José Luis Rivera Parra

José Luis Rivera Parra (Dr.) is Chair of the Biology Department at Escuela Politécnica Nacional (EPN) in Ecuador. In this role, he is in charge of coordinating the research and other academic duties of several researchers in a wide range of disciplines within the biological sciences. Moreover, he is currently President of the Committee for the Sustainability of the Campus, where he coordinates, at the institutional level, different processes to improve the environmental performance of EPN, including use of resources, reducing carbon footprint, and conducting research for sustainability. He is also an active researcher focused on studying the effects of extractive industries (mining and petroleum production) on the natural habitat at various scales.

Preface to "Artisanal and Small-Scale Gold Mining (ASGM) Related Environmental and Health Problems"

Artisanal and small-scale gold mining (ASGM) activity has been undertaken in developing countries for socio-economic reasons; however, it accounts for the largest anthropogenic source of mercury (Hg) emission, releasing over 700 metric tons into the atmosphere, in addition to approximately 800 metric tons into land and water annually, exposing both ASGM miners and nearby communities to the mercury. Moreover, heavy metals other than Hg have been released into the environment as a result of the ASGM process. Therefore, ASGM causes environmental disruption and health problems not only in ASGM communities but across the whole nation too because of the risks it poses.

Several studies related to Hg pollution caused by ASGM activity and its impact on the environment and health have been published; yet, there is a need to explore the hazards of ASGM due to the nature of this global problem.

Therefore, we initiated this Special Issue to explore research related to the environmental and health problems posed by ASGM across the world, as well as its economic impacts and the proposed sustainable solutions to ASGM problems. A total of 20 papers are published and introduced in this book, covering regions including Southeast Asia, South America, and South Africa, in addition to papers covering more general ASGM issues.

Southeast Asia

Indonesia is known as the country with the most prevalent ASGM activities in Southeast Asia (SEA). The first paper describes the ASGM profile of Indonesia in a narrative literature review, including the Hg problem, occupational hazards, the environmental and health impacts caused by ASGM, and recommendations on the management of the ASGM sector of Indonesia.

The second and third papers investigate the transformations of roving camp-type ASGM activities in Katingain Regency, Central Kalimantan Province, using the Sentinel-1 time-series dataset, with the results are expected to contribute to environmental governance actions.

Additionally, this Special Issue included seven papers on research conducted using multiple approaches to analyze ASGM in Gorontalo Province, Indonesia. Papers four and five assess the transformation of camp-type ASGM in Bone Bolango Regency using remote sensing imagery and field surveys. These findings shed light on the community's socio-environmental pollution, allowing for better environmental governance. According to the findings of paper six's household survey of rural villagers residing close to ASGM in Bone Bolango Regency, diverse choices about work prospects are found to be useful in delinking reliance on ASGM, decreasing health hazards and improving the rural villagers' livelihoods. Paper seven reports on river basin sediment contamination with toxic elements at ASGM sites, such as Hg, zinc (Zn), lead (Pb), and arsenic (As). Paper eight suggests *P. vittate* as a bioindicator for assessing the environmental pollution caused by Pb and Hg. Moreover, paper nine examines the atmospheric Hg attachment to tree bark and finds that the tree types employed in the study may be used as biomonitors of atmospheric Hg pollution in ASGM regions. Finally, paper 10 explores geomedical science using geogenic samples, hair samples, and health evaluations, and finds that geogenic concentrations of specific components match their accumulation.

Myanmar is also a SEA country with abundant ASGM activities and high Hg usage due to poor national ASGM sector management and little environmental and health research, aggravating the problems. Paper 11 presents a case study conducted during the COVID-19 pandemic, providing the first transdisciplinary online health assessment of the ASGM community of Mandalay Region, Myanmar, which is suggested as an effective regular long-term health assessment of the residents following the physical health surveys in ASGM areas.

Paper 12 reviews the Hg pollution in ASGM areas of Myanmar and some other SEA countries, exploring the severe Hg contamination characteristic of the ASGM process, specifically the gold amalgamation stage. Paper 13 summarizes the ASGM profiles of SEA countries and the consequent environmental, community health, and socioeconomic problems, introducing a sustainable approach to alleviate severe ASGM issues and improve environmental governance and community health in ASGM communities in SEA.

South America

Four studies focused on various regions of South America—Ecuador, Brazil, and Peru—are included in this Special Issue. Paper 14 investigates the possible composition and distribution of toxic element contamination in various metallic deposits, primarily from ASGM districts throughout Ecuador and a major oilfield in the Ecuadorian Amazon basin, as well as how mismanagement of local mines and petroleum exploitation projects results in environmental degradation. Paper 15 evaluates the sustainability of cocoa growing in the mining district of Ponce Enriquez, Ecuador, in connection to mining operations, which may pose health concerns to humans and the environment.

Paper 16 proposes a methodology to quantify the average economic health impact of the extraction of gold by ASGM in Brazil, converting it into monetary losses. This method can assess ASGM's environmental impact.

Paper 17 analyzes the amount of methylmercury that people in the headwaters of the Manu River and Manu National Park in southwestern Peru are exposed to. The paper also looks into the link between people's exposure to methylmercury and the functioning of their brains.

South Africa

Paper 18 represents a case report derived from a cross-sectional study of Pb exposure in incarcerated juveniles in greater Johannesburg, South Africa, and argues that illegal gold mining has a negative impact on the livelihoods of young males, with particular attention paid to education, health, and violent behavior.

General

Paper 19 is a bibliometric analysis on artisanal and small-scale mining (ASM), showing that the current research from 2010 onwards has focused on (i) the social factors affecting ASM, (ii) the effects of ASM on the environment, (iii) Hg contamination and its effect on health and the environment, and (iv) ASM as a livelihood. Ecotourism in artisanal mining regions has also become a major trend in the last decade, promoting sustainable development as well as the preservation and utilization of geological and mining heritage. In paper 20, which is a systematic review, the authors argue that occupational Hg exposure seriously affects miners' health and wellbeing, possibly resulting in neuro-psychological disorders, including ataxia, tremor, or memory problems; however, the major reported symptoms of the previous studies were largely unspecific, such as hair loss or pain.

To summarize, these Special Issue studies posit that: (1) ASGM activity using toxic elements, such as Hg and cyanide due to the lack of legal enforcement on ASGM sector, is still widely practiced as an alternative livelihood in rural areas in many countries around the world; (2) ASGM has a negative impact on environmental ecosystems, causes occupational health problems for miners, and causes chronic health disorders in both ASGM communities and those living far from ASGM areas. However, some studies explore the specific potential bioindicators of environmental damage caused by ASGM to suggest sustainable solutions to its severe environmental and social problems.

We are grateful to all the authors for their important contributions, as well as Darija Žilić at MDPI for facilitating the publication of this Special Issue. Our thanks go out to everyone who took part in the studies and to the workplaces that allowed us to study.

Masayuki Sakakibara, Win Thiri Kyaw, and José Luis Rivera Parra
Editors

International Journal of
Environmental Research and Public Health

MDPI

Review

Indonesian Artisanal and Small-Scale Gold Mining—A Narrative Literature Review

Ami A. Meutia [1,*], Royke Lumowa [2] and Masayuki Sakakibara [1,3,4]

[1] Research Institute for Humanity and Nature, Kyoto 603-8047, Japan; sakaki@chikyu.ac.jp
[2] School of Environmental Sciences, Universitas Indonesia, Jakarta 10430, Indonesia; roykevanlumowa@gmail.com
[3] Department of Earth Science, Graduate School of Science and Engineering, Ehime University, Matsuyama 790-8577, Japan
[4] Faculty of Collaborative Regional Innovation, Ehime University, Matsuyama 790-8577, Japan
*   Correspondence: ami.meutia@chikyu.ac.jp; Tel.: +81-75-707-2390

Abstract: Indonesia is host to a long history of gold mining and is responsible for a significant contribution to world gold production. This is true not only with regard to large gold mining companies but also to small-scale mining groups comprised of people and enterprises that participate in the gold industry of Indonesia. More than two thousand gold mining locations exist in present day Indonesia. Artisanal and small-scale gold mining (ASGM) sites are spread out across thirty provinces in Indonesia, and have provided work opportunities and income for more than two million people. However, the majority of ASGM activities use rudimentary technologies that have serious impacts upon the environment, public health, and miners' safety, which in turn generate socio-economic impacts for people residing around the mine sites. Moreover, many ASGMs are not licensed and operate illegally, meaning that they are immune to governmental regulation, and do not provide income to the regions and states via taxes. The possibility for more prudent management of ASGM operations could become a reality with the involvement and cooperation of all relevant parties, especially communities, local government, police, and NGOs.

Keywords: Indonesian gold mining; ASGM; illegal mining; environmental and health impact

Citation: Meutia, A.A.; Lumowa, R.; Sakakibara, M. Indonesian Artisanal and Small-Scale Gold Mining—A Narrative Literature Review. *Int. J. Environ. Res. Public Health* **2022**, *19*, 3955. https://doi.org/10.3390/ijerph19073955

Academic Editor: Paul B. Tchounwou

Received: 19 February 2022
Accepted: 22 March 2022
Published: 26 March 2022

Publisher's Note: MDPI stays neutral with regard to jurisdictional claims in published maps and institutional affiliations.

1. Introduction

Community mining has been practiced throughout Indonesia for hundreds of years and is facing an even greater increase in activity in the present day. Indonesia is included in the top ten gold producers worldwide (with production in December 2020 alone reaching 130 tons) [1,2] and is host to large number of artisanal and small-scale gold mining (ASGMs) businesses and operational facilities. The definition offered by Wiriosudarmo (1999) in Prabawa (2020) [3] states that ASGM can be interpreted as any mining operation whose investors and participants are common people or belong to the local community themselves. Simultaneously, the Ministry of Environment and Forestry (KLHK) regulation defines ASGM as gold mining which is conducted by individual miners or small enterprises with limited capital investment and production. ASGM actors generally operate without a license and exploit marginal gold reserves located in remote areas with hard-to-reach access, such as protected or preserved forests, or even conservation areas. In some regions, however, the activities of ASGMs have also been found to be conducted in the middle of residential areas [4].

There are two types of ASGMs in Indonesia: one is licensed ASGM and the other is unlicensed ASGM. Most ASGM activity in Indonesia still operates illegally, because they do not possess a license permit from the government. Illegal ASGM is considered detrimental to the state because they are unlicensed, do not pay royalties, cannot be regulated, and contribute to damage to the environment, as well as adverse health impacts caused by the use of mercury.

The existence of ASGM also has an impact on micro-businesses' development. ASGM activities are considered as occupying the informal sector, and as a means to provide job opportunities and sources of income [5–8], especially in light of the current COVID-19 pandemic [9,10]. At present, approximately two million people depend on ASGM activities for their living expenses, spread out across thousands ASGM locations in various regions of Indonesia [9]. The Ministry of Environment and Forestry states that such mining activities do not require special training. Many rural communities choose to make a living as miners or they combine farming and mining in order to subsidize their income. It is difficult for miners who earn money from processed gold taken from mining activities to move to other livelihoods. Such mining practices are becoming more widespread, strengthening in line with democratization and improvements in human rights awareness [11].

Most of the ASGMs in Indonesia use a traditional method utilizing mercury amalgamation in the gold extraction process. However, the use of toxic mercury results in severe environmental and health problems. Many alternatives to the harmful mercury amalgamation have already been studied [12].

Sustainable development in gold mining has received a great deal of attention in the last decade, including the legitimate Community Mining Area/WPR (*Wilayah Pertambangan Rakyat*) promoted by the government. However, implementing sustainable development in this industry has become a complex dilemma. Research on the potential for other types of business income is needed to ensure a legal, but sound livelihood for the miners. In order to minimize the amount of illegal ASGM activity, it is considered effective to organize mining actors into co-operatives and provide financial assistance. This would involve supporting them in switching to other types of businesses, whilst also providing environmental education with regard to the negative effects of mercury use and adapting appropriate technology to be used in its place.

In this article, we will set out a review of ASGM in Indonesia based on the papers that have been published in Indonesia and other countries regarding ASGM in Indonesia. In the last section, the author will propose that the problem of illegal mining and the dangers of mercury use that accompanies it can be resolved by the various parties concerned.

2. History, Location and Production

Indonesia has a very long history of gold mining (Table 1).

For quite some time, many local people have been engaged with gold mining in Indonesia in line with customary law; that being that rights related to gold mining are combined with the rights to possess land. Historically, inhabitants of the land offered payments to the chiefs, who provided permits to their communities to cultivate the land and develop forests. Gold mining rights took the same form as this. However, it was the 1899 Dutch East Indies Law that denied these customary mining rights (Table 1). This made it possible for European companies to operate by establishing a mining concession, proclaiming that the rights of landowners do not extend to the mineral reserves.

The customary idea that land right holders also have mining rights was continuously denied even after independence. For example, the 1967 Basic Mining Law stipulates that all underground reserves are controlled by the state (Table 1). ASGM managed by local communities, formed of small, mutually supported businesses that use simple tools for their livelihood, are stipulated to only gain rights to proceed with their activities upon being granted a license from the government. The dispute which has arisen between the customary idea that land right holders can mine and the national law that mining is allowed only when in possession of a government-issued license persists to the present day. Even now, a large number of ASGM actors are not licensed, but they believe that mining rights have been customarily granted. As such, this disagreement could be described as being at the heart of the widespread problem of illegal mining conducted by many ASGM actors.

Table 1. History of Active Mining and Law in Indonesia.

Year	Location and mining operation/activity	Law
8th century AD (Hindu Period)	* West Kalimantan [13,14] " Sumatera * Java	
1669–1928	* West Sumatera: Salido Ketek	-
1760–1880	* West Kalimantan: Landak (China District) by Chinese immigrant [15]	-
1850–1899	* North Sulawesi: Bukit Mesel (1850) *Bengkulu: Lebong (1896) by Lebong Goud Syndicaat	* 1899: A mining law called the Indische Mijnwet Staatsblad was issued [16].
1910–1939	* Bengkulu: Simau (1910), Salida (1914), Lebong Simpang (1921) and Tambang Sawah (1923) [17,18] * West Sumatera: Tambang Manggani (1913), Belimbing, Gunung Arum (1935), Muarasipongi (1936) * West Java: Cikotok (1940)	-
1940–1941	* Riau: Logas, Kuantan Singingi [19] * North Sulawesi: Tapaibekin & Ratatotok (1940) [20] * Aceh: Meulaboh (1941)	-
1950–1959	Most of the people's mining is active: * Bengkulu [21] * Kalimantan * North Sulawesi Mining company (State Company Antam): * South Banten: Cikotok * Riau: Logas	* 1959: Law Number 10 of 1959 was issued regarding the cancellation of mining rights.
1960–1967	-	* 1960: Indische Mijnwet was replaced by Government Regulation in Lieu of Law Number 37 of 1960 concerning Mining [22]. * 1967: the Foreign Capital Investment Law was introduced, together with the New Mining Law Number 11.
1972–1989	* Papua: Ertsberg, (1972 & 1980), Grasberg (1989) * Banten: Cirotan, (1978) * Central Kalimantan Ampalit (1988)	-
1991–1998	* North Maluku: Lerokis (1991), Kali Kuning (1994) * East Kalimantan: Kelian (1992) * Central Kalimantan: Gunung Muro (1994), Mirah (1995) * West Java: Pongkor (1994) * North Sulawesi: Messel (1995) * North Sumatera: Martabe (1997) * NTB: Batu Hijau (1998) [23]	* 1992: Government Regulation Number 79 of 1992 concerning mining permits from the Minister of Mines was issued [24].
2000–2020	* North Maluku: Gosowong (2005), Buru Island [25,26] * East Java: Tambang Tujuh Bukit, Banyuwangi (2017). * West Nusa Tenggara: Prabu Village, Lombok Regency (2011) [27]	* 2001 & 2004: Government Regulation Number 75 of 2001 and Law Number 32 of 2004 on local government explaining the mining management authority is on the local government. * 2009: Law number 4 of 2009 was issued regarding mining [28]. * 2014: The enactment of Law number 23 of 2014 concerning the revocation of mining management authority at the district/city level. * 2020: Law number 3 of 2020 concerning the transfer of mining management authority from the provincial government to the central government [29].

We would argue that it is necessary to approach this problem by respecting the opinions of the inhabitants and recognize customary mining rights, rather than convicting ASGM actors as illegal miners. In fact, the 2009 Mineral and Coal Mining Act also states that mining activities conducted by communities in the area will be given priority as a Community Mining Area, as well as proposing to promote safe mining through cooperatives. In this way, what is needed is an approach that grants advantages to the local communities, paving the way for the regulated development of ASGM that would allow communities to make the best use of the reserves present in the area.

Gold mining takes place in almost all of the islands comprising Indonesia (Table 1). This is in part due to the existence of bountiful gold deposits along the western segment of the Sunda-Banda Neogene Arc, extending from Sumatra in the north, across Java, and continuing to Maluku in the east [30].

ASGM activities have continued to increase quantitatively and qualitatively since the reforms for democratization and decentralization were introduced in 1998. According to data from the Directorate General of Mineral and Coal of the Indonesian Ministry of Energy and Mineral Resources [31], the number of illegal ASM (artisanal and small-scale Mining) including gold mining operations as of August 2021 amounts to 2645 locations spread across thirty provinces in Indonesia, with the number varying per province (Figure 1).

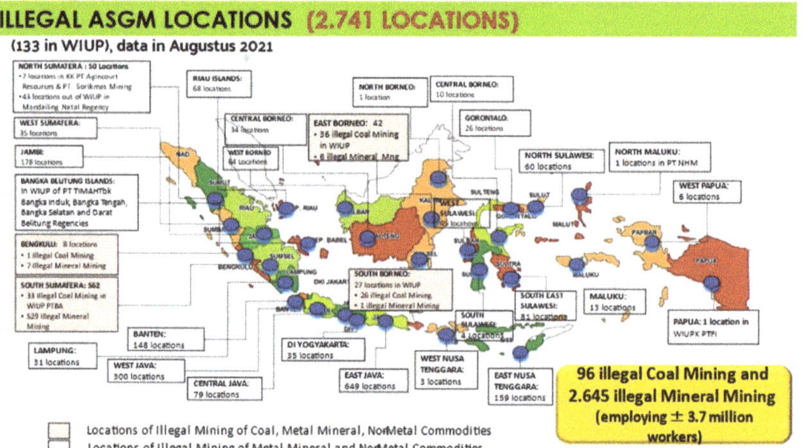

Figure 1. Locations of Illegal ASGM in Indonesia. Source: Directorate of Mineral and Coal of the Indonesian Ministry of Energy and Mineral Resources (2021).

Table 2 shows data from companies that are still actively producing with their production, exports, and domestic consumption displayed. Figure 2 provides data of Indonesian gold production from Statistics Indonesia and Ministry of Energy and Mineral Resources (Figure 3), so that the graphs can complement each other, such as data in the year 2016 which is not available in Statistics Indonesia. Vice versa, data before 2015 can only be obtained from Statistics Indonesia.

Table 2. Top 15 Gold Companies.

Company Name	Production [ton]	Export [ton]	Domestic [ton]
Antam Co. (UBPP Logam Mulia)	44.13	17.60	13.70
Freeport Indonesia	28.01	11.63	19.51
Agincourt Resources	12.17	11.93	0.00
Tambang Tondano Nusajaya	6.8	7.03	0.00
Nusa Halmahera Minerals	5.1	5.55	0.00
J Resources Bolaang Mongondow	2.6	2.78	0.00
Indo Muro Kencana	1.92	1.87	0.00
Amman Mineral Nusa Tenggara	1.73	0.83	1.17
Bumi Suksesindo	1.56	1.56	0.00
Antam Co. (UBPE Pongkor)	1.42	1.05	0.00
Meares Soputan Mining	1.33	1.34	0.00
Natarang Mining	0.9	0.74	0.00
Kasongan Bumi Kencana	0.86	0.86	0.00
Sago Prima Pratama	0.49	0.49	0.00
Sultan Rafli Mandiri	0.01	0.00	0.00

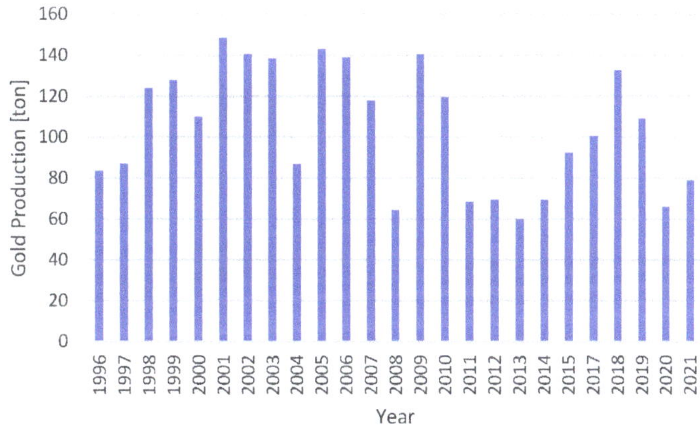

Figure 2. Indonesian Gold Production (Source: Statistics Indonesia).

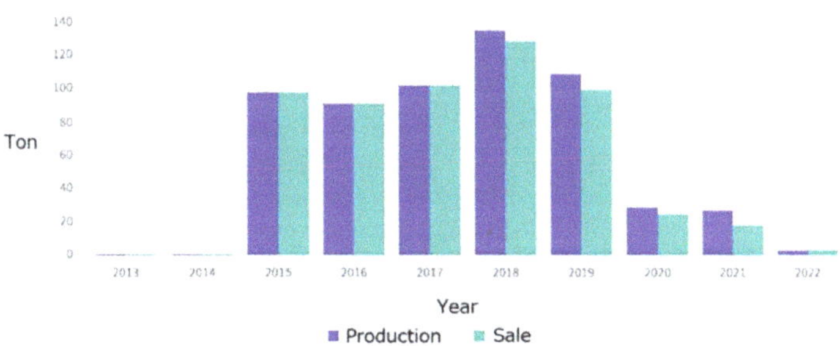

Figure 3. Gold production in Indonesia (Source: Ministry of Energy and Mineral Resources).

The oldest producer in Java, ANTAM Co., was established as a state-owned enterprise in 1968 through the merger of several national projects and mining companies. Lately, there are also local governments that conduct gold mining activities together with private companies via share ownership, for example the South Tapanuli Regency Government and the North Sumatra Provincial Government in gold mining areas around Martabe, North Sumatra.

Information on the production costs of ASGM in 2012 and 2021 can be found in Table 3. It appears that the selling price at the location increased in line with the increase in world gold prices.

Table 3. Gold Cost Production of ASGM.

Year	Cost Production [Rp./gram]	Gold Price [Rp./gram]	Profit [Rp./gram]	1 US Dollar (December)	Reference
2012	78,400	250,000	171,600	9670	Anang Suherman [32]
2021	100,000	500,000	400,000	14,266	Prabawa [3]

3. Illegal ASGM

Unlicensed ASGMs, also known as PETIs in the Indonesian language, are operated by individuals or groups of people, sometimes conducted as a side business (Figure 4). One study shows, however, that due to increasing economic needs, illegal mining ventures are gradually being transformed from secondary businesses into main businesses [15]. One of the driving forces behind the increase in illegal ASGM is the excessive proliferation of gold mining companies competing in the same location [20,33].

Figure 4. Illegal ASGM location in Buru Island, Maluku Province.

Illegal ASGM is illegal according to Article 158 and Article 160 of the Law of The Republic of Indonesia number 4 of 2009 on Mineral and Coal Mining [28]. However, enforcement against illegal mining presents a significant dilemma for law enforcement officers. Raids and crackdowns on illegal gold mining have been conducted, but have not been shown to cause a deterrent effect for the actors. This is partly due to a lack of awareness surrounding the impact caused by illegal gold mining on the environment and human health. Moreover, the existence of illegal ASGM is directly related to the social and economic problems of the impoverished communities living around mining areas. In total, 77% of illegal ASGM miners reported that they experienced an increase in their general welfare thanks to illegal mining activities [6].

Throughout many generations, there has arisen a number of factors that have enabled the existence and increasing expansion of illegal ASGM practices. Firstly, the capital required to participate is relatively small. The businesses are conducted using simple (traditional) technologies, namely mercury or a combination process utilizing mercury and cyanide. Socio-economic factors such as limited employment opportunities, as well as the significant financial benefits offered by ASGM provide an attractive incentive for miners to engage in such illegal modes of business. These businesses have several characteristics in common, such as non-compliance with mining laws and regulations, lack of supervision frameworks, and resistance to law enforcement. Businesses often act under the knowledge that the process of obtaining a mining license is one fraught with complicated bureaucratic procedures, tending to incur high costs [6,15,34].

On the other hand, the unregulated development of illegal ASGM results in several negative consequences, such as: environmental degradation as a result of mercury use, changes in landscapes, forest/soil damage, water pollution, poor handling of mining waste due to the rudimentary skills used, issues related to health, mining accidents, high-interest illicit banking, monopolistic illicit trade, loss of state revenue, violation of the official taxation system, social vulnerability, disturbance of the community, legal abuse, unfavorable investment climates, security disturbances [15,35–38], and threats towards biodiversity [39,40]. Examples of such impacts, among others, have occurred in Mount Botak (Maluku), Lebak (Banten), Solok (West Sumatra), Poboya (Central Sulawesi), and in other regional locations in North Sulawesi, such as Ratatotok, Lanut, Lolayan, and Dumoga [5,41,42]. The large scale of illegal mining activity has been shown to pose serious threats to the health and socio-economic living conditions of communities situated around the mining sites [3,43,44]. Such threats have presented themselves continuously since the establishment of illegal ASGM sites.

According to Khotami (2020) [45], illegal ASGM also results in damage to food resources, as well as to breakdowns of established social order and discipline. However, the cooperative relationship between mining workers and investors is quite close, mutual trust, responsibility, tolerance, and prioritizing the attitude of kinship and mutual need for each other [46]. Additionally, in mining areas, located far from the reach of government supervision, the level of community income remains low and such conditions have been inherited from generation to generation due to mining activities that do not prioritize aspects of sustainable development [45,47].

The issues surrounding illegal mining is very complex in nature, so ideally any policy-based attempts to regulate illegal ASGM should be directed through a social approach that acknowledges the needs of the community, in conjunction with the need to enforce law. Extension programs or guidance administered to local communities are needed to overcome this problem. The level of control conducted by the authorities must be intensified so that the illegal miners (*gurandils* in local language), money brokers, owners, and investors (*cukongs* in local language) who operate in areas rich in gold mining do not run rampant. Illegal mining provides many benefits for the laborers, but even more so for the owners, as the exploitation of illegal miners offers significant profit opportunities for such secondary and tertiary parties. According to the accounts of several laborers, finances are often handled by unscrupulous persons, so that their mines can be operated securely, beyond the reach of the law. For each deposit, an allowance of two sacks of stone containing gold is allotted per hole. The value of each, however, does not necessarily depend on the gold content itself.

However, law enforcement and the forced closure of illegal ASGMs has had a significant impact on social security. According to residents of Lebak Regency, Banten, almost 99 percent of the people in their village work as an illegal miner. According to accounts in the community, since the closure of illegal mining facilities cases of motorcycle theft and other types of petty crime began to rise within the village. Apparently, this increase in crime is related to the population of illegal miners who have been left unemployed following the dismantling of unlawful sites.

From the aspect of law enforcement of illegal ASGM, this has become a serious dilemma, as it spans multidimensional factors such as social, economic, and legal aspects involving employment and poverty, as well as violations of law, so it is necessary to have policies that are able to accommodate such complex and wide-ranging problems [48,49]. Redi (2016) states that illegal mining law enforcement policy solutions should make policies based on cost and benefit analysis to ensure the achievement of the policy. Law enforcement officers and related officials must prioritize non-penal policies through monitoring and coaching small-scale miners towards a more legal form of mining [6].

Some more specific solutions include social outreach and education programs about the impact of illegal gold mining activities on the community, in conjunction with cooperation and coordination between related agencies and community support. The regulation of illegal gold mining can be conducted gradually and continuously, in tandem with strict law enforcement against illegal miners (workers and investors), prioritizing the implementation of deterrents towards local police officers suspected of being involved in extortion or bribery [7].

The police must also play an active role in implementing repressive, preemptive, and preventive measures towards the handling of illegal ASGM activity. However, it must also be recognized that law enforcement faces numerous constraints in this regard, held back by legislation, bureaucratic obstacles, lack of organizational infrastructure, as well as factors related to community legal culture [50].

As an example, by utilizing sufficient preventive and repressive measures, the Landak Police Department stated that the handling of illegal gold mining by the Landak Resort Police was met with some levels of success [43]. Similarly, the investigation into the criminal activity of illegal ASGM in South Solok Regency carried out by the Regional Police Special Criminal Investigation Directorate West Sumatra, based on the data obtained, was deemed to be moderately effective [51].

Based on one of the author's experiences in collecting primary data as the head of the West Papua police region in 2016–2017, the head of the Maluku police region in 2018–2019, and the head of the North Sulawesi police region in 2020, it was found that there are four levels of capability designated to a police region in the handling of illegal mining (Table 4).

Table 4. Designated levels granted to police regions in their combat against illegal ASGM.

Level	Leniency	Bribery	Enforcement Success	Future Outlook
Level 1	Permits illegal ASGM activity	Accepts illegal levies	N/A	Unchanged
Level 2	Does not permit illegal ASGM activity	Does not accept illegal levies	Poor	Little change
Level 3	Actively enforces against illegal ASGM activity	Does not accept illegal levies	Successful	Illegal sites closed, but no steps towards legalization of sites
Level 4	Actively enforces against illegal ASGM activity	Does not accept illegal levies	Successful	Groundwork laid for the transformation of illegal sites into legal, community-operated ones

As shown in the table above, the effectiveness of a police region's measures in enforcing against illegal AGSM can vary widely and is measured based on a number of interlinking factors.

Given the wide range of socio-economic factors that underlie it, one could argue that illegal ASGM activity will continue indefinitely. Millions of people depend on the labor it provides [52,53] and many regions rely on the large revenues from the mining business, including the state's revenue generated from the acquisition of mineral resources. Illegal mining continues to operate in the Merangin, Sarolangun, Bungo, and Tebo mining areas in the Jambi Province, because the communities there view it as an expeditious method for generating profit. Such labor is easily accommodated by those who stand to benefit in the gold industry and is rendered secure by officers who are bribed to provide security and government officials who enforce regulations loosely [54]. As a result, more radical strategies are urgently needed to reorganize unlicensed mining activities, through improved management and environmentally friendly mining sites.

4. Impact of ASGM Activities

4.1. Mercury Problems

For ASGMs throughout various countries, the primary method by which gold is recovered during the mining process is via the use of mercury, usually in very high quantities. More than 1000 tons of mercury used in mining activities are released into the environment every year, and an estimated 10–19 million people are at risk from mercury exposure globally [55]. This problem is particularly exacerbated in many developing countries.

Almost all ASGMs in Indonesia use mercury to separate gold from ore in the amalgamation process. The mercury and gold precipitate to form a mercury–gold amalgam, which is then heated at high temperatures. Finally, the gold is extracted via evaporation of the mercury. Primary gold processing via the mercury method is less efficient, as it can only distill 10–40% of the contained gold [56]. Due to this, many miners conduct further processing using the cyanide method to extract the remaining gold from the initial processing waste. Very often in this way, miners in several areas process waste twice with cyanide, in order to maximize the produced quantity of gold. In several locations, it was noted that many miners believe that cyanide is capable of increasing the quantity of gold, whereas mercury increases its quality.

The traditional processing methods used by miners are generally only able to extract a small amount of gold with high levels of mercury and dust (Figure 5), causing environmental and health problems not only among the active miners involved in the processing, but also among the surrounding communities that are not actually involved in mining activities. For instance, mercury concentrations have been found in the hair of the heads of people living in the Tulabolo sub-watershed [57]. In addition to the elimination of mercury use, the treatment process could also benefit from improvement, so that its productivity can be increased and the processing waste produced can meet the accepted threshold value [30].

Figure 5. Process of grinding primary rock deposits (hard rock).

The United Nations Environmental Program (UNEP) 2018 states that global mercury emissions in 2015 amounted to 2220 tons, of which 49% came from Southeast Asia, 18% from South America and 16% from Africa. The small-scale gold mining industry is the largest contributor to mercury emissions with a total of 38% globally from 2010 to 2015 [48]. Indonesia is referred to by the United Nations as the third largest mercury emitter in the world, after China and India. Mercury pollution by the ASGM sector in Indonesia has increased significantly over the last two decades. For practical reasons, nearly 90% of small-scale gold mines in Indonesia still use mercury in their processing methods. Citing research by Ismawati (2013) in relation to mercury pollution's effects on health, in Indonesia alone, around 195 tons of mercury were identified to be released into the environment per year. This amount represents 20% of global mercury emissions. Of this amount, about 57.5% of the mercury is released into the air, 15.5% into water, and 14% into soil or sediment [58].

Mercury pollution is exposed to humans via a variety of different routes. In the mining sector, it is poisoning due to direct exposure of workers to mercury. Outside of this, mercury contamination makes its way into the surroundings by spreading through rivers, agricultural damage, bioaccumulation of mercury in plants [59] which can produce harmful agricultural products (such as rice) [60], as well as mercury pollution in the ocean and mercury accumulation in fish [61] and seafood which is harmful to consumer health. Most ASGM sites in Indonesia possess no reliably safe method by which to dispose of mercury waste, resulting in miners generally depositing the waste into nearby rivers [26]. However, the main route of exposure is from the atmosphere. People can be exposed to mercury by breathing contaminated air produced during the amalgam smelting process, consuming contaminated food and through direct absorption via the skin. Mercury can cause digestive, respiratory, skin, and kidney problems, the effects of which can be very dangerous, even in minute quantities. At high doses, mercury can cause permanent damage to the brain and nervous system, kidneys [62], impaired fetal development [54,63], and lung damage [64].

In Indonesia, mercury has been designated as a hazardous and toxic substance (B3) by government regulations. Its use has been banned, and various measures have been taken to crack down on its employment in illegal mining. The Indonesian government signed the Minamata Convention on International Treaty in October 2013 and has ratified The Minamata Convention on Mercury into a domestic law by Law Number 11 of 2017 in September 2017 [65]. This commitment of the Indonesian government is carried out through the 2019 Presidential Regulation Number 21 in the National Action Plan for Mercury Reduction and Elimination (RAN-PPM), which aims to ban the use of mercury in ASGM as a National Priority Program and reduce/eliminate mercury contamination in an integrated and sustainable manner [66]. This is reinforced by the Regulation of the Minister of Energy and Mineral Resources of the Republic of Indonesia Number 16 of 2020 concerning Strategic Plan of the Ministry of Energy and Mineral Resources for 2020–2024 [67]. The Indonesian government has drawn up a national action plan for the reduction and elimination of mercury by 2030 [68]. This commitment was proven to be effective in Indonesia through real statistical examples. Until 2020, based on the RAN-PPM report [69], efforts to decrease the use of mercury in the ASGM sector resulted in a reduction of 10.45 tons of mercury. This was achieved not only through the curbing of ASGM that use mercury, but also by endeavors into the development of non-mercury-based gold processing methods.

Another important statistic to pay attention to is the increasing circulation and distribution of mercury sales in various regions. Scientists and environmental activists hope that global restrictions on mercury trade will raise mercury prices and reduce the mercury use and pollution involved in ASGM. However, in Indonesia, despite restrictions on global mercury trade, increased domestic supplies of mercury have conversely made mercury cheaper and more widely available. This phenomenon has the opposite effect of increasing environmental pollution and exposure to mercury by miners [58]. Consequentially, more far-reaching steps are needed by the Indonesian government to completely halt the illegal trade of mercury entering various regions. For that purpose, it has been argued that the government should play a more vital role in the reduction and use of mercury in small holder mining [70].

Amalgamation and cyanidation are the two main gold extraction methods that are currently in common use. However, these methods have a harmful impact on the environment and the health of miners and local people. Significant research on alternative gold processing outside the use of mercury has been carried out both on a global scale [71] and nationally within Indonesia [12], for example, the hydrometallurgical use of chemical solvents such as thiocyanate, thiourea, and thiosulphate. The advantages of the leaching technique using thiosulfate reagents include less environmental damage compared to commonly used methods and a faster dissolving process of gold compared to cyanide solution [72,73].

While there are other methods for extracting gold, the onus largely rests upon the miners. If the miners consider this method to be more efficient, cheap, and effective, they will naturally move from a dangerous method to using a more environmentally friendly and economical method. On the other hand, several problems arise in relation to alternative methods, such as the difficulty of changing the method because new equipment must be brought in to replace the equipment from the former method; an issue made all the more difficult due to the remoteness of the areas in which gold mining operates. Additionally, new methods are not easy for miners to learn and the necessary chemicals are not easily obtained by miners.

Alternative technologies to the traditional equipment in mining that uses mercury and other harmful materials is being sought by the government of Indonesia. Despite the fact that mercury is a material that is prohibited from being used in mining, the use of mercury is still ongoing even today, because switching to alternative technologies would incur large costs in procuring equipment that is difficult to carry out by illegal ASGM actors. Therefore, as a solution, the author proposes the concept of Willingness to Pay, so that the replacement to this alternative technology can be carried out immediately and the use of mercury in illegal ASGM can be stopped. Another idea is that more licenses will be issued by the government if simplification of the process to obtain a license is implemented on the condition that miners must use alternative mercury-free technologies. With the existence of TDCoP (transdisciplinary communities of practices) groups that carry out transformative learning [74] in good mining practice, awareness of the dangers of mercury and environmental safety and health, it is hoped that public awareness will increase.

In terms of solving mercury pollution, the author once again reiterates the use of the concept of WTP (Willingness to Pay). This can be successful due to the fact that it is not only miners and mining workers who suffer from the negative impacts of mercury use, but also people indirectly related to mining, due to the effects of environmental pollution. Taking that into account, the impact of mercury intoxication not only places an additional burden on the victim's family, but also on the community in general. If we only rely on the concept of "those who pollute must pay for it", the problem of removing mercury in mining will not be solved; it will persist as it has up until now. However, if the cost of eliminating the use of mercury in gold mining and the shift to mercury-free technologies is shared by all communities in a certain region, the problem of eliminating the use of mercury will be quickly resolved.

4.2. Environmental Impacts

Environmental pollution represents one of the most tangible adverse impacts of gold mining activities. Illegal gold mining in particular poses a remarkably higher threat to the environment, due to the increased danger to health and influence upon the occurrence of natural disasters it is directly responsible for. With regard to the latter, the effects of illegal gold mining have the potential to cause environmental damage in the long term, taking the form of negative changes to the landscape, landslides, and erosion, as well as water pollution in/around mining sites. A number of studies have reported upon the contribution of ASGM to land degradation [19,42,75], river water pollution [18,74,76,77], and soil/sediment pollution [18,25,26,36,57,78], whose conditions do not meet established quality standards. Indonesia Government Regulation 82/2001 for mercury in river water is 0.002 mg/L and in drinking water or in the water supply it is 0.001 mg/L.

Over time, the illegal mining that has been carried out traditionally (Figure 6) in Indonesia has resulted in damage to the agricultural land surrounding mine sites. The stagnation of water flows often leads to irrigation channels on agricultural land becoming a breeding ground for mosquitoes, whilst also polluting rivers and aquatic biota, ultimately altering the soil structure present around the mines [54].

Figure 6. Gold processing sites.

As a result of mercury being distributed in the water sediments along Kayeli Bay, the aquatic ecosystem of Kayeli Bay has been contaminated with mercury as its bioconcentration (accumulation) was found in the leaves and rhizomes of the seagrass *Enhalus acoroides* [79]. Likewise, the concentration of mercury in sediments along the Tulabolo River, Gorontalo Province was deemed unsatisfactory by European Safety Standards and the water quality was found to be too close to the Government Regulation, PP82 threshold, 2001 [57]. The water and sediment mercury concentration of the Sekonyer River in the village of Aspai, Central Kalimantan was also found to have exceeded acceptable thresholds; a consequence of the boom in unlicensed gold mining which has occurred there since 1990. Additionally, the accumulation of mercury in samples of several types of fish and shrimp from the river exceeded acceptable EPA standards [78]. Similarly, the existence of gold mines along the Batanghari River, Dharmasraya Regency, West Sumatra has led to severe damage to the ecosystem around the river, causing the extinction of living things in the river and the cloudiness of the river water reaching levels that render it dangerous to consume [36].

A study which took place in the Batang Asai District, Sarolangun Regency, Jambi showed that the negative impacts of illegal gold mining activities resulted in not only a decrease in the quality of water, river, sediment, and soil but also to an increase in noise pollution, dust, forest conversion, river silting, the emergence of large holes, soil abrasion, and the disappearance of the meranti (*Shorea* sp.) and damar (*Agathis damara*) species of flora. Furthermore, a decline in the population of the semah fish (*Tor* sp.), once considered a common species to the area, was also measured [80]. Identical conditions related to the emergence of ground holes, loss of the semah fish population, forest vegetation decline, river silting, and cloudy river water were also noted in Muara Mensao Village, Jambi Province [47]. Mercury is not only found in fish and plants, livestock such as cattle in mining areas are also contaminated with mercury [81]. Finally, a high level of mercury contamination has been reported in various other ASGM communities throughout various provinces, such as in West Java [82], West Nusa Tenggara [62], Gorontalo [83], Southeast Sulawesi [84], and Buru Island, Mollucas [85].

As described above, environmental degradation from illegal gold mining exerts a significant influence upon river water pollution that extends to irrigation dams, causing the pollution of productive agricultural land and fisheries. Due to this, communities around the watershed often have difficulty accessing clean water for their daily needs and fishermen struggle to sustain their livelihoods. As such, the existence of illegal gold mining has led to numerous social conflicts. Likewise, incidents of landslides near mining sites have also added to the list of environmental damages and resulting social tensions [21,50].

The damage inflicted by ASGM operations has also been deemed a threat to world heritage geological sites along the Marupa and Kahayan Rivers in Indonesia, Central Kalimantan. Indonesia boasts a high level of biodiversity, with its tropical rainforests and unique geological characteristics, including the presence of forests in watersheds and mountains [40]. ASGM actors also frequently intrude and conduct activities within many national parks [20,39]. However, illegal gold mining activities greatly imperil the country's biodiversity, as well as the world's geological heritage as a whole.

From the various studies above, it is very clear that environmental safety is very important in mining. Mining sites containing hazardous chemicals such as mercury, etc. pose a risk to the environment. Mercury not only contaminates local mining areas, but can also be transported by rivers into the sea and can be carried by the wind to make its ways into plants and animals. According to several studies, traces of mercury have been detected not only in plants and animals around mining areas, but also in the bodies of fish, shellfish, and seaweed far from the mining site. Environmental safety is inseparably linked to public health. Even people who are not in direct contact with mining activities will also be affected if they eat fish or come into contact with objects which have been indirectly contaminated by mercury. Environmental safety is also an important concern for post-mining sites, because such land usually cannot be repurposed as agricultural, plantation, and fishery land for some time.

In order to combat the acidity introduced to water reserves, limestone can be added into reservoirs so that the pH of the water in the pond increases. In addition, for mining that has a clear post-operation plan, land surface arrangement and revegetation are conducted, such as the closure of pyrite material at a mining site with good soil material and planting vegetation on it so that the concentration of acid in the water is reduced.

4.3. Economic and Social Impacts

According to UNDP data, ASGM is an important source of income for Indonesia's estimated 300,000 miners [86]. The gold procured by ASGM producers in rural areas can earn as much as 70% or more of the standard international price. With these considerations in mind, people living in remote areas often view small-scale gold mining as a way out of poverty [87]. The actors who participate in gold mining include agricultural and fishery workers who work part-time and require additional income in order to sustain their livelihoods. Meanwhile, for some communities, AGSM activity is the main source of income by which to support their daily lives.

In two provinces of Nusa Tenggara (NTB & NTT), especially in Sekotong Island-Lombok and Taliwang-Sumbawa, it is found that ASGM leads to an increase in economic activity, employment, income, and opportunities, meaning that they are considered as a form of positive social development [8]. However, over the generations, the lack of socio-economic benefits received by communities from illegal ASGM activities has been measured in the North Lebong District, Lebong Regency, Bengkulu Province [14]. In Kuantan Singingi, it has been noted that illegal gold mining tends to bring long-term benefits only to a limited number of actors, such as investors, and local police officers who accept bribes [7]. The findings in the province of East Nusa Tenggara and Sulawesi revealed however that existing mining sites there did not provide major long-term economic benefits to local communities [44].

In a similar nature, a study into the management of the gold mine in Bakan Village, Bolaang Mongondow Regency, North Sulawesi Province showed that the management of the gold mine brought about both positive and negative impacts. The positive impacts included the improvement of the miners' immediate economic welfare, while the negative impacts included elements of danger related to health and the environment [5].

Ultimately, though migrant miners and those involved in the mining network benefited from AGSM activity, residents who are a part of the local community do not stand to profit from the mining operations, and instead suffer from the damage inflicted by destruction of their environment. In other words, any economic advantages earned from AGSM activity comes with the price of environmental degradation, as well as social vulnerability and injustice [21,88], including social security disturbances, and corrupt behavior [44]. It could therefore be argued that in order to avoid the negative impacts of AGSM, miners must operate only when in possession of a license. With a license, miners can work within strict safety standards that are maintained by work facility management [5].

The labor management system in illegal small-scale mining varies diversely and has no objective standards, as found in research related to small-scale gold mining in Pongkor, West Java. The absence of a defined labor system arrangement that is in accordance with labor standards puts mining workers at a major disadvantage. As a result, the laborers are not entitled to equal work and employment guarantees because there is no health insurance, safety guarantees, or regulations related to working hours [89].

From an economic point of view, mining was shown to improve the overall economy of Dharmasraya Regency, West Sumatera communities, especially those who were in possession of customary land rights along the banks of the river. On the other hand, the existence of mining facilities led to increased wealth disparity between the rich and the poor because only the owners of the capital and means of production stood to fully reap the benefits [36].

One phenomenon that is also commonly found is social conflict involving communities surrounding mining areas. Conflicts between communities and private companies began to occur in the Bogani Nani Wartabone National Park area, Bone Bolango Regency in Gorontalo Province after the government offered work contracts to private companies to mine gold. Even following the end of the Dutch occupation in 1940, traditional community gold mining activities are still illegal here, even though the land that has been owned by the community for generations will be passed on to their children and grandchildren [17].

Mining sites attract migrant miners from various regions, as seen in the example of the mining site of Talawaan-Tatelu, North Sulawesi Province, which is visited by miners from all over the province of North Sulawesi [78]. Disputes, or even fights between miners, as well as factions of miners involving inter-ethnic groups, almost always occur around mining sites. Such conditions lead to a poor sense of security and public order in the villages around the mining sites [51,54,88].

However, the illegal mining activity in Muara Mensao Village, Jambi Province has been shown to contribute significantly to the local economy. This is because rises in income can lead to positive impacts in the education sector. The education levels of the younger generation in the area revealed promising results [47].

In Perentak Village, Merangin Regency, Jambi Province, the existence of illegal gold mining conducted by Perentak villagers has had positive implications for household economies in the short term, but in the long term, environmental impacts caused by an increase in unlicensed gold mining has in turn brought negative implications for the average household economy. Any positive aspects brought about are offset by the fact that serious cooperation is needed between the community, government, non-governmental organizations, and the scientific community to prevent damage to natural resources in the future. Post-treatment of land following illegal gold mining activities is one method which could help to restore sustainable natural conditions [47]. Another would be the empowering of the traditional community by granting a platform for local wisdom to express itself, especially with regard to the economic disparity introduced to the community via AGSM. Accordingly, it is necessary to pay attention to the economic needs of the community, finding ways to provide benefits that grant advantages for the local villages [48].

ASGM is one of the economic activities that has been hit hardest by the COVID-19 pandemic. However, the COVID-19 outbreak can be used as an opportunity by which to reconfigure public opinion towards ASGM activities, instead allowing us to view it as a safety net during economic crises, that can act as a fallback for local economies that have been unable to find alternative methods of income [9].

In conclusion, ASGM needs to be recognized as a formal economic sector with solid legal foundations, in order to maximize the benefits and minimize adverse impacts brought about by the sector. An approach which integrates the idea of sustainable mining is one method by which to move forward, counteracting the current exploitation of natural resources and unsustainable resource management [7].

4.4. Impact on Health and Education

Illegal ASGM is very harmful to the safety and health of workers. Every year, several accidents occur at illegal ASGM sites. Figure 7 shows total data from registered mining companies for all types of mining. As such, it cannot reveal data related to illegal ASGM, only displaying statistics related to legally operated mining.

Table 5 shows that accidents that occur in illegal ASGM are very frequent every year, caused by landslides, oxygen deprivation, and toxic gas inhalation. Therefore, it is very important to legalize ASGM by giving illegal miners licenses so that they can be taught how to conduct safe mining practice and minimize mining accidents in TDCoP groups through transformative learning.

Figure 7. Total number of accidents in mining companies including gold mining (Source: Ministry of Energy and Mineral Resources).

Table 5. Victims of illegal mining over the last 5 years (Source: online news).

Month, Year	Location	Dead, Injured [Number of Persons]	Cause
January 2022	Atoga Timur Village, Bolaang Mongondow Timur District, North Sulawesi	2, not mentioned	poison gas
October 2021	Tumbang Torung, Kotawaringin Timur, Central Kalimantan	6, not mentioned	landslide
October 2021	Gapit Village, Sumbawa Regency, Nusa Tenggara Barat	4, not mentioned	poison gas
July 2021	Tambang Saweak, Lebong, Bengkulu	1, not mentioned	landslide
May 2021	Timbahan Nagari Abai, Solok Selatan, West Sumatera	8, 9	landslide
February 2021	Buranga Village, Parigi Moutong Regency, Central Sulawesi	6, 16	landslide
December 2020	Taman Nasional Gunung Halimun Salak (TNGHS) Lebak, Banten	4, 2 (missing)	landslide
November 2020	Sungai Seribu, Kotawaringin Barat, Central Kalimantan	10, not mentioned	landslide
October 2020	Sekatak Buji Village, Bulungan Regency, North Kalimantan	5, not mentioned	landslide
December 2019	Pulau Baru, Merangin Regency, Jambi	4, 2 (missing)	landslide
July 2019	Bakan Village, Bolaang Mongodow Regency, North Sulawesi	2, not mentioned	landslide
May 2019	Gunung Pongkor, Bogor, West Java	5, 15	landslide
March 2019	Bakan Village, Bolaang Mongodow Regency, North Sulawesi	16,18	landslide
August 2018	Gunung Botak, Wamsait Village, Buru Regency, Maluku	2, 2	landslide
June 2018	Bakan Village, Bolaang Mongodow Regency, North Sulawesi	5, not mentioned	landslide
June 2018	Gunung Suge, West Lombok	7, 6	poison gas

Buru Regency is one of the main areas in Maluku which experience health problems as a direct result of the use of mercury and cyanide in the amalgamation process of traditional gold mining [26].

Illegal mining has caused many cases of health deterioration [83,84,90] and disease in the community [80], as well as amongst the miners themselves, stemming from the identification of mercury traces in the miners' bodies [91–93]. This is revealed by the research that was conducted in Ratatotok, Southeast Minahasa Regency, North Sulawesi Province in 2019 [58] and the accounts of a number of residents living around Mount Botak mining area on Buru Island [85]. In North Lebong District, Lebong Regency, Bengkulu Province, the community felt that illegal ASGM practices caused nearly 57% of all health issues, such as coughing, lung problems, and tuberculosis [14]. Moreover, the risk of soil and plant contamination through mining activities has led to high concentrations of mercury building up in the bodies of local residents within a short period of time [8].

The impact of mining activities is serious, especially in children and women, the most vulnerable, who participate in mining or live around the mining areas of Merangin, Sarolangun, Bungo and Tebo, Jambi Province. As illustrated in Figure 8, female laborers in the mines working without using gloves is a common sight (Figure 8). The immediate influence on health problems includes impaired growth and development of children living in locations around the mine, with long-term impacts including the threat of permanent disability and malignancy to children and the overall community [54,63].

Figure 8. Female worker at an illegal ASGM, carrying out gold processing work without gloves.

Health problems and diseases that arise in gold miners come in the form of chronic and acute diseases. Chronic diseases caused by mercury intoxication in gold miners include the occurrence of liver dysfunction, decreased leukocytes, paralysis of limbs, numbness, and tremors (Parkinson's disease). Tremors is a condition where the hands and feet are always shaking, while the facial muscles and lips often move unconsciously. Other health problems that arise include a lack of passion for activities (depression), difficulty sleeping, emotional turbulence, poor memory, cramps during weather conditions, cold, and feelings of anxiety. While the acute diseases that arise are acute poisoning, diarrhea, upper respiratory tract infections, eye diseases, vertigo, miscarriage, and skin diseases.

In the case of childbirth, there are inhibitors to fetal development because organic mercury from the methyl mercury form can enter the placenta and inhibit fetal development in women who are pregnant. This can cause birth defects in the baby, damage DNA, and interfere with blood flow to the brain, causing damage to brain tissue.

Table 6 shows symptoms of miners/inhabitants caused by mercury intoxication. Data regarding children born under the influence of mercury intoxication are the results of reports conducted by BaliFokus and Medicuss Foundation in Cisitu (Lebak Regency, Banten Province), Bombana Regency (Southeast Sulawesi Province), and Sekotong (West Lombok Regency, West Nusa Tenggara Province). They found several severe mercury intoxication suspects in adults and children. The studies were supplemented by measuring mercury levels in hair or blood. The results showed that in some samples the internal mercury levels exceeded the permissible standard values. There are some photos of children suffering from synostosis, seizures, hypersalivation, short size hand, kriptorkismus, labiognatopalatoschizis, congenital talipes, cataract juvenile, deafness, anokuli, and high myopia [94]. Unfortunately, it is quite difficult to obtain data from the local health office, as reports and previous studies are often incomplete and accurate.

Table 6. Mercury intoxication and its effects on health.

Location	Clinical Symptoms	Ranges Mercury Concentration	Remark	Reference
Hulawa and Ilangata, North Gorontalo, Gorontalo	Bluish gums, babinski reflex, labial reflex and tremor	2.1–144.8 µg/g in hair WHO standard 1–2 mg/kg	44 miners and inhabitants got symptoms.	Arifin, Y. et al. (2015) [83]
Bolaang Mongondow, North Sulawesi	Signs of bluish discoloration of gums, rigidity and ataxia (walking or standing), alternating movements or dysdiadochokinesia, irregular eye movements or nystagmnus, Field of vision, knee jerk reflex, biceps reflex, sensory examination, tremor	0.51–79.27 µg/g in hair WHO standard 1–2 mg/kg	50 miners and inhabitants	Arifin, Y. et al. (2017) [91]
Kurun, Gunung Mas, Central Kalimantan	Easy fatigue, headache, shaking/shivering, and stiff joints	0.5178–10.4682 µg/g in hair WHO standard 1–2 mg/kg	80.5% miners got mercury contamination	Lestarisa, T. (2010) [95]
Talakiak Village, Sangir, South Solok, West Sumatera	Stiff joint disease, muscle pain, rheumatism, aches, foot/hand joints feel tingling, achy, tired, shivering/shaking, fever, sore waist and chest pain), and skin diseases of itching/itching/allergy	No data	22 miners (39%) got symptoms	Putri, G. E. (2017) [96]
Cisarua Village, Nangung, Bogor, West Java	Tremor, frequent tingling, stiff facial muscles, eye irritation, metallic taste in the mouth, muscle aches and spasms, thickened skin on the palms and soles, and headaches	0.28–68 µg/g in hair WHO standard 1–2 mg/kg	24 miners (60%) have mercury intoxication.	Junita, N. R. (2013) [97]
Lebaksitu Village, Banten	No data	0.00–188.28 µg/L in blood WHO standard (5–10 µg/L)	77.9% respondents mercury in blood more than 10 µg/L.	Kristianingsih, Y. (2018) [98]
Bulawa, Bone Bolango Regency, Gorontalo	No data	2.92–378.90 µg/L in blood Standard 8.0 µg/L 0.48–260.20 µg/g in hair Standard 2.0 µg/g	52 respondents have mercury content in the blood that exceeds the standard and 57 respondents have mercury content in their hair that exceeds the standard.	Singga, S. (2013) [99]
Kayeli Village, Gunung Botak, Buru Regency, Maluku	No data	0.10–3.25 ppm in hair Standard 0.5 ppm	Repondents are inhabitants	Rumatoras, et al. (2016) [85]

Education of children is often neglected at illegal mining sites, and there are often no schools or educational facilities in many mining districts. In addition, the presence of sex workers who make a living around mining sites holds the potential for causing a myriad of other social problems. Overall, living conditions and general wellbeing of residents around mining areas are often poor, meaning that a comprehensive approach would be required to tackle the wide variety of problems being faced there.

One form of good mining practice is an environmental education and the teaching of how to make mining tunnels that are safe from landslides, oxygen deprivation, and toxic gases. Miners and mining workers require the use of personal protective equipment (PPE) such as rubber gloves, masks, shoes, and helmets. Meanwhile, for owners/funders of illegal mining, the use of PPE compounds the difficulty, because there are many cases of gold loss both in excavation and processing which are actually hidden in PPE equipment by workers (interview result with owner). The use of alternative technologies prevents the community from being affected by mercury intoxication. However, during the transition period, while they are still using these hazardous chemicals, it is necessary to conduct periodic health checks for miners. This routine health test should be carried out by the Community Health Center (Puskesmas) or Health Office in the province. Our project has established the Health Village (*Kampung Tangguh Kesehatan*, KTK) in the East Suwawa, Gorontalo Province in collaboration with the local government in planning to carry out health checks for the miners. Small groups of TDCoPs can increase communities' awareness by teaching more about the dangers of mining using mercury.

5. Government Role and Policy/Customary Law Involvement

5.1. Government Role and Policy

Natural resource management is inseparable from the issue of licensing illegal mining activities. Every mining actor, both individual and enterprise alike, is required to obtain a license as regulated in Regulation Number 4 of 2009 [28] regarding Mineral Mining in conjunction with Government Regulation Number 23 of 2010 [100] about the Implementation of Mineral Mining. The rise of unlicensed gold mining makes it difficult for the government to supervise. All existing regulations and laws work together to regulate the actors of gold mining activities, in a manner which seeks to prevent permanent damage to the environment. The combined functions of law enforcement are based on the rules contained in Law Number 4 of 2009 with regard to Mineral and Coal Mining, UU Number 32 of 2009 [101] pertaining to Environmental Protection and Management, Government Regulation Number 23 of 2010 [100] about the Implementation of Mineral and Coal Mining Business Activities, and Regulation of the State Minister for the Environment Number 23 of 2008 [102] concerning Technical Guidelines for the Prevention of Pollution and/or Environmental Damage Due to Community Gold Mining. Law enforcement is carried out against parties involved in illegal gold mining using a number of both preventive and repressive measures. However, when it comes to the implementation of the above policies, the laws are very much subject to violation, especially regarding licensing issues.

Following the enactment of Law Number 23 of 2014 [103] concerning Regional Government, effective from October 2016, the designated roles and policies to be enacted by local governments became far more palpable. Law Number 23 of 2014 concerning Regional Governments emphasizes that regents and mayors are no longer authorized to assign mining business areas and mining business permits to enterprises. That authority now belongs only to the governor and to the central government. Prior to 2 October 2016, local governments had the authority to prohibit and control mining activities. However, after the new law was instated, the local government was relinquished of any authority over mining activities in the area, which was now in the hands of the provincial government. In reality however, far from being more regulated, mining in Indonesia is actually becoming increasingly less controlled, and is instead experiencing rapid expansion. The latest Law Number 3 of 2020 [29] introduced a shift in authority from the provinces to the central government (Table 1).

Environmental damage and mercury waste pose a severe threat to the people of Prabu Village, Central Lombok Regency, West Nusa Tenggara. The government closed this village's gold mining site, resulting in conflict between the government and the mining communities [27]. The closure of this mine was ultimately ineffective, since the residents continued to carry out mining activities in secret.

Another definitive example can be found in the case of illegal gold mining in Banyumas, which has also been subject to numerous issues in relation to governmental intervention. The majority of mining that takes place in Banyumas is of a traditional, and consequentially illegal nature. In this manner, illegal mining could be said to be a form of vertical conflict between the community and the local government, causing dilemmas for the local government in the prohibition and control of mining operations [88,104–106]. The law enforcement of illegal gold mining in Banyumas Regency is implemented by forcefully stopping mining activities in the field, sealing mining sites and seizing goods related to mining activities [105]. In order to guarantee its success, strong cooperation between the government, the community, and the private sector must continue and should be executed under strict supervision.

Furthermore, support from non-governmental organizations (NGOs) is needed to handle the ongoing crisis of unlicensed gold mining. The role of NGOs as an intermediate actor allows them to act as a balancing force, offering greater empowerment to the unlicensed miners [85,107].

Observing how institutions deal with illegal gold mining in Kuantan Singingi Regency, Riau Province, it has been shown that the leadership role holds little sway, both formally and informally, in supervising the activities conducted by capital owners and mining actors. In addition, there were no regional regulations implemented that could act as a legal umbrella over community mining. Limited resources in terms of costs and facilities, and the remote distance of mining sites make it difficult for police to enforce control. Cooperation is needed in institutional development between provincial and district governments, the Regional House of Representatives and the community in overcoming the problem of illegal gold mining in Kuantan Singingi Regency [45].

Actions taken by the government up until now include the closing of many mining sites operated by illegal ASGM, in collaboration between the local government and the local police. Although the closure of mining sites is not an easy task because of resistance from the mining community, limited government officials or police, locations deep in the jungle or on top of mountains that make it difficult due to limited access, and the presence of people who protect the existence of illegal mining activities.

The government has a responsibility to take anticipatory steps in preventing health problems caused by mercury and other heavy metals in the human body, especially children living in proximity to illegal mining areas. It is important to advocate for policy makers to construct policies that protect people's health and prioritize environmental conservation [54].

The problem of stopping illegal mining in Indonesia does not only concern one governmental department. Although there are already clear regulations, it seems that changing regulations and shifting powers would allow for various opportunities. For the government authorities, the authors believe that cooperation between relevant departments, police, and between local and central governments is the best solution to stop illegal mining. Meanwhile, from the community point of view, they need certainty for livelihoods, security and justice in obtaining natural resources around their area. The viewpoint shared amongst illegal miners is that they do not want to be spectators in their area, where their area's natural resources are being exploited by companies, most of which are not from their area or even from abroad (interview result). Due to this, illegally operated mining can commonly be found being conducted around/adjacent to large-scale official (company) mining areas, resulting in conflicts with the mining business license holders [15].

The current reality is that many traditional mines are located in mining company locations. Therefore, it is necessary to make additional rules that accommodate both in the same location.

5.2. Customary Law

Under the Indonesian Mining Law, unlicensed mining activities are considered criminal acts regulated by Law Number 4 of 2009 concerning Mineral and Coal Mining. In spite of formal legal attempts to curb unlicensed mining, ASGM activities operating beyond the reach of the law are still ongoing. There has in the past been cases where instances of unlicensed gold mining were resolved using alternative, community-based legal instruments, in the form of customary sanctions in Kualan Hulu Village, Simpakng Hulu

District, Ketapang Regency [108]. These customary legal sanctions against illegal mining actors in Kualan Hulu Village were considered effective in instilling a sense of justice, whilst achieving the objectives of the applicable law. The application of customary law in implementing sanctions against illegal mining actors is something which could also be adopted into the governmental mining law system in Indonesia [108].

Perhaps one of the reasons why customary law could function effectively in combatting ASGM, is related to the trust and social capital that community leaders command, in stark contrast to the lack of trust garnered towards government officials. Resistance can be established from a bottom-up perspective, allowing for new internal norms to be instilled, that can be adhered to by the community [41]. The disparity in trust between community and government was made apparent in the granting of a mining contract of work permit in the Central Sulawesi Grand Forest Park to a private company. This resulted in clashes of perspective between the government and the community which ended up with the development of an illegal mining site managed by the community.

In several villages in Indonesia, customary law (*adat*) plays a role in governing their communities. Traditional customary institutions were previously able to regulate how to manage the environment. Several villages have proven to be able to conserve the environment by prioritizing the functions of customary institutions. In the illegal gold mining areas of Kuantan Singingi Regency, the function of traditional customary institutions in preventing environmental damage, however, has gradually weakened. Therefore, the revitalization of customary and legal institutions is an urgent task in the battle for environmental protection [109].

6. Prudent Management of ASGM

As described above, there are two major impacts of illegal mining on the community, positive impacts such as increasing community economic welfare and negative impacts, namely environmental damage and decreased health quality due to mercury intoxication. Meanwhile, the impact to the government is the loss of income tax from mining activities and environmental damage, the repair of which requires government expenses.

One way to tackle these effects would be through prudent management. Prudent management of ASGM is an alternative approach which should be taken immediately in order to prevent and reduce severe damages to the environment and food resources. This must be realized together through a strong and integrated collaboration between the government, the community, and the private sector. Furthermore, a strict, consistent and non-discriminatory form of law enforcement should take place in the handling of ASGM [36]. Such enforcement must intimately involve the relevant government institutions and policing authorities [7,105]. This is due to the wide range of participating actors and the complex socio-economic processes and structures which ASGM involves [109]. The inhibiting factors faced in constructing a sustainable form of mining development are primarily a result of the lack of legal awareness in mining communities, as well as the rampant corruption taking place on both an individual level [43,105] and an organizational level, especially with regard to the behavior of officers [44] who accept illegal levies.

Indeed, such incidents must be stopped and sensible management of mining must be implemented, so that these conditions in which severe environmental damage and the overexploitation of natural resources are permitted to occur do not persist. Mining management begins with the regulation of unlicensed miners, which deters them from engaging in activities at the sites and allows for the closure of the unlawful mining sites [3].

The formalization of ASGM is a demand that must be followed up immediately [9,49] by the relevant agencies. This change in system or regulation can bridge the needs of community mining with the government that has the authority to issue and supervise mining permits [9,49,109]. Such efforts, on the one hand, would allow miners to obtain permits through cooperatives or associations. In addition, a strict environmental management on the safety and health of miners, as well as environmental quality around the mine site can also be maintained. However, the processes of legalizing community mining must account for strict environmental considerations and be supervised by impartial and fair institutions [8,9,48]. Such processes must also take into consideration the perspectives of miners and local communities [44,49] in a balanced way.

Therefore, in the case of legalizing community mining, the customary rights of indigenous miners must be recognized. Additionally, companies that wish to operate in a community mining location must be able to co-exist with community miners who have worked there for decades.

At the community level, TDCoPs (transdisciplinary communities of practices) of small groups can be used so that the community's perspective on good mining practice and health can be changed through transformative learning. TDCoPs can be applied throughout various layers of society. In the Suwawa Timur region, in collaboration with the local government, there is a Health Village (a combination of several TDCoPs) which aims to raise public awareness of good mining practices, mercury impacts, environmental concerns, and community safety and health. At the Health Village, the community can learn environmental education, especially concerning mercury and mining. This project is available for children, elementary and junior high school students, as well as women (Pateda, personal communication) [110].

7. Discussion

The ASGM problem in Indonesia is already an exceedingly complex one. It is however becoming further complicated and distinct compared to the activities taking place in other countries, because illegal ASGM in Indonesia has specific problems related to history and makes it difficult for the government to control through regulations alone. Traditional people's mining that has operated since hundreds of years ago is not recognized by the government and is considered illegal, with regulations made later and applied retroactively. However, people who have been mining for generations or have customary rights to their land have different opinions from the government. According to the author's opinion, mining customary rights of local people should be respected, and mining licenses (IPR) should be granted to those who were actively mining before the regulations were made, and exercised their customary rights. Miners should be organized as cooperatives that function via rules which are compatible with government regulations, and must follow the rules and regulations made by the government. The neglect of local people's rights is the most important problem that is protracting the issue of illegal mining in Indonesia. Miners working via cooperatives would be able to employ alternative technologies that avoid contamination from mercury. Simplifying the process of granting mining licenses would actually be very beneficial for the government, because the government can find out the exact number of miners who operate illegally, can urge the miners to protect the environment and health, as well as gain significant revenue in the form of taxes. For mining workers with low education, it is necessary to provide training to find other forms of livelihood through transformative learning in a TDCoP group.

Even though illegal ASGM activity has been prohibited amidst a flurry of changing regulations, community actors continue to conduct illegal ASGM. If these circumstances do not change, the environment will suffer from further deterioration, and in the end, no parties will benefit. The economic situation of the community around mining areas is not improving, whilst at the same time the government cannot control the illegal ASGM.

The government cannot obtain tax revenue from illegal ASGM, since illegal ASGM actors are not officially registered as an individual, cooperative, or company. The mining sector is one of the biggest non-taxed sources of income for Indonesia. It is estimated that the potential loss of state revenue in 2019 is at least USD 908,544,000 per year [111]. In addition, the post-mining environmental damage that remains after gold has been fully extracted from the land, without any rejuvenation measures enacted by illegal ASGM actors, will end up as the burden of the local government. The conditions of the people who are sick and suffer due to mercury poisoning will also become a burden to the community and will indirectly result in losses to the local government. If the tax from mining can be procured, it can be used in a sustainable manner for regional development and environmental conservation.

The above Table 7 shows a comparison between illegal ASGM and mining companies in Indonesia. It is shown that illegal ASGM results in more losses than benefits. Therefore, to replace current mining practices, legalization of ASGM is an urgent need. The closure of illegal ASGM sites alone is not enough because it will eliminate people's livelihoods. Thus, the local government should cooperate with the central government to find the best way to simplify the process of ASGM legalization. ASGM legalization will introduce multiple benefits, both socio-economic and environmental. However, the reality is that not all illegal ASGM locations can be legalized, such as activity taking place in national parks or on land owned by mining companies. For this reason, it is necessary to provide alternative livelihood solutions for the people affected by the illegal mining closures. One way is by transformative learning through TDCoP (transdisciplinary communities of practices).

Table 7. Comparison between illegal ASGM and mining companies in Indonesia.

Aspect	Illegal ASGM	Mining Company
Legality	unlicensed	licensed
Number	uncountable	countable
Production	unrecorded	well recorded
Number of workers	unrecorded	recorded
Safety	uncontrolled	controlled
Health insurance	without health insurance	with health insurance
Environmental safety	unconcerned	concerned
Post-mining reclamation	no	conducted
Community safety	unconcerned	concerned
Surrounding community economic impact	direct	indirect
Social impact	high	moderate
Taxable	no	yes
Conflict	high	high

Currently, what the government is doing to control illegal mining is the closure of ASGM illegal mining sites and the requirement of anyone who wants to partake in mining as ASGM to obtain a license directly from the central government (new law Number 3 of 2020). Previously, the granting of licenses was the authority of provincial governments. The function of local governments now is to submit applications for a Community Mining Area (WPR) to the central government. After a Community Mining Area is established, the individual or cooperative can apply for a mining license (IPR). Many people are reluctant to apply for a mining license because the community mining location does not contain significant gold (information directly from miners). Especially if there is a Mining Business Permit Area (WIUP) owned by a mining company in that area (which actually came after the community started mining). The company is granted the area by the central government. Thus, the Community Mining Area cannot take place in the Mining Business Permit Area. As a result, the community ends up with land that is less suitable for gold mining and of no interest to miners. Therefore, it is necessary to make additional rules to allow the community and mining companies to coexist in adjoining locations. There are also people who are reluctant to apply for a license because it is a complicated, expensive, and time-consuming process to obtain a mining license. Therefore, even though the number of Community Mining Areas is large in quantity, the number of license holders is still low. To date, only 100 licenses have been issued throughout the 3329 Community Mining Areas. In fact, there has been no new licenses issued after Law No. 3 2020 (personal communication) [112]. Simplification of the time-consuming, difficult, and costly licensing process will lead to an increase in legally operated ASGM.

Another problem, however, is that mining companies who do obtain permits from the government often expel traditional (illegal) miners from their regular working sites. This has led to conflict between companies and local communities. Therefore, there must be a solution in the form of new regulation or an intermediary regulation that accommodates the presence of traditional miners in locations adjacent to license-operated mining companies.

The next step is to charge the cost of eliminating the use of mercury for all communities in a region. The benefits of improving the environment, public health, and removing mercury are enormous. If the costs for these benefits are shared, it will relieve the miners so that they will not use mercury in traditional people's mining. For that purpose, we should apply a methodology of evaluation of environmental hazard by ASGM and mercury contamination. For example, with the concept of WTP (Willingness to Pay), we can evaluate the negative impacts on the environmental ecosystem in a particular region, and the local government can charge a cost, in order to offer alternative ways of mining, such as applying intermediate technology to avoid mercury contamination, offering alternative sources of livelihood for the miners, managing the cooperatives for ASGM small miners, environmental education and so on.

Above all, resource utilization should use the best possible recovery approach. In line with this, we offer the following suggestions:

a. People's customary rights to mine should be respected. Though difficult, time-consuming, and costly, simplifying the legalization process is considered to be one possible way. Another would be the making of additional or complementary regulations, so that traditional people who have been mining in a location for a long time are not expelled due to the arrival of companies that have mining permits in that area. Alternatively, compensation money could also be offered.

b. Dialogue with the community and raising public awareness of good mining practice without mercury, improved safety and health through transformative learning by creating small TDCoP (transdisciplinary communities of practices) groups.

c. Alternative and appropriate technologies should be supplied to avoid the use of mercury. Miners would be organized in the form of cooperatives where such alternative technologies would be readily supplied, as well as forward-thinking measures such as credit, environmental education, and the offering of alternative livelihoods.

d. It is also necessary to introduce other alternative livelihoods for the community that are valuable and in line with sustainable development through transformative learning.

e. In order to cover the cost of operating the cooperatives, the author proposes to apply the concept of Willingness to Pay (WTP). In this way, people who live in provinces or regencies that suffer from the environmental damages caused by mercury use via illegal ASGM activities could share the cost to eliminate the use of mercury in illegal gold mining. This method has merits, as ultimately, almost all residents in the province/regency will face the negative impact caused by environmental degradation. The funds collected must be managed prudently by the government and be used to replace traditional equipment using mercury with alternative technology, offer training in the use of this new equipment and how to conduct good mining practices, restoration of post mining areas, environmental conservation, and routine health checks for miners, so as to achieve sustainable development. This can only be achieved if the funds are treated according to their purpose.

8. Summaries

Based on the review of the article above, it can be summarized that:

1. Artisanal and small-scale gold mining (ASGM) in Indonesia is conducted in more than two thousand locations in thirty provinces and most of them operate without a license. The mining activities are practiced using simple technologies that employ mercury and cyanide, which is not environmentally friendly.

2. It is an urgent task for the government to regulate ASGMs immediately, in order to stop further environmental degradation, overexploitation, impairment of environmental health, and reduction in the socio-economic wellbeing of surrounding communities.
3. Wise management of ASGM can only begin by cracking down on miners who work unlicensed and by closing illegal mining sites, followed by the legalization of Community Mining Areas (WPR) where miners and their associations can apply for Community Mining Licenses (IPR) in order to work legally.
4. The formalization of ASGM is necessary and should be enacted immediately, in a way which bridges the needs of actors in the mining industry and the priorities of the government that has the authority to issue licenses and supervise activities. The procedures of legalizing community mining should be framed by strict environmental considerations and supervised by impartial institutions, whilst also including the participation of local people, considering the perspective of local knowledge and traditions of miners and communities.
5. The community must be educated to increase their awareness in terms of good mining practice, mining safety and health, as well as the dangers of mercury and its consequences through transformative learning in TDCOP (transdisciplinary communities of practices). The community must also be provided with opportunities for other forms of livelihood that are more valuable and sustainable than gold mining—an unsustainable industry which will eventually come to an end in the future.
6. By using the concept of WTP (Willingness to Pay), the negative impacts on the environmental ecosystem in a particular region can be reduced and alternative technologies without mercury can be more widely disseminated.

Author Contributions: Conceptualization, supervision, and funding acquisition, M.S.; resources, R.L.; writing—original draft preparation, A.A.M. All authors have read and agreed to the published version of the manuscript.

Funding: This research was funded by The Research Institute for Humanity and Nature (RIHN: a constituent member of NIHU), Project No. 14200102.

Institutional Review Board Statement: Not Applicable.

Informed Consent Statement: Not Applicable.

Data Availability Statement: No data.

Acknowledgments: The authors give thanks to the anonymous referees for their perceptive comments and recommendations.

Conflicts of Interest: The authors declare no conflict of interest.

References

1. Indonesia Gold Production. Available online: https://www.ceicdata.com/ (accessed on 11 January 2022).
2. Gold Mine Production. 2021. Available online: https://www.gold.org/goldhub/data/historical-mine-production (accessed on 11 January 2022).
3. Prabawa, F.Y. *Pemodelan Sistem Dinamis Paramagnon Emas Rakyat Menuju Pertambangan Berkelanjutan, Studi Penambangan Emas rakyat Desa Kertajaya Kecamatan Simpenan Kabupaten Sukabumi Jawa Barat*; Disertasi Sekolah Ilmu Lingkungan Universitas Indonesia: Jakarta, Indonesia, 2020.
4. Kementerian Lingkungan Hidup dan Kehutanan. Grand Design, Pengurangan dan Penghapusan Merkuri Pada Pertambangan Emas Skala Kecil. 2017. Available online: http://sib3pop.menlhk.go.id/index.php/articles/view?slug=pertambangan-emas-skala-kecil-pesk (accessed on 20 December 2021).
5. Dondo, S.M.; Kiyai, B.; Palar, N. Dampak sosial pengelolaan tambang emas di Desa Bakan Kabupaten Bolaang Mongondow. *J. Jar. Adm. Pembang.* **2021**, *7*, 63–72.
6. Redi, A. Dilema Penegakan Hukum Penambangan Mineral dan Batubara Tanpa Izin pada Pertambangan Skala Kecil. *J. Rechts Vinding* **2016**, *5*, 399–420.
7. Nopriadi. Solution on handling of illegal gold mining activities in Kuantan Singingi Region. *Int. J. Appl. Environ. Sci.* **2016**, *11*, 1213–1235.

8. Krisnayanti, B.D. ASGM Status in West Nusa Tenggara Province, Indonesia. *J. Degrad. Min. Lands Manag.* **2018**, *5*, 1077–1084. [CrossRef]
9. Nugroho, H. Pandemi COVID-19: Tinjau ulang kebijakan mengenai PETI (Pertambangan Emas Tanpa Izin) di Indonesia. *Indones. J. Dev. Plan.* **2020**, *4*, 117–125. [CrossRef]
10. Perks, R.; Schneck, N. COVID-19 in artisanal and small-scale mining communities: Preliminary results from a global rapid data collection exercise. *Environ. Sci. Policy* **2021**, *121*, 37–41. [CrossRef]
11. Bansah, K.J. From diurnal to nocturnal: Surviving in a chaotic artisanal and small-scale mining sector. *Resour. Policy* **2019**, *64*, 101475. [CrossRef]
12. Aisyah Syafei, A. Penyediaan Alternatif Teknologi Pengolahan Emas Non Merkuri. 2019. Available online: https://sitkb3.menlhk.go.id/infomerkuri/?p=4652 (accessed on 25 January 2022).
13. Nafsiatun; Saptomo, P.; Najib, W.; Hartini. Characteristics of Environmental Conflicts Caused by Illegal Gold Mining in West Kalimantan, Indonesia. *J. Humanit. Soc. Sci.* **2019**, *3*, 1–3.
14. Trimiska, L.; Wiryono, W.; Suhartoyo, H. Kajian Penambangan Emas Tanpa Izin (PETI) di Kecamatan Lebong Utara, Kabupaten Lebong. *Naturalis* **2018**, *7*, 67–76. [CrossRef]
15. Herman, D.Z. Pertambangan Tanpa Izin (PETI) dan Kemungkinan Alih Status Menjadi Pertambangan Skala Kecil. Buletin Sumber Daya Geologi. Available online: http://psdg.geologi.esdm.go.id/buletin_pdf_file (accessed on 11 January 2022).
16. Darmono, D. *Mineral dan Energi Kekayaan Bangsa: Sejarah Pertambangan dan Energi Indonesia*; Darmono, D., Ed.; Departemen Energi dan Sumber Daya Mineral: Jakarta, Indonesia, 2009.
17. Rahmana, S. Pengaruh Pendirian Perusahaan Pertambangan Emas Kolonial Belanda di Lebong Tahun 1897–1930. *J. Aghinya Stiesnu Bengkulu* **2018**, *1*, 74–86.
18. Mulyadi, I.; Zaman, B.; Sumiyati, S. Mercury Concentrations of River Water and Sediment in Tambang Sawah Village Due to Unlicensed Gold Mining. *J. Ilm. Tek. Kim.* **2020**, *4*, 93–97.
19. Buchori, M.I.E. Kerusakan Lahan Akibat Kegiatan Penambangan Emas Tanpa Izin di Sekitar Sungai Singingi Kabupaten Kuantan Singingi. *J. Pembang. Wil. Dan Kota* **2019**, *15*, 174–188.
20. Rahim, S. Konflik Pemanfaatan Ruang Akibat Penambangan Emas Tanpa Ijin (PETI) di Kawasan Hutan Produksi Terbatas. *J. GeoEco* **2017**, *3*, 17–25.
21. Andriyanto, R.; Fitrisia, A. Eksplorasi dan eksploitasi penambangan emas Lebong Donok (Bengkulu) tahun 1897–1942. *Kronologi* **2019**, *1*, 10–21.
22. Available online: https://peraturan.bpk.go.id/Home/Details/53563/perpu-no-37-tahun-1960 (accessed on 11 January 2022).
23. Carlile, J.C.; Mitchell, A.H.G. Magmatic Arcs and Associated Gold and Copper Mineralization in Indonesia. *J. Geochem. Explor.* **1994**, *50*, 91–142. [CrossRef]
24. Available online: https://peraturan.bpk.go (accessed on 11 January 2022).
25. Male, Y.T.; Reichelt-Brushett, A.J.; Pocock, M.; Nanlohy, A. Recent mercury contamination from artisanal gold mining on Buru Island, Indonesia—Potential future risks to environmental health and food safety. *Mar. Pollut. Bull.* **2013**, *77*, 428–433. [CrossRef] [PubMed]
26. Mariwy, A.; Male, Y.T.; Manuhutu, J.B. Mercury (Hg) Contents Analysis in Sediments at Some River Estuaries in Kayeli Bay Buru Island. In *IOP Conference Series: Materials Science and Engineering*; IOP Publishing: Bristol, UK, 2019; Volume 546, p. 022012. [CrossRef]
27. Agus, A.; Dwimawanti, I.H. Rationality Conflict Between the Government and The Community: A Case Study on Illegal Gold Mining in Prabu Village, West Nusa Tenggara Province. In Proceedings of the 4th International Conference on Indonesian Social and Political Enquiries, Semarang, Indonesia, 21–22 October 2019; p. 9.
28. Law of the Republic of Indonesia Number 4 of 2009 on Mineral and Coal Mining. Available online: http://www.apbi-icma.org/uploads/files/old/2013/11/uu_no_4_2009_en.pdf (accessed on 11 January 2022).
29. Available online: https://peraturan.bpk.go.id/Home/Details/138909/uu-no-3-tahun-2020 (accessed on 11 January 2022).
30. Aziz, M. The model of traditional gold mining and its environmental management in the Paningkaban Village, Gumelar District, Banyumas Regency, Central Java. *Din. Rekayasa* **2014**, *10*, 20–28.
31. Kementerian Energi dan Sumber Daya Mineral. Perkembangan Kebijakan Sub Sektor Pertambangan Mineral dan Batubara. 2021. Available online: http://www.minerba.esdm.go.id (accessed on 11 January 2022).
32. Suherman, A. Usaha Tambang Emas. 2012. Available online: https://www.academia.edu/6849960/USAHA_TAMBANG_EMAS (accessed on 11 January 2022).
33. Zulkarnain, I. Mengenal Fenomena PETI di Kawasan Pertambangan Emas Pongkor. 2006. Available online: http://lipi.go.id/berita/mengenal-fenomena-peti-di-kawasan-pertambangan-emas-pongkor/233LIPI (accessed on 11 January 2022).
34. Hasibuan, O.; Tjakraatmadja, J.H.; Sunitiyoso, Y. Finding workable and mutually beneficial solution to eradicate illegal gold mining. *Bisnis Birokrasi J. Ilmu Adm. Dan Organ.* **2021**, *28*, 5. [CrossRef]
35. Kasworo, Y. Pertambangan Emas Tanpa Izin (PETI), Dapatkah Ditanggulangi? *J. Rechtsvinding*. 2015. Available online: https://rechtsvinding.bphn.go.id/jurnal_online/PETI_YERICO.pdf (accessed on 11 January 2022).
36. Nuzul, A.Y. Dampak Pertambangan Emas Ilegal di Aliran Sungai Batanghari Kabupaten Dharmasraya Sumatera Barat. *Res. Gate* **2018**. Available online: https://www.researchgate.net/publication/325310760 (accessed on 25 January 2022).

37. Santoso, D.H.; Gomareuzzaman, M. Kelayakan teknis penambang emas pada wilayah pertambangan rakyat (Studi kasus: Desa Kalirejo, Kecamatan Kokap, Kabupaten Kulon Progo. *J. Sci. Technol.* **2018**, *4*, 19–28. [CrossRef]
38. Sittadewi, E.H. Mitigasi lahan terdegradasi akibat penambangan melalui revegetasi. *J. Sains Dan Teknol. Mitigasi Bencana* **2016**, *11*, 51–60. [CrossRef]
39. Puluhulawa, U.F.; Harun, A.A. Biodiversity protection from the impact of illegal gold mining for sustainability. In *IOP Conference Series: Earth and Environmental Science*; IOP Publishing: Bristol, UK, 2019; Volume 519, p. 012031. [CrossRef]
40. Bruno, E.D.; Rubane, D.A.; Tiess, G.; Perrone, B.; Perrota, P.; Mikhailenko, A.; Ermolev, V.; Yashalova, N. Artisanal and small-scale gold mining, meandering tropical rivers, and geological heritage: Evidence from Brazil and Indonesia. *Sci. Total Environ.* **2019**, *715*, 136907. [CrossRef] [PubMed]
41. Amelia, N.R.; Kartodihardjo, H.; Sundawati, L. Peran modal sosial masyarakat penambang emas dalam mempertahankan tambang ilegal di Taman Hutan Raya Sulawesi Tengah. *J. Sylva Lestari* **2019**, *7*, 255–266. [CrossRef]
42. Wawo, A.R.H.; Widodo, S.; Jafar, N.; Yusuf, F.N. Analisis pengaruh penambangan emas terhadap kondisi tanah pada pertambangan rakyat Poboya Palu, Provinsi Sulawesi Tengah. *J. Geomine* **2017**, *5*, 116–119. [CrossRef]
43. Basuki, A. Penegakan Hukum Terhadap Tindak Pidana Penambangan Tanpa Izin oleh Polres Landak (Tinjauan Yuridis-Sosiologis). *J. Nestor Magister Huk.* **2018**, *1*.
44. Erb, M.; Mucek, A.E.; Robinson, K. Exploring a social geology approach in eastern Indonesia: What are mining territories? *Extr. Ind. Soc.* **2021**, *8*, 89–103. [CrossRef]
45. Khotami. Institution building dalam mengatasi persoalan pertambangan emas tanpa izin di Kabupaten Kuantan Singingi Provinsi Riau. *Nakhoda* **2020**, *19*, 17–37. [CrossRef]
46. Triyana, I.; Ikhwan, I. Dinamika Sosial dan Ekonomi Pekerja Tambang Emas Pasca Ditutupnya Tambang Emas Ilegal di Nagari Palangki Kabupaten Sijunjung. *Cult. Soc. J. Anthropol. Res.* **2019**, *1*, 84–89. [CrossRef]
47. Susanti, T.; Hidayat; Sartika, D.; Utami, W.; Viktres, R.H.; Novallyan, D. The influence of illegal gold mining (IGM) on environmental, economic, and educational sectors of Muara Mensao Village, Jambi. *Int. Conf. Basic Sci. Its Appl.* **2019**, *2019*, 307–322. [CrossRef]
48. Solichin, E. Implications of Illegal Gold Mining on the Household Economy and the Environment. *Saudi J. Bus. Manag. Stud.* **2020**, *5*, 70–73. [CrossRef]
49. Hasibuan, O.P.; Tjakraatmadja, J.H.; Sunitiyoso, Y. Illegal gold mining in Indonesia: Structure and causes. *Int. J. Emerg. Mark.* **2020**, *17*, 177–197. [CrossRef]
50. Rosjadi, D.; Taufiq, M. Efektivitas peranan kepolisian dalam menertibkan penambangan emas tanpa izin (PETI) yang dilakukan oleh masyarakat di lahan penambangan PT AntamTbk dari sisi pembangunan berkelanjutan. *J. Huk. De'rechtsstaat* **2019**, *5*, 119–128. [CrossRef]
51. Pamungkas, P.D. Efektivitas penyidikan tindak pidana penambangan emas tanpa izin di Kabupaten Solok Selatan (Studi pada Direktorat Reserse Kriminal Khusus Kepolisian Daerah Sumatera Barat). *UNES Law Rev.* **2018**, *1*, 134–147. [CrossRef]
52. Arif, I. *Emas Indonesia*; Gramedia Pustaka Utama: Jakarta, Indonesia, 2020; 325p.
53. Ismawati, Y. Gold, Mercury, and the Next Minamata. *Indones. J. Leadersh. Policy World Aff.* **2014**. Available online: https://ipen.org/news/gold-mercury-and-next-minamata-article-yuyun-ismawati-ipen-steering-committee-member-and-lead (accessed on 11 January 2022).
54. Guswahyuni, S.M. Pertambangan Emas Tanpa Izin dan Masalah Public Health Dari Anak-Anak Yang Tinggal di Sekitar. In Proceedings of the 3rd UGM Public Health Symposium, Yogyakarta, Indonesia, 7–9 May 2018. [CrossRef]
55. Bank, M.S. The mercury science-policy interface: History, evolution and progress of the Minamata Convention. *Sci. Total Environ.* **2020**, *722*, 137832. [CrossRef]
56. Krisnayanti, B.D.; Probiyantono, A.S. (Eds.) *Teknologi Pengolahan Emas Pada Pertambangan Emas Skala Kecil di Indonesia, buku 4*; 2020, 17p. Available online: https://goldismia.org/sites/default/files/2020-12/View%20Buku%204%20%281%29.pdf. (accessed on 11 January 2022).
57. Mahmud, M. Model Sebaran Spasial Temporal Konsentrasi Merkuri Akibat Penambangan Emas Tradisional Sebagai Dasar Monitoring dan Evaluasi Pencemaran di Ekosistem Sungai Tulabolo Provinsi Gorontalo. Ph.D. Thesis, Gajahmada University, Yogyakarta, Indonesia, 2012.
58. Gundo, S.D.I.; Polii, B.J.V.; Umboh, J.M.L. Kandungan merkuri pada penambang emas rakyat. *Indones. J. Public Health Community Med.* **2020**, *1*, 13–18.
59. Prasetia, H.; Sakakibara, M.; Omori, K.; Laird, J.S.; Sera, K.; Kurniawan, I.A. Mangifera indica as bioindicator of mercury atmospheric contamination in an ASGM area in North Gorontalo Regency, Indonesia. *Geosciences* **2018**, *8*, 31. [CrossRef]
60. Zhang, H.; Feng, X.; Larssen, T.; Qiu, G.; Vogt, R.D. In inland China, rice, rather than fish, is the major pathway for methylmercury exposure. *Env. Health Perspect* **2010**, *118*, 1183–1188. [CrossRef] [PubMed]
61. Junaidi, M.; Krisnayanti, B.D.; Juharfa, J.; Anderson, C. Risk of Mercury Exposure from Fish Consumption at Artisanal Small-Scale Gold Mining Areas in West Nusa Tenggara, Indonesia. *J. Health Pollut.* **2019**, *9*, 190302. [CrossRef]
62. Ekawanti, A.; Krisnayanti, B.D. Effect of Mercury Exposure on Renal Function and Hematological Parameters among Artisanal and Small-scale Gold Miners at Sekotong, West Lombok, Indonesia. *Health Pollut.* **2015**, *5*, 25–32. [CrossRef] [PubMed]

63. Ismawati, Y. *Children's Exposure to Mercury in Artisanal and Small-Scale GOLD Mining Areas in Indonesia and in More than 70 Countries*; UN Comminttee on the Rights of the Child: Geneva, Switzerland, 2016.
64. Afrifa, J.; Opoku, Y.K.; Gyamerah, E.O.; Ashiagbor, G.; Sorkpor, R.D. The clinical importance of the mercury problem in artisanal small-scale gold mining. *Front. Public Health* **2019**, *7*, 131. [CrossRef]
65. Undang-Undang Nomor 11 tahun 2017 Tentang Pengesahan Minamata Convention on Mercury. Available online: https://peraturan.bpk.go.id/Home/Details/53614#:~:text=UU%20No.%2011%20Tahun%202017,Merkuri) (accessed on 11 January 2022).
66. Lembaran Negara Republik Indonesia. Available online: https://peraturan.go.id/common/dokumen/ln/2019/ps21-2019.pdf (accessed on 11 January 2022).
67. Kementerian Energi dan Sumber Daya Mineral. 2021. Perkembangan Kebijakan Sub Sektor Pertambangan Mineral dan Batubara. Available online: https://jdih.esdm.go.id/storage/document/Permen%20ESDM%20Nomor%2016%20Tahun%202020.pdf (accessed on 11 January 2021).
68. Djatmiko, A.; Purwendah, E.K.; Pudyastiwi, E. Benefits of Indonesia Ratification of Minamata Convention on Mercury. *Int. J. Bus. Econ. Law* **2019**, *18*, 1–6.
69. Available online: https://www.goldismia.org/sites/default/files/2021-06/Annual%20Report%202020-8.pdf (accessed on 11 January 2022).
70. Spiegel, S.J.; Agrawal, S.; Mikha, D.; Vitamerry, K.; Billon, P.L.; Veiga, M.; Konolius, K.; Paul, B. Phasing Out Mercury? Ecological Economics and Indonesia's Small-Scale Gold Mining Sector. *Ecol. Econ.* **2018**, *144*, 1–11. [CrossRef]
71. Aylmore, M.G.; Muir, D.M. Thiosulfate leaching of gold- A. Review. *Miner. Eng.* **2001**, *14*, 135–174. [CrossRef]
72. Yustanti, E.; Guntara, A.; Wahyudi, T. Ekstraksi Bijih Emas Sulfida Tatelu Minahasa Utara Menggunakan Reagen Ramah Lingkungan Tiosulfat. *Teknika* **2018**, *12*, 97–106. [CrossRef]
73. Wahyudin, P.; Mubarok, M.Z. Perilaku Adsorpsi Emas dari Larutan Ammonium Thiosulfat dengan Karbon Aktif dan Resin Penukar Ion. *Metalurgi* **2016**, *2*, 69–78.
74. Matsumoto, Y.; Kasamatsu, H.; Sakakibara, M. Challenges in Forming Transdisciplinary Communities of Practice for Solving Environmental Problems in Developing Countries. *World Futures* **2022**. [CrossRef]
75. Hindratmo, B.; Rita, R.; Masitoh, S.; Kusumardhani, M.; Junaedi, E. Kandungan Logam Berat Merkuri (Hg) pada Area Bekas Penambangan Emas Skala Kecil (PESK): Studi Kasus di Gunung Botak, Kabupaten Buru, Provinsi Maluku. *Ecolab* **2019**, *13*, 127–132. [CrossRef]
76. Indah, M.F.; Agustina, N.; Ariyanto, E. Analysis of Mercury Levels, Degree of Acidity and Health Risk Factors for Unlicensed Gold Miners in Cempaka Sub-District. Bul. *Penelit. Kesehat.* **2020**, *48*, 281–290. [CrossRef]
77. Palapa, T.M.; Maramis, A.A. Heavy Metals in Water of Stream Near an Amalgamation Tailing Ponds in Talawaan –Tatelu Gold Mining, North Sulawesi, Indonesia. *Procedia Chem.* **2015**, *14*, 428–436. [CrossRef]
78. Hidayanti, K. Distribusi logam berat pada air dan sedimen serta potensi bioakumulasi pada ikan akibat penambangan emas tanpa izin (Studi kasus: DAS Sekonyer, Kalimantan Tengah). *Media Ilm. Tek. Lingkung.* **2019**, *4*, 24–33. [CrossRef]
79. Fakaubun, F.R.; Male, Y.T.; Selanno, D.A.J. Bioconcentration and Bioaccumulation of Mercury (Hg) in Seagrass *Enhalus Acoroides* in Kayeli Bay, Buru Regency, Maluku Province. *Indones. J. Chem. Res.* **2020**, *8*, 159–166.
80. Susanti, T.; Utami, W.; Hidayat, H. The negative impact of illegal gold mining on the environmental sector in Batang Asai, Jambi. *Sustinere J. Environ. Sustain.* **2018**, *2*, 108–167. [CrossRef]
81. Sakakibara, M.; Sera, K.; Kurniawan, I.A. Mercury contamination of cattle in artisanal and small-scale gold mining in bombana, Southeast Sulawesi, Indonesia. *Geosciences* **2017**, *7*, 133.
82. Harianja, A.H.; Saragih, G.S.; Fauzi, R.; Hidayat, M.Y.; Syofyan, Y.; Tapriziah, E.R.; Kartiningsih, S.E. Mercury Exposure in Artisanal and Small-Scale Gold Mining Communities in Sukabumi, Indonesia. *J. Health Pollut.* **2020**, *10*, 201209. [PubMed]
83. Arifin, Y.; Sakakibara, M.; Sera, K. Impacts of artisanal and Small-Scale Gold Mining (ASGM) on environment and human health of Gorontalo Utara Regency, Gorontalo Province, Indonesia. *Geosciences* **2015**, *5*, 160–176. [CrossRef]
84. Sakakibara, M.; Sera, K. Current mercury exposure from artisanal and small-scale gold mining in Bombana, Southeast Sulawesi, Indonesia—future significant health risks. *Toxics* **2017**, *5*, 7.
85. Rumatoras, H.; Taipabu, M.I.; Lesiela, L. Analysis of mercury (Hg) content on hair villagers Kayeli, ilegal gold mining result in Botak Mountain Area, Buru Regency-Maluku Province. *Ind. J. Chem. Res.* **2016**, *3*, 290–294.
86. Planet Gold Gold-Ismia. Available online: https://www.planetgold.org/indonesia (accessed on 11 January 2022).
87. Program Emas Rakyat Sejahtera. Quaterly Email Newsletters. 2019. Available online: https://pers.no-hg.org/wordpress/wp-content/uploads/2019/12/e-Newsletter-PERS_Bahasa-Version_Final_Final.pdf (accessed on 12 December 2021).
88. Zuhdi, S.; Wahyudi, B.; Munawwaroh, T. The role local goverment in gold mine conflict handling in Trenggalek Regency, East Jawa Province. *J. Prodi Damai Dan Resolusi Konflik* **2018**, *4*, 45–71.
89. Libassi, M. Matthew Libassi Mining heterogeneity: Diverse labor arrangements in an Indonesian informal gold economy. *Extr. Ind. Soc.* **2020**, *7*, 1036–1045.
90. Abbas, H.H.; Sakakibara, M.; Sera, K.; Arma, L.H. Mercury Exposure and Health Problems in Urban Artisanal Gold Mining (UAGM) in Makassar, South Sulawesi, Indonesia. *Geosciences* **2017**, *7*, 44. [CrossRef]
91. Arifin, Y.I.; Sakakibara, M.; Sera, K. Heavy metals concentrations in scalp hairs of ASGM miners and inhabitants of the Gorontalo Utara regency. *IOP Conf. Ser. Earth Environ. Sci.* **2017**, *71*, 012028. [CrossRef]

92. Zaharani, F. Salami IRS. Kandungan Merkuri Pada Urin Dan Rambut Sebagai Indikasi Paparan Merkuri Terhadap Pekerja Tambang Emas Tanpa Izin (Peti) Di Desa Pasar Terusan Kecamatan Muara Bulian Kabupaten Batanghari–Jambi. *J. Teh Lingkung* **2015**, *21*, 168–179.
93. Sumantri, A.; Laelasari, E.; Junita, N.R.; Nasrudin, N. Logam Merkuri pada Pekerja Penambangan Emas Tanpa Izin. *Kesmas Natl. Public Health J.* **2014**, *8*, 398–403. [CrossRef]
94. Ismawati, Y. Preliminary Environmental Health Report from 3 ASGM Hotspots in Indonesia: Bombana, Sekotong and Cisitu. 2015. Available online: https://www.academia.edu/16013226/Preliminary_environmental_health_report_from_3_ASGM_Hotspots_in_Indonesia_Bombana_Sekotong_and_Cisitu (accessed on 25 January 2022).
95. Lestarisa, T. Faktor-Faktor Yang Berhubungan Dengan Keracunan Merkuri (Hg) Pada Penambang Emas Tanpa Ijin (PETI) di Kecamatan Kurun, Kabupaten Gunung Mas, Kalimantan Tengah. Master's Thesis, Diponegoro University, Semarang, Indonesia, 2010.
96. Putri, G.E. Gejala Kesehatan yang diderita penambang emas akibat proses penambangan emas menggunakan merkuri (Hg). *Med. St.* **2017**, *8*, 69–78. [CrossRef]
97. Junita, N.R. Resiko keracunan merkuri (Hg) pada pekerja penambang emas tanpa izin (PETI) di Desa Cisarua Kecamatan Nanggung Kabupaten Bogor Tahun 2013. Bachelor Thesis, Syarif Hidayatullah Islamic State Univeristy, Jakarta, Indonesia, 2013. Available online: https://repository.uinjkt.ac.id/dspace/bitstream/123456789/24270/1/NITA%20RATNA%20JUNITA-fkik.pdf (accessed on 11 January 2022).
98. Kristianingsih, Y. Bahaya merkuri pada masyarakat di Pertambangan Emas Skala Kecil (PESK) Lebaksitu. *J. Ilm. Kesehat.* **2018**, *10*, 32–38. [CrossRef]
99. Singga, S. Analisis resiko kesehatan pajanan pada masyarakata Kecamatan Bulawa, Kabupaten Bone Bolango Provinsi Gorontalo. *J. MKMI* **2013**, *9*, 21–28.
100. Available online: https://jdih.esdm.go.id/storage/document/PP%20No.%2023%20Thn%202010.pdf (accessed on 5 March 2022).
101. Available online: https://jdih.esdm.go.id/storage/document/UU%2032%20Tahun%202009%20(PPLH).pdf (accessed on 5 March 2022).
102. Available online: https://peraturan.bpk.go.id/Home/Details/4835/pp-no-23-tahun-2008#:~{}:text=PP%20No.%2023%20Tahun%202008,Penanggulangan%20Bencana%20%5BJDIH%20BPK%20RI%5D (accessed on 5 March 2022).
103. Available online: https://peraturan.bpk.go.id/home/Details/38685/uu-no-23-tahun-2014 (accessed on 5 March 2022).
104. Muslihudin, M.; Santosa, I.; Setyoko, P.I.; Bahtiar, R.A. Local Government's Role and Policy on Illegal Mining (case Study of gold Mining in Banyumas Indonesia). *Am. J. Humanit. Soc. Sci. Res.* **2020**, *4*, 275–282. Available online: https://www.researchgate.net/publication/339445336 (accessed on 25 January 2022).
105. Muryani, E. Sinergisitas Penegakan Hukum Pada Kasus Pertambangan Emas Tanpa Izin di Kabupaten Banyumas, Jawa Tengah. *J. Best.* **2019**, *7*, 84–92. [CrossRef]
106. Nainggolan, P. Resistensi Penambang Ilegal: Studi Kasus eksploitasi Tambang Galian B (Emas) di Desa Sayurmatua Kecamatan Naga Juang Kabupaten Mandailing Natal. *J. Buana* **2018**, *2*, 870–881. [CrossRef]
107. Yani, R.F.; Asrinaldi, A.; Rahmadi, D. Peran WALHI Sumbar dalam Investigasi Tambang Emas Ilegal di Kota Padang. *J. Demokr. Dan Polit. Lokal* **2019**, *1*, 88–100. [CrossRef]
108. Damaryanti, H.; Yenny, A.S. Implementation of Customary Sanction "Pengocek Torun" in Dayak Simpakng Community to Settlement of Illegal Gold Mining Ketapang Regency. *Adv. Soc. Sci. Educ. Humanit. Res.* **2020**, *477*, 308–311. [CrossRef]
109. Tinov, M.T. Strengthening institutions in the effort adat costumary law enforcement in illegal gold mining areas affected. *J. Niara* **2019**, *12*, 19–28. [CrossRef]
110. Pateda, S.; Gorontalo State University, Golontalo, Indonesia. Personal communication, 2021.
111. Krisnayanti, B.D.; Probiyantono, A.S. (Eds.) Penggunaan merkuri dan dampaknya terhadap lingkungan serta sebaran lokasi pertambangan emas skala kecil, buku 2. 2020, 13p. Available online: https://www.goldismia.org/sites/default/files/2021-01/View%20Buku%202.pdf (accessed on 11 January 2022).
112. Wafid, M.A.N.; Ministry of Energy and Mineral Resources, Jakarta, Indonesia. Personal communication, 2022.

International Journal of
Environmental Research and Public Health

MDPI

Article

Investigation of Long-Term Roving Artisanal and Small-Scale Gold Mining Activities Using Time-Series Sentinel-1 and Global Surface Water Datasets

Satomi Kimijima [1,*], **Masayuki Sakakibara** [1,2] and **Masahiko Nagai** [3,4]

[1] Research Institute for Humanity and Nature, Kyoto 603-8047, Japan; sakaki@chikyu.ac.jp
[2] Graduate School of Science & Engineering, Ehime University, Matsuyama 790-8577, Japan
[3] Graduate School of Science and Technology for Innovation, Yamaguchi University, Ube 755-8611, Japan; nagaim@yamaguchi-u.ac.jp
[4] Center for Research and Application of Satellite Remote Sensing, Yamaguchi University, Ube 755-8611, Japan
* Correspondence: kimijima@chikyu.ac.jp

Abstract: Artisanal and small-scale gold mining (ASGM) is a significant source of gold production globally despite the sector being informal and illegal. The rapid increase in the number of roving mining camps has negatively impacted the surrounding environment; however, the formation and transformation of roving mining camps have not been well studied. This study investigated the long-term trends and significant hotspots of roving camp-type ASGM (R-C-ASGM) in Katingain Regency, Central Kalimantan Province, Indonesia, from 1988 to 2020 using remotely sensed data, including Sentinel-1 time-series, global surface water (GSW), and world landcover datasets. Results show that several active R-C-ASGM sites existed in the Galangan and Kalanaman areas in 2017/2018. According to the GSW dataset, the Galangan area was estimated to be formed earlier, whereas the Kalanaman areas were recently formed and were associated with the Kalanaman river expansion. Notably, the center of Galangan was still a significant R-C-ASGM hotspot. The findings of this study broaden our understanding of R-C-ASGM transformation and identify significant R-C-ASGM hotspots over a long period. This study contributes to the development of timely and appropriate interventions for strengthening environmental governance.

Keywords: active mining; alluvial mining; artisanal and small-scale gold mining; Indonesia; remote sensing; SAR; surface water occurrence

Citation: Kimijima, S.; Sakakibara, M.; Nagai, M. Investigation of Long-Term Roving Artisanal and Small-Scale Gold Mining Activities Using Time-Series Sentinel-1 and Global Surface Water Datasets. *Int. J. Environ. Res. Public Health* **2022**, *19*, 5530. https://doi.org/10.3390/ijerph19095530

Academic Editor: Paul B. Tchounwou

Received: 30 March 2022
Accepted: 22 April 2022
Published: 2 May 2022

Publisher's Note: MDPI stays neutral with regard to jurisdictional claims in published maps and institutional affiliations.

1. Introduction

Artisanal and small-scale gold mining (ASGM) is a significant source of gold production globally. ASGM is the world's largest employer in the gold mining sector, employing 70–80% of informal small-scale workers [1]. This has been carried out continuously in more than 80 countries as a tool for poverty alleviation and socioeconomic development [2,3]. In Indonesia, both active and inactive ASGM sites have been identified in 93 regencies in 30 of the country's 34 provinces, with more than 1200 hotspots estimated in 2017 [4] and 250,000–300,000 miners [5]. Although this sector has provided economic benefits at various levels, the substantial harmful environmental and health risks associated with mercury pollution are devastating [6–10].

Kalimantan Island is an ASGM hotspots with alluvial operations [4]. Illegal mining activities are widespread on this island, even in conserved areas, negatively affecting biodiversity and human health [4]. Alluvial-based ASGM activities affect waterbodies and the surrounding environment, resulting in deforestation; high mercury contamination; and changes in geomorphological processes, biogeographic conditions, hydrological regime, and river courses [11,12].

The ASGM sector can be categorized into "travel-type" and "camp-type." For the former, miners commute daily from their local residences to the mining sites. For the latter (hereafter called C-ASGM), miners live and conduct mining activities at informal worksites [13]. C-ASGM can be further categorized as either roving or non-roving practices. Both in terms of size and workforce, the ASGM sector has grown in tandem with increases in gold prices since 2000 [14]. Studies of C-ASGM in Indonesia have demonstrated the sector's growth and the magnitude of its activities [13,15,16]. Accordingly, the sector's rapid growth is anticipated to accelerate the environmental and health risks at various levels and on a wider scale.

As an adopter of the Minamata Convention on Mercury (MCM) initiated by the United Nations Environment Programme, Indonesia is attempting to reduce mercury use in the ASGM sector. The MCM is a global treaty that protects human health and the environment from anthropogenic emissions, mercury releases, and its compounds [17] and was adapted and implemented in October 2013 and August 2017, respectively [18]. The MCM's Article 7 focuses primarily on the ASGM sector, facilitating the formalization of action plans and various regulations at a country level among the ratifying nations [19]. However, the policy formalization has often been hindered by insufficient institutional frameworks, capacities, and funds [20,21].

In addition to ASGM being informal, illegal, and unregulated, the uncontrollability of the sector by law allows for increased mercury use, endangering the environment and human health. Furthermore, geographical characteristics of the C-ASGM sector, for instance, the fact that most C-ASGM sites are located in remote rural areas, restrict the collection of information, such as the sector's status and transformation, making it difficult to monitor them [13]. Remote sensing technologies enable the monitoring of time-series spatial changes in the C-ASGM sector to gain a better understanding of the sector. Satellite observations have been used to quantify the C-ASGM sector [13,15,16]. Cloud-free data sets are difficult to obtain in areas prone to heavy rainstorms. However, synthetic aperture radar (SAR), an active independent Earth observation system [22], can be a suitable alternative to measure optical data [23], facilitating the qualitative and comprehensive understanding of the C-ASGM sector.

Previous studies on the transformation of roving camp-type ASGM (R-C-ASGM) have investigated the active and inactive status of R-C-ASGM and its changes using Sentinel-1 (S-1) time-series datasets [16]. However, the availability of S-1 datasets is restricted to 2014. To investigate long-term R-C-ASGM practices, tracking surface water occurrence (SWO), which represents the frequency of land surface water from 1984 to 2020, associated with the spatial distributions of active R-C-ASGM sites may be key to recognizing the formation periods of mines in this sector. Therefore, this study primarily investigated a long-term trend and significant hotspots of R-C-ASGM from 1988 to 2020 in Katingan Regency, Central Kalimantan Province, Indonesia, using S-1 time series, global surface water (GSW), and world land cover datasets.

2. Materials and Methods

2.1. Overall Methodological Workflow

The methodological workflow used in this study is depicted in Figure 1. This workflow comprised four main steps to achieve its primary objective of investigating long-term R-C-ASGM trends and significant R-C-ASGM hotspots. First, the R-C-ASGM status from 2015 to 2020 was identified using the S-1 temporal time series. Second, the surface water extents observed from 1988 to 2020 were extracted. Third, targeted SWOs were extracted, along with a landcover map. Fourth, significant R-C-ASGM hotspots were identified by overlaying the results generated in steps 1 and 3. The results from these steps improve our understanding of the significant long-term trend of R-C-ASGM practices in the study site. This paper presents a discussion based on the findings described above. The methods employed in each step are explained in the following sections.

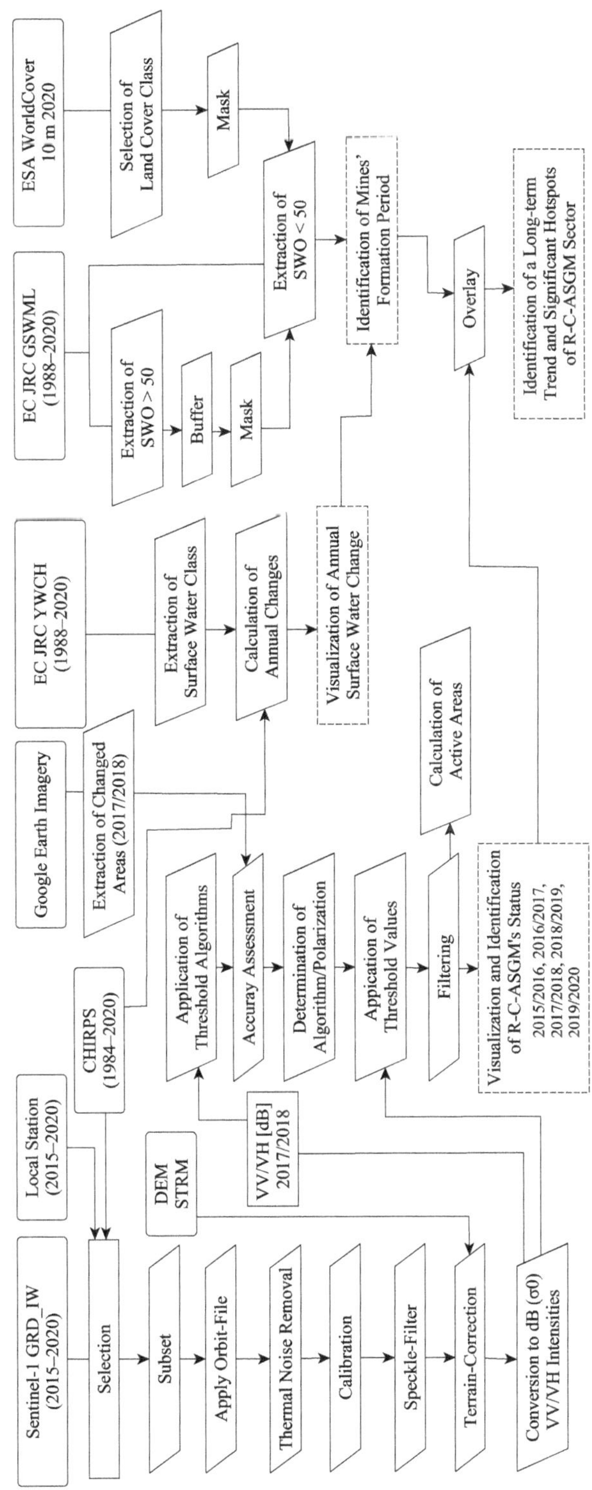

Figure 1. Overall methodology.

2.2. Study Area

Indonesia is a well-mineralized metallogenic region with significant gold mineralization associated with quartz veins in andesite-hosted epithermal settings. Gold-bearing alluvial soils in Central Kalimantan, a significant ASGM hotspot, have attracted several ASGM-targeted migrants from Java and South Kalimantan [24]. In this study, we focused on mining activities in Galangan, the center of Katingan Regency, Central Kalimantan Province, Indonesia, where the alluvial-based mining method is employed (Figure 2). In the early 1990s, the Galangan mining region rapidly developed and was designated as the geographical and historical center of land-based mining areas [25]. Especially, Hampalit town served as a base for mining activities for both indigenous miners and gold companies [24,26]. To date, migrated miners have continuously practiced R-C-ASGM to greater extents in various areas ranging from Kalanaman, Pundu, and Galangan to explore newer locations with more outstanding gold production by season [24]. Particularly, alluvial-based ASGM in this region employs various techniques, such as open pits, deep excavation pits, and floating pumps. For example, the open-pit method involves removing all soils and vegetation landscape from the surface, creating a barren wasteland [24,26]. Moreover, the floating pump method disturbs riverbanks and increases sediment volumes [24].

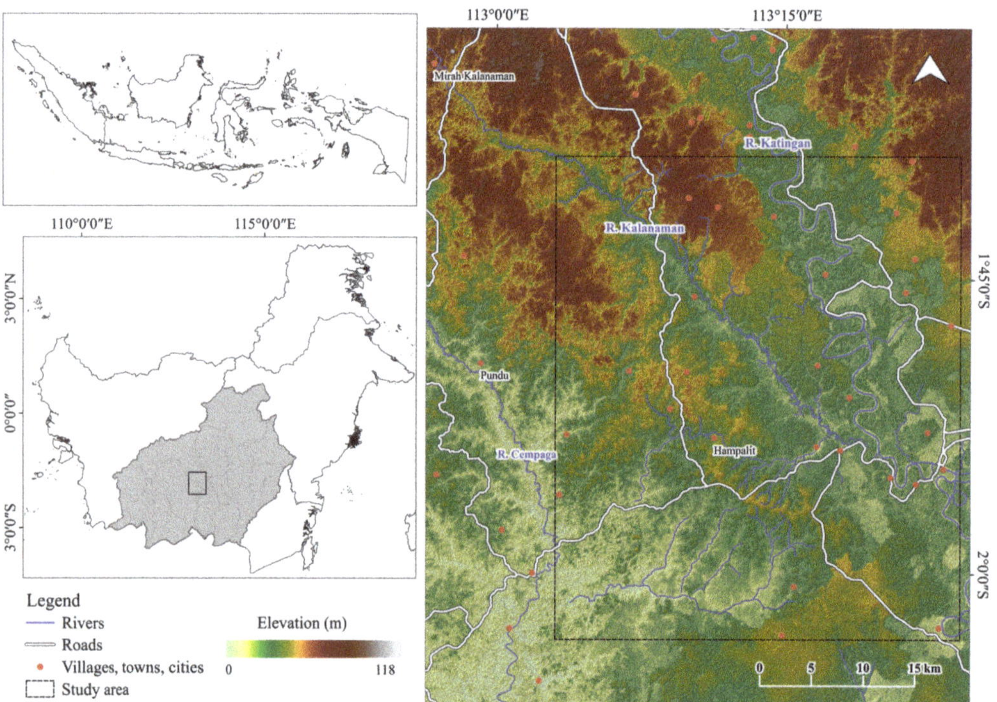

Figure 2. Study area.

2.3. Identification of Active and Inactive Status of R-C-ASGM Sites

2.3.1. S-1 Imagery

Six level-1 grand range detected (GRD) S-1 datasets covering 2015–2020 downloaded from the European Space Agency (ESA) were used to investigate the status of R-C-ASGM practice and its change. Through the EU/ESA Copernicus program, the S-1 mission (S-1A and S-1B) provides an exceptional combination of high spatial (10 m) and temporal (6 days) resolution data by operating two polar-orbiting radar imaging systems working with C-band (~5.7 cm wavelength). The main operational mode is the Interferometric Wide swath mode (IW) with VV/VH polarizations, which are freely and routinely available [27]. By referring to the Climate Hazards Group InfraRed Precipitation with Station (CHIRPS) and local station data, we used datasets acquired between July and August for the study period with a relative orbit number of 3 for a better comparison of the backscatter intensity of each image. Table 1 summarizes the main specifications of the databases used in this study.

Table 1. Main specification of satellite imagery used in this study.

Satellite	Type	Acquisition Date	Spatial Resolution	Image Number	Polarization	Wavelength
Sentinel-1	C-SAR	20 July 2015 7 August 2016 21 July 2017 4 July 2018 11 July 2019 10 August 2020	10 m	3	Descending (VV, VH)	C band

2.3.2. Image Preprocessing

The ESA's open-source software, Sentinel Application Platform (version 8.0.0), was used for image preprocessing. The following five steps were implemented in the S-1 Tool-box: (1) orbit correction, (2) thermal noise removal, (3) radiometric calibration, (4) speckle filtering with 5×5 windows, and (5) terrain correction using the 3-arcsec digital elevation model (DEM) from the Shuttle Radar Topography Mission (SRTM) [28]. As a result, all imagery was converted to the digital pixel value of S-1 images, resulting in an image intensity value of σ^0.

2.3.3. Selection of Threshold and Detection of Changed Extents in Time Series

After image preprocessing, we identified optimized threshold values by the VV/VH polarizations acquired in 2017 and 2018. Sixteen different automatic global thresholding algorithms [29–43] were used in this process to identify mining-induced areas using an open-source Java image processing package, namely, Fiji (version 2.1.0) software (https://imagej.net/software/fiji/ accessed on 1 March 2022). Moreover, a supervised classification method, such as histogram intersection, was used. To determine the best separability for change detections, results were validated using high-resolution images obtained on 9 June 2017, and 23 September 2018, through Google Earth Pro. One hundred points were randomly selected from the datasets. Consequently, the best combination of algorithm and polarization was applied to all datasets. Next, the post-classification of a majority filter with a moving window size of 5×5 pixels was used to remove isolated pixels. After extracting the differences between the two target years, the areas observed in the river buffers were further eliminated to remove mudflats, possibly caused by the changes in the annual precipitation volume between the years. Consequently, the annual changes in the extent of illegal mining were computed for the following five temporal series: 2015/2016, 2016/2017, 2017/2018, 2018/2019, and 2019/2020. The correlation between detected active areas and the Indonesian gold price was evaluated statistically at the 95% confidence level.

2.4. Identification of SWO

In this process, European Commission (EC) Joint Research Centre (JRC) Yearly Water Classification History, v1.3 (YWCH, 1988–2020), EC JRC Global Surface Water Mapping Layers v1.3 (GSWML, 1988–2020), and ESA WorldCover 10 m 2020 (WC2020, 2000) datasets were used to determine a long-term change in surface water extent and its occurrence. The YWCH dataset contains yearly classifications of the seasonality of water detected throughout the year [44]. The GSWML dataset contains different facets of surface water data. Both datasets were generated on the basis of Landsat 5, 7, and 8 with 30 m ground resolution. Further, the WC2020 dataset provides a global land cover map of 2020 generated on the basis of S-1 and Sentinel-2 datasets with 10 m ground resolution.

First, seasonal and permanent water classes were extracted from the YWCH dataset, and long-term changes in surface water extents were identified. Trends of annual permanent and seasonal surface water were evaluated statistically using Sen's Slope test with significance at the 95% confidence level. Meanwhile, the trend of monthly precipitation was calculated from 1985 to 2020 to validate the obtained results. Second, the occurrence band of the GSWML dataset, representing the frequency of water from 1984 to 2020, was primarily used to investigate specific surface water extents, which R-C-ASGM activities may have caused. Here, SWO greater and less than 50% (SWO > 50 and SWO < 50) were primarily considered permanent surface water and temporal- and mining-induced surface water extents, respectively. Third, a 50 m buffer was applied to SWO > 50 images. Fourth, the barren/sparse vegetation class among 11 land classes was extracted from the WC2020 dataset. Fifth, SWO < 50 images were masked by the buffered SWO > 50 and the barren/sparse vegetation class extracted from the WC2020 dataset. Here, buffering was applied to avoid possible errors, such as over-extraction of SWO < 50 at river edges, probably caused by changes in annual precipitation volume.

2.5. Identification of Long-Term Trends and Hotspots of R-C-ASGM

Long-term trends of R-C-ASGM and its hotspots were investigated by overlapping the results generated from Sections 2.3 and 2.4.

3. Results

3.1. Determination of Threshold and Polarization Channels

Both VV/VH polarizations acquired in 2017 and 2018 were primarily used to derive the best combination of algorithm and polarization to identify illegal mining-induced landcover changes. Changed areas identified from each result were validated using features extracted from high-resolution Google Earth images, as mentioned in Section 2.5. Thus, this study found the best locally sensitive algorithm and polarization combination to be the IJ_Isodata algorithm and VH polarization, achieving 76.0% accuracy. Subsequently, the following image-specific thresholding values were generated for the final classification, probably leading to better results in detecting annual changes caused by active R-C-ASGM activities: -20.88 (2015), -19.95 (2016), -21.47 (2017), -20.16 (2018), -20.36 (2019), and -20.76 (2020).

3.2. Transformation of R-C-ASGM in Time Series

The occurrences of active mining sites in the five periods (2015/2016, 2016/2017, 2017/2018, 2018/2019, and 2019/2020) are shown in Figure 3. The possible active mining areas were estimated to be 18.2 km^2 (2015/2016), 6.5 km^2 (2016/2017), 26.2 km^2 (2017/2018), 14.5 km^2 (2018/2019), and 4.8 km^2 (2019/2020). The 2017/2018 period exhibited the peak change; meanwhile, smaller changes were observed in 2016/2017 and 2019/2020 in the study area. The detected areas were primarily located in the center of the Galangan region and the west of the Kalanaman River. After the implementation of the MSM in 2018, the mining activities in these detected areas decreased. Additionally, a negative correlation of -0.51 was found between the detected active areas and the Indonesian gold price from 2016 to 2020.

3.3. Surface Water Extents in Time Series

An increase in yearly surface water extent was found in the study area, from 20.4 km^2 in 1998 to 33.4 km^2 in 2020 (Figure 4). The analysis indicated that an increase in the amount of seasonal water was observed from 2000. According to the statistical test described in Section 2.4, positive increase trends were found in both permanent and seasonal water. Furthermore, a higher slope of 0.30 was found in seasonal water than in permanent water

(0.12). In comparison, no trends were statistically identified from the monthly precipitation during 1985/2020 in the study area.

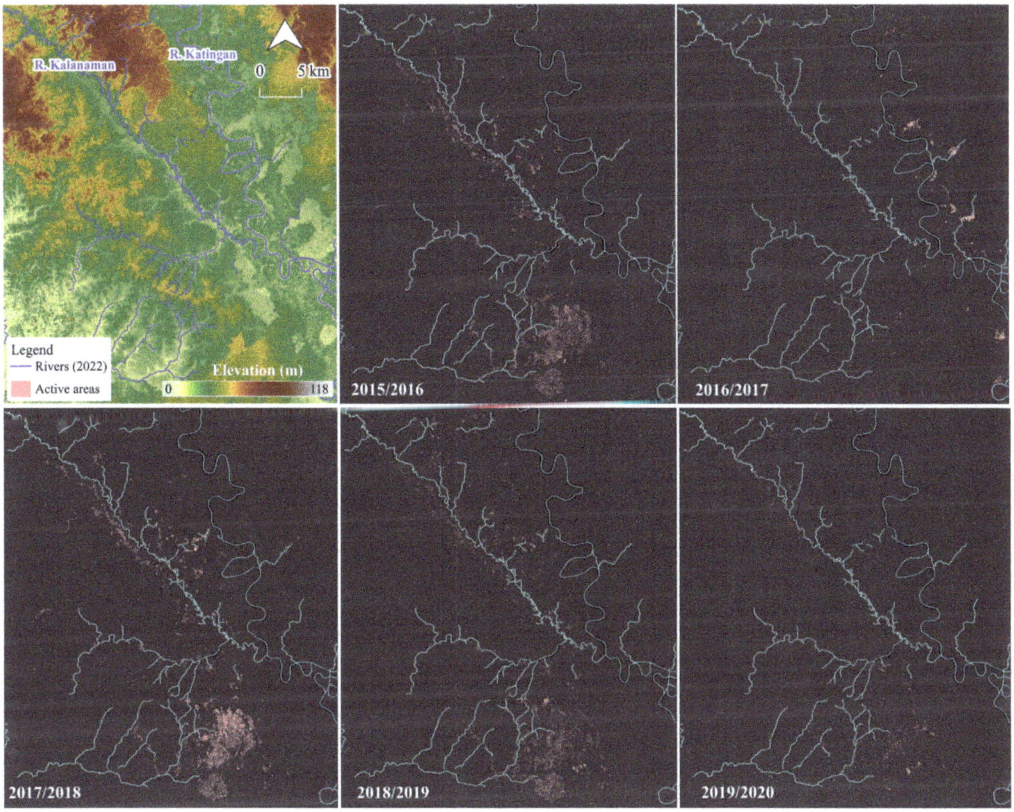

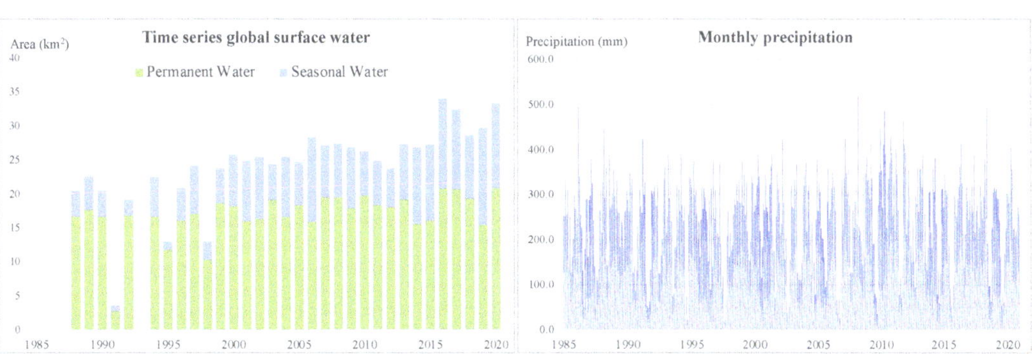

Figure 3. Transformation of active R-C-ASGM sites detected from IJ_Isoda algorithm and VH polarization.

Figure 4. Trends of global surface water and monthly precipitation.

Changes in the yearly amount of surface water, particularly in 1988, 2000, 2010, and 2020, are shown in Figure 5. There were no changes in the Katingan River, whereas in the Kalanaman River, some changes were observed in 2000 and 2010, and the river's extent expanded toward the northwestern parts of the Kalanaman area, forming the river (B in Figure 5). Moreover, surface water was also observed at the center of the Galangan area between 2000 and 2010 (A in Figure 5). This site is located further away from the main river networks but expanded its extent until 2020.

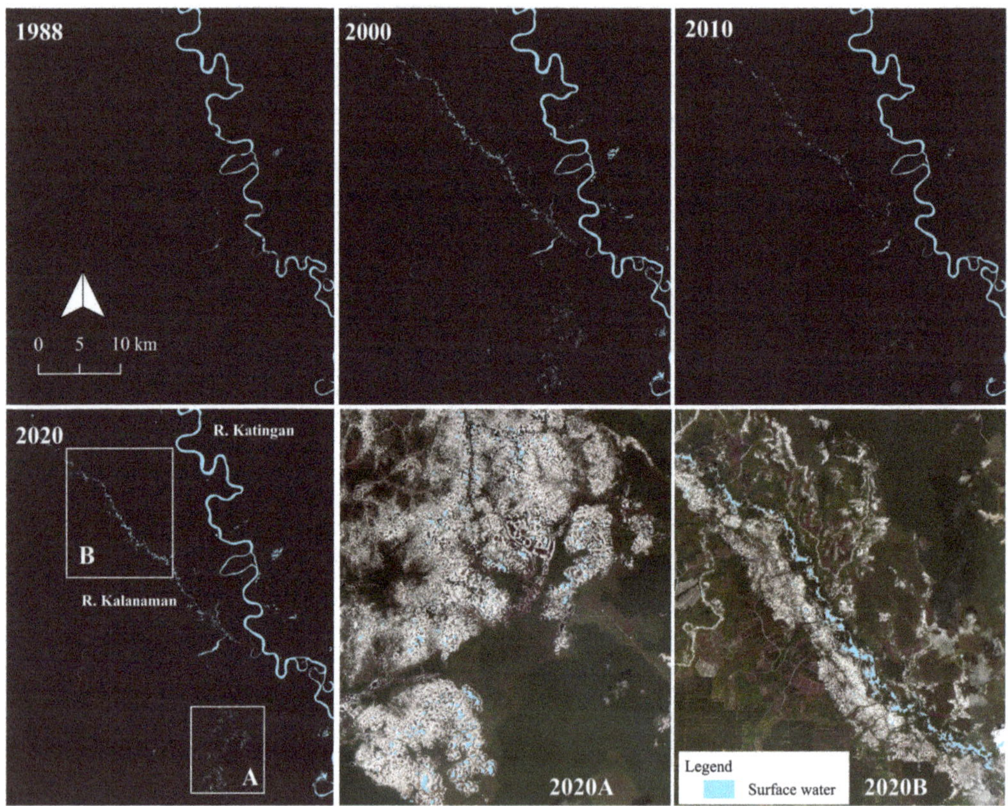

Figure 5. Changes in yearly surface water.

3.4. Surface Water Occurrence Associated with R-C-ASGM Activities

The SWO < 50 observed during 1984–2020 was extracted, as described in Section 2.4, showing a total extent of 0.25 km^2 in the study area. The SWOs in the Galangan area were as follows: 1–9% (75.1%), 10–19% (19.8%), 20–29% (4.1%), 30–39% (0.6%), and 40–49% (0.3%). Moreover, the SWOs in the Kalanaman area were as follows: 1–9% (64.8%), 10–19% (22.7%), 20–29% (9.1%), 30–39% (2.6%), and 40–49% (0.9%). Furthermore, the results were overlayed on the possible active R-C-ASGM sites found in Section 3.1 (Figure 6). Notably, a higher density of SWO was observed in the central of the Galangan area (A in Figure 6). In comparison, a lower density of SWO was observed in surrounding areas. Their distributions were toward the eastern, northern, and southern areas from the center of the Galangan area. Water areas separately identified in the southern part exhibited relatively lower SWOs. In comparison, relatively lower SWOs were observed along the Kalanaman River in the Kalanaman area (B in Figure 6). Slightly higher SWOs were found at approximately 3 km intervals along the river. Conversely, lower SWOs were found in the northern and southern parts of the river. Fewer pixels were observed, primarily in the river's northern part. According to the overlay analysis, notably, possible hotspots of active mines were mostly observed in similar areas with SWO < 50 during the study period.

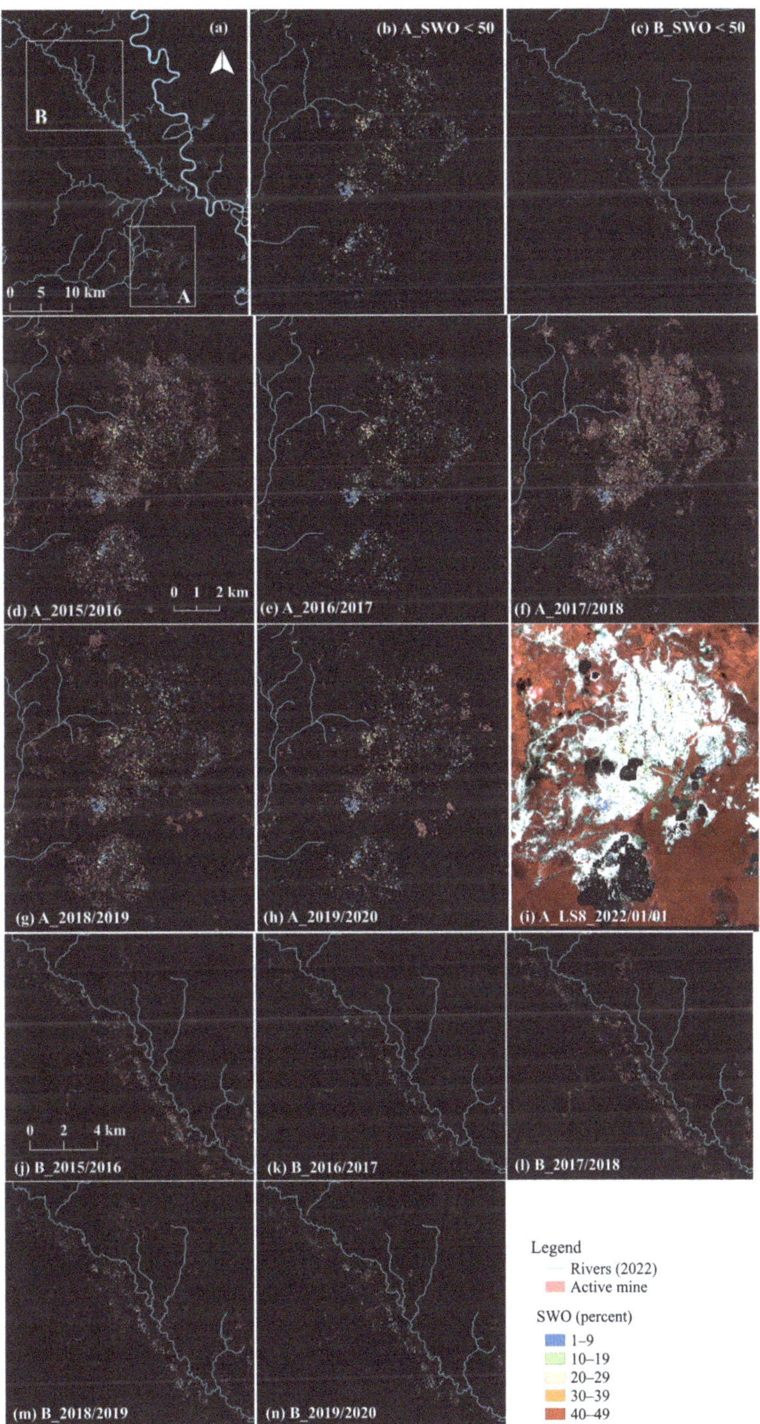

Figure 6. R-C-ASGM hotspots revealed by overlaying with the identified active mine sites and SWOs: (**a**) regional overview of SWO < 50 and waterbodies; (**b**) Galangan area (**A**); (**c**) Kalana-man area (**B**);

(**d–h**) SWO < 50 and detected active sites in Galangan area in 2015/2016, 2016/2017, 2017/2018, 2018/2019, and 2019/2020, respectively; (**i**) overlaying SWO < 50 on Landsat8 imagery acquired on 1 January 2022; (**j–n**) SWO < 50 and detected active sites Kalanaman area in 2015/2016, 2016/2017, 2017/2018, 2018/2019, and 2019/2020, respectively.

4. Discussion

4.1. Time-Series Analysis of R-C-ASGM

The time-series analysis contributed to revealing a pattern of R-C-ASGM activities, which helps predict future activity trends. According to the results, the most active period of R-C-ASGM during the study period occurred in 2017/2018. Despite being an informal sector, it is predicted that an increase in global gold prices accelerates a massive entry of immigrants from different islands into mining sites. The gold price in Indonesia has risen since 2007 (5,722,115 IDR/oz. in January), with an especially steady rise since 2017 (15,907,804 IDR/oz. in January), with the price nearly doubling by 2020 (26,257,748 IDR/oz. in December) [45]. Previous studies have indicated a strong relationship between the development of the C-ASGM sector and increases in gold prices [13,15]; however, this study shows a moderate negative correlation. This trend may be attributed to the coronavirus pandemic affecting mining activities and the gold market rather than the adaptation of the MCM. Although the gold market price increased during the pandemic, mining operation costs increased because of the disruption in labor, supply chains, and cash flow, increasing the gold price [46,47].

4.2. Tracing Mines' Formation Period and Hotspots in R-C-ASGM

While GSW datasets generated from the Landsat series may have weather-related effects, the SWO datasets can be a significant indicator of transformations of R-C-ASGM activities at a regional level. A long-term quantitative analysis of R-C-ASGM broadens our understanding of the scale and pattern of their transformation over time, as well as tracking its responses to global factors, such as the MCM and the gold price. Furthermore, estimating the formation period of mines and recognizing significant R-C-ASGM hotspots is essential to identify a significant source of high pollution, which may lead to significant socio-environmental destruction at the local and community levels.

The results of this study quantified significant SWOs resulting from R-C-ASGM activities and demonstrated their hotspots, along with the status of R-C-ASGM mines. Most notably, in 2017/2018, an active area peak was identified, which was largely concentrated at the center of the Galangan area and along the Kalanaman River. Similarly, higher SWOs were observed in the central of the Galangan area. Conversely, lower SWOs were observed in their surrounding areas and along the Kalanaman River. This trend may indicate that the central Galanga area was formed earlier and is still a significant hotspot. However, the Kalanaman area was recently formed, expanding toward the northeast along the river. The alluvial-based R-C-ASGM practices have accelerated fluvial changes with this expansion [11]. To date, only a few studies have quantified the transformation of the R-C-ASGM sector using remote technology. The application of SAR technology enables the monitoring of R-C-ASGM changes. Previously, [16] investigated the active and inactive status of R-C-ASGM practices from 2015 to 2021 using the S-1 temporal series. Conversely, our work quantified the transformation of R-C-ASGM practice for a longer time frame while also using global surface datasets. The utilization of SWO associated with the spatial distributions of active R-C-ASGM helps in estimating the formation periods of R-C-ASGM mines. Furthermore, overlaying the R-C-ASGM status contributes to detecting mining hotspots, which may be a significant source of high pollution, leading to the destruction of the surrounding environment and increasing health risks.

4.3. Limitations

The results of this study have some limitations in terms of the quality of input data. First, precipitations that occurred before the acquisition time can decrease the backscatter intensity in polarizations, overestimating illegal mining extents. Second, some smaller areas were undetected because of the spatial resolution of the datasets used.

5. Conclusions

In this study, the long-term trend and significant hotspots of R-C-ASGM in Katingan Regency, Central Kalimantan Province, were investigated using time-series S-1, YWCH, GSWML, and WC2020 datasets. The results show a massive occurrence of active R-C-ASGM sites with 2017/2018 as the peak period, primarily at the center of the Galangan area and along the Kalanaman River. With the combination of SWO datasets, the Galangan area was estimated to have formed earlier than other study areas, and its central area was still a significant hotspot. Conversely, Kalanaman areas were recently formed, and their sites expanded with the creation of the Kalanaman river. Therefore, the long-term trend of R-C-ASGM and its significant hotspots can be detected from a combination of time-series datasets. These quantitative analysis results broaden our understanding of R-C-ASGM distributions, transformation, mine occurrence periods, and significant hotspots over a long period. Recognizing long-term R-C-ASGM transformation and identifying significant R-C-ASGM hotspots are also essential to tracking R-C-ASGM responses to global factors/events, such as the MCM and gold prices. This further helps predict the magnitude of environmental destruction at the local and regional levels. These findings are expected to assist in developing rapid and appropriate interventions for strengthening environmental governance by involving various stakeholders.

Author Contributions: S.K. contributed to designing the research, data analysis, and data visualization. M.S. provided comments. M.N. provided technical advice and critical comments. All authors have read and agreed to the published version of the manuscript.

Funding: This research was financially supported by the Research Institute for Humanity and Nature (RIHN: a constituent member of NIHU). Project No. RIHN 14200102.

Institutional Review Board Statement: Not applicable.

Informed Consent Statement: Not applicable.

Data Availability Statement: Not applicable.

Acknowledgments: We thank Ujaval Gandhi and Santhosh M, Spatial Thoughts, for the kind advice on coding. We thank Gunawan Pratama Yoga and Fika Rofiek Mufakhir, National Research and Innovation Agency, Indonesia, for sharing information.

Conflicts of Interest: The authors declare no conflict of interest.

References

1. International Institute for Sustainable Development. *Global Trends in Artisanal and Small-Scale Mining (Asm): A Review of Key Numbers and Issues*; International Institute for Sustainable Development: Winnipeg, MB, Canada, 2018.
2. United Nations Environment Programme. *Estimating Mercury Use and Documenting Practices in Artisanal and Small-scale Gold Mining (ASGM)*; United Nations Environment Programme: Geneva, Switzerland, 2017. [CrossRef]
3. Wilson, M.L.; Renne, E.; Roncoli, C.; Agyei-Baffour, P.; Tenkorang, E.Y. Integrated assessment of artisanal and small-scale gold mining in—Part 3: Social sciences and economics. *Int. J. Environ. Res. Public Health* **2015**, *12*, 8133–8156. [CrossRef] [PubMed]
4. BaliFokus Foundation. *Mercury Country Situation Report Indonesia*; BaliFokus Foundation: Bali, Indonesia, 2018.
5. Agrawal, A.W.S.; Bebbington, A.J.; Imhof, A.; Jebing, M.; Royo, N.; Sauls, L.A.; Sulaiman, R.; Toumbourou, T. *Impacts of Extractive Industry and Infrastructure on Forests: Indonesia*; Climate and Land Use Allaince: San Francisco, CA, USA, 2018.
6. World Health Organization. *Artisanal and Small-Scale Gold Mining and Health*; World Health Organization: Geneva, Switzerland, 2016.
7. Saldarriaga-Isaza, A.; Villegas-Palacio, C.; Arango, S. The public good dilemma of a non-renewable common resource: A look at the facts of artisanal gold mining. *Resour. Policy* **2013**, *38*, 224–232. [CrossRef]

8.   Macháček, J. Typology of environmental impacts of artisanal and small-scale mining in African Great Lakes Region. *Sustainability* **2019**, *11*, 3027. [CrossRef]

9.   United Nations Environment Programme. *Technical Background Report to the Global Mercury Assessment 2018*; United Nations Environment Programmet: Tromsø, Norway, 2018.

10.   Taux, K.; Kraus, T.; Kaifie, A. Mercury Exposure and Its Health Effects in Workers in the Artisanal and Small-Scale Gold Mining (ASGM) Sector—A Systematic Review. *Int. J. Environ. Res. Public Health* **2022**, *19*, 2081. [CrossRef]

11.   Macháček, J. Alluvial artisanal and small-scale mining in a river stream-Rutsiro case study (Rwanda). *Forests* **2020**, *11*, 762. [CrossRef]

12.   Mestanza-Ramón, C.; Cuenca-Cumbicus, J.; D'Orio, G.; Flores-Toala, J.; Segovia-Cáceres, S.; Bonilla-Bonilla, A.; Straface, S. Gold Mining in the Amazon Region of Ecuador: History and a Review of Its Socio-Environmental Impacts. *Land* **2022**, *11*, 221. [CrossRef]

13.   Kimijima, S.; Sakakibara, M.; Nagai, M.; Gafur, N.A. Time-series Assessment of Camp-type Artisanal and Small-scale Gold Mining Sector with Large Influxes of Miners using LANDSAT Imagery. *Int. J. Environ. Res. Public Health* **2021**, *18*, 9441. [CrossRef]

14.   GoldPrice.Org. Gold Price in USD. 2021. Available online: https://goldprice.org/spot-gold.html (accessed on 19 July 2021).

15.   Kimijima, S.; Sakakibara, M.; Nagai, M. Detection of artisanal and small-scale gold mining activities and their transformation using earth observation, nighttime light, and precipitation data. *Int. J. Environ. Res. Public Health* **2021**, *18*, 954. [CrossRef]

16.   Kimijima, S.; Sakakibara, M.; Nagai, M. Characterizing Time-series Roving Artisanal and Smallscale Gold Mining Activities in Indonesia Using Sentinel-1 Data. *Int. J. Environ. Res. Public Health* **2022**, *under submitted*.

17.   United Nations Environment Programmet. *Developing National ASGM Formalization Strategies within National Action Plans*; United Nations Environment Programmet: Geneva, Switzerland, 2018.

18.   United Nations Environment Programme. *Global Mercury Assessment 2018*; United Nations Environment Programmet: Geneva, Switzerland, 2018.

19.   United Nations Environment Programme. *Developing a National Action Plan to Reduce and, Where Feasible, Eliminate Mercury Use in Artisanal and Small-Scale Gold Mining*; United Nations Environment Programmet: Geneva, Switzerland, 2017.

20.   Hilson, G.; Zolnikov, T.R.; Ortiz, D.R.; Kumah, C. Formalizing artisanal gold mining under the Minamata convention: Previewing the challenge in Sub-Saharan Africa. *Environ. Sci. Policy* **2018**, *85*, 123–131. [CrossRef]

21.   Kinyondo, A.; Huggins, C. *Promoting Environmental Sustainability in the Artisanal and Small-Scale Mining Sector in Tanzania*; The United Nations University World Institute for Development Economics Research (UNU-WIDER): Helsinki, Finland, 2021. [CrossRef]

22.   *The National Aeronautics and Space Administration. What is Synthetic Aperture Radar?* The National Aeronautics and Space Administration: Washington, DC, USA, 2022.

23.   Forkuor, M.G.G.; Ullmann, T. Mapping and Monitoring Small-Scale Mining Activities in Ghana using Sentinel-1 Time Series 2015–2019. *Remote Sens.* **2007**, *12*, 911. [CrossRef]

24.   Agrawal, S. *Community Awareness on Hazards of Exposure to Mercury and Supply of Equipment for Mercury-Cleaner Gold Processing Technologies in Galangan, Central Kalimantan, Indonesia*; United Nations Industrial Development Organization: Galangan, Indonesia, 2007.

25.   Telmer, K.H.; Stapper, D. *Evaluating and Monitoring Small Scale Gold Mining and Mercury Use: Building a Knowledge-Base with Satelite Imagery and Field Work*; United Nations Industrial Development Organization: Victoria, BC, Canada, 2007.

26.   Bose-O'Reilly, S.; Drasch, G.; Beinhoff, C.; Rodrigues-Filho, S.; Roider, G.; Lettmeier, B.; Maydl, A.; Maydl, S.; Siebert, U. Health assessment of artisanal gold miners in Indonesia. *Sci. Total Environ.* **2009**, *408*, 713–725. [CrossRef] [PubMed]

27.   European Space Agency. *Frencem Sentinel-1*; European Space Agency (ESA): Paris, France, 2022.

28.   Filipponi, F. Sentinel-1 GRD Preprocessing Workflow. *Proceedings* **2019**, *18*, 11. [CrossRef]

29.   Huang, L.K.; Wang, M.J.J. Image thresholding by minimizing the measures of fuzziness. *Pattern Recognit.* **1995**, *28*, 41–51. [CrossRef]

30.   Prewitt, J.M.S.; Mendelsohn, M.L. The Analysis of Cell Images. *Ann. N. Y. Acad. Sci.* **2006**, *128*, 1035–1053. [CrossRef] [PubMed]

31.   Otsu, N. A Threshold Selection Method from Gray-Level Histograms. *IEEE Trans. Syst. Man. Cybern.* **1979**, *9*, 62–66. [CrossRef]

32.   Doyle, W. Operations Useful for Similarity-Invariant Pattern Recognition. *J. ACM* **1962**, *9*, 259–267. [CrossRef]

33.   Shanbhag, A.G. Utilization of Information Measure as a Means of Image Thresholding. *CVGIP Graph. Model. Image Process.* **1994**, *56*, 414–419. [CrossRef]

34.   Zack, S.A.; Rogers, W.G.; Latt, W.E. Automatic measurement of sister chromatid exchange frequency. *J. Histochem. Cytochem.* **1977**, *25*, 741–753. [CrossRef]

35.   Jui-Cheng Yen, S.C.; Chang, F.-J. A new criterion for automatic multilevel thresholding. *IEEE Trans. Image Process.* **1995**, *4*, 370–378. [CrossRef]

36.   Ridler, T.W.; Calvard, S. Picture Thresholding Using an Interactive Selection Method. *IEEE Trans. Syst. Man Cybern.* **1978**, *8*, 630–632. [CrossRef]

37.   Li, C.H.; Lee, C.K. Minimum cross entropy thresholding. *Pattern Recognit.* **1993**, *26*, 617–625. [CrossRef]

38.   Li, C.H.; Tam, P.K.S. An iterative algorithm for minimum cross entropy thresholding. *Pattern Recognit. Lett.* **1998**, *19*, 771–776. [CrossRef]

39. Sezgin, B.S.M. Survey over image thresholding techniques and quantitative performance evaluation. *J. Electron. Imaging* **2004**, *13*, 146. [CrossRef]
40. Kapur, J.N.; Sahoo, P.K.; Wong, A.K.C. A new method for gray-level picture thresholding using the entropy of the histogram. *Comput. Vis. Graph. Image Process.* **1985**, *29*, 273–285. [CrossRef]
41. Glasbey, C.A. An Analysis of Histogram-Based Thresholding Algorithms. *Graph. Model. Image Process.* **1993**, *55*, 532–537. [CrossRef]
42. Kittler, J.; Illingworth, J. Minimum error thresholding. *Pattern Recognit.* **1986**, *19*, 41–47. [CrossRef]
43. Tsai, W.H. Moment-preserving thresholding: A new approach. *Comput. Vis. Graph. Image Process.* **1985**, *29*, 377–393. [CrossRef]
44. Pekel, J.F.; Cottam, A.; Gorelick, N.; Belward, A.S. High-resolution mapping of global surface water and its long-term changes. *Nature* **2016**, *540*, 418–422. [CrossRef]
45. World Gold Council. Gold Prices. 2021. Available online: https://www.gold.org/goldhub (accessed on 6 June 2021).
46. PlanetGOLD. The Prevalence of COVID-19 within GOLD-ISMIA's Project Locations. 2020. Available online: https://www.planetgold.org/sites/default/files/2020-04/planetGOLD_Indonesia_Factsheet_COVID19_ISMIA.pdf (accessed on 5 March 2022).
47. PlanetGOLD. The Gold Price Trend within GOLD-ISMIA's Project Locations amid the COVID-19 Pandemic. 2020. Available online: https://www.planetgold.org/sites/default/files/2020-04/GOLD-ISMIAFactSheet_GoldPriceTrend.pdf (accessed on 5 March 2022).

International Journal of
Environmental Research and Public Health

Article

Characterizing Time-Series Roving Artisanal and Small-Scale Gold Mining Activities in Indonesia Using Sentinel-1 Data

Satomi Kimijima [1,*], Masayuki Sakakibara [1,2] and Masahiko Nagai [3,4]

[1] Research Institute for Humanity and Nature, Kyoto 603-8047, Japan; sakaki@chikyu.ac.jp
[2] Graduate School of Science & Engineering, Ehime University, Matsuyama 790-8577, Japan
[3] Graduate School of Science and Technology for Innovation, Yamaguchi University, Ube 755-8611, Japan; nagaim@yamaguchi-u.ac.jp
[4] Center for Research and Application of Satellite Remote Sensing, Yamaguchi University, Ube 755-8611, Japan
* Correspondence: kimijima@chikyu.ac.jp

Abstract: The rapid growth of roving mining camps has negatively influenced their surrounding environment. Although artisanal and small-scale gold mining (ASGM) is a major source of gold production, the mining activities and their activeness are not well revealed owing to their informal, illegal, and unregulated characteristics. This study characterizes the transformations of roving camp-type ASGM (R-C-ASGM) activities in Central of Katingan Regency, Central Kalimantan Province, Indonesia, from 2015 to 2021 using remotely sensed data, such as the time-series Sentinel-1 dataset. The results show that the growth of active R-C-ASGM sites was identified at the center of the Galangan mining region with expansions to the northwest part along the Kalanaman River, especially in 2021. Hence, these approaches identify the transformations of roving mining activities and their active or nonactive status even in tropical regions experiencing frequent heavy traffic rainstorms. They provide significant information on the socioenvironmental risks possibly caused at local and regional levels. Our results also inform the design of timely interventions suited to local conditions for strengthening environmental governance.

Keywords: alluvial mining; artisanal and small-scale gold mining; Indonesia; landcover change; remote sensing; synthetic aperture radar

Citation: Kimijima, S.; Sakakibara, M.; Nagai, M. Characterizing Time-Series Roving Artisanal and Small-Scale Gold Mining Activities in Indonesia Using Sentinel-1 Data. *Int. J. Environ. Res. Public Health* **2022**, *19*, 6266. https://doi.org/10.3390/ijerph19106266

Academic Editors: Paul B. Tchounwou and Wei Song

Received: 21 April 2022
Accepted: 18 May 2022
Published: 21 May 2022

Publisher's Note: MDPI stays neutral with regard to jurisdictional claims in published maps and institutional affiliations.

1. Introduction

The rapid growth of the rove-type mining sector has negatively influenced their surrounding environments. Therefore, detecting such occurrences, determining their development rate, and identifying their active or nonactive status should provide significant insights into identifying possible socioenvironmental problems caused at local and regional levels. This may also allow environmental governance to be promoted at various levels.

Artisanal and small-scale gold mining (ASGM) is a major source of gold production using rudimentary technology at individual or community levels despite being informal, illegal, and unregulated [1]. This sector has the largest employer in gold mining at the global level comprising 70% to 80% of informal small-scale workers [2]. Mercury is commonly used to increase the gold extraction process, resulting in highly toxic environmental and health risks due to mercury pollution throughout its emissions and release into water and the atmosphere, respectively [3–5]. Such mercury pollution has largely been observed in South America, Africa, and Asian regions. Indeed, environmental impacts, such as deforestation, geomorphic and hydrological changes [6–11], and health problems, such as mercury intoxication-oriented movement disorders and various injuries associated with the ASGM activities have been reported [12,13]. Despite its significant socioenvironmental impacts, more than 80 countries have continuously employed ASGM to alleviate poverty for their socioeconomic development [14,15].

Continuous growth has been observed in Indonesia. Active and nonactive ASGM practices have been placed in 93 regencies of 30 of the 34 provinces, estimating 250,000–300,000 miners [16] in more than 1200 hotspots in 2017 [17]. Furthermore, the country has been the fastest increase in polluted sites in the last 20 years on a global scale [2]. In Kalimantan island, one of the ASGM hotspots with alluvial operations, many illegal mining activities have been widespread even in conservation areas, impacting biodiversity and human health [17].

The ASGM sector can be classified into the following two types: "travel-type," in which the miners commute from their local residences to the mining sites, and "camp-type," in which the miners live and conduct mining activities on informal worksites [18] (hereafter referred to as C-ASGM). In the C-ASGM sector, both roving and non-roving practices are observed. The scale of the workforce in the ASGM sector has expanded with the increasing gold prices since 2000 [19]. The strong relationship between ASGM increases and the high price of gold has been confirmed in the literature [7,18,20].

Remote-sensing technologies have been widely used to characterize natural features and physical objects and monitor their spatial changes over time. Additionally, this technology provides a wide variety of continuous data with temporal, spatial, and spectral resolutions. Freely available satellite remote-sensing data, such as the Landsat series, have provided long-term Earth observation data since the 1970s and have been widely used for land cover detection and monitoring [21–24]. Despite the development in geoinformation technology, few studies have focused on the ASGM sector for quantitative assessments experiencing the harmful environmental and health risks caused by mercury pollution. Even [6–11,25–27] demonstrated time-series assessments in deforestation, mining area detection, and geomorphic and hydrological changes; however, they mainly examined the travel-type mining sites. To investigate the closed C-ASGM sites, Ref. [18] recently conducted a quantitative time-series analysis of the growth in C-ASGM sites using satellite remote-sensing imagery. Furthermore, Ref. [28] analyzed the transformation of C-ASGM activities by integrating nighttime light (NTL) intensities as a magnitude of mining activities. Although a time-series assessment of the closed C-ASGM sector with non-roving practices has been conducted by [18,28], a roving C-ASGM sector (hereafter R-C-ASGM) has not yet been discovered. The major challenges, such as acquiring an optical cloud-free time-series dataset [18,28,29], lead to further difficulty in understanding the R-C-ASGM sector, operated at a larger scale in tropical regions experiencing frequent heavy traffic rainstorms.

The use of the synthetic aperture radar (SAR), an active independent Earth observation system from solar illumination or day–night cycles [30], is an alternative suited tool for optical data [31]. Further, Ref. [32] reviewed the optical and SAR data for monitoring ASGM sites and ensured results between the datasets. Previous studies have revealed the potential of SAR data usage in mining-induced area detection using SAR sensitivities of radar systems to surface roughness and dielectric properties of materials [27,31,32]. Therefore, SAR data are a powerful tool to overcome weather-related limitations mainly found with optical sensors. This helps detect and monitor closed R-C-ASGM sectors to obtain a qualitative and comprehensive understanding.

This study primarily assesses the transformation of the R-C-ASGM activities from 2015 to 2021 in Katingan Regency, Central Kalimantan Province, Indonesia, where active alluvial-based R-C-ASGM activities have been conducted. This study's results are expected to contribute to the understanding of R-C-ASGM development spread in remote rural areas, the prediction of the level of socioenvironmental pollution, and strengthening environmental governance at the regional level.

2. Materials and Methods

2.1. Overall Methodological Workflow

The methodological workflow used in this study is demonstrated in Figure 1. This workflow employed three main steps to achieve its primary objective of assessing the transformation of the R-C-ASGM activities. First, the S-1 backscattering coefficients (σ^0) were calculated with vertical–vertical (VV) and vertical–horizontal (VH) polarizations. Second, selections of algorithm/polarization were performed to detect the most locally sensitive values. Third, the changes in the R-C-ASGM occurrences during 2015–2021 were calculated based on the S-1 temporal series. This evidence allowed us to understand the historical transformation of the R-C-ASGM activities at the study site. This study presents a discussion based on all the findings described above. The methods used in each step are explained in the following sections.

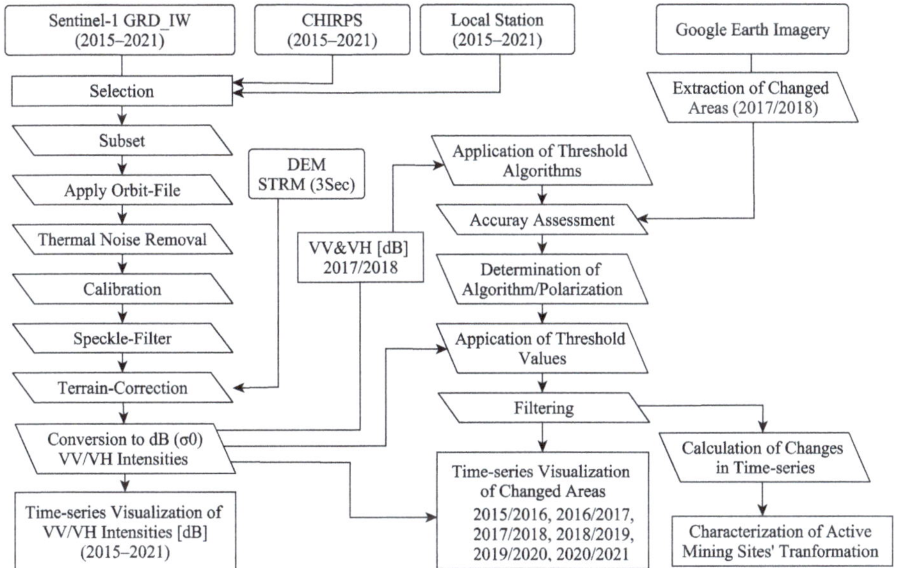

Figure 1. Overall methodology.

2.2. Study Area

Indonesia is a well-mineralized metallogenic region with significant gold mineralization, associated with quartz veins in andesite-hosted epithermal settings. One of the major ASGM hotspots in Central Kalimantan with gold-bearing alluvial soils has attracted large ASGM-targeted migrants from Java and South Kalimantan [33]. The Galangan mining region in Central Kalimantan is the geographical and historical center of the land-based mining area, which developed rapidly in the early 1990s [32]. The Hampalit town, especially, was a base for active mining activities for both indigenous miners and a gold company, namely PT Hampalit Mas Perdahana, which closed during the financial crash of 1997. Company-initiated mining activities extracted heavy minerals through an openpit method, digging deep excavation pits. Thus, removing all the soil and vegetation landscape on the surface creates a barren wasteland [13,33]. However, indigenous ASGM communities have extracted gold along with river systems by floating pumps, resulting in disturbances of riverbanks and an increase in sediment volumes [33]. After the company's closure, the lands were taken by migrated miners. They have continuously traveled to the greater areas, from Kalanaman, Pundu to Galangan, to explore newer locations with greater gold production by seasons [33].

This study targets Galangan mining (Central of Katingan Regency, the Central Kalimantan Province, Indonesia), utilizing the alluvial-based mining method (Figure 2). In this mining region, the Katingan River, one of the major river basins in Southeast Asia, flows north to south.

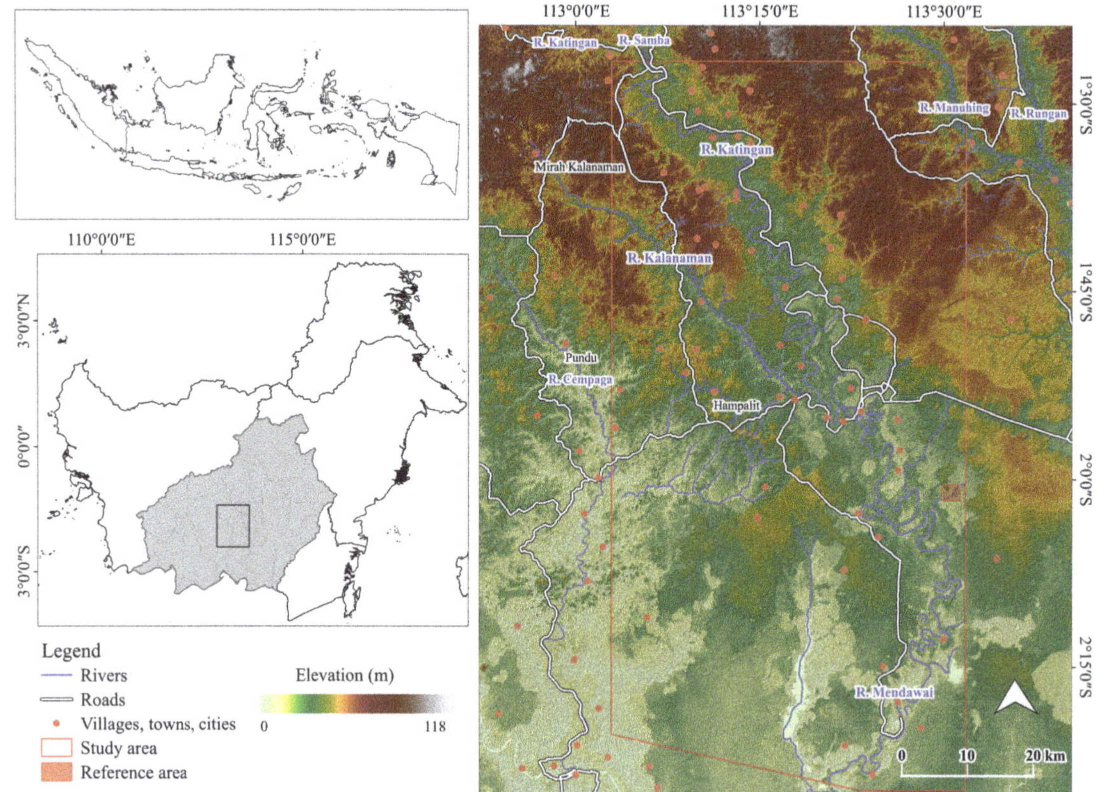

Figure 2. Study area.

2.3. S-1 Imagery

A total of seven level-1 grand range detected (GRD) Sentinel-1 datasets, covering 2015–2021, downloaded from the European Space Agency (ESA), were utilized to extract and calculate time-series changes of the ASGM occurrences. Through the EU/ESA Copernicus program, the S-1 mission (S-1A and S-1B) provides an exceptional combination of high spatial (10 m) and temporal (6 days) resolution data by operating two polar-orbiting radar imaging systems working with the C band (~5.7 cm wavelength). The main operational mode is interferometric wide swath mode (IW) with VV and VH polarizations, and images are freely and routinely available [34].

To reduce atmospheric effects, which reduced the quality of images, Climate Hazards Group InfraRed Precipitation with Station (CHIRPS) data was referred to using Google Earth Engine to target months experiencing less rain with local weather station data. Thus, this study focused on July to August from 2015 to 2021.

All datasets were acquired from the descending track with relative orbit number 3 of each image's backscatter intensity to better the image. The available S-1 dense time-series offers a unique opportunity to monitor ASGM activities, especially in tropical regions experiencing the magnitude of frequent rainstorms.

2.4. Image Preprocessing

The preprocessing workflow is based on ESA's open-source software, ESA named sentinels application platform (version 8.0.0), and its functionalities. The following steps were implemented in the S-1 Toolbox: orbit correction, thermal noise removal, radiometric calibration, speckle filtering with 5×5 windows, and terrain correction using the 3-arcsec digital elevation model (DEM) from the shuttle radar topography mission (SRTM) [35]. Here, the radiometric calibration aims to convert the digital pixel value of the S-1 images into an image intensity value of σ^0. The data were projected to the World Geodetic System 1984, Universal Transverse Mercator Zone 49 South. Terrain-corrected σ^0 intensities of the VV and VH were used for further analysis.

2.5. Selection of Threshold and Detection of Changed Areas in Time-Series

After image preprocessing, optimized threshold values were identified based on the VV and VH polarizations acquired in 2017 and 2018. Sixteen automatic global thresholding algorithms and binary image classifications using one-dimensional feature space were applied to extract mining-induced areas. In this process, Fiji (version 2.1.0) software (https://imagej.net/software/fiji/, accessed on 1 March 2022), an open-source Java image processing package, was used to determine each algorithm's threshold values. Huang's fuzzy [36], Internodes [37], Isodata [38], IJ_Isodata, Li's Minimum Cross-Entropy [39–41], Maximum Entropy [42], Mean [43], Minimum Error [44], Minimum [37], Moments [45], Otsu's [46], Percentile [47], Renyi's Entropy [42], Shanbhag's [48], Triangle [49], and Yen's [50] threshold algorithms were separately performed. This study also tested a supervised classification method, such as histogram intersection, applied by [31]. Subsequently, the results were validated using reference data to examine the best separability for the change detection. The reference data for the accuracy assessment were derived from high-resolution images obtained on 9 June 2017 and 23 September 2018, using Google Earth Pro.

Owing to heavy cloud coverage in the study area, the acquisition of the scenes was extremely limited only to the abovementioned data. However, these images identified mining activities along the Katingan River. According to human visual image interpretation, areas affected by mining activities were separately digitized, and the changed areas were identified by overlaying. Third, 100 points were randomly selected from the datasets to determine the best suitability by polarizations. Fourth, the determined best combination of algorithm and polarization was applied to all datasets post-classification of a majority filter with a moving window size of 5×5 pixels to remove isolated pixels. Furthermore, the detected areas observed in the river buffers were eliminated to remove the mudflats in the rivers, possibly caused by changes in the magnitude of precipitation between the acquired years. Consequently, the annual changes in the extent of illegal mining were calculated for the following six temporal series: 2015/2016, 2016/2017, 2017/2018, 2018/2019, 2019/2020, and 2020/2021.

In previous studies, mining areas in the Central of Katingan Regency were estimated to cover ~400 km^2 in 2007 [32]. Hence, the long-term trends in R-C-ASGM sites could be observed from satellite imagery even with a 10-m ground resolution. We summarized the main specifications of the databases used in Table 1.

Table 1. Main specification of satellite imagery used in the study.

Satellite	Type	Acquisition Date	Spatial Resolution	Image Number	Polarization	Wavelength
Sentinel-1	C-SAR	20 July 2015 7 August 2016 21 July 2017. 4 July 2018. 11 July 2019 10 August 2020 24 July 2021	10 m	3	Descending (VV, VH)	C band

3. Results

3.1. Visualization of Time-Series Color Composites of VV and VH Polarizations

The processed VV and VH polarizations were displayed in RGB color composites in six temporal series: 2015/2016, 2016/2017, 2017/2018, 2018/2019, 2019/2020, and 2020/2021, as shown in Figure 3. In this visualization process, the older years were assigned red, and the newer years were assigned green and blue, which detects changes in land covers between two different periods. VV polarizations show slightly brighter intensities compared to that of VH's. VH polarization can detect significant landcover changes along the Katingan and Kalanaman Rivers in 2016/2017 and 2020/2021.

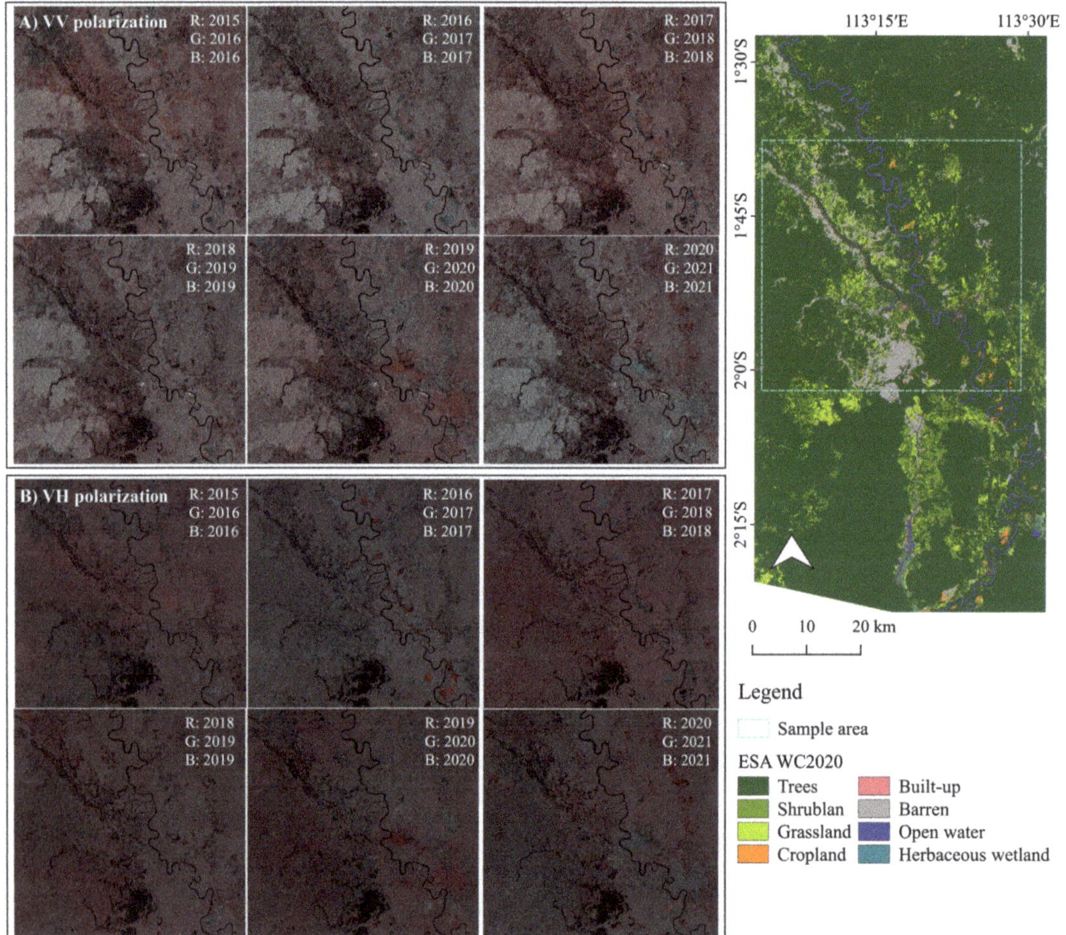

Figure 3. Time-series color composites by (**A**) VV and (**B**) VH polarization channels.

3.2. Determination of Threshold

Both VV and VH polarizations acquired in 2017 and 2018 were primarily used to derive the best combination of algorithm and polarization. The changed areas identified from each result were validated using features extracted from high-resolution Google earth images (GEI), as mentioned in Section 2.5. The most sensitive algorithm and polarization channel indicate changes in illegal mining extents.

After processing the optimized thresholding, the locally sensitive methods were found only in the IJ_Isodata and Yen algorithms. Therefore, those were applied both to VV & VH polarization channels. Table 2 presents the identified thresholds. The threshold values identified by the IJ_Isodata showed lower intensities: −15.07 dB (2017 VV), −21.47 dB (2017_ VH), −14.84 dB (2018_VV), and −20.16 dB (2018_VH). The Yens were: −15.07 dB (2017_VV), −20.16 dB (2017_VH), 13.32 dB (2018_VV), and −20.16 dB (2018_VH). The results show no significant value differences in both VV and VH polarizations (IJ_Isodata algorithm). The same values were generated in 2017_VV and 2018_VH from the Yen algorithm; however, 2018_VV showed a larger difference between the two periods. As new mining areas are usually associated with land cover changes from vegetation to bare areas or water, the magnitude of intensity in such areas is expected to be lower intensities in VV and VH polarizations. Thus, the IJ_Isodata algorithm was more sensitive to finding mining activity-induced land landcover changes than the Yen algorithm in this study.

Table 2. Threshold values by algorithm and polarizations.

	2017		2018	
Algorithm	**VV**	**VH**	**VV**	**VH**
IJ_Isodata	−15.07 dB	−21.47 dB	−14.84 dB	−20.16 dB
Yen	−15.07 dB	−20.16 dB	13.32 dB	−20.16 dB

Figure 4 shows detected areas induced by R-C-ASGM activity during 2017/2018, based on human visual interpretation of GEI and thresholding results by VV and VH polarizations optimized by the IJ_Isodata algorithm. After different threshold values, similar intensities were found in both VV and VH polarization in the identified areas. For example, an average of −18.03 dB (standard deviation (STDEV) of 1.14 dB)) and −24.23 dB (STDEV of 1.51 dB) was observed for 2017 VV and VH, respectively. Furthermore, −18.11 dB (STDEV of 1.55 dB) and −23.89 dB (STDEV of 2.22 dB) were observed during 2018. By comparing the results, some areas in the middle part were not detected by VV and VH polarization; however, the visual comparison indicates that areas induced by R-C-ASGM activities can be detectable in both time-series features.

3.3. Detection of Newly Expanded R-C-ASGM Areas

Using the results in Section 3.1, the accuracy assessment was performed to judge their sensitivity. The results show 73.3% and 76.0% for the VV and VH polarizations, respectively. We recalculated the accuracy by omitting the points found at boundaries due to high spectral resolution sensitivity resulting from mixed pixels. As a result, we found 76.7% and 82.1% accuracies for the VV and VH polarizations, respectively. The best combination was found with the IJ_Isodata algorithm with VH polarization. The particular threshold values for each VH polarization were generated for the final classification, possibly leading to better detection of active R-C-ASGM activities (Table 3).

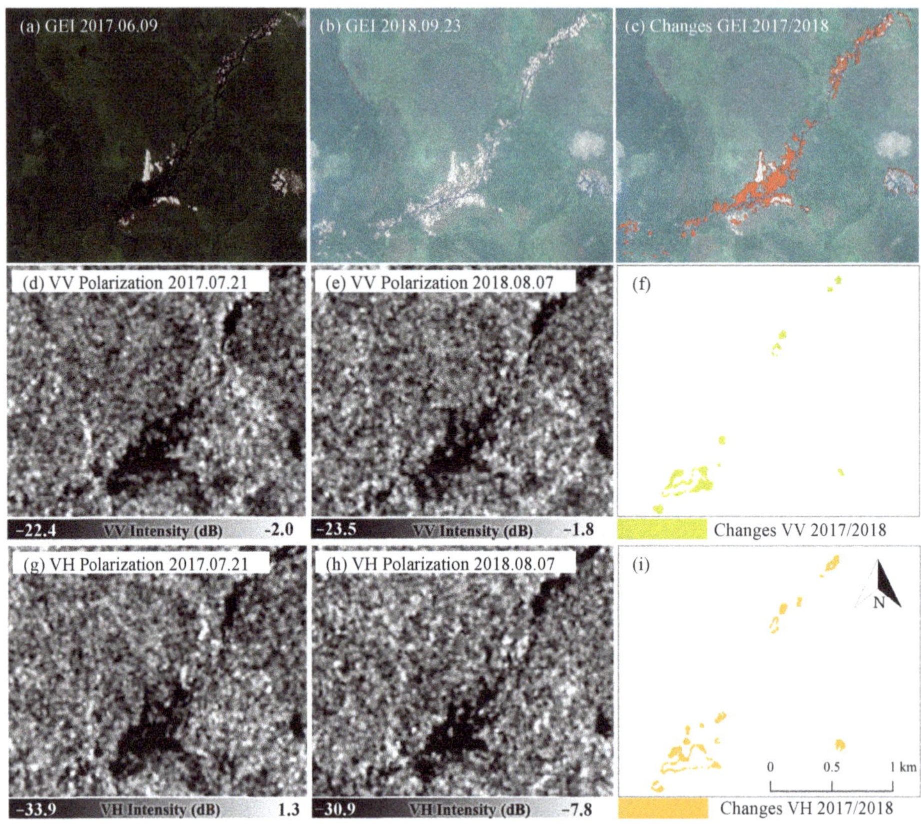

Figure 4. GEI 2017 (**a**), 2018 (**b**), detected changes from GEI 2017–2018 (**c**). VV polarization in 2017 (**d**), 2018 (**e**), detected changes from VV 2017–2018 after applying the threshold values (**f**). VH polarization in 2017 (**g**), 2018 (**h**), detected changes from VH 2017–2018 after applying the threshold values (**i**).

Table 3. Threshold values for time-series VH polarizations.

Threshold (IJ_Isodata Algorithm)	2015	2016	2017	2018	2019	2020	2021
Intensities (dB)	−20.88	−19.95	−21.47	−20.16	−20.36	−20.76	−19.8

Figure 5 shows the occurrence of active mining sites for the six periods (2015/2016, 2016/2017, 2017/2018, 2018/2019, 2019/2020, and 2020/2021), overlaying on the European Space Agency (ESA) WorldCover 10 m 2020 (WC2020). The occurrence of R-C-ASGM-induced areas exhibited 25.0 km^2 (2015/2016), 28.0 km^2 (2016/2017), 32.1 km^2 (2017/2018), 20.3 km^2 (2018/2019), 7.4 km^2 (2019/2020), and 47.9 km^2 (2020/2021), respectively. The magnitude of the occurrences was found in 2015/2016–2017/2018; however, fewer occurrences were observed in 2019/2020. Simultaneously, the largest occurrence was again observed in 2020/2021 along the river. The detected areas were concentrated in the center of the Galangan region and along the Kalanaman River, where LC is classified as barren in ESA WC2020. The magnitudes of the occurrences were particularly observed in 2020/2021 in the northwestern parts of the study area along the Kalanaman River. The pattern of occurrences is observed mostly along with the river networks.

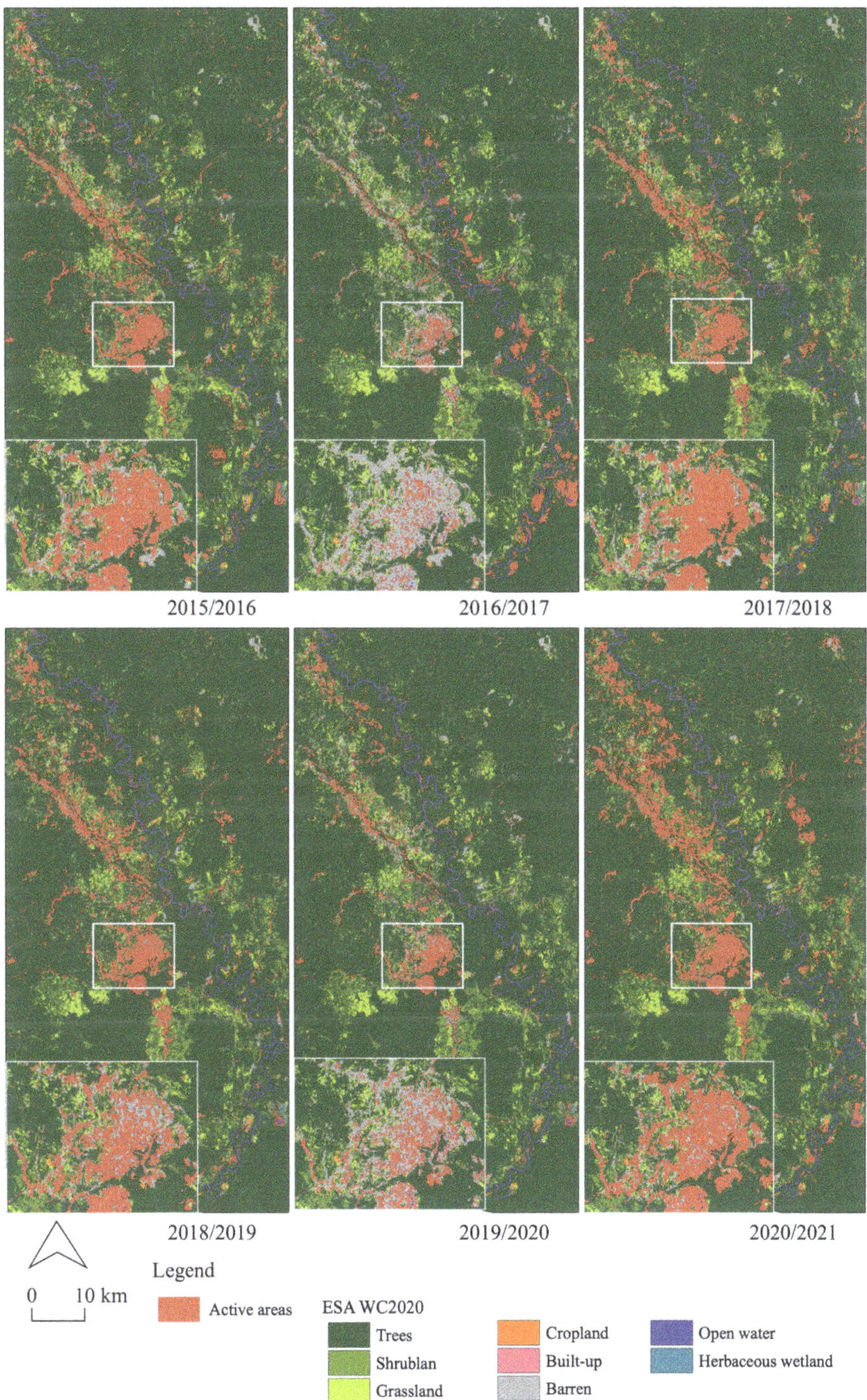

Figure 5. Occurrence of active mining sites detected by VH polarizations and their overlay on the ESA WC2020.

4. Discussion

4.1. Contributions

We studied the transformations of the R-C-ASGM activities from 2015 to 2021 using the S-1 time series. A quantitative time-series analysis of the R-C-ASGM sectors can help better understand the rate and pattern of development of such mining activities over time. Detecting such occurrences and their patterns in tropical regions experiencing the magnitude of frequent rainstorms can provide significant information or estimation on the potential rates and levels of socioenvironmental pollution and its human risk resulting from mercury use at R-C-ASGM sites. Understanding the characteristics of R-C-ASGM practices helps strengthen environmental governance at various levels.

As described, the establishment of new mining areas is usually associated with changes in land cover from vegetation to bare/water areas. We employed a change detection method based on generating the binary masks using a threshold defined by the image, optimized by the IJ_Isodata algorithm (Table 2). The analysis reveals that VH polarization was more sensitive than VV polarization, resulting in better separation of areas induced by mining activity (Table 3). While the classification accuracy was 76.0%, a higher accuracy (82.1%) was found with omission of random points at boundary. For 24.0% of errors, influence factors would not be caused by algorithm matter. Instead, the following factors can be considered for this misclassification; SAR specific errors such as foreshortening and layover in mountainous areas owing to the side looking of SAR; differences in data acquired time; weather conditions before the data acquired time; and spatial resolution of data. Previous studies using SAR datasets in the mining sectors only achieved 52.0% [51], 84.9% (producer accuracy), and 72.4% (user accuracy) [31]. Our study does not focus on generating a high-accuracy map of active mining, which can be a replacement for a field survey. Instead, we aim to provide information that leads to and supports the initial survey and social implementation at a local level. Without any field data, we cannot target any destinations for surveying, resulting in huge loss of time and cost. Thus, the generated possible active map helps plan field survey. Our study is not an alternative tool to a field survey; therefore, 76.0% (82.1% highest) accuracy is sufficient for this study.

This study demonstrated the transformation of active sites in the R-C-ASGM sector from 2015 to 2021 in the Galangan region, Central Kalimantan, Indonesia, where active alluvial-based R-C-ASGM activities have been historically conducted. This study detected the active mines and their various transformation forms using a quantitative analysis over time (Figure 5), as described in Section 3.2. Few studies have quantified R-C-ASGM practices with satellite imagery data. A recent study by [18,28] conducted a quantitative time-series analysis of the closed Non-R-C-ASGM sites, employing the vertical tunnel method (shaft) of mining, in Golontato, Indonesia, using optical satellite remote-sensing imagery. However, this work further quantified R-C-ASGM sites where activities are operated at a large scale in tropical regions experiencing frequent heavy rainstorms.

Few studies have utilized satellite data to reveal the volume of illegal mining activities. Especially, a recent study by [28] quantified that the extent of illegal mining sites and the magnitude of mining activities in the camps experienced 4.8- and 3.8-fold increases, respectively, from 2014 to 2020. Although the study areas, mining type, and indicator of transformations differ from this study, a similar trend in the occurrence of active illegal mining activities was found in their results. Further, Ref. [32] focused on a study area that is comparable to ours. Their study found an annual expansion of 8 km^2 through the Landsat series in the Galangan region from 1999 to 2002 [32]. This study found a higher magnitude of the occurrence rate during 2015/2021. The possible reason for this may be associated with an increase in the global gold price since 2006. Similarly, the gold price in Indonesia has increased since 2007, with an especially steady increase since 2017, which approximately doubled at the end of 2020 [18]. The magnitude of occurrences was found especially in 2020/2021 when a rapid increase in the gold price was observed globally and nationally, while the lower rate of occurrence observed in 2019/2020 could be due to the globally spread influence of the coronavirus pandemic, which affects the workforce, mining activities, markets' supply chain, and cash flow [52,53].

For the shifts in the occurrence patterns, the result showed that the magnitudes of active areas were found in the western Galangan region and along with the river networks in 2020/2021. Even migrated miners have continuously roved the greater areas by season [33]. Their main mining target sites can be shifted along the Kalanaman Rivers to explore greater gold production. The possible reason for this shift can be associated with the expansions of river extents. From the European Commission Joint Research Centre Yearly Water Classification History dataset [54], which contains yearly water classifications from 1988 to 2020, water extent along the Kalanaman River areas exhibited 1.83 (2015), 2.90 (2016), 2.91 (2017), 2.76 (2018), 2.59 (2019), and 3.40 km^2 (2020), respectively. Thus, alternative R-C-ASGM sites were expected to be further developed that were associated with the expansion of water extents along the Kalanaman River after 2020. Moreover, Ref. [32] previously revealed shifts in mining direction with PALSAR (June-September 2006). The authors found the various shifts in the active area of the western Galangan region. However, a time-series analysis of the study further identifies the detailed characteristics such as volumes, mine status, and trends of active mins' shifting directions in the hidden R-C-ASGM sectors, representing a more comprehensive understanding of R-C-ASGM sectors across the region.

The R-C-ASGM sector can operate successfully due to its high productivity of gold. However, it is estimated that approximately 270 tons of mercury are annually released only from Central Kalimantan to the Sea of Java as of 2007 [32]. Furthermore, severe mercury contamination (sediment, local fish, and hair samples) and typical symptoms resulting from mercury intoxications (ataxia, tremor, and dysdiadochokinesia) among workers with high ASGM activities have been observed in the Galangan region [13]. Despite its status as an informal sector, the increase in global gold prices accelerated the massive entry of immigrants into the mining sector, resulting in its massive growth. The growth in those sectors further accelerate to cause mercury-related environmental pollution and health problems at the stages of mining and amalgamation. Therefore, detecting such rapidly developing hidden R-C-ASGM sectors can provide significant insights into the potential rates and the levels of socioenvironmental pollution. This would also strengthen the environmental governance with the participation of different stakeholders at various levels.

4.2. Limitations

This study has certain limitations associated with the characteristics of SAR data. Although SAR data helps in the active independent observation of weather, it causes foreshortening and layover in mountainous areas owing to the side looking of SAR, leading to misclassification. Further, precipitation before the acquisition can decrease the backscatter intensity in polarizations, overestimating illegal mining extents. Moreover, some smaller and complex areas are undetectable due to the used datasets (10 × 10 km grid cell). Finally, because of the S-1 series' operation period, the methodologies applied in this study are limited only to the period after 2014.

Although the proposed method cannot detect the existing mining areas before 2015, it identifies the occurrences of R-C-ASGM-related active sites and their changing patterns.

5. Conclusions

This study assessed the transformation of the R-C-ASGM sector in Katingan Regency, Central Kalimantan Province, Indonesia, using S-1 time-series data. The results presented herein show the massive occurrence of the active R-C-ASGM sites. In particular, a magnitude of occurrence was found in the center of the Galangan region and along the Kalanaman River in 2021. Therefore, it can be concluded that the active mining sector undertaken by the R-C-ASGM method can be detected from a set of time-series datasets. These results extend our understanding of the transformations of the mining site and the status of their activeness in the hidden R-C-ASGM sectors. Subsequently, it also provides significant insight into the potential for further socioenvironmental problems at the regional level. These findings are expected to assist in developing rapid and appropriate interventions for strengthening environmental governance by involving various stakeholders.

Author Contributions: S.K. contributed to designing the research, data analysis, and data visualization. M.S. provided comments. M.N. provided technical advice and critical comments. All authors have read and agreed to the published version of the manuscript.

Funding: This research was financially supported by the Research Institute for Humanity and Nature (RIHN: a constituent member of NIHU). Project No. RIHN 14200102.

Institutional Review Board Statement: Not applicable.

Informed Consent Statement: Not applicable.

Data Availability Statement: Not applicable.

Acknowledgments: We thank Gunawan Pratama Yoga and Fika Rofiek Mufakhir, National Research and Innovation Agency, Indonesia for sharing information.

Conflicts of Interest: The authors declare no conflict of interest.

References

1. De Lobo, F.L.; Costa, M.; de Moraes Novo, E.M.L.; Telmer, K. Distribution of Artisanal and Small-Scale Gold Mining in the Tapajós River Basin (Brazilian Amazon) over the Past 40 Years and Relationship with Water Siltation. *Remote. Sens.* **2016**, *8*, 579. [CrossRef]
2. International Institute for Sustainable Development. *Global Trends in Artisanal and Small-Scale Mining (Asm): A Review of Key Numbers and Issues*; International Institute for Sustainable Development: Winnipeg, MB, Canada, 2018.
3. World Health Organization. *Artisanal and Small-Scale Gold Mining and Health*; World Health Organization: Geneva, Switzerland, 2016.
4. Saldarriaga-Isaza, A.; Villegas-Palacio, C.; Arango, S. The public good dilemma of a non-renewable common resource: A look at the facts of artisanal gold mining. *Resour. Policy* **2013**, *38*, 224–232. [CrossRef]
5. United Nations Environment Programmet. *Technical Background Report to the Global Mercury Assessment 2018*; United Nations Environment Programmet: Tromsø, Norway, 2018.
6. Espejo, J.C.; Messinger, M.; Román-Dañobeytia, F.; Ascorra, C.; Fernandez, L.E.; Silman, M. Deforestation and foarest degradation due to gold mining in the peruvian amazon: A 34-year perspective. *Remote Sens.* **2018**, *10*, 1903. [CrossRef]
7. Swenson, J.J.; Carter, C.E.; Domec, J.-C.; Delgado, C.I. Gold mining in the peruvian amazon: Global prices, deforestation, and mercury imports. *PLoS ONE* **2011**, *6*, e18875. [CrossRef] [PubMed]
8. Emel, J.; Plisinski, J.; Rogan, J. Monitoring geomorphic and hydrologic change at mine sites using satellite imagery: The Geita Gold Mine in Tanzania. *Appl. Geogr.* **2014**, *54*, 243–249. [CrossRef]
9. Gallwey, J.; Robiati, C.; Coggan, J.; Vogt, D.; Eyre, M. A Sentinel-2 based multispectral convolutional neural network for detecting artisanal small-scale mining in Ghana: Applying deep learning to shallow mining. *Remote Sens. Environ.* **2020**, *248*, 111970. [CrossRef]
10. Owusu-Nimo, F.; Mantey, J.; Nyarko, K.; Appiah-Effah, E.; Aubynn, A. Spatial distribution patterns of illegal artisanal small scale gold mining (Galamsey) operations in Ghana: A focus on the Western Region. *Heliyon* **2018**, *4*, e00534. [CrossRef]
11. Schmid, T.; Rico, C.; Rodríguez-Rastrero, M.; Sierra, M.J.; Díaz-Puente, F.J.; Pelayo, M.; Millán, R. Monitoring of the mercury mining site Almadén implementing remote sensing technologies. *Environ. Res.* **2013**, *125*, 92–102. [CrossRef]
12. Macháček, J. Typology of Environmental Impacts of Artisanal and Small-Scale Mining in African Great Lakes Region. *Sustainability* **2019**, *11*, 3027. [CrossRef]
13. Bose-O'Reilly, S.; Drasch, G.; Beinhoff, C.; Rodrigues-Filho, S.; Roider, G.; Lettmeier, B.; Maydl, A.; Maydl, S.; Siebert, U. Health assessment of artisanal gold miners in Indonesia. *Sci. Total Environ.* **2010**, *408*, 713–725. [CrossRef]
14. United Nations Environment Programme. *Estimating Mercury Use and Documenting Practices in Artisanal and Small-Scale Gold Mining (ASGM)*; United Nations Environment Programme: Geneva, Switzerland, 2017. [CrossRef]
15. Wilson, M.L.; Renne, E.; Roncoli, C.; Agyei-Baffour, P.; Tenkorang, E.Y. Integrated Assessment of Artisanal and Small-Scale Gold Mining in Ghana—Part 3: Social Sciences and Economics. *Int. J. Environ. Res. Public Health* **2015**, *12*, 8133–8156. [CrossRef] [PubMed]
16. Agrawal, A.W.S.; Bebbington, A.J.; Imhof, A.; Jebing, M.; Royo, N.; Sauls, L.A.; Sulaiman, R.; Toumbourou, T. *Impacts of Extractive Industry and Infrastructure on Forests: Indonesia*; Climate and Land Use Allaince: San Francisco, CA, USA, 2018.
17. BaliFokus Foundation. *Mercury Country Situation Report Indonesia*; BaliFokus Foundation: South Jakarta, Indonesia, 2018.
18. Kimijima, S.; Sakakibara, M.; Nagai, M.; Gafur, N.A. Time-Series Assessment of Camp-Type Artisanal and Small-Scale Gold Mining Sectors with Large Influxes of Miners Using LANDSAT Imagery. *Int. J. Environ. Res. Public Health* **2021**, *18*, 9441. [CrossRef] [PubMed]

19. GoldPrice.Org. Gold Price in USD. 2021. Available online: https://goldprice.Org/spot-gold.html (accessed on 19 July 2021).
20. Asner, G.P.; Llactayo, W.; Tupayachi, R.; Luna, E.R. Elevated rates of gold mining in the Amazon revealed through high-resolution monitoring. *Proc. Natl. Acad. Sci. USA* **2013**, *110*, 18454–18459. [CrossRef] [PubMed]
21. Kimijima, S.; Nagai, M. Study of Urbanization Corresponding to Socio-Economic Activities in Savannaket, Laos Using Satellite Remote Sensing. *Malaysisan J. Remote Sens. GIS* **2014**, *3*, 71–75. [CrossRef]
22. Kimijima, S.; Sakakibara, M.; Amin, A.; Nagai, M.; Arifin, Y.I. Mechanism of the Rapid Shrinkage of Limboto Lake in Gorontalo, Indonesia. *Sustainability* **2020**, *12*, 9598. [CrossRef]
23. Alam, A.; Bhat, M.S.; Maheen, M. Using Landsat satellite data for assessing the land use and land cover change in Kashmir valley. *GeoJournal* **2020**, *85*, 1529–1543. [CrossRef]
24. Pericak, A.A.; Thomas, C.J.; Kroodsma, D.A.; Wasson, M.F.; Ross, M.R.; Clinton, N.E.; Campagna, D.J.; Franklin, Y.; Bernhardt, E.S.; Amos, J.F. Mapping the yearly extent of surface coal mining in Central Appalachia using Landsat and Google Earth Engine. *PLoS ONE* **2018**, *13*, e0197758. [CrossRef]
25. Xiao, W.; Deng, X.; He, T.; Chen, W. Mapping Annual Land Disturbance and Reclamation in a Surface Coal Mining Region Using Google Earth Engine and the LandTrendr Algorithm: A Case Study of the Shengli Coalfield in Inner Mongolia, China. *Remote Sens.* **2020**, *12*, 1612. [CrossRef]
26. Isidro, C.M.; McIntyre, N.; Lechner, A.M.; Callow, I. Applicability of Earth Observation for Identifying Small-Scale Mining Footprints in a Wet Tropical Region. *Remote Sens.* **2017**, *9*, 945. [CrossRef]
27. Ammirati, L.; Mondillo, N.; Rodas, R.A.; Sellers, C.; Di Martire, D. Monitoring Land Surface Deformation Associated with Gold Artisanal Mining in the Zaruma City (Ecuador). *Remote Sens.* **2020**, *12*, 2135. [CrossRef]
28. Kimijima, S.; Sakakibara, M.; Nagai, M. Detection of Artisanal and Small-Scale Gold Mining Activities and Their Transformation Using Earth Observation, Nighttime Light, and Precipitation Data. *Int. J. Environ. Res. Public Health* **2021**, *18*, 10954. [CrossRef] [PubMed]
29. Kimijima, S.; Sakakibara, M.; Nagai, M. Investigation of Long-Term Roving Artisanal and Small-Scale Gold Mining Activities Using Time-Series Sentinel-1 and Global Surface Water Datasets. *Int. J. Environ. Res. Public Health* **2022**, *19*, 5530. [CrossRef]
30. The National Aeronautics and Space Administration. What Is Synthetic Aperture Radar? The National Aeronautics and Space Administration: Washington, DC, WA, USA, 2022.
31. Forkuor, G.; Ullmann, T.; Griesbeck, M. Mapping and Monitoring Small-Scale Mining Activities in Ghana using Sentinel-1 Time Series 2015–2019. *Remote Sens.* **2020**, *12*, 911. [CrossRef]
32. Telmer, K.H.; Stapper, D. *Evaluating and Monitoring Small Scale Gold Mining and Mercury Use: Building a Knowledge-Base with Satelite Imagery and Field Work*; United Nations Industrial Development Organization: Victoria, BC, Canada, 2007.
33. Agrawal, S. *Community Awareness on Hazards of Exposure to Mercury and Supply of Equipment for Mercury-Cleaner Gold Processing Technologies in Galangan, Central Kalimantan, Indonesia*; United Nations Industrial Development Organization: Central kalimantan, Indonesia, 2007.
34. European Space Agency. *Sentinel-1*; European Space Agency: Harwell, UK, 2022.
35. Filipponi, F. Sentinel-1 GRD Preprocessing Workflow. *Multidiscip. Digit. Publ. Inst. Proc.* **2019**, *18*, 11. [CrossRef]
36. Huang, L.K.; Wang, M.-J.J. Image thresholding by minimizing the measures of fuzziness. *Pattern Recognit.* **1995**, *28*, 41–51. [CrossRef]
37. Prewitt, J.M.S.; Mendelsohn, M.L. THE ANALYSIS OF CELL IMAGES*. *Ann. N.Y. Acad. Sci.* **1966**, *128*, 1035–1053. [CrossRef] [PubMed]
38. Ridler, T.W.; Calvard, S. Picture Thresholding Using An Interactive Selection Method. *IEEE Trans. Syst. Man Cybern.* **1978**, *8*, 630–632. [CrossRef]
39. Li, C.H.; Lee, C.K. Minimum cross entropy thresholding. *Pattern Recognit.* **1993**, *26*, 617–625. [CrossRef]
40. Li, C.H.; Tam, P.K.S. An iterative algorithm for minimum cross entropy thresholding. *Pattern Recognit. Lett.* **1988**, *19*, 771–776. [CrossRef]
41. Sankur, B. Survey over image thresholding techniques and quantitative performance evaluation. *J. Electron. Imaging* **2004**, *13*, 146–168. [CrossRef]
42. Kapur, J.N.; Sahoo, P.K.; Wong, A.K.C. A new method for gray-level picture thresholding using the entropy of the histogram. *Comput. Vis. Graph. Image Process.* **1985**, *29*, 273–285. [CrossRef]
43. Glasbey, C. An Analysis of Histogram-Based Thresholding Algorithms. *Graph. Model. Image Process.* **1993**, *55*, 532–537. [CrossRef]
44. Kittler, J.; Illingworth, J. Minimum error thresholding. *Pattern Recognit.* **1986**, *19*, 41–47. [CrossRef]
45. Tsai, W.-H. Moment-preserving thresolding: A new approach. *Comput. Vision Graph. Image Process.* **1985**, *29*, 377–393. [CrossRef]
46. Otsu, N. A threshold selection method from gray-level histograms. *IEEE Trans. Syst. Man Cybern.* **1979**, *9*, 62–66. [CrossRef]
47. Doyle, W. Operations Useful for Similarity-Invariant Pattern Recognition. *J. ACM* **1962**, *9*, 259–267. [CrossRef]
48. Shanbhag, A. Utilization of Information Measure as a Means of Image Thresholding. *CVGIP Graph. Model. Image Process.* **1994**, *56*, 414–419. [CrossRef]
49. Zack, G.W.; Rogers, W.E.; Latt, S.A. Automatic measurement of sister chromatid exchange frequency. *J. Histochem. Cytochem.* **1977**, *25*, 741–753. [CrossRef]

50. Yen, J.C.; Chang, F.-J.; Chang, S. A new criterion for automatic multilevel thresholding. *IEEE Trans. Image Process.* **1995**, *4*, 370–378. [CrossRef]
51. Nicolau, A.P.; Flores-Anderson, A.; Griffin, R.; Herndon, K.; Meyer, F.J. Assessing SAR C-band data to effectively distinguish modified land uses in a heavily disturbed Amazon forest. *Int. J. Appl. earth Obs. Geoinf. ITC J.* **2020**, *94*, 102214. [CrossRef]
52. PlanetGOLD. The Prevalence of COVID-19 within GOLD-ISMIA' s Project Locations. 2020. Available online: https://www.planetgold.org/sites/default/files/2020-04/planetGOLD_Indonesia_Factsheet_COVID19_ISMIA.pdf (accessed on 5 March 2022).
53. PlanetGOLD. The Gold Price Trend within GOLD-ISMIA' s Project Locations amid the COVID-19 Pandemic. 2020. Available online: https://www.planetgold.org/sites/default/files/2020-04/GOLD-ISMIAFactSheet_GoldPriceTrend.pdf (accessed on 5 March 2022).
54. European Commission Joint Research Centre. Yearly Water Classification History. 2022. Available online: https://global-surface-water.appspot.com/ (accessed on 10 April 2022).

International Journal of
Environmental Research and Public Health

MDPI

Article

Time-Series Assessment of Camp-Type Artisanal and Small-Scale Gold Mining Sectors with Large Influxes of Miners Using LANDSAT Imagery

Satomi Kimijima [1,*], **Masayuki Sakakibara** [1,2], **Masahiko Nagai** [3] **and Nurfitri Abdul Gafur** [4]

[1] Research Institute for Humanity and Nature, Kyoto 603-8047, Japan; sakaki@chikyu.ac.jp
[2] Graduate School of Science & Engineering, Ehime University, Matsuyama 790-8577, Japan
[3] Graduate school of Science and Technology for Innovation, Yamaguchi University, Ube 755-8611, Japan; nagaim@yamaguchi-u.ac.jp
[4] Bappeda-Litbang Bone Bolango, Suwawa 96113, Indonesia; vivinurv3@hotmail.com
* Correspondence: kimijima@chikyu.ac.jp

Abstract: Mining sites development have had a significant impact on local socioeconomic conditions, the environment, and sustainability. However, the transformation of camp-type artisanal and small-scale gold mining (ASGM) sites with large influxes of miners from different regions has not been properly evaluated, owing to the closed nature of the ASGM sector. Here, we use remote sensing imagery and field investigations to assess ASGM sites with large influxes of miners living in mining camps in Bone Bolango Regency, Gorontalo Province, Indonesia, in 1995–2020. Built-up areas were identified as indicators of transformation of camp-type ASGM sites, using the Normalized Difference Vegetation Index, from the time series of images obtained using Google Earth Engine, then correlated with the prevalent gold market price. An 18.6-fold increase in built-up areas in mining camps was observed in 2020 compared with 1995, which correlated with increases in local gold prices. Field investigations showed that miner influx also increased after increases in gold prices. These findings extend our understanding of the rate and scale of development in the closed ASGM sector and the driving factors behind these changes. Our results provide significant insight into the potential rates and levels of socio-environmental pollution at local and community levels.

Keywords: artisanal and small-scale gold mining; gold price; Indonesia; influxes of miners; landcover change; mining camp; remote sensing

Citation: Kimijima, S.; Sakakibara, M.; Nagai, M.; Gafur, N.A. Time-Series Assessment of Camp-Type Artisanal and Small-Scale Gold Mining Sectors with Large Influxes of Miners Using LANDSAT Imagery. *Int. J. Environ. Res. Public Health* **2021**, *18*, 9441. https://doi.org/10.3390/ijerph18189441

Academic Editor: Daniela Varrica

Received: 7 July 2021
Accepted: 31 August 2021
Published: 7 September 2021

Publisher's Note: MDPI stays neutral with regard to jurisdictional claims in published maps and institutional affiliations.

1. Introduction

The development of mining sites has had a significant impact on local socioeconomic conditions, the environment, and their sustainability. These developments have harmful socio-environmental consequences. Therefore, understanding the speed and scale of the development of mining sites and the factors driving these changes should provide significant insight into the potential rates and levels of socio-environmental pollution at local and community levels. This may also allow problems to be avoided and alternative sustainable strategies to be developed by involving various stakeholders at different levels.

The artisanal and small-scale gold mining (ASGM) sector is a significant gold-producing sector and is the largest employer in gold mining throughout the world. This sector represents ~20% (400–600 T/year) of the gold production and 90% of the gold-producing workforce on the global stage [1]. 'ASGM' refers to the mining practiced with rudimentary technology by an individual, group, or community [2]. The sector can be generally characterized as informal, unregistered, and illegal [3]. In this practice, gold is commonly extracted with mercury at the stage of amalgamation, causing extremely harmful environmental and occupational health hazards as a result of mercury pollution [4–6]. Mercury emissions into the atmosphere and its releases into water from ASGM are significant, and the ASGM was the dominant sector emitting mercury (37.7%) into the air in 2015 [5], mainly in South America, Africa, and Asia [4,7]. Other health problems, such as silicosis, methyl-orthophosphate-related poisoning, and various injuries, also occur during the mining process [8]. Despite the negative socio-environmental consequences, ASGM activities are still undertaken, predominantly in rural areas of >80 countries, as a significant poverty-alleviation mechanism, and to drive their economic development [1,3]. Indonesia, where the national poverty line is 9.4% as of 2019 [9], shows the continuous growth of ASGM across the country [10,11].

The Minamata Convention on Mercury, a global treaty protecting human health and the environment from anthropogenic emissions and the releases of mercury and its compounds [12], was adapted and came into force in October 2013 and August 2017, respectively [7]. Article 7 of the Convention focuses especially on ASGM sector, advocating the reduction of mercury use in the sector, and has been strategically taken up in national action plans and various national regulations among the ratifying nations [13]. However, the formalization of this policy has often been impeded by political issues, such as insufficient institutional frameworks, capacities, and funds [14,15]. In Indonesia, despite the formalization of the Convention and the development of a national action plan, the alternative supply of mercury has been domestically produced [16,17], resulting in a significant increase in informal mercury imports [18]. Therefore, it is expected that the generation of an alternative supply of mercury at a lower cost will accelerate ASGM and allow operators to expand their activities and the use of mercury beyond the regulated levels, as the price of gold on the global market increases [19]. A relationship between increases in ASGM and in gold price has been reported in the literature [20,21].

The ASGM sector can be categorized into two types: the 'travel-type', in which miners commute from their local residences to mining sites, and the 'camp-type', in which miners live and conduct their mining activities at mining camps. Recent research has focused mainly on the environmental and human health effects of mercury pollution caused by the travel-type ASGM sites spread over a large number of areas [2,8,22–29], largely employing the alluvial mining method. Although camp-type ASGM sites have also been studied [30], they have only been recorded as points on maps, and there has been no quantitative analysis of the changes in the camp-type mining sector over time.

As remote sensing technologies have developed, they have been widely used to characterize natural features and physical objects, allowing spatial changes in these to be monitored over time. Remote sensing also provides diverse continuous data with temporal, spatial, and spectral resolutions. Freely available satellite remote sensing data, such as Landsat series, have provided long-term datasets of Earth observations since the 1970s, and are extensively used to detect and monitor landcover [31–34]. The use of such long-term satellite datasets has allowed the development of qualitative and comprehensive understanding of various changes, including ASGM development [2,27,34,35]. Using remote sensing technologies, several studies have assessed ASGM-related qualitative spatiotemporal changes, such as deforestation, the extent of mining areas, and geomorphic and hydrological changes [21,27–29,35,36]. However, those studies were limited to ASGM areas with long and well-known mining histories, and mainly examined the travel-type part of the sector.

However, as the gold price has increased, artificially developed camp-type ASGM sites, spread across small areas, have developed in remote rural areas, with significant influxes of miners from neighboring regions [16,37,38]. The movement of such invisible informal influxes of miners to artificial camps causes informal communities to form and expand, accelerating the severe socio-environmental pollution inside the camps. However, this camp-type ASGM, with large influxes of miners, has never been quantitatively investigated in depth. Because camp-type ASGM in remote rural areas is associated with influxes of miners who live at the mining camps, the spatial distribution of built-up areas can be a significant indicator of the transformation of otherwise invisible ASGM communities.

In this study, our primary objective was to assess the transformation of the ASGM sectors by large influxes of miners living at mining camps in Bone Bolango Regency, Indonesia. Our specific objectives were: (1) to assess the landcover changes in 1995–2020, using remotely sensed imagery, such as a Landsat series; and (2) to correlate the ASGM-directed transformations with the prevalent gold price. The results of this study contribute to our understanding of the spread of camp-type ASGM activities spread across a small remote rural area and allow the rate and level of socio-environmental pollution in its train to be predicted.

2. Materials and Methods

2.1. Overall Methodological Workflow

Figure 1 shows the methodological workflow used in this study. We focused on three significant steps to achieve our primary objective of assessing the transformation of ASGM with large influxes of miners into mining camps. First, the built-up areas in the mining camp in 1995–2020 were calculated from Landsat data. Second, the relationship between the identified built-up areas and the historical gold price was assessed. Third, a field survey was conducted to investigate the characteristics of the ASGM camps. Together, this evidence allowed us to understand the transformation of the ASGM sector by large influxes of miners living at mining camps. We present a discussion based on all the findings described. The methods used in each step are explained in the following sections.

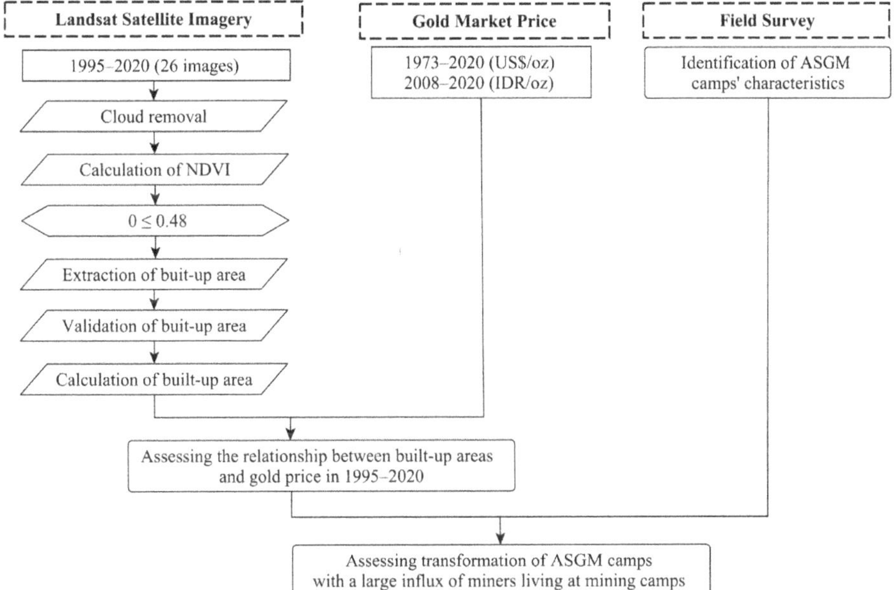

Figure 1. Overall methodology.

2.2. Study Area

The North Sulawesi, Indonesia, is a well-mineralized metallogenic region, with significant gold mineralization associated with quartz veins, in a variety of porphyry and epithermal settings. The vertical tunnel method (shaft) of mining is predominantly used in the country, and the gold is extracted with mercury amalgamation in almost all of Indonesia's ASGM hotspots [11], including in Sulawesi.

The Motomboto ASGM area is located ~30 km southeast of the city of Gorontalo in Bone Bolango Regency, Gorontalo Province, Indonesia. This area is categorized as having high-sulfidation epithermal deposits of copper, gold, and silver [39]. Gold mining in Bone Bolango Regency, Gorontalo began in the Dutch era (18th century) [40], and later, mining activity in the West Motomboto and Tulabolo areas was developed by Tropic Endeavour Indonesia in 1988 [41]. However, these mining sites were closed in 1991 because they intruded upon the Bogani Nani Wartabone National Park development [41]. The closure of the former mining site has triggered the entry of residents to the area to undertake mining activities [41]. In 2013, more than 9000 small-scale miners were reported in the Bogani Nani Wartabone National Park [38]. The majority of them came from neighboring regions, including Bolaang Mongondow and Minahasa in North Sulawesi, Indonesia [38].

In this study, we examined the Motomboto ASGM area, and divided it into mining camps 1, 2, and 3 (Figure 2).

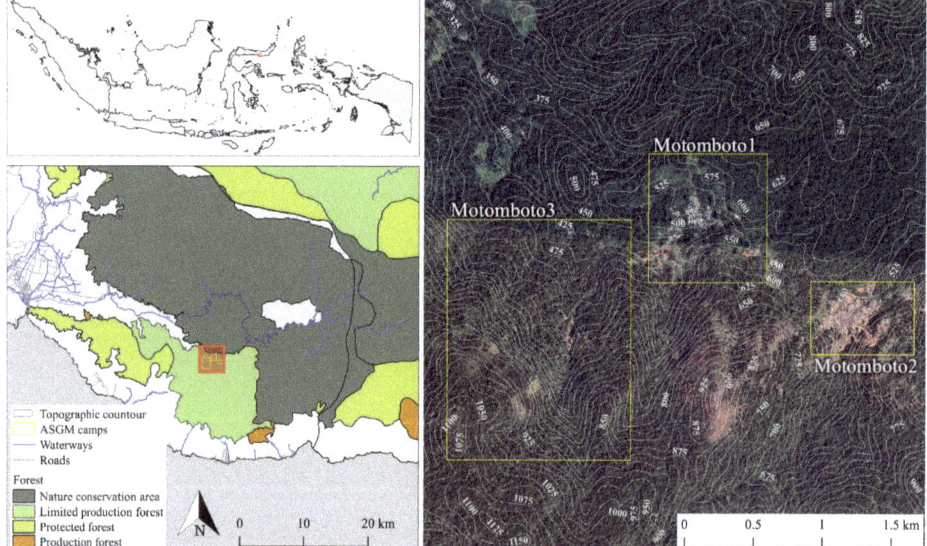

Figure 2. Study area.

2.3. Satellite Imagery

Atmospherically corrected cloud-free Landsat data from the Thematic Mapper (TM), Enhanced Thematic Mapper Plus (ETM+) and Operational Land Imager (OLI), satellite images, available from the United States Geological Survey (USGS) through the Google Earth Engine, were used to extract a time series of built-up areas. The images were chosen based on cloud coverage and satellite data availability to minimize any potential influencing factors. Consequently, imagery from March–May was primarily selected. For those years in which data for the target months were not available, imagery collected in adjacent months was used. With these provisos, satellite imagery acquired from 1995 to 2020, with a ground resolution of 30 m in the World Geodetic System 84 (WGS84) geographic coordinate reference system, was used to detect and analyze the built-up areas in the study region.

In previous studies, the mining areas in Bone Bolango Regency were estimated to cover 0.62 km^2 in 2012 [30]. Therefore, the long-term trend in ASGM sites could be detected with satellite imagery, even at a ground resolution of 30 m. The main specifications of the sensors used in this study are summarized in Table 1.

Table 1. Main specification of satellite imagery in the study.

Acquisition Date	Sensor	NIR (μm)	Red (μm)	Green (μm)
30 May 1995 26 May 1997 11 May 2000 4 April 2001 25 May 2002	Landsat 5 TM	0.70–0.80	0.60–0.70	0.50–0.60
27 March 2004 10 February 2005 17 March 2006, 25 September 2006 23 May 2007, 30 October 2007 26 April 2009 24 February 2010, 4 September 2010 10 January 2011, 26 November 2011 27 March 2013 24 April 2014	Landsat 7 ETM+	0.76–0.90	0.63–0.69	0.52–0.60
8 July 2015 5 April 2016 24 April 2017 10 March 2018 4 August 2019 15 March 2020	Landsat 8 OLI	0.85–0.88	0.64–0.67	0.53–0.59

2.4. Extraction and Calculation of Built-Up Areas

Landsat satellite images acquired in 1995–2020 were used. Since the transformation by ASGM in remote rural areas is associated with influxes of miners who live at mining camps, as described in the Introduction, the spatial distribution of built-up areas was extracted as a significant indicator of the transformation of the ASGM camps resulting from large influxes of miners. Built-up areas can be defined by their physical aspects, such as predominantly human-constructed elements [42], as in this study. A number of spectral indices, including the Urban Index (UI) [43], Normalized Difference Built-up Index (NDBI) [44], Index-based Built-up Index [45], Built-up Area Extraction Method [46], Enhanced Built-up and Bareness Index [47], Band Ratio for Built-up Area [48], Built-up Index [49], Normalized Difference Vegetation Index (NDVI) [34], and Automated Built-up Extraction Index [50], have been developed to extract built-up areas from satellite imagery. Furthermore, human visual interpretation was also used [42]. Previous studies found that NDBI [51,52] and UI [53,54] are most sensitive in retrieving built-up areas, although these have mainly been used in urban studies. As NDBI and UI are incapable of efficiently separating built-up areas from bare land [44], the separation of these two land types in rural areas is more complicated. Therefore, in this study, we used NDVI, as used elsewhere [34], to analyze remote mining areas over long timescales. The value of NDVI in the built-up areas was calculated with Equation (1).

$$NDVI = (NIR - Red)/(NIR + Red) \qquad (1)$$

NDVI, ranging from -1 to 1, shows a high value for dense vegetation and a low value for desert or unvegetated areas [55]. In this study, we further restricted the built-up areas using the NDVI threshold, $0 \leq NDVI \leq 0.48$, to exclude vegetated areas on the land surface. The value was determined based on comparisons of the accuracy levels by referring high-resolution satellite data. In this way, the built-up areas were identified, and results were visualized as a time series. To assess the accuracy of the results, 100 points were randomly selected in the study area and validated using a high-resolution image obtained on 8 February 2017 using Google Earth Pro. Because no images were available in Google Earth Pro for the same dates as the Landsat imagery acquired from USGS, images acquired on the closest date (24 April 2017) were used. In this study, we applied the validated accuracy to all of the classification results, owing to the unavailability of reference data.

The relationship between the built-up areas and the gold price was also assessed. The global gold prices [56] and gold prices in Indonesian rupiah [57] were obtained for 1973–2020 and 2008–2020, respectively. They were then graphed against the total built-up area in the mining camps across time, and the correlation between the two parameters was calculated.

2.5. Investigation of ASGM Camps

Field observations of the Motomboto ASGM camps were made on 6 February 2020. Settlements, trommel machines, pools for immersing the materials, and miners' camps were investigated. Interviews were also conducted with key miners in the mining camps.

3. Results

3.1. Expansion of Built-Up Areas in the Mining Camps

To detect the changes in the landcover surrounding the ASGM camps, NDVI was primarily calculated for 1995–2020. The appearance of changes and the rate of their development varied across the camps. Figures 3–5 show the changes in NDVI in each camp over the 25-year study period. Motomboto ASGM camp 1 was first identified around 2011. However, it did not show significant expansion until 2020. In comparison, camp 2 existed before 1995 and developed gradually after 2015. Camp 3 was newly identified in 2019 and expanded rapidly into the eastern and southern areas.

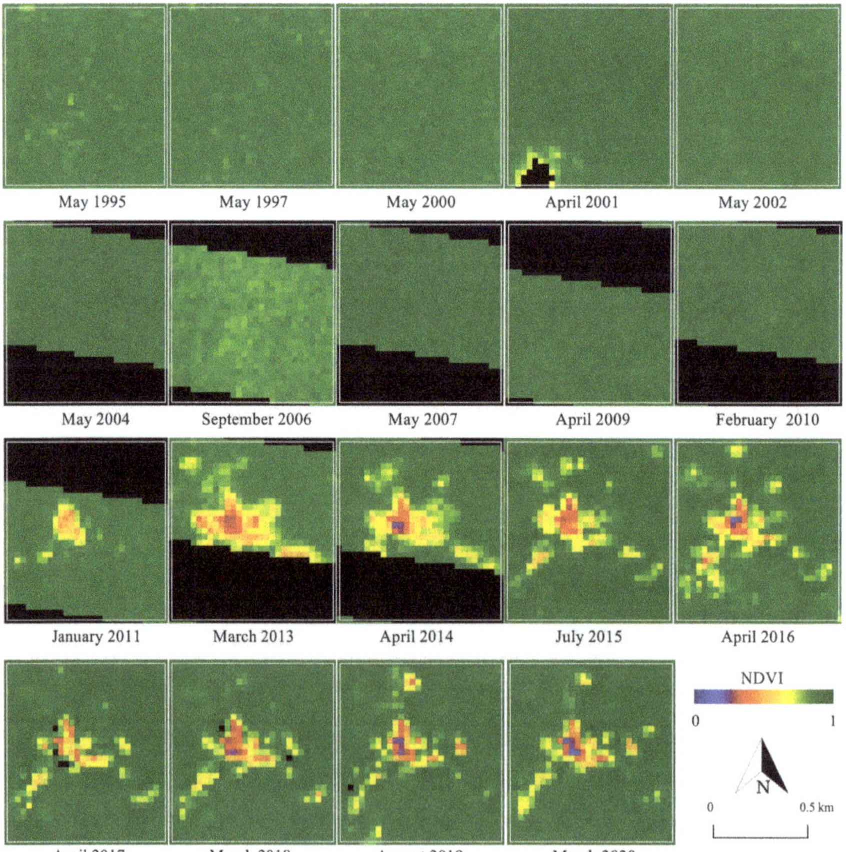

Figure 3. Changes in NDVI in Motomboto ASGM camp 1.

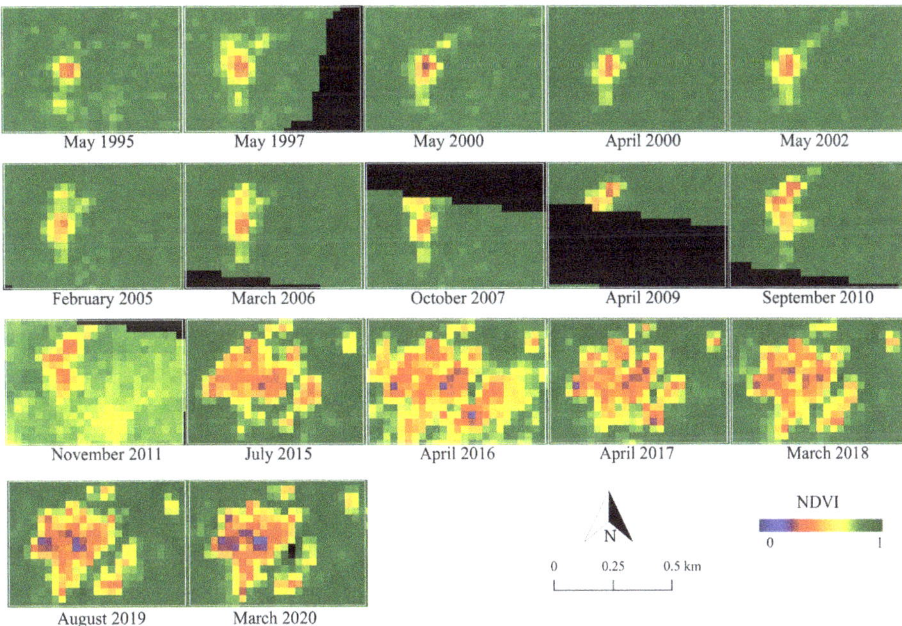

Figure 4. Changes in NDVI in Motomboto ASGM camp 2.

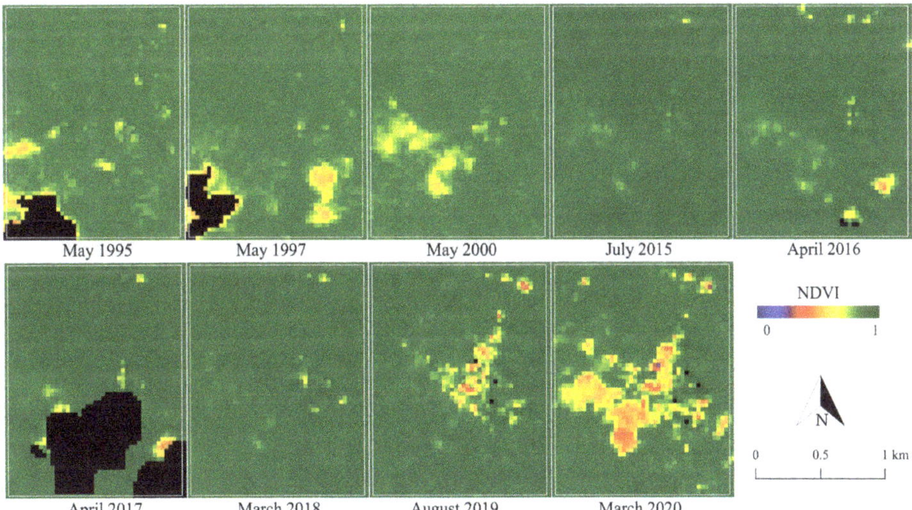

Figure 5. Changes in NDVI in Motomboto ASGM camp 3.

From the results of NDVI, the built-up areas in the ASGM camps were extracted as described in Section 2.4. Together with various landcover changes, built-up areas were detected in all of the ASGM camps, with an accuracy of 96% (Figure 6). Built-up areas were identified from 2011, 1995, and 2016 in Motomboto ASGM camps 1, 2, and 3, respectively. The built-up area in camp 1 developed largely in 2013, but the area remained around the same size. Camp 2 mainly developed around 2015, and the largest area was detected in 2016. Camp 3 was clearly distinguishable from the others and showed a continuous rapid increase in size even after the Minamata Convention on Mercury was brought into force in 2017. Notably, in 2020, ASGM camp 3 showed a 23-fold increase in size over that in 2017. The built-up areas in the mining camps tended to be detected approximately 1 year after the landcover changes described in the previous paragraph were identified.

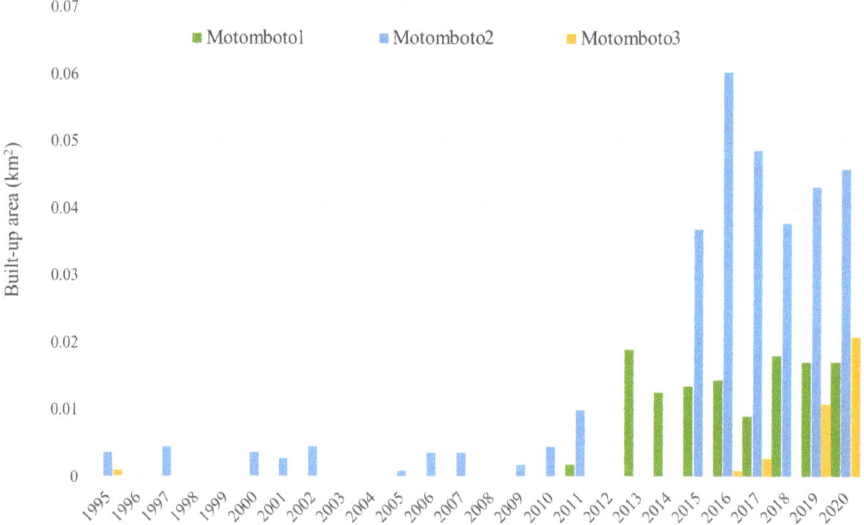

Figure 6. Built-up areas by ASGM camp.

3.2. Investigation of ASGM Camps

Motomboto ASGM camp is located 4–6 h from the center of East Suwawa. There are no paved roads, and the camps are only accessible by motorcycle. The road conditions are poor, and miners are required to cross several rivers by motorcycles to reach the camps. The basic settlements in the camps are composed of tin roofs or are covered by tarpaulins and are spread over in the small valley, forming village-like settlements (Figure 7a). In these simple settlements, all the processes required for the gold extraction are performed, and the incoming miners also stay in these mining camps.

Based on our field investigations, the total number of miners in the Suwawa area was approximately 10,000. Because the miners tended to be replaced frequently, they were difficult to count accurately; however, 75–85% of the incoming miners were from rural areas, where they engaged in agricultural or fishery industries. At the ASGM sites, more than two teams were organized per tunnel, working in 24 h shifts (as reported by a local miner).

The activities conducted in the Motomboto ASGM camps are shown in Figure 7b–e. Mercury and cyanide are required in the process of gold amalgamation in the tailings. Figure 7f shows the residences of the incoming miners and their families, where the miners' families generally operate small restaurants or grocery stores to service the influx miners and their families' daily needs. The ASGM activity in this area has increased rapidly since 2017, with influxes of miners after the gold price increased (as reported by a local miner).

Figure 7. Motomboto ASGM camp. (**a**) Settlements in Motomboto ASGM 3 taken by an unmanned aerial vehicle. (**b**) An active mine hole. (**c**) Trommels for milling materials mined from holes. (**d**) A pool of water mixed with hydrogen peroxide for immersing the milled materials. (**e**) A pool of mercury mixed with cyanide for immersing the materials. (**f**) A common settlement where incoming miners and their families stay and manage a grocery shop.

3.3. Relationship between the Built-Up Areas and the Gold Price

The relationship between the built-up areas in the ASGM camps and the gold price was assessed (Figure 8). The built-up areas identified at the three mining sites were combined. The global gold price increased rapidly from 2006 to 2012. It then decreased and remained steady till early 2019, when it increased again until 2021. The gold price in Indonesia has increased since 2007, with an especially steadily increase since 2017, approximately doubling by late 2020. Although there are differences in the Indonesian and global gold prices, similar tends were observed. From 2007 to 2020, the increase in the built-up areas was significantly associated with the gold price ($R^2 = 0.91$).

Figure 8. Global gold price, Indonesian gold price, and built-up areas in the camps.

4. Discussion and Limitations

4.1. Discussion

The built-up areas developed in ASGM camps are indicators of transformations at the ASGM sites and, in this study, they provided a developmental time series that correlated with the changes in gold prices. A quantitative analysis of the ASGM sector extends our understanding of the rate of development of mining sites and their transformation over time. Understanding the status of ASGM is essential to tracking its responses to global factors, such as the gold price and the Minamata Convention on Mercury (2017), and to predicting the rate and level of socio-environmental destruction at the local and community levels.

Identification of proposed built-up areas, as an indicator of the growth of the camp-type ASGM, using Landsat series with 30 m ground resolution, demonstrated the transformation of camp-type ASGM over decades (Figures 3–6), as reported previously [28]. With a quantitative time-series analysis, we detected various forms of built-up areas in the mining camps, indicative of the camps' characteristics. For example, Motomboto camp 1 was identified as a newly opened area in 2011, whereas the mining activities at camp 2 expanded rapidly in 2015. Although, the development of camp 3 is considered recent, it developed rapidly after 2016. This expansion and development resulted from new influxes of miners and the weak enforcement of their regulation. Illegal immigrant miners from neighboring regions, including Bolaang Mongondow and Minahasa (North Sulawesi), have been reported [38]. The population of East Suwawa, where the targeted mining camps are located, did not show a significant change with the population in 2007–2019 [58–68]. Because local residents travel from their own villages near the mining camp (field interview), the development of the built-up areas in a mining camp reflects the increased influx of miners into that mining camp. Possible factors encouraging their entry include the weak regulation of the ASGM sector resulting from its informal, illegal, and closed nature [3]; limited governments resources and administrative capacity to provide adequate technical assistance or to enforce compliance [69]; and the remote locations of the mining sites [35]. The identification of policy problems and the development of planning and management solutions in remote rural locations further impede the proper management of the sector [70]. Because the Motomboto ASGM camps are located in remote rural areas in a national park, the regulation and monitoring of miner influxes from other regions and mining activities are more complex than those at easily accessible sites.

Our finding of a significant relationship between the developmental scale represented by built-up areas and the gold price in Indonesian rupiah is consistent with those previous studies [20,21]. Although mining types differ, similar trends have been detected between these two factors.

The living conditions in the mining camps can have negative social and health effects, resulting from the population's exposure to high levels of mercury vapor. The emission of mercury from the camps into the atmosphere and its releases into Bone River, the main water supply for the city of Gorontalo, further harm both human and environment health at the community and regional levels. Other problems are expected inside the mining camps, such as a high incidence of infection among children, an increased prevalence of tropical diseases such as Dengue fever, and a lack of access to health care, education for children, safe water, wastewater treatment, and sanitation supplies, as reported in previous studies [71–74]. However, artificial camp-type AGSM sites, including the Motomboto sites do not generally contain any basic infrastructures, resulting in rapid environmental pollution owing to poor waste treatment.

4.2. Limitations

This study had several limitations associated with the quality of the input data. First, cloud-free Landsat series and complete images of Landsat7 were limited due to various factors, including scanline errors. Second, differences in the spatial resolution of the datasets used resulted in mixed pixels, possibly causing the overestimation or miscalculation of built-up areas. Third, the methodology used in this study is only applicable to similar camp-type mining sectors.

5. Conclusions

In this study, the transformation of the ASGM sector, with large influxes of miners living at mining camps in Bone Bolango Regency, Indonesia, was assessed with remote sensing imagery and field investigations. The results presented here show that the total built-up area in the target ASGM sites identified by Landsat series in 2020 had increased 18.6-fold relative to that in 1995 and correlated with the increase in the gold price in Indonesian rupiah. Furthermore, the large influx of miners living in mining camps paralleled the increase in the market price of gold. Therefore, we conclude that the spread of camp-type ASGM across a small remote area with large influxes of miners is detectable by monitoring the built-up areas in those mining camps. These results extend our understanding of the rate and scale of the development of the closed ASGM sector and provide significant insight into the potential for environmental pollution at the local and community levels. This will allow precautions to be taken and alternative sustainable strategies to be developed at the local, community, and regional levels.

Author Contributions: S.K. contributed to design the research, data analysis, and data visualization. M.S. and M.N. provided conceptual advice and critical comments. N.A.G. conducted the field investigation. All authors have read and agreed to the published version of the manuscript.

Funding: This research was financially supported by the Research Institute for Humanity and Nature (RIHN: a constituent member of NIHU).

Institutional Review Board Statement: Not applicable.

Informed Consent Statement: Not applicable.

Acknowledgments: This research was financially supported by The Research Institute for Humanity and Nature (RIHN: a constituent member of NIHU), project no. 14200102. We are grateful to the Indonesian State Ministry of Research and Technology (RISTEK) for providing research permits in Indonesia.

Conflicts of Interest: The authors declare no conflict of interest.

References

1. United Nations Environment Programme. *Estimating Mercury Use and Documenting Practices in Artisanal and Small-Scale Gold Mining (ASGM)*; UN Environment: Geneva, Switzerland, 2017.
2. Lobo, F.D.L.; Costa, M.; Novo, E.M.L.D.M.; Telmer, K. Distribution of Artisanal and Small-Scale Gold Mining in the Tapajós River Basin (Brazilian Amazon) over the Past 40 Years and Relationship with Water Siltation. *Remote Sens.* **2016**, *8*, 579. [CrossRef]
3. Wilson, M.L.; Renne, E.; Roncoli, C.; Agyei-Baffour, P.; Tenkorang, E.Y. Integrated Assessment of Artisanal and Small-Scale Gold Mining in Ghana—Part 3: Social Sciences and Economics. *Int. J. Environ. Res. Public Health* **2015**, *12*, 8133–8156. [CrossRef] [PubMed]
4. World Health Organization. *Artisanal and Small-Scale Gold Mining and Health*; World Health Organization: Geneva, Switzerland, 2016.
5. United Nations Environment Programmet. *Technical Background Report to the Global Mercury Assessment 2018*; United Nations Environment Programme: Tromsø, Norway, 2018.
6. Isaza, A.S.; Villegas-Palacio, C.; Arango, S. The public good dilemma of a non-renewable common resource: A look at the facts of artisanal gold mining. *Resour. Policy* **2013**, *38*, 224–232. [CrossRef]
7. United Nations Environment Programme. *Global Mercury Assessment*; United Nations Environment Programme: Geneva, Switzerland, 2018.
8. Macháček, J. Typology of Environmental Impacts of Artisanal and Small-Scale Mining in African Great Lakes Region. *Sustainability* **2019**, *11*, 3027. [CrossRef]
9. World Bank Group. Poverty Headcount Ratio at National Poverty Lines (% of Population)-Indonesia. Available online: https://data.worldbank.org/indicator/SI.POV.NAHC?locations=ID (accessed on 6 June 2021).
10. BALIFOKUS. *Mercury Contaminated Sites in ASGM Sites of Indonesia*; BALIFOKUS: Bali, Indonesia, 2019.
11. Ismawati, J.; Petrlik, Y.; DiGangi, J. Mercury Hotspots in Indonesia ASGM Sites: Poboya and Sekotong in Indonesia. Available online: https://ipen.org/hgmonitoring/pdfs/indonesia-report-en.pdf (accessed on 6 June 2021).
12. United Nations Environment Programmet. *Developing National ASGM Formalization Strategies within National Action Plans*; United Nations Environment Programme: Geneva, Switzerland, 2018.
13. United Nations Environment Programmet. *Developing a National Action Plan to Reduce and, Where Feasible, Eliminate Mercury Use in Artisanal and Small-Scale Gold Mining*; United Nations Environment Programme: Geneva, Switzerland, 2017.
14. Hilson, G.; Zolnikov, T.R.; Ortiz, D.R.; Kumah, C. Formalizing artisanal gold mining under the Minamata convention: Previewing the challenge in Sub-Saharan Africa. *Environ. Sci. Policy* **2018**, *85*, 123–131. [CrossRef]
15. Kinyondo, A.; Huggins, C. Promoting environmental sustainability in the artisanal and small-scale mining sector in Tanzania. *WIDER Work. Pap.* **2021**, *119*, 1–25.
16. International Institute for Sustainable Development. *Global Trends in Artisanal and Small-Scale Mining (Asm): A Review of Key Numbers and Issues*; IISD: Winnipeg, Canada, 2018.
17. Spiegel, S.J.; Agrawal, S.; Mikha, D.; Vitamerry, K.; Le Billon, P.; Veiga, M.; Konolius, K.; Paul, B. Phasing Out Mercury? Ecological Economics and Indonesia's Small-Scale Gold Mining Sector. *Ecol. Econ.* **2018**, *144*, 1–11. [CrossRef]
18. Ismawati, Y. ASGM: The Production of Social and Environmental Suffering Gold, mercury and the next Minamata tragedy. *Bali Fokus. Denpasar.* **2014**, *2009*, 1–14.
19. World Gold Council. Gold Prices. Available online: https://www.gold.org/goldhub (accessed on 6 June 2021).
20. Asner, G.P.; Llactayo, W.; Tupayachi, R.; Luna, E.R. Elevated rates of gold mining in the Amazon revealed through high-resolution monitoring. *Proc. Natl. Acad. Sci. USA* **2013**, *110*, 18454–18459. [CrossRef]
21. Swenson, J.J.; Carter, C.E.; Domec, J.-C.; Delgado, C.I. Gold Mining in the Peruvian Amazon: Global Prices, Deforestation, and Mercury Imports. *PLoS ONE* **2011**, *6*, e18875. [CrossRef]
22. Macháček, J. Alluvial Artisanal and Small-Scale Mining in a River Stream—Rutsiro Case Study (Rwanda). *Forests* **2020**, *11*, 762. [CrossRef]
23. Kahhat, R.; Parodi, E.; Larrea-Gallegos, G.; Mesta, C.; Vázquez-Rowe, I. Environmental impacts of the life cycle of alluvial gold mining in the Peruvian Amazon rainforest. *Sci. Total Environ.* **2019**, *662*, 940–951. [CrossRef] [PubMed]
24. Nakazawa, K.; Nagafuchi, O.; Kawakami, T.; Inoue, T.; Yokota, K.; Serikawa, Y.; Cyio, B.; Elvince, R. Human health risk assessment of mercury vapor around artisanal small-scale gold mining area, Palu city, Central Sulawesi, Indonesia. *Ecotoxicol. Environ. Saf.* **2016**, *124*, 155–162. [CrossRef] [PubMed]
25. Bose-O'Reilly, S.; Drasch, G.; Beinhoff, C.; Rodrigues-Filho, S.; Roider, G.; Lettmeier, B.; Maydl, A.; Maydl, S.; Siebert, U. Health assessment of artisanal gold miners in Indonesia. *Sci. Total Environ.* **2010**, *408*, 713–725. [CrossRef]
26. Wyatt, L.; Ortiz, E.J.; Feingold, B.; Berky, A.; Diringer, S.; Morales, A.M.; Jurado, E.R.; Hsu-Kim, H.; Pan, W. Spatial, Temporal, and Dietary Variables Associated with Elevated Mercury Exposure in Peruvian Riverine Communities Upstream and Downstream of Artisanal and Small-Scale Gold Mining. *Int. J. Environ. Res. Public Health* **2017**, *14*, 1582. [CrossRef]
27. Schmid, T.; Rico, C.; Rodríguez-Rastrero, M.; Sierra, M.J.; Díaz-Puente, F.J.; Pelayo, M.; Millán, R. Monitoring of the mercury mining site Almadén implementing remote sensing technologies. *Environ. Res.* **2013**, *125*, 92–102. [CrossRef]
28. Espejo, J.C.; Messinger, M.; Román-Dañobeytia, F.; Ascorra, C.; Fernandez, L.E.; Silman, M. Deforestation and Forest Degradation Due to Gold Mining in the Peruvian Amazon: A 34-Year Perspective. *Remote Sens.* **2018**, *10*, 1903. [CrossRef]

29. Emel, J.; Plisinski, J.; Rogan, J. Monitoring geomorphic and hydrologic change at mine sites using satellite imagery: The Geita Gold Mine in Tanzania. *Appl. Geogr.* **2014**, *54*, 243–249. [CrossRef]
30. Puluhulawa, F.; Harun, A.A. Policy formalization of Artisanal and Small-Scale Gold Mining (ASGM) post-ratification of Minamata Convention for Sustainability (case study of ASGM Gorontalo). *E3S Web Conf.* **2019**, *125*, 02006. [CrossRef]
31. Kimijima, S.; Nagai, M. Study of urbanization corresponding to socio-economic activities in Savannaket, Laos using satellite remote sensing. *Malaysisan J. Remote Sens. GIS* **2014**, *3*, 71–75.
32. Kimijima, S.; Sakakibara, M.; Amin, A.; Nagai, M.; Arifin, Y.I. Mechanism of the Rapid Shrinkage of Limboto Lake in Gorontalo, Indonesia. *Sustainability* **2020**, *12*, 9598. [CrossRef]
33. Alam, A.; Bhat, M.S.; Maheen, M. Using Landsat satellite data for assessing the land use and land cover change in Kashmir valley. *GeoJournal* **2020**, *85*, 1529–1543. [CrossRef]
34. Pericak, A.A.; Thomas, C.J.; Kroodsma, D.A.; Wasson, M.F.; Ross, M.R.; Clinton, N.E.; Campagna, D.J.; Franklin, Y.; Bernhardt, E.S.; Amos, J.F. Mapping the yearly extent of surface coal mining in Central Appalachia using Landsat and Google Earth Engine. *PLoS ONE* **2018**, *13*, e0197758. [CrossRef]
35. Gallwey, J.; Robiati, C.; Coggan, J.; Vogt, D.; Eyre, M. A Sentinel-2 based multispectral convolutional neural network for detecting artisanal small-scale mining in Ghana: Applying deep learning to shallow mining. *Remote Sens. Environ.* **2020**, *248*, 111970. [CrossRef]
36. Owusu-Nimo, F.; Mantey, J.; Nyarko, K.; Appiah-Effah, E.; Aubynn, A. Spatial distribution patterns of illegal artisanal small scale gold mining (Galamsey) operations in Ghana: A focus on the Western Region. *Heliyon* **2018**, *4*, e00534. [CrossRef] [PubMed]
37. Lampost.co. Tambang Liar di Register 20 Gunung Bunder Tak Pernah Berakhir. Available online: https://www.lampost.co/berita-tambang-liar-di-register-20-gunung-bunder-tak-pernah-berakhir.html (accessed on 10 July 2021).
38. Hatu, R.A. Socio-economic Conditions in The Illegal Gold Miners Tulabolo Village, Gorontalo-in Indonesian. *Asian J. Appl. Sci.* **2016**, *9*, 97–105. [CrossRef]
39. PT Bumi Resources Minerals Tbk. *Laporan Tahunan 2019 Annual Report*; PT Bumi Resources Minerals Tbk: Jakarta, Indonesia, 2019.
40. Van Bemmelen, R.W. *The Geology of Indonesia. General Geology of Indonesia and Adjacent Archipelagoes*; Government Printing Office: The Hague, The Netherlands, 1949; pp. 545–547, 561–562.
41. Kesatuan Pengelotaan Hutan. *Rencana Pengelolaan Hutan Jangka Panjang Kphp Unit Vii Bone Bo-Lango Tahun 2016–2025*; Bone Bolango Regency: Suwawa, Indonesia, 2016.
42. Kimijima, S.; Nagai, M. Human Mobility Analysis for Extracting Local Interactions under Rapid Socio-Economic Transformation in Dawei, Myanmar. *Sustainability* **2017**, *9*, 1598. [CrossRef]
43. Kawamura, Y.; Jayamana, M.; Tsujiko, S. Relation between social and environmental conditions in colombo sri lanka and the urban index estimated by satellite remote sensing data. *Int. Arch. Photogramm. Remote Sens.* **1996**, *31*, 321–326.
44. Zha, Y.; Gao, J.; Ni, S. Use of normalized difference built-up index in automatically mapping urban areas from TM imagery. *Int. J. Remote Sens.* **2003**, *24*, 583–594. [CrossRef]
45. Xu, H. A new index for delineating built-up land features in satellite imagery. *Int. J. Remote Sens.* **2008**, *29*, 4269–4276. [CrossRef]
46. Bhatti, S.S.; Tripathi, N.K. Built-up area extraction using Landsat 8 OLI imagery. *GIScience Remote Sens.* **2014**, *51*, 445–467. [CrossRef]
47. As-Syakur, A.R.; Adnyana, I.W.S.; Arthana, I.W.; Nuarsa, I.W. Enhanced Built-Up and Bareness Index (EBBI) for Mapping Built-Up and Bare Land in an Urban Area. *Remote Sens.* **2012**, *4*, 2957–2970. [CrossRef]
48. Waqar, M.M.; Mirza, J.F.; Mumtaz, R.; Hussain, E. Development of New Indices for Extraction of Built-Up Area & Bare Soil. *Open Access Sci. Rep.* **2012**, *1*, 1–4.
49. Kaimaris, D.; Patias, P. Identification and Area Measurement of the Built-up Area with the Built-up Index (BUI). *Int. J. Adv. Remote Sens. GIS* **2016**, *5*, 1844–1858.
50. Firozjaei, M.K.; Sedighi, A.; Kiavarz, M.; Qureshi, S.; Haase, D.; Alavipanah, S.K. Automated Built-Up Extraction Index: A New Technique for Mapping Surface Built-Up Areas Using LANDSAT 8 OLI Imagery. *Remote Sens.* **2019**, *11*, 1966. [CrossRef]
51. Villa, P.; Mousivand, A.; Bresciani, M. Aquatic vegetation indices assessment through radiative transfer modeling and linear mixture simulation. *Int. J. Appl. Earth Obs. Geoinf.* **2014**, *30*, 113–127. [CrossRef]
52. Zhou, G.; Ma, Z.; Sathyendranath, S.; Platt, T.; Jiang, C.; Sun, K. Canopy Reflectance Modeling of Aquatic Vegetation for Algorithm Development: Global Sensitivity Analysis. *Remote Sens.* **2018**, *10*, 837. [CrossRef]
53. Jaskuła, J.; Sojka, M. Assessing Spectral Indices for Detecting Vegetative Overgrowth of Reservoirs. *Pol. J. Environ. Stud.* **2019**, *28*, 4199–4211. [CrossRef]
54. Villa, P.; Bresciani, M.; Braga, F.; Bolpagni, R. Comparative Assessment of Broadband Vegetation Indices Over Aquatic Vegetation. *IEEE J. Sel. Top. Appl. Earth Obs. Remote Sens.* **2014**, *7*, 3117–3127. [CrossRef]
55. Japan Association on Remote Sensing. *Remote Sensing Note*; Japan Association on Remote Sensing: Tokyo, Japan, 1993.
56. Kogyo, T.K. Gold Price Change. Available online: https://gold.tanaka.co.jp/commodity/souba/m-gold.php (accessed on 11 June 2021).
57. Bullion Rates. Gold Price History in Indonesian Rupiahs (IDR). Available online: https://www.bullion-rates.com/ (accessed on 11 June 2021).

58. Statistics of Bone Bolango Regency. *Bone Bolango Dalam Angka 2010*; BPS- Statistics of Bone Bolango Regency: Suwawa, Indonesia, 2010.
59. Statistics of Bone Bolango Regency. *Bone Bolango Dalam Angka 2011*; BPS- Statistics of Bone Bolango Regency: Suwawa, Indonesia, 2011.
60. Statistics of Bone Bolango Regency. *Bone Bolango Dalam Angka 2020*; BPS- Statistics of Bone Bolango Regency: Suwawa, Indonesia, 2020.
61. Statistics of Bone Bolango Regency. *Bone Bolango Dalam Angka 2012*; BPS- Statistics of Bone Bolango Regency: Suwawa, Indonesia, 2012.
62. Statistics of Bone Bolango Regency. *Bone Bolango Dalam Angka 2013*; BPS- Statistics of Bone Bolango Regency: Suwawa, Indonesia, 2013.
63. Statistics of Bone Bolango Regency. *Bone Bolango Dalam Angka 2014*; BPS- Statistics of Bone Bolango Regency: Suwawa, Indonesia, 2014.
64. Statistics of Bone Bolango Regency. *Bone Bolango Dalam Angka 2015*; BPS- Statistics of Bone Bolango Regency: Suwawa, Indonesia, 2015.
65. Statistics of Bone Bolango Regency. *Bone Bolango Dalam Angka 2016*; BPS- Statistics of Bone Bolango Regency: Suwawa, Indonesia, 2016.
66. Statistics of Bone Bolango Regency. *Bone Bolango Dalam Angka 2017*; BPS- Statistics of Bone Bolango Regency: Suwawa, Indonesia, 2017.
67. Statistics of Bone Bolango Regency. *Bone Bolango Dalam Angka 2018*; BPS- Statistics of Bone Bolango Regency: Suwawa, Indonesia, 2018.
68. Statistics of Bone Bolango Regency. *Bone Bolango Dalam Angka 2019*; BPS- Statistics of Bone Bolango Regency: Suwawa, Indonesia, 2019.
69. Sousa, R.N.; Veiga, M.M.; Meech, J.; Jokinen, J.; Sousa, A.J. A simplified matrix of environmental impacts to support an intervention program in a small-scale mining site. *J. Clean. Prod.* **2011**, *19*, 580–587. [CrossRef]
70. Corbett, T.; O'Faircheallaigh, C.; Regan, A. 'Designated areas' and the regulation of artisanal and small-scale mining. *Land Use Policy* **2017**, *68*, 393–401. [CrossRef]
71. Gafur, N.A.; Sakakibara, M.; Sano, S.; Sera, K. A Case Study of Heavy Metal Pollution in Water of Bone River by Artisanal Small-Scale Gold Mine Activities in Eastern Part of Gorontalo, Indonesia. *Water* **2018**, *10*, 1507. [CrossRef]
72. Long, R.N.; Renne, E.P.; Basu, N. Understanding the Social Context of the ASGM Sector in Ghana: A Qualitative Description of the Demographic, Health, and Nutritional Characteristics of a Small-Scale Gold Mining Community in Ghana. *Int. J. Environ. Res. Public Health* **2015**, *12*, 12679–12696. [CrossRef] [PubMed]
73. Basu, N.; Renne, E.P.; Long, R.N. An Integrated Assessment Approach to Address Artisanal and Small-Scale Gold Mining in Ghana. *Int. J. Environ. Res. Public Health* **2015**, *12*, 11683–11698. [CrossRef] [PubMed]
74. Rajaee, M.; Obiri, S.; Green, A.; Long, R.; Cobbina, S.J.; Nartey, V.; Buck, D.; Antwi, E.; Basu, N. Integrated Assessment of Artisanal and Small-Scale Gold Mining in Ghana—Part 2: Natural Sciences Review. *Int. J. Environ. Res. Public Health* **2015**, *12*, 8971–9011. [CrossRef] [PubMed]

International Journal of
*Environmental Research
and Public Health*

Article

Detection of Artisanal and Small-Scale Gold Mining Activities and Their Transformation Using Earth Observation, Nighttime Light, and Precipitation Data

Satomi Kimijima [1,*], Masayuki Sakakibara [1,2] and Masahiko Nagai [3]

[1] Research Institute for Humanity and Nature, Kyoto 603-8047, Japan; sakaki@chikyu.ac.jp
[2] Graduate School of Science & Engineering, Ehime University, Matsuyama 790-8577, Japan
[3] Graduate School of Science and Technology for Innovation, Yamaguchi University, Ube 755-8611, Japan; nagaim@yamaguchi-u.ac.jp
* Correspondence: kimijima@chikyu.ac.jp

Abstract: The rapid growth of artificially constructed mining camps has negatively impacted the camps' surrounding environment and the informal communities that have developed inside the camps. However, artisanal and small-scale gold mining (ASGM) is generally informal, illegal, and unregulated; thus, transformations of the mining activities and potential social-environmental problems resulting from these changes are not revealed. This study assesses the transformation of mining activities in camp-type ASGM sectors in Gorontalo, Indonesia, during 2014–2020 using remotely sensed data, such as Landsat series, nighttime light, and precipitation data obtained through Google Earth Engine. Results show that the combined growth of the built-up areas increased 4.8-fold, and their annual mean nighttime light increased 3.8-fold during 2014–2019. Furthermore, diverse increases in the sizes of area and nighttime light intensity were identified from the mining camps. Among the studied camps, since 2017, Motomboto camp 3 showed a particularly rapid change in activity regardless of the season of the year. Hence, these approaches are capable of identifying rapid transformations in the mining activities and provide significant insight into the socio-environmental problems originating from the closed and vulnerable camp-based ASGM sector. Our results also contribute to developing rapid and appropriate interventions and strengthening environmental governance.

Keywords: artisanal and small-scale gold mining; environmental governance; Indonesia; landcover change; mining camp; nighttime light; remote sensing

Citation: Kimijima, S.; Sakakibara, M.; Nagai, M. Detection of Artisanal and Small-Scale Gold Mining Activities and Their Transformation Using Earth Observation, Nighttime Light, and Precipitation Data. *Int. J. Environ. Res. Public Health* **2021**, *18*, 10954. https://doi.org/10.3390/ijerph182010954

Academic Editor: Paul B. Tchounwou

Received: 6 September 2021
Accepted: 16 October 2021
Published: 18 October 2021

Publisher's Note: MDPI stays neutral with regard to jurisdictional claims in published maps and institutional affiliations.

1. Introduction

Rapid growth of artificially constructed mining camps has negatively impacted their surrounding environments and the informal communities that have developed inside them. The communities that have developed in such camps may face severe socio-environmental problems at various levels owing to their informal status. Therefore, detecting such camps and determining their rate of development as well as the transformations of their activities should provide significant insights into and the identification of possible social-environmental problems that originate in these vulnerable mining communities. This may also enable environmental governance to be promoted at various levels.

The artisanal and small-scale gold-mining (ASGM) sector, which is characterized as informal, unregistered, and illegal, is a significant gold-producing sector that uses rudimentary technology at the individual, group, or community levels [1]. In this sector, 70–80% of small-scale miners are informal workers [2]. The United Nations Environment Programme reported that ASGM is the largest employer in gold mining throughout the world, representing approximately 20% (400–600 T/year) of the worldwide gold production and 90%

of the global gold-mining workforce, respectively [3]. In the process of gold extraction, mercury is commonly used at the stage of amalgamation, resulting in substantial harmful environmental and health risks owing to mercury pollution [4–6], such as mercury emissions into the atmosphere and release into water. Such mercury pollution has mainly been reported in South America, Africa, and Asia [4,7]. Additionally, other health problems—such as silicosis, methyl orthophosphate-oriented poisoning, and various injuries—also occur during the mining process [8]. Despite the high socio-environmental risks, ASGM has been undertaken continuously in more than 80 countries as a tool for poverty alleviation during their socioeconomic development [3,9]. In Indonesia, a continuous growth in ASGM has been observed across the country. Both active and non-active ASGM practices have been located in 93 regencies in 30 out of 34 provinces in Indonesia, estimating more than 1200 hotspots in 2017 [10] with 250,000–300,000 miners [11]. Furthermore, Indonesia's fastest rise in the number of polluted sites has been reported in the past 20 years at the global level [2]. In Gorontalo province, which shows the fifth highest poverty rate of 25.9% in 2019 [12], many informal mining activities have been widespread even in national park areas, affecting biodiversity and human health [10].

The ASGM sector can be categorized into two types: "travel-type," in which miners commute daily from their local residences to the mining sites, and "camp-type," in which miners live and conduct mining activities at the worksites [13]. The camp-type ASGM (hereafter referred to as C-ASGM) sites are artificially constructed, basic settlements—in general, with poor infrastructure—resulting in the formation of an informal society in each camp [13]. The scale and workforce of the ASGM sector has been expanding along with the increase in gold prices since 2000 [14]. A relationship between the increases in ASGM and the high price of gold has been confirmed in the literature [15,16].

Recent research has focused mainly on the environmental and health assessments of mercury pollution originating in the ASGM sector [1,8,17–24]. Several studies have focused on the C-ASGM sites, but they have been limited to point-based, time-cross-sectional analyses [25,26]. Thus, quantitative analyses of the time-series of the transformations of mining activities have not been dealt with in-depth. Furthermore, due to the development of geoinformation technology, several studies have conducted ASGM-related time-series assessments of associated features, such as deforestation, mining-area detection, and geomorphic and hydrological changes [16,22–25,27], but they have mainly examined travel-type ASGM sites. To investigate the closed C-ASGM sites, [13] recently conducted a quantitative time-series analysis using satellite remote-sensing imagery of the growth of the built-up areas at the C-ASGM sites. However, the authors only captured the growth of the mining camps represented by the built-up areas, and the detailed changes and volumes of mining activities in these camps are not well understood.

However, miners living in the C-ASGM sites may face severe social risks within these communities owing to their informal, illegal, unregulated, and vulnerable natures. Thus, a better understanding of C-ASGM sites is required to reveal hidden, severe social problems. For this reason, [28] investigated the economic outputs of small-scale areas with low economic densities using remote-sensing-based light measurements. Applying such data may provide a key to understanding how remote rural ASGM camps have developed and how their mining activities have been transformed. Thus, tracing the nighttime light (NTL) and weather data associated with the spatial distributions of the built-up areas may provide better indicators of activity transformations in the mining camps located in remote rural areas.

This study primarily assesses the transformation of the ASGM activities during 2014–2020 in Bone Bolango Regency, Gorontalo Province, Indonesia, where active C-ASGM activities have been conducted. Specifically, our objectives were: (1) to assess the built-up areas in the mining camps using the Landsat series and (2) to characterize the mining activities by associating the detected built-up areas with the NTL data obtained using the Visible Infrared Imaging Radiometer Suite (VIIRS) from the National Oceanic and Atmospheric Administration (NOAA) and the Climate Hazards Group InfraRed Precipitation with Station (CHIRPS) data. The results of this study are expected to contribute to identifying potential socio-environmental problems originating in vulnerable mining communities, potentially resulting in the strengthening of environmental governance.

2. Materials and Methods

2.1. Overall Methodological Workflow

Figure 1 shows the methodological workflow used in this study. This workflow employed three main steps to achieve its primary objective of assessing the rapid transformation of ASGM activities. First, the areas built-up in the mining camps during 2014–2020 were identified using Landsat series data. Second, the NTL intensities for those areas were calculated using VIIRS–NOAA data. Third, the amounts of precipitation in those areas were obtained using CHIRPS data. Then, the relationships between the built-up areas was identified, and the volume of NTL and the amount of precipitation were assessed. Together, this evidence enabled us to understand the rapid transformation of the ASGM activities at the mining camps. In this report, we present a discussion based on all the findings described above. The methods utilized in each step are explained in the following sections.

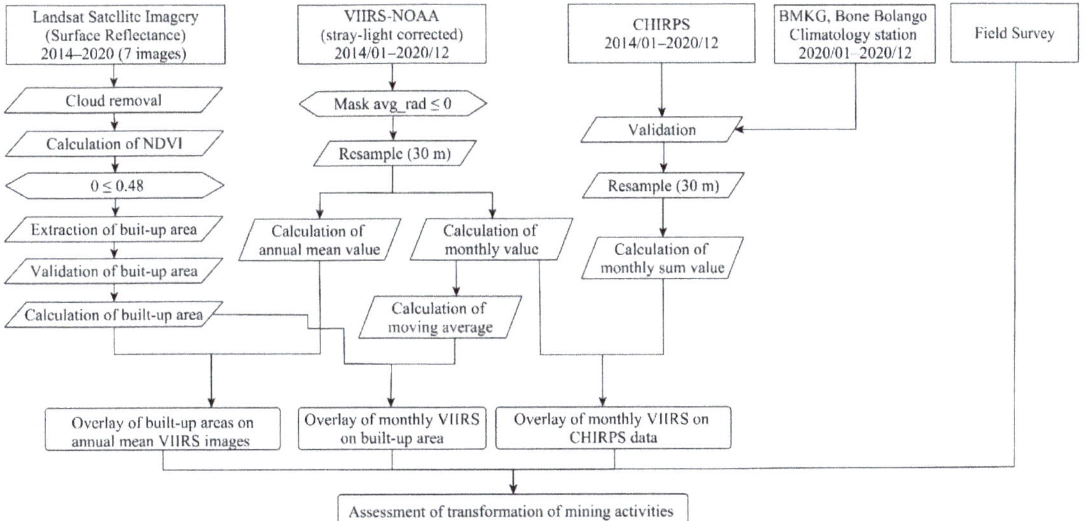

Figure 1. Overall methodology.

2.2. Study Area

North Sulawesi, Indonesia, is a well-mineralized metallogenic region with significant gold mineralization associated with quartz veins in andesite-hosted epithermal settings. The East Suwawa ASGM area is located in Bogani Nani Wartabone National Park, Bone Bolango Regency, approximately 30 km southeast of the city of Gorontalo, Gorontalo Province, Indonesia. This East Suwawa region is categorized as a high-sulfidation epithermal setting containing copper, gold, and silver [29].

The first mining activities in Bone Bolango Regency occurred in the Dutch era (18th century) [30]. Much later, mining activity in the West Motomboto and Tulabolo areas was developed by Tropic Endeavour Indonesia in 1988 [31]. However, these mining sites were closed in 1991 because they overlapped the area being developed into the Bogani Nani Wartabone National Park [31]. The closure of the former mining site triggered the entry of residents to carry out mining activities [31]. In 2013, more than 9000 small-scale miners were reported in the Bogani Nani Wartabone National Park [32].

In this study, the Mohutango and Motomboto ASGM camps 1, 2, and 3 in East Suwawa in Bone Bolango Regency, Gorontalo Province, Indonesia—each of which utilizes the shaft-based method of mining—are targeted (Figure 2). Those camps are located 4–6 h away from the center of East Suwawa. Access to the camps is very poor; they are only accessible by motorcycle and require the crossing of several rivers in the mountains [13]. The basic settlements in these C-ASGM sites consist of tin roofs covered by tarpaulins, and they are spread across small valleys, forming village-like settlements. All the gold-mining activities are conducted in these simple camps using 24-h shift operations [13].

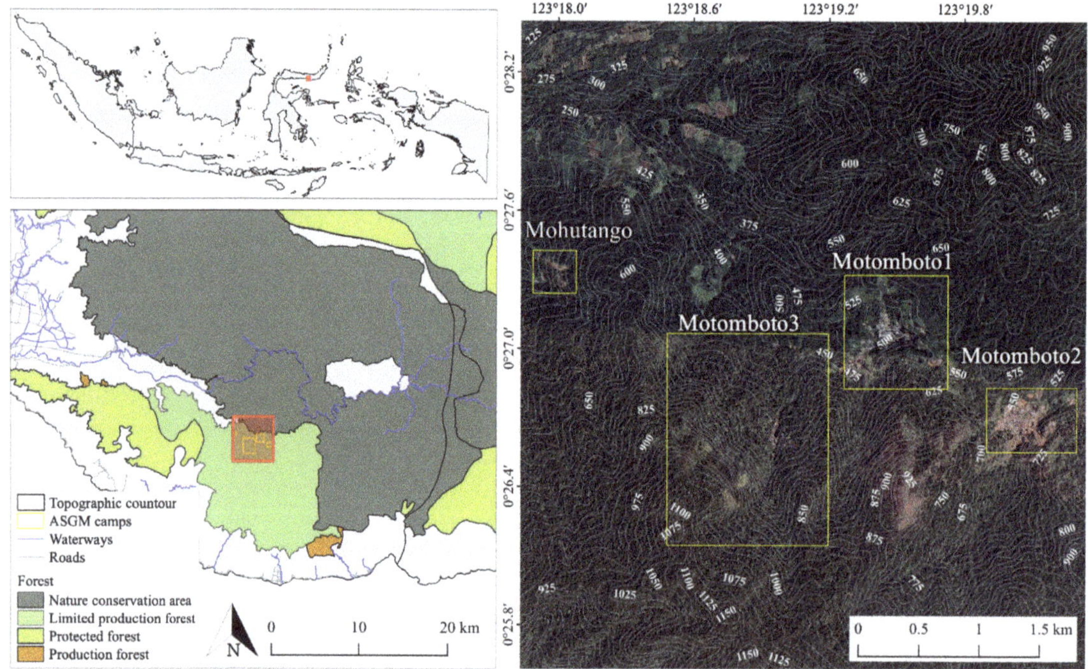

Figure 2. Study area.

2.3. Satellite Imagery

Atmospherically corrected, cloud-free Landsat data from the Enhanced Thematic Mapper Plus (ETM+) and Operational Land Imager (OLI)'s surface-reflectance products, available from the United States Geological Survey, together with the VIIRS–NOAA and CHIRPS data products, were used. Those datasets are available in Google Earth Engine, and they can be used to extract and calculate time-series of the built-up areas, the NTL intensities, and the amounts of precipitation. Therefore, the Landsat series from 2014 to 2020, which has a ground resolution of 30 m in the World Geodetic System 84 (WGS84) geographic-coordinate reference system and applies the cloud-removal function, was utilized to extract the built-up areas. Furthermore, VIIRS (stray-light-corrected)–NOAA datasets acquired during 2014–2020 were masked, with the NTL values masked to be greater than or equal to 0. Then, annual mean and monthly values were calculated by applying resampling to a spatial resolution of 30 m. The CHIRPS datasets acquired for 2014–2020 were used to calculate monthly-sum values by resampling them to the same scale. Finally, the resulting data for the built-up areas, annual mean NTL, monthly NTL, and monthly precipitation were overlaid.

In previous studies, the mining areas in Bone Bolango Regency were estimated to cover a total of 0.62 km^2 in 2012 [26]. However, a continuous expansion of the built-up areas in the camps that employ 24-h shift operations has since been reported [13]. Hence, the long-term trends in ASGM camps could be observed from satellite imagery even with a 30-m ground resolution. The main specifications of the databases used in this study are summarized in Table 1.

Table 1. Main specification of satellite imagery in the study.

Satellite	Acquisition Date	Temporal Resolution	Spatial Resolution
NOAA VIIRS	2014.01.01–2020.12.31	Monthly	15 arc seconds
CHIRPS	2014.01.01–2020.12.31	Daily	0.05 degrees
Landsat 7 ETM+	2014.04.24	16 days	30 m
Landsat 8 OLI	2015.07.08 2016.04.05 2017.04.24 2018.03.10 2019.08.04 2020.03.15	16 days	30 m

2.4. Extraction of Built-Up Areas, NTL, and Precipitation Data

Satellite-based observational data—such as Landsat, VIIRS, and CHIRPS—acquired in 2014–2020 were used. Because the transformation of ASGM activities in remote rural areas is associated with the miners living at the worksites, a combination of the growth of the built-up areas and associated changes in the NTL can provide significant indicators for assessing the detailed activity in the camps. Furthermore, as described in Section 2.2, the basic settlements in the studied C-ASGM sites are made of poor materials; thus, their mining activities have higher sensitivity to weather conditions. To clarify their changes, we further assessed them along with the changes in precipitation amount.

In this study, the built-up areas were defined based on their physical aspects, such as a built-up environment consisting mostly of human-constructed elements [33]. A number of spectral indices, together with human visual interpretation [33–41], were employed to detect built-up areas using remote-sensing technology. Previous studies found that the Normalized Difference Built-up Index (NDBI) [42,43] and the Urban Index (UI) [44,45] had high sensitivities for retrieving built-up areas; however, these have been employed mainly in urban studies. As the NDBI and UI are incapable of separating built-up areas from bare land effectively [35], separating the two in rural areas can be expected to be more complicated. Therefore, in this study, the Normalized Difference Vegetation Index (NDVI) was applied, as employed by [13,46], to detect remote rural mining areas over long timescales. The value of NDVI in the built-up areas was calculated using Equation (1):

$$NDVI = (NIR - Red)/(NIR + Red) \tag{1}$$

NDVI, which ranges from -1 to $+1$, has a high value for denser vegetation, while it is lower for desert or non-vegetation areas [47]. In this study, the NDVI was further restricted to the range $0 \leq NDVI \leq 0.48$ in order to exclude vegetated areas on the land surface from the built-up areas. This threshold value was determined based on comparisons to the accuracy levels for high-resolution satellite data. In this way, the results were visualized in time-series. A hundred points were randomly selected within the study area, and the accuracy of the results was assessed using a high-resolution image obtained on 8 February 2017 from Google Earth Pro. Because images were not available on the same date for which the Landsat imagery had been acquired, images acquired on the closest date, 24 April 2017, were used. In this study, the validated accuracy was applied to all classification results owing to the unavailability of reference data.

NTL data were acquired from the VIIRS, representing radiance values from −1.5 to 193,564.92, for the period 2014–2020. The negative radiance is generated by the airglow effect in unpopulated regions, where the probability of illumination is zero or very low [48]. For this study, values less than 0 were excluded from the whole dataset. After summarizing the monthly radiance values, a 12-month moving average was calculated for 2014–2020. Meanwhile, an annual mean radiance value was calculated per year to generate a time-series of annual maps, which were overlaid against the detected built-up areas.

Precipitation data was acquired from CHIRPS for the period 2014–2020. Before the CHIRPS data were analyzed, data consistency was evaluated using reference data observed at Bone Bolango climatology station provided by the Indonesian Agency for Meteorology, Climatology and Geophysics (BMKG) [49]. Precipitation data provide by CHIRPS were accumulated in the form of monthly data to match the monthly data provided by BMKG. After the data validation, the sums of the amounts of precipitation by month were calculated for each camp. These results were graphed together with the detected monthly NTL.

2.5. Investigation of ASGM Camps

Field observation was conducted on 6 February 2020 to investigate the ASGM camps. Additionally, interviews were conducted with key informant miners on the worksites.

3. Results

3.1. Growth of Built-Up Areas in the Mining Camps

To detect the land-cover changes surrounding the ASGM camps during the period of 2014–2020, the calculated NDVI was primarily used. Using these NDVI results, the built-up areas in the ASGM camps were calculated with an accuracy of 99%, as described in Section 2.4. Figure 3 shows how the built-up areas have developed over time in the camps. The built-up areas in the Mohutango and Motomboto ASGM camps 1 and 2 were identified beginning early 2014, and camp 3 was identified early in 2016, and the camps showed various types of growth. The growth of all the built-up areas combined exhibited a 4.8-fold increase during 2014–2020. While the Mohutango and Motomboto ASGM camp 1 remained similar in extent from 2014 to 2020, the Motomboto camps 2 and 3 developed substantially in extent in 2015 and 2019, respectively. Among these mining camps, the growth of camp 3 is clearly distinguishable from the cases mentioned above, showing continuous and rapid annual growth of the built-up areas to the southern part. While the growth of the Mohutango camp and Motomboto camp 1 showed only 1.1- and 1.4-fold increases, respectively, during 2014–2020, and camp 2 showed a 1.2-fold increase during 2015–2020, camp 3 showed a remarkable 23.1-fold increase during 2016–2020. Through the field observations, we confirmed that the identified built-up areas were either residences of miners or settlements for mining activities where trommel machines and pools for immersing the materials were placed. Furthermore, the ASGM activity in this area has rapidly increased since 2017 after the gold price increased (according to interview with a local miner).

3.2. Relationship between Built-Up Areas and NTL Intensity in the Mining Camps

The built-up areas were further overlaid against the corresponding annual mean NTL images as an indicator of the volume of mining activity in time-series (Figure 3). The total annual mean NTL showed a 3.8-fold increase during 2014–2020, while the individual Mohutango and Motomboto camps 1, 2, and 3 showed 3.0-, 2.8-, 2.6-, and 5.4-fold increases, respectively. The highest NTL increase occurred in mining camp 3, where the rapid growth of built-up areas was observed (Section 3.1). On the contrary, the Mohutango and Motomboto camps 1 and 2, which were already identified in 2014, showed lower growths of NTL intensity in comparison to Motomboto camp 3. Based on our field observations, electricity was generated using diesel generators and distributed across the camp (as reported by a local miner).

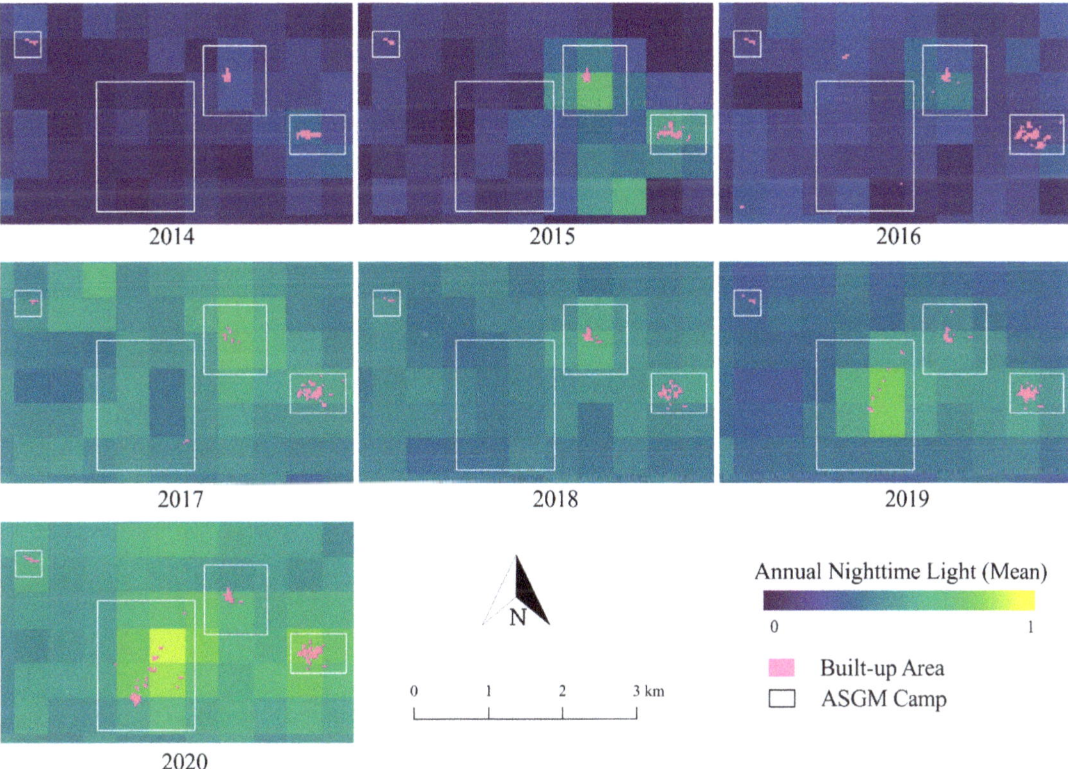

Figure 3. Overlay of built-up areas on annual mean NTL images in time-series.

To deepen our understanding of the mining activities in the camps represented by the NTL intensity, the monthly NTL intensity and the 12-month moving average of NTL were graphed against the built-up areas (Figure 4). Even though the NTL intensities vary by camps, the study found a similar trend of the NTL intensity by season. For example, lower NTL intensities were found in the rainy seasons—i.e., during April–June and November–December—while higher intensities were found in the other months, during the dry seasons. There was no significant change in the maximum values in the Mohutango and Motomboto camps 1 and 2; however, the NTL from Motomboto camp 3 showed a 2.0-fold increase for 2014–2020. Furthermore, notable increases in the NTL were found in Motomboto camp 3 even in the rainy season, along with the growth of built-up areas since 2017. The 12-month moving average for Motomboto camp 3 also showed a continuous increase.

3.3. Relationship between the NLT Intensity and Precipitation by Month

As described in Section 2.2, the basic settlements in the studied C-ASGM sites are made of tin roofs covered by tarpaulins; consequently, their mining activities may have higher sensitivity to weather conditions. We therefore expanded our analysis of the NTL intensity described in Section 3.1 by associating it with the amount of precipitation, with a correlation of 89%. As described above, higher NTL intensities have been observed since 2017, even in the rainy season, especially in Motomboto camp 3; thus, we summarized the monthly precipitation and NTL intensities by camp in a graph (Figure 5). The annual precipitation in 2014–2020 was 1333, 887, 1437, 1760, 1170, 918, and 1643 mm, respectively. Even though the annual amount of precipitation varied, the NTL volumes in 2020 were the highest during the entire study period. Despite the increase in the volume of precipitation, the

NTL intensity increased during the rainy season even in the simple settlements, especially after 2017.

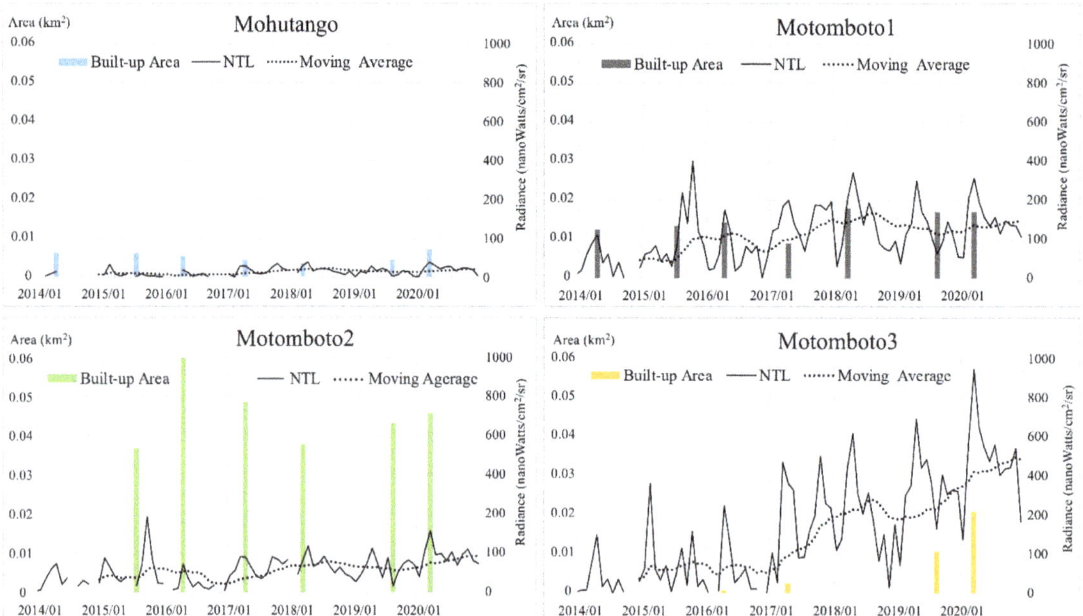

Figure 4. Built-up areas and monthly NTL intensities for the ASGM camps.

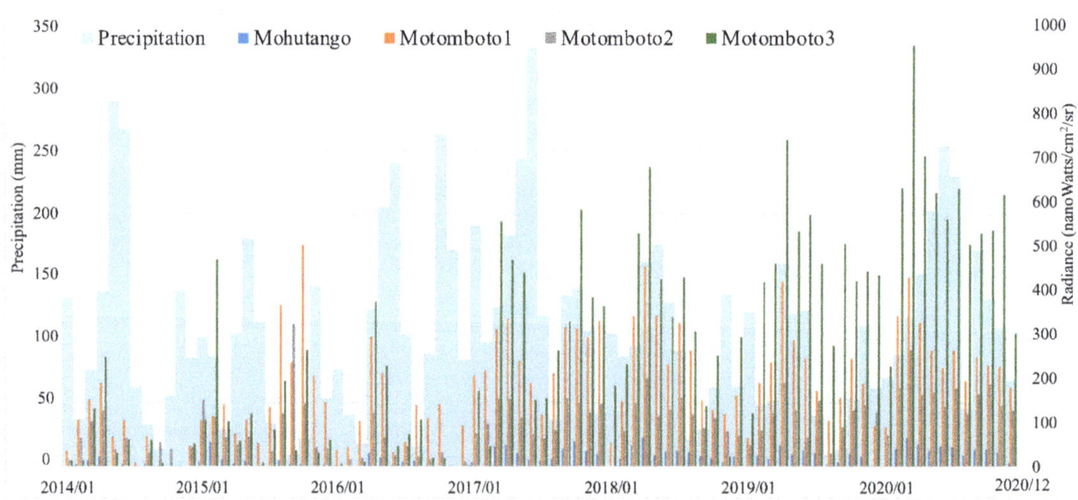

Figure 5. Monthly precipitation (sum) and NTL intensity by mining camps.

4. Discussion and Limitations

4.1. Discussion

We studied the rapid transformation of activities in the C-ASGM sector from 2014–2020 using time-series associated with the built up areas, the NTL, and precipitation data. A quantitative time-series analysis of the artificially constructed C-ASGM sectors can help to achieve better understanding of the rate of development of such mining activities and their transformations across time. Detecting such rapid transformations can provide significant insights into hidden, severe social problems inside these vulnerable mining communities, potentially resulting in the strengthening of environmental governance at various levels.

By combining the extraction of indicators of the growth of the built-up areas of C-ASGMs, the NTL intensity as an indicator of the mining activities, and the precipitation data, this study demonstrated the transformations of the mining activities undertaken in the C-ASGM sectors over a significant fraction of a decade (Figures 3–5). Using a quantitative analysis over the time, this study detected the various forms of built-up areas and NTL intensities in the mining camps. For example, Motomboto camp 3 was identified in 2016, and it showed a more rapid and extensive growth of the built-up areas and the NTL intensity than the other camps during 2016–2020, with 23.1- and 2.0-fold increases, respectively, as described in Section 3.1. We found notable increases in Motomboto camp 3 since 2017, which is in accordance with a previous study [13]. As the study sites are remote, rural C-ASGMs, a significant source of the growth identified in this area may be due to large influxes of miners from neighboring regions, such as Bolaang Mongondow and Minahasa in North Sulawesi [32]. This huge entry may have been encouraged by weak regulations resulting from the informal, illegal, and closed nature of this sector [9]; limited government resources and administrative capacity to provide adequate technical assistance or enforce compliance [50]; and the remote locations of the mining sites [27], which could further result in large socioeconomic problems.

The transformation of the working pattern was also identified since 2017 regardless of the season. This may imply that local villagers near the mining sites previously engaged in the mining activity as an additional income-generating activity during the agricultural off-season. However, the increased influx of miners from neighboring regions that has occurred since 2017 appears to have resulted in workers staying at the mining camps continuously, becoming occupational miners throughout the year. As discussed above, [13] previously revealed the expansion of the C-ASGM sector; however, the increase of NTL, in association with the precipitation data, enables the further identification of the detailed volumes of and changes in the hidden mining activities.

The C-ASGM sector can operate successfully owing to its high productivity of gold. Despite its status as an informal sector, large influxes of miners have entered the camps continuously, resulting in their rapid growth. These influxes of large populations into artificially constructed spaces, which lack fundamental infrastructures and systems, definitely cause and accelerate socioeconomic and environmental problems relating to children, family, education, health, sexual health, sanitation, garbage, and water usage, as reported elsewhere as well [51–54]. However, their informal and illegal status limits the power of law over and control of the camps, resulting in severe situations at some camps. In particular, this may be the case for the rapidly growing C-ASGM sector, as has been observed at Motomboto camp 3. Thus, detection of such rapidly developing, hidden C-ASGM sectors can definitely contribute to strengthening environmental governance by attracting and involving various stakeholders at various levels.

Many of ASGM-related assessments are limited to a particular discipline; thus, interdisciplinary researches involving health and environmental impact assessments will be carried out in different regions to prove the effectiveness of remote sensing analysis in the future.

4.2. Limitations

The results of this study have some limitations associated with the quality of the input data. First, the presence of small negative-radiance values caused by the airglow effect in uninhabited regions [48] can lead to underestimates of the NTL intensity. Second, differences in the spatial resolution of the utilized datasets results in mixed pixels, which can cause the overestimation or miscalculation of factors such as built-up areas and NTL intensities. Third, the methodology used in this study is applicable only to similar mining sectors that employ 24-h operations.

5. Conclusions

The rapid growth of artificially constructed mining camps has negatively impacted the environment surrounding them and the informal societies that have formed inside them. In this study, the transformations of the ASGM activities in Bone Bolango Regency, Gorontalo Province, Indonesia, were assessed using remote-sensing data. The results presented herein show that the growth of the built-up areas and annual mean NTL experienced 4.8- and 3.8-fold increases, respectively, during 2014–2020. In particular, rapid changes in the working patterns were found in Motomboto camp 3 regardless of the season. Therefore, it can be concluded that transformations of the activities undertaken in the closed C-ASGM sites can be determined by combining observations of the built-up areas and NTL in the mining camps with the precipitation volumes. These results extend our understanding of the transformations of mining activities in the hidden C-ASGM sectors and provide significant insight into the potential for social problems that can occur in vulnerable informal mining communities. These findings are expected to assist in developing rapid and appropriate interventions for strengthening environmental governance by involving various stakeholders.

Author Contributions: S.K. contributed to design the research, data analysis, and data visualization. M.S. and M.N. provided conceptual advice and critical comments. All authors have read and agreed to the published version of the manuscript.

Funding: This research was financially supported by the Research Institute for Humanity and Nature (RIHN: a constituent member of NIHU).

Institutional Review Board Statement: Not applicable.

Informed Consent Statement: Not applicable.

Data Availability Statement: Not applicable.

Acknowledgments: This research was financially supported by The Research Institute for Humanity and Nature (RIHN: a constituent member of NIHU), project no. 14200102. We are grateful to the Indonesian State Ministry of Research and Technology (RISTEK) for providing research permits in Indonesia.

Conflicts of Interest: The authors declare no conflict of interest.

References

1. De Lobo, F.L.; Costa, M.; de Novo, E.M.L.M.; Telmer, K. Distribution of Artisanal and Small-Scale Gold Mining in the Tapajós River Basin (Brazilian Amazon) over the Past 40 Years and Relationship with Water Siltation. *Remote Sens.* **2016**, *8*, 579. [CrossRef]
2. International Institute for Sustainable Development. *Global Trends in Artisanal and Small-Scale Mining (Asm): A Review of Key Numbers and Issues*; International Institute for Sustainable Development: Winnipeg, MB, Canada, 2018.
3. United Nations Environment Programme. *Estimating Mercury Use and Documenting Practices in Artisanal and Small-scale Gold Mining (ASGM)*; International Institute for Sustainable Development: Geneva, Switzerland, 2017.
4. World Health Organization. *Artisanal and Small-Scale Gold Mining and Health*; World Health Organization: Geneva, Switzerland, 2016.
5. United Nations Environment Programmet. *Technical Background Report to the Global Mercury Assessment 2018*; United Nations Environment Programmet: Tromsø, Norway, 2018.
6. Saldarriaga-Isaza, A.; Villegas-Palacio, C.; Arango, S. The public good dilemma of a non-renewable common resource: A look at the facts of artisanal gold mining. *Resour. Policy* **2013**, *38*, 224–232. [CrossRef]
7. United Nations Environment Programme. *Global Mercury Assessment 2018*; United Nations Environment Programme: Geneva, Switzerland, 2018.
8. Macháček, J. Typology of environmental impacts of artisanal and small-scale mining in African Great Lakes Region. *Sustainability* **2019**, *11*, 3027. [CrossRef]

9. Wilson, M.L.; Renne, E.; Roncoli, C.; Agyei-Baffour, P.; Tenkorang, E.Y. Integrated assessment of artisanal and small-scale gold mining in Ghana—Part 3: Social sciences and economics. *Int. J. Environ. Res. Public Health* **2015**, *12*, 8133–8156. [CrossRef] [PubMed]
10. BaliFokus Foundation. *Mercury Country Situation Report Indonesia*; BaliFokus Foundation: Bli, Indonesia, 2018.
11. Agrawal, A.W.; Anthony, S.; Bebbington, J.; Imhof, A.; Jebing, M.; Royo, N.; Sauls, L.A.; Sulaiman, R.; Toumbourou, T. *Impacts of Extractive Industry and Infrastructure on Forests: Indonesia*; Climate and Land Use Allaince: San Francisco, CA, USA, 2018.
12. Statistics of Bone Bolango Regency. *Bone Bolango Dalam Angka 2010*; BPS-Statistics of Bone Bolango Regency: Suwawa, Indonesia, 2010.
13. Kimijima, S.; Sakakibara, M.; Nagai, M.; Gafur, N.A. Time-series Assessment of Camp-type Artisanal and Small-scale Gold Mining Sector with Large Influxes of Miners using LANDSAT Imagery. *Int. J. Environ. Res. Public Health* **2021**, *18*, 9441. [CrossRef] [PubMed]
14. GoldPrice.Org. Gold Price in USD. 2021. Available online: https://goldprice.org/spot-gold.html (accessed on 19 July 2021).
15. Asner, G.P.; Llactayo, W.; Tupayachi, R.; Luna, E.R. Elevated rates of gold mining in the Amazon revealed through high-resolution monitoring. *Proc. Natl. Acad. Sci. USA* **2013**, *110*, 18454–18459. [CrossRef]
16. Swenson, J.J.; Carter, C.E.; Domec, J.C.; Delgado, C.I. Gold mining in the peruvian amazon: Global prices, deforestation, and mercury imports. *PLoS ONE* **2011**, *6*, e18875. [CrossRef]
17. Macháček, J. Alluvial artisanal and small-scale mining in a river stream-Rutsiro case study (Rwanda). *Forests* **2020**, *11*, 762. [CrossRef]
18. Kahhat, R.; Parodi, E.; Larrea-Gallegos, G.; Mesta, C.; Vázquez-Rowe, I. Environmental impacts of the life cycle of alluvial gold mining in the Peruvian Amazon rainforest. *Sci. Total Environ.* **2019**, *662*, 940–951. [CrossRef]
19. Nakazawa, K.; Nagafuchi, O.; Kawakami, T.; Inoue, T.; Yokota, K.; Serikawa, Y.; Basir-Cyio, M.; Elvince, R. Human health risk assessment of mercury vapor around artisanal small-scale gold mining area, Palu city, Central Sulawesi, Indonesia. *Ecotoxicol. Environ. Saf.* **2016**, *124*, 155–162. [CrossRef] [PubMed]
20. Bose-O'Reilly, S.; Drasch, G.; Beinhoff, C.; Rodrigues-Filho, S.; Roider, G.; Lettmeier, B.; Maydl, A.; Maydl, S.; Siebert, U. Health assessment of artisanal gold miners in Indonesia. *Sci. Total Environ.* **2009**, *408*, 713–725. [CrossRef]
21. Wyatt, L.; Ortiz, E.J.; Feingold, B.; Berky, A.; Diringer, S.; Morales, A.M.; Jurado, E.R.; Hsu-Kim, H.; Pan, W. Spatial, Temporal, and Dietary Variables Associated with Elevated Mercury Exposure in Peruvian Riverine Communities Upstream and Downstream of Artisanal and Small-Scale Gold Mining. *Int. J. Environ. Res. Public Health* **2017**, *14*, 1582. [CrossRef] [PubMed]
22. Schmid, T.; Rico, C.; Rodríguez-Rastrero, M.; José Sierra, M.; Javier Díaz-Puente, F.; Pelayo, M.; Millán, R. Monitoring of the mercury mining site Almadén implementing remote sensing technologies. *Environ. Res.* **2013**, *125*, 92–102. [CrossRef]
23. Espejo, J.C.; Messinger, M.; Román-Dañobeytia, F.; Ascorra, C.; Fernandez, L.E.; Silman, M. Deforestation and forest degradation due to gold mining in the Peruvian Amazon: A 34-year perspective. *Remote Sens.* **2018**, *10*, 1–17. [CrossRef]
24. Emel, J.; Plisinski, J.; Rogan, J. Monitoring geomorphic and hydrologic change at mine sites using satellite imagery: The Geita Gold Mine in Tanzania. *Appl. Geogr.* **2014**, *54*, 243–249. [CrossRef]
25. Owusu-Nimo, F.; Mantey, J.; Nyarko, K.B.; Appiah-Effah, E.; Aubynn, A. Spatial distribution patterns of illegal artisanal small scale gold mining (Galamsey) operations in Ghana: A focus on the Western Region. *Heliyon* **2018**, *4*, e00534. [CrossRef] [PubMed]
26. Puluhulawa, F.; Harun, A.A. Policy formalization of Artisanal and Small-Scale Gold Mining (ASGM) post-ratification of Minamata Convention for Sustainability (case study of ASGM Gorontalo). *E3S Web Conf.* **2019**, *125*, 02006. [CrossRef]
27. Gallwey, J.; Robiati, C.; Coggan, J.; Vogt, D.; Eyre, M. A Sentinel-2 based multispectral convolutional neural network for detecting artisanal small-scale mining in Ghana: Applying deep learning to shallow mining. *Remote Sens. Environ.* **2020**, *248*, 111970. [CrossRef]
28. Chen, X.; Nordhaus, W. A test of the new VIIRS lights data set: Population and economic output in Africa. *Remote Sens.* **2015**, *7*, 4937–4947. [CrossRef]
29. PT Bumi Resources Minerals Tbk. *Laporan Tahunan 2019 Annual Report*; PT Bumi Resources Minerals Tbk: Jakarta, Indonesia, 2019.
30. Van Bemmelen, R.W. The Geology of Indonesia. General Geology of Indonesia and Adjacent Archipelagoes. *Gov. Print. Off. Hague* **1949**, *545–547*, 561–562.
31. Kesatuan Pengelotaan Hutan. *Rencana Pengelolaan Hutan Jangka Panjang Kphp Unit Vii Bone Bolango Tahun 2016–2025*; Bone Bolango Regency: Suwawa, Indonesia, 2016.
32. Hatu, R.A. Socio-economic conditions in the illegal gold miners Tulabolo Village, Gorontalo-in Indonesian. *Asian J. Appl. Sci.* **2016**, *9*, 97–105. [CrossRef]
33. Kimijima, S.; Nagai, M. Human mobility analysis for extracting local interactions under rapid socio-economic transformation in Dawei, Myanmar. *Sustainability* **2017**, *9*, 1598. [CrossRef]
34. Kawamura, M.; Jayamana, S.; Tsujiko, Y. Relation between social and environmental conditions in colombo sri lanka and the urban index estimated by satellite remote sensing data. *Int. Arch. Photogramm. Remote Sens.* **1996**, *31*, 321–326.
35. Zha, Y.; Gao, J.; Ni, S. Use of normalized difference built-up index in automatically mapping urban areas from TM imagery. *Int. J. Remote Sens.* **2003**, *24*, 583–594. [CrossRef]
36. Xu, H. A new index for delineating built-up land features in satellite imagery. *Int. J. Remote Sens.* **2008**, *29*, 4269–4276. [CrossRef]

37. Bhatti, S.S.; Tripathi, N.K. Built-up area extraction using Landsat 8 OLI imagery. *GIScience Remote Sens.* **2014**, *51*, 445–467. [CrossRef]
38. As-syakur, A.R.; Adnyana, I.W.S.; Arthana, I.W.; Nuarsa, I.W. Enhanced built-UP and bareness index (EBBI) for mapping built-UP and bare land in an urban area. *Remote Sens.* **2012**, *4*, 2957–2970. [CrossRef]
39. Waqar, M.M.; Mirza, J.F.; Mumtaz, R.; Hussain, E. Development of New Indices for Extraction of Built-Up Area & Bare Soil. *Open Access Sci. Reports* **2012**, *1*, 1–4.
40. Kaimaris, D.; Patias, P. Identification and Area Measurement of the Built-Up Area with the Built-Up Index (bui). Available online: https://www.semanticscholar.org/paper/Identification-and-Area-Measurement-of-the-Built-up-Kaimaris-Patias/8af141e5f616488d238384111b69137bb54b1ec9 (accessed on 1 August 2021).
41. Firozjaei, M.K.; Sedighi, A.; Kiavarz, M.; Qureshi, S.; Haase, D.; Alavipanah, S.K. Automated built-up extraction index: A new technique for mapping surface built-up areas using LANDSAT 8 OLI imagery. *Remote Sens.* **2019**, *11*, 1966. [CrossRef]
42. Villa, P.; Mousivand, A.; Bresciani, M. Aquatic vegetation indices assessment through radiative transfer modeling and linear mixture simulation. *Int. J. Appl. Earth Obs. Geoinf.* **2014**, *30*, 113–127. [CrossRef]
43. Zhou, G.; Ma, Z.; Sathyendranath, S.; Platt, T.; Jiang, C.; Sun, K. Canopy reflectance modeling of aquatic vegetation for algorithm development: Global sensitivity analysis. *Remote Sens.* **2018**, *10*, 837. [CrossRef]
44. Jaskuła, J.; Sojka, M. Assessing spectral indices for detecting vegetative overgrowth of reservoirs. *Polish J. Environ. Stud.* **2019**, *28*, 4199–4211. [CrossRef]
45. Villa, P.; Bresciani, M.; Braga, F.; Bolpagni, R. Comparative Assessment of Broadband Vegetation Indices Over Aquatic Vegetation. *IEEE J. Sel. Top. Appl. Earth Obs. Remote Sens.* **2014**, *7*, 3117–3127. [CrossRef]
46. Pericak, A.A.; Thomas, C.J.; Kroodsma, D.A.; Wasson, M.F.; Ross, M.R.V.; Clinton, N.E.; Campagna, D.J.; Franklin, Y.; Bernhardt, E.S.; Amos, J.F. Mapping the yearly extent of surface coal mining in central appalachia using landsat and google earth engine. *PLoS ONE* **2018**, *13*, 1–15. [CrossRef]
47. Japan Association on Remote Sensing. *Remote Sensing Note*; Japan Association on Remote Sensing: Tokyo, Japan, 1993.
48. Jeswani, R.; Anurag, K.; Gupta, P.K.; Srivastav, S.K. Evaluation of the consistency of DMSP-OLS and SNPP-VIIRS night-time light datasets Reshma. *J. Geomatics* **2019**, *13*, 98–105.
49. Indonesian Agency for Meteorology Climatology and Geophysics. Daily Data of Bone Bolango Climatology Station. 2021. Available online: https://dataonline.bmkg.go.id/ (accessed on 10 October 2021).
50. Sousa, R.N.; Veiga, M.M.; Meech, J.; Jokinen, J.; Sousa, A.J. A simpli fi ed matrix of environmental impacts to support an intervention program in a small-scale mining site. *J. Clean. Prod.* **2011**, *19*, 580–587. [CrossRef]
51. Gafur, N.A.; Sakakibara, M.; Sano, S.; Sera, K. A case study of heavy metal pollution in water of Bone River by Artisanal Small-Scale Gold Mine Activities in Eastern Part of Gorontalo, Indonesia. *Water* **2018**, *10*, 1507. [CrossRef]
52. Long, R.N.; Renne, E.P.; Basu, N. Understanding the social context of the asgm sector in ghana: A qualitative description of the demographic, health, and nutritional characteristics of a small-scale gold mining community in Ghana. *Int. J. Environ. Res. Public Health* **2015**, *12*, 12679–12696. [CrossRef]
53. Basu, N.; Renne, E.P.; Long, R.N. An integrated assessment approach to address artisanal and small-scale gold mining in Ghana. *Int. J. Environ. Res. Public Health* **2015**, *12*, 11683–11698. [CrossRef]
54. Rajaee, M.; Obiri, S.; Green, A.; Long, R.; Cobbina, S.J.; Nartey, V.; Buck, D.; Antwi, E.; Basu, N. Integrated Assessment of Artisanal and Small-Scale Gold Mining In Ghana—Part 2: Natural Sciences Review. *Int. J. Environ. Res. Public Health* **2015**, *12*, 8971–9011. [CrossRef] [PubMed]

International Journal of
Environmental Research and Public Health

Article

Examining Preference Heterogeneity in Best-Worst Scaling: Case of Preferences for Job Opportunities in Artisanal Small-Scale Gold Mining (ASGM) Communities in Indonesia

Satoru Komatsu [1],*, Yayu Isyana D. Pongoliu [2], Masayuki Sakakibara [3,4] and Taro Ohdoko [5]

[1] Graduate School of Global Humanities and Social Sciences, Nagasaki University, Nagasaki 852-8521, Japan
[2] Faculty of Economics, State University of Gorontalo, Gorontalo 96128, Indonesia; yayuidp@gmail.com
[3] Research Institute for Humanity and Nature, Kyoto 603-8047, Japan; sakaki@chikyu.ac.jp
[4] Graduate School of Science and Engineering, Ehime University, Matsuyama 790-8577, Japan
[5] Faculty of Economics, Dokkyo University, Saitama 340-0042, Japan; ohdoko@dokkyo.ac.jp
* Correspondence: satoru.komatsu@gmail.com; Tel.: +81-95-819-2915

Abstract: This research empirically examines the preferences for job-related attributes among rural villagers living close to artisanal and small-scale gold mining (ASGM) in Indonesia. Based on hypothetical scenarios in which a private company collaborates with the local government to establish a food processing industry in these villages, a questionnaire survey designed with best–worst scaling (BWS) was administered to households to determine their preferences. Additionally, the heterogeneity among the villagers was examined by applying a latent class logit (LCL) model. The main household survey was conducted in 2019 in Bone Bolango Regency, Gorontalo Province. The estimation results revealed that villagers are separated into four classes, and each class has different and unique preferences. Creating more job opportunities for society is a highly evaluated attribute; however, the preference for skill acquisition differs among groups. The results indicate that accounting for heterogeneous preferences regarding job opportunities is helpful to delink dependency on ASGM and health hazards and improve the livelihoods of rural villagers. The study yields key information to substantially reduce environmental and health hazards in the poverty-plagued ASGM community by facilitating job opportunities in Indonesia.

Keywords: ASGM; best-worst scaling; job opportunities; preferences; Indonesia

Citation: Komatsu, S.; Pongoliu, Y.I.D.; Sakakibara, M.; Ohdoko, T. Examining Preference Heterogeneity in Best-Worst Scaling: Case of Preferences for Job Opportunities in Artisanal Small-Scale Gold Mining (ASGM) Communities in Indonesia. *Int. J. Environ. Res. Public Health* **2022**, *19*, 306. https://doi.org/10.3390/ijerph19010306

Academic Editor: Paul B. Tchounwou

Received: 26 November 2021
Accepted: 23 December 2021
Published: 28 December 2021

Publisher's Note: MDPI stays neutral with regard to jurisdictional claims in published maps and institutional affiliations.

1. Introduction

The release of mercury and its components into the atmosphere has had serious impacts on human health and the local environment. One of the major activities that emits mercury (Hg) is artisanal and small-scale gold mining (ASGM), an activity in which individuals or small groups extract gold using simple tools and methods. Commonly, mercury is used to extract gold from gold ores to make an amalgam, which is then burned to distill mercury and isolate gold. Mercury is released to the air, water and soil during the vaporization and tailing process and is diffused into the environment.

ASGM is the predominant source of anthropogenic mercury emissions, accounting for 37.7% of the total mercury emissions in 2015 [1]. ASGM plays a major role in gold extraction; it is estimated to constitute 17–20% of global gold production [2]. The release of Hg from ASGM is higher in countries with lower levels of technology [3].

ASGM practices have expanded, especially in many low- and middle-income countries, and have caused occupational mercury intoxication in local societies. Approximately 15 million people, including approximately 3 million women and children, engage in ASGM activities in developing countries [4]. Seccatore et al. estimated that over 16 million miners were involved in gold mining worldwide [3]. Gold extraction in ASGM is commonly informal, and ASGM causes widespread mercury contamination for miners and surrounding villagers. Globally, 3.3 to 6.5 million miners suffer from moderate levels of chronic metallic mercury vapor intoxication [5]. Given the impacts of ASGM on the rural economy and the mercury consumption in this sector, the sustainable management of the ASGM sector is crucially important to achieve sustainable economic growth while preventing environmental degradation and reducing health risk.

Engagement in this environmentally and occupationally hazardous activity is closely associated with chronic poverty in rural societies, where income-generating opportunities are limited. In places where job opportunities are limited and unstable, obtaining cash through extracting gold is crucial for sustaining day-to-day livelihood. Even if villagers are not directly engaged in mining, mining operations have effects on other industries, such as shops, restaurants, driving and agriculture. Therefore, the local economy where ASGM is located is connected to the mining sector.

There are two ways to reduce poverty and to improve local livelihoods in existing ASGM communities. The first way is to minimize the extensive environmental and health hazards of ASGM. Various previous studies have attempted to replace mercury amalgamation or cyanide processing, which also poses safety and health concerns (refer to [6], for example). Alternative technologies can address safety and health concerns, but they require users to have prior knowledge and technical training [6]. Major technological transformations usually require continuous effort to develop the capacity of miners in the long run. Alternative technologies should be affordable, cost-effective, and available when necessary. Moreover, technological development cannot directly overcome the underlying reality of rural villages, i.e., pervasive poverty.

Another way is an alternative livelihood approach. Alternative livelihood approaches in the context of mining assume that farmers can cease or reduce their reliance on ASGM if livelihood-based policy interventions, such as the development of plantations, are effective in generating sufficient cash income. If such interventions are successful, alternative job opportunities that generate cash income are promising to eradicate chronic poverty and to reduce dependency on ASGM.

Alternative livelihood approaches have extensively been applied in the context of biodiversity conservation to convert conventional livelihoods that have substantial impacts on biodiversity into sustainable livelihoods that reduce resource consumption. Examples of alternative livelihoods are ecotourism in Nepal [7] and the provision of microcredit for villager groups in Indonesia [8]. Based on an extensive review on the effects of alternative livelihood initiatives on biodiversity conservation, Roe et al. found mixed outcomes; nine initiatives reported positive conservation outcomes, another nine reported no change, and the remaining three reported negative outcomes [9].

However, to date, various studies have pointed out that alternative livelihood approaches in the field of mining fail to incentivize the abandonment of mineral extraction. Empirical findings on the expansion of agriculture as a livelihood intervention tool in Myanmar concluded that mining and agriculture are unlikely to be substitute livelihood options for villagers; rather, they are complementary options [10]. Agriculture and mining were found to be complementary in Sierra Leone, where inhabitants engage in farming in rainy seasons and mining in dry seasons [11].

How can we achieve the success of alternative livelihood approaches to attain the multifaceted goals of eliminating rural poverty and minimizing environmental hazards? Answering such questions is difficult; however, we assert that bottom-up approaches that incorporate the latent preferences of miners and villagers are crucially important. Alternative livelihoods are presumed to be a promising opportunity that can absorb local labor forces and yield sufficient income, in addition to meeting the latent demands of potential local workers. Collecting and incorporating possible job seekers' attitudes and opinions on alternative job opportunities are the basis for formulating effective policies. Alternative livelihoods are effective if a series of income-earning activities can prevent people from pursuing employment in ASGM and are attractive for individuals already engaged in mining activities [12]. Therefore, for proposing attractive alternative livelihoods and addressing local public health issues, it is critical to clarify the latent demand for job opportunities of rural villagers who may otherwise work at ASGM sites or surrounding industries.

To date, a limited number of case studies have evaluated preferences for job opportunities in the context of ASGM. A conference presentation appears to be the first attempt to analyze citizens' preferences for possible job opportunities in Indonesia [13]. The study conducted household surveys to elicit job preferences through best–worst scaling (BWS) and applied a mixed logit model to examine preferences for job opportunities [13].

This paper applies the data set used in [13] and extends it to examine the source of preference heterogeneity from various socioeconomic variables (household income, dependency on mining, education, residency, etc.). If the source of heterogeneity is successfully and rigorously identified, this information can support policymakers in designing alternative livelihood policies.

BWS is frequently applied to examine the relative importance of attributes in the field of health economics and marketing and is increasingly applied in environmental and resource economics. BWS was first introduced by Louviere and Woodworth in a working paper in 1990, while Finn and Louviere published the first application of the technique [14]. Commonly, respondents are asked to select their most preferred choice and their least preferred choice in a choice set. Extensive literature has examined the relative importance of policy measures and opinions on environmental attributes as well as latent attributes among stakeholders.

This research empirically examines the preferences for job-related attributes among rural villagers living close to ASGM in Gorontalo Province of Indonesia. Based on hypothetical scenarios in which a private company collaborates with the local government to establish a food processing industry in the villages, the research applies BWS techniques to determine potential workers' latent demands for job opportunities. Given the heterogeneous livelihoods of the villagers, the study also examines preference heterogeneity by applying a latent class logit (LCL) model. This research aims to expand the discussion and provide new insights into job opportunities to reduce dependency on ASGM and improve public health in poverty-plagued villages in Indonesia.

2. ASGM Operations in Indonesia

Indonesia emits the second largest amount of mercury in the world [15]. ASGM operations have spread widely in various provinces in Indonesia. There are limited reliable sources of data, but the estimated number of workers in artisanal and small-scale mining in Indonesia is 109,000 [16]. Note that this number includes workers engaged in extracting a broad range of minerals, including gold. Ten percent of workers involved in artisanal and small mining operations are women and children [16]. Extensive previous studies have described the economic benefits accrued from mining operations and noted the mercury contamination in ASGM and surrounding areas as well as safety and health hazards. Research comparing development benefits of ASGM and large-scale mining in North Sulawesi found that ASGM brought substantial benefits for the local community in the form of a decreased unemployment rate, increased income, better infrastructure, and increased market diversification [17]. At the same time, a high level of mercury contamination has been reported in ASGM communities in various provinces, such as in West Java [18], West Nusa Tenggara [19], Buru Island, and Mollucas [20].

Gorontalo Province, located on northern Sulawesi Island, is one of the typical provinces where ASGM activities have deeply penetrated the economy. Similar to the literature examining other provinces, studies conducted in Gorontalo have pointed out adverse effects of ASGM on the environment and human health. Mining operations in Hulawa villages in Gorontalo Utara Regency date back to the 19th century, whereas mining operations in Ilangata villages began just 15 years ago [15]. Inhabitants living close to ASGM sites experience higher hair mercury contamination than those living in non-ASGM sites [21]. Based on the results of spirometry tests, miners suffer from lung disorders after exposure to evaporated mercury [22]. Part of the context of these ASGM activities is the limited number of prospective industries that can absorb labor in Gorontalo, hindering sustainable development, particularly in rural villages.

Bone Bolango Regency, located in the Gorontalo Province in Indonesia, is a location where many ASGM extraction sites are located. The livelihoods of the inhabitants of the Bone Bolango Regency depend on the Bone River. The inhabitants use the river's water as a source of drinking water and for cooking and agricultural production. Fisheries are common along the Bone River; inhabitants commonly consume fish and shrimp caught from the river. A recent survey noted heavy metal contamination in the Bone River. Concentrations of As (arsenic), Hg, and Pb (lead) in water samples were found to exceed the water safety standards defined by the World Health Organization [23]. Contamination by heavy metals is due to ASGM activities [23].

In rural Bone Bolango, opportunities to obtain cash income are limited to agriculture, fisheries, chauffeuring, and retail work, among other activities. In particular, households with low educational attainment face limited options for obtaining cash income; they are more dependent on mining, probably due to the limited job opportunities in the formal sector in Gorontalo [24]. Generating alternative livelihoods that can incentivize rural villagers to avoid mining activities is key to achieving multifaceted development goals, namely, avoiding environmental pollution, mitigating adverse health effects, and improving rural livelihoods.

This research assumes that the food processing industry in Gorontalo is promising to absorb the labor force and effectively utilize local agricultural production. In Gorontalo, agriculture is the second largest economic sector [25]. Currently, the economy of Gorontalo relies mainly on agriculture, which produces primary commodities (rice, maize). Corn is a commodity that is commonly planted and dominates dryland agricultural activity in Gorontalo [25]. Research notes that the development of the food processing industry can facilitate more production and increase the value added of cultivated commodities. Gorontalo has the potential to develop small and medium enterprises for maize products given the availability and affordability of the labor force and raw materials [26]. The development and use of technologies that can be adopted for the corn processing industry can benefit the economic profile of Gorontalo [27]. Considering the economic profile and prospects of the food processing industry, it is important to examine the potential of this industry from the perspective of the potential labor supply.

The objectives of this research are to examine job-related attributes that are highly evaluated by rural villagers in Gorontalo, Indonesia. Based on hypothetical scenarios that present opportunities to work in the food processing industry, the research examines the preferences of villagers in relation to the following seven attributes: contribution to local environmental quality, frequency of payments, employment of friends in the same company, occupation-related health risk, creating more job opportunities for society, reputation of the company, and acquisition of new skills. Estimation results obtained from the collected information are expected to support the encouragement of alternative livelihoods and minimize dependency on mining.

3. Methodology

3.1. Survey Areas and Questionnaire Design

The survey was conducted in Bulawa and Suwawa Timur districts in the Bone Bolango Regency in Gorontalo Province in September 2019 (refer to [13] for the specific names of the surveyed villages of Bulawa and Suwawa Timur districts). The households were visited based on the map created from satellite imagery to cover households in the surveyed districts and to minimize sampling biases. In the field survey, not every identified household could be the location of a survey because these buildings are not always utilized for residential purposes and sometimes they house governments or serve other purposes. The research collected data from newly identified residential households via field observations that were not identified from satellite imagery. During the survey, field investigators identified residential properties whose household members were not at home. Investigators also identified households whose adult members were not at home. Since the household survey was conducted beginning in the early morning, if respondents were not at home, the households were revisited at night. We did not collect data on rejection rate; however, most of the household members were very responsive and kind to the investigators if they understood the objectives of the survey; thus, rejection was rare. In this sense, sampling biases due to rejection were limited.

The survey was administered to 503 villagers, and a sample of 91 villagers who were the household heads served as the sample for this paper. Most of the answers were collected from the wife of the household, who was unlikely to be a job seeker. To examine the consistency of the results, the authors checked that similar estimation results could be obtained when answers from non-household heads were incorporated.

Prior to administering the survey, the investigators received training to ensure that they understood the survey objectives and questionnaire design. Since the literacy rate of rural households was considered low due to lower educational attainment, answers from respondents were collected through individual face-to-face interviewer-administered elici-tations. To obtain reliable and representative answers from households, the investigators sought respondents who were the head of their household or served as the household's decision maker.

Table 1 shows the attributes selected to elicit job preferences. To examine the validity of each attribute, this research refers to a previous survey on job preferences conducted in 2017 in rural areas of Gorontalo. Accounting for the local characteristics in the ASGM region, environmental and health attributes were incorporated into the survey. Income-based attributes on job preferences, such as a periodic increase in salary and higher income than that offered by other companies, are generally evaluated higher than other attributes. Income-based attributes were not included in the questionnaire to optimize identification of additional relevant attributes affecting job preferences.

Table 1. Selected Preference Attributes.

Attributes	Description
Environment	Contribution to better local environmental quality
Frequency	Frequency of payments (e.g., daily, weekly, monthly)
Friend	Your friends are working in the same company
Health	Occupation-related health risk
Job	Creating more job opportunities for society
Reputation	Reputation of the company
Skill	Obtaining new skills

Respondents were presented with the following hypothetical scenarios.

Suppose that a private company in collaboration with the local government establishes a food processing industry near the market of Kecamatan. The company is going to hire local people for food processing duty. Monthly payments are equivalent to those of the other companies in the same industry in Gorontalo Province, but you are expected to receive more compensation as you gain experience. Suppose that you receive a job opportunity in that company. If you decide to work there, you must work there as a full-time employee, meaning that you cannot work full-time for other enterprises. Which factors (attributes) do you think are the most important and least important in deciding to accept a new job opportunity?

One of the sample choice sets is shown in Figure 1. Each respondent was asked to choose what he or she considered the best and worst of the four attributes. Each respondent repeated this exercise seven times with different choice sets. Respondents were instructed to select "I don't know" if they could not understand the question or decide on their answers.

Q. Please tick the most important and least important factors (attributes) for each question. Tick "I don't know" if you cannot decide.

Most Important	Attributes (Factors)	Least Important
☐	Obtaining new skills	☐
☐	Frequency of payments (e.g. daily, weekly, monthly etc.)	☐
☐	Your friends are working in the same company	☐
☐	Creating more job opportunities for society	☐

☐   I don't know

Figure 1. Sample questions.

To construct BWS questions, this research employed a balanced incomplete block design (BIBD) to ensure that each alternative appeared an equal number of times and was equally paired with each of the other alternatives across all choice sets (refer to Street and Burgess [28] and Hinkelmann and Kempthorne [29]).

3.2. Estimation Procedures

To examine latent preferences obtained from BWS, maximum difference (maxdiff) models are commonly applied, and simple counting methods are used to obtain the best-minus-worst score [30,31]. The basic theory of the maxdiff model is random utility maximization [32,33]. Suppose a random utility of choosing i as the best item and i' as the worst $U_{ii'} = [V(i) - V(i')] + \varepsilon_{ii'} / \varphi$, where $V(\cdot)$ denotes the deterministic component of indirect utility, $\varepsilon_{ii'}$ is the *iid* (independently and identically distributed) error component, and φ is the scale parameter, which is inversely proportional to the standard deviation of the error component. When the error component $\varepsilon_{ii'}$ depends on an independent standard Gumbel or type I extreme value distribution, an individual choice probability can be formed from the well-known form of a multinomial or conditional logit model, as shown in Equation (1), representing the maxdiff model:

$$P(ii', i \neq i' | M) = \frac{exp(\varphi(V(i) - V(i')))}{\sum_{\substack{j, j' \in M \\ j \neq j'}} exp(\varphi(V(j) - V(j')))} \tag{1}$$

where M denotes the options, which consist of the choice set provided to the respondents.

To examine data collected from BWS that contain "best" and "worst" answers, a random utility function that assumes a linear-in-parameter form, as shown in Equations (2) and (3), has been frequently used:

$$U(ii', i \neq i') = \alpha_i x_i - \alpha_{i'} x_{i'} + \varepsilon_{ii'} / \varphi \tag{2}$$

$$U^*(ii', i \neq i') = \varphi U(ii', i \neq i') = \beta_i x_i - \beta_{i'} x_{i'} + \varepsilon_{ii'}{}^* = \beta_i x_i + \beta_{i'}(-x_{i'}) + \varepsilon_{ii'}{}^* \tag{3}$$

where x_i and $x_{i'}$ denote dummy variables that take a value of 1 when the item is chosen by the respondent and 0 otherwise. α_i and $\alpha_{i'}$ are the true marginal utility parameters. β_i and $\beta_{i'}$ are the parameters where the true marginal utility parameters, α_i and $\alpha_{i'}$, and the scale parameter, φ, are jointly estimated; these 7 arameters are difficult to be estimated separately. In coding the data, we set the most important item as 1, the least important as -1, and all other items as 0.

When utilizing a multinomial logit model Equation (1), there are several strict assumptions to relax: preference homogeneity, a property of independence of irrelevant alternatives (IIA), and an uncorrelation of marginal utility parameters. We employ an LCL model to relax these model assumptions because of the anticipated heterogeneity in surveyed preferences [34–36]. Since any form of correlation can be permitted under LCL [37], we can control parameter correlations caused by behavioral phenomena and scale heterogeneity when we employ LCL. Refer to Appendix A for further discussion.

For the LCL model, the number of classes is specified exogenously. This research incorporates a number of variables in the membership functions, including dependency on mining, household income (natural log), age, education, residency, and duration of residency in the current location. Several information criteria have been utilized in previous papers. Boxall and Adamowicz [34] utilized the minimum Akaike information criterion (AIC) and the Bayesian information criterion (BIC). Hynes et al. [38] proposed employing AIC3 and corrected AIC (crAIC). We decided to utilize these four criteria to determine the number of classes with every combination of covariates to be considered.

4. Estimation Results

Table 2 shows the socioeconomic variables of the households. The average number of household members is 4.3. Annual income is IDR 36.542 million (USD 2583) per household, or IDR 10.65 million (USD 740) per capita, indicating that villagers are not wealthy. The share of agricultural income and mining income is approximately 20%, and other income sources include working in a shop, working as a driver, or engaging in freelance activities.

Table 2. Socioeconomic Variables of the Households (data collected from the household head only).

	Variable	Sample Sizes	Mean	Std. Dev.	Min.	Max.
	Number of household members (num.)	91	4.308	1.872	1	9
Household income	Total (per year, in million Rupiah)	91	36.542	45.147	0	270.06
	Per capita (per year, in million Rupiah)	91	10.465	16.079	0	115
	Share of agricultural income	88	0.195	0.340	0	1
	Share of mining income	88	0.172	0.322	0	1
Mining	Whether household has miner (1 = yes, 0 = no)	91	0.286	0.454	0	1
	Whether household head is miner (1 = yes, 0 = no)	91	0.220	0.416	0	1
	Number of miners per household (num.)	91	0.341	0.619	0	3
Head's demographics	Age (years)	88	46.466	12.641	21	79
	Years of education (years)	89	7.921	3.076	3	17
	Status of residency in the current place of residence (1 = living in the household except occasional trip, 0 = living outside from household)	87	0.885	0.321	0	1
	Duration of stay in the current residential place (1 = 10 years or more, 0 = less than 10 years)	89	0.809	0.395	0	1

Note: 1 USD is equivalent to 14,148 Rupiah in 2019 [39]. Refer to [40,41] about descriptive statistics for codes (0, 1).

The estimation results of the conditional logit model are shown in Table 3. Reputation (reputation of the company) is set as a reference attribute; therefore, respective parameters should be interpreted as the relative importance compared with the reputation attribute. Estimation results indicate that job (creation of more job opportunities for society) is evaluated highest, and the friend attribute, i.e., employment of friends in the same company, is evaluated as least important. Based on the estimation results of the CL (Conditional Logit) model, villagers seek more job opportunities for society.

Table 4 shows the estimation results of the LCL model that assumes preference heterogeneity among different classes. Based on the information criterion, four class models with a constant membership function model were identified. The estimation results that included socioeconomic variables in the membership functions provided lower explanatory power than the estimation model presented in Table 4. The class share is 20.0% for class 1, 16.0% for class 2, 27.5% for class 3, and 36.5% for class 4.

Table 3. Estimation Results of CL (Conditional Logit) model.

Factors	Coef.
Job	1.388 ***
	(0.114)
Frequency	1.046 ***
	(0.111)
Environment	1.008 ***
	(0.111)
Health	0.836 ***
	(0.110)
Skill	0.756 ***
	(0.109)
Friend	0.094
	(0.107)
Log-likelihood	−1351.604
Number of Observations	7164

Note: Standard errors are in parentheses. The symbol *** denotes statistical significance at the 1% levels.

Table 4. Estimation Results of LCL model.

	Class 1	Class 2	Class 3	Class 4
Attributes				
Environment	1.462 ***	1.856 ***	−0.035	2.136 ***
	(0.399)	(0.428)	(0.222)	(0.278)
Frequency	0.620	4.461 ***	0.005	1.950 ***
	(0.438)	(0.744)	(0.229)	(0.291)
Friend	−0.160	−0.283	0.243	0.247
	(0.324)	(0.361)	(0.211)	(0.218)
Health	2.682 ***	3.926 ***	−0.296	0.620 **
	(0.551)	(0.693)	(0.261)	(0.261)
Job	1.809 ***	2.774 ***	0.028	3.302 ***
	(0.460)	(0.636)	(0.237)	(0.348)
Skill	1.299 ***	0.444	0.732 ***	1.038 ***
	(0.371)	(0.345)	(0.237)	(0.250)
Membership Function (reference = class 4)				
constant	−0.602	−0.828	−0.283	
	(0.408)	(0.371)	(0.325)	
Membership Probability				
	0.200	0.160	0.275	0.365
Log-likelihood		−1225.222		
Number of Observations		7164		

Note: Standard errors are in parentheses. The symbols *** and ** denote statistical significance at the 1%, and 5 % levels, respectively.

Compared with the estimation results of the LCL model, Table 4 explains distinct features in the estimation results. Class 1 prioritized occupation-related health risks, followed by job opportunities, better local environmental quality, and skill acquisition. The employment of friends in the company and the frequency of payments were evaluated as less important. In contrast to class 1, class 2 evaluated frequency of payments highest, followed by health consciousness, job opportunities and better local environmental quality. Among the four classes, class 2 is the only class that did not positively evaluate skill acquisition. The employment of friends in the company is also evaluated as less important by this class.

Class 3 is quite unique. Skill acquisition is the only attribute that is positively evaluated by this class. Other attributes are not statistically significant with respect to the baseline attribute (reputation of the company).

Class 4 evaluated job opportunity highest, followed by better local environmental quality, frequency of payments, skill acquisition and health consciousness. The results are similar to those from the conditional logit model presented in Table 3. Since this class was the largest and the majority of respondents were in this category, this class reflected the average characteristics and its constant was applied as the reference for the membership function of the other three classes.

5. Discussion

The estimation results in Table 3 show that respondents attribute high importance to more local job prospects, flexibility in payment schemes, the company's environmental consciousness, the company's consciousness of occupation-related health risks, and opportunities to obtain skills. The estimation results of the LCL model indicate the need to account for preference heterogeneity. In Table 4, the most prevalent class is class 4, which comprises 36.5% of respondents, and the estimation results are similar to those in Table 3. However, the proportion of the respondents (27.5%) who are more likely to be categorized into class 3, which places importance on only skill acquisition, is not negligible, according to critical mass theory in sociology [42–44]. Therefore, even though job opportunities are highly evaluated by classes other than class 3, the unique features of class 3 need to be taken into account in managing ASGM issues in Gorontalo, Indonesia.

An attribute that is relatively highly evaluated overall is the creation of more job opportunities for society. This implies that limited job opportunities are a serious concern and an urgent issue for rural villagers in Bone Bolango. Although no classes evaluate environmental consciousness as the most important attribute, class 4 (the largest class) ranks it the second most important attribute. Therefore, heavy metal contamination is recognized by a non-negligible number of individuals as a serious local environmental issue that Bone Bolango faces.

Although class 2 is estimated to represent only 16.0% of the total, payment frequency received a great deal of attention from this class, which implies that these villagers want prospective companies to provide flexible salary payment options.

Contrasting features are found in the skill acquisition attribute. The estimation results indicate that skill acquisition at work received relatively low interest from villagers categorized in classes 1, 2, and 4. This may indicate that rural villagers have little experience receiving incremental salaries based on their acquired skills since jobs that require professional skills are scarce there. In contrast, class 3 evaluates only skill acquisition highly, meaning that members of this category evaluate whether opportunities to obtain skills are available. These data suggest that to some, the opportunity to acquire skills is crucially important to boosting income and improving living standards above all other preferences. It is important to examine this distinction in order to raise awareness of the importance of skill development for poverty eradication in rural Gorontalo.

The results for occupation-related health risk in Table 4 indicate unique characteristics. Villagers categorized into class 1 (20.0%) and class 2 (16.0%) prefer companies with consciousness of occupation-related health risks; however, respondents in class 4 (36.5%) place less importance on this attribute. There are no statistically significant differences for class 3 (27.5%), meaning that occupation-related health risks are not important compared with the baseline attribute (reputation of the company). If alternative livelihood approaches are implemented to alleviate health risks, such practices may not be widely accepted by potential employees.

The results obtained from the LCL model yield important policy implications for managing ASGM and improving public health in the survey regions because they revealed preference heterogeneity. This indicates that a single policy does not apply to all; several packages that assume heterogeneous preferences should be provided to meet the latent demands of villagers.

The research notes two limitations of generalizing results for establishing policies in ASGM communities to reduce environmental degradation. The first limitation is the generalizability of the findings. Our sampling procedures attempted to minimize sampling biases but are not considered random sampling because complete sample frames could not be prepared before the survey. The research cautions that existing sampling errors may influence the generalizability of the results. Second, the sampling could not focus on potential job seekers due to the lack of formal job markets. If sampling was conducted for only job seekers, the estimation results would yield better policy implication.

Since the research was conducted to examine latent preferences for job-related attributes, the level of attributes was not examined. For example, respondents would provide different opinions and preferences regarding the level of occupation-related health risks and possible permanent damages. It is necessary to conduct a stated preference survey to identify preferences on the level of those attributes, which can help formulate more detailed policy to minimize dependency on ASGM.

This research includes various sociodemographic variables in the LCL model membership function to find determinants segregating different classes of workers. If current miners are identified in alignment with specific classes, this allows for interpretation of heterogeneity of preferences for job opportunities and the consequent dependency on mining. Other observable socioeconomic variables, such as the duration of residency in the current location, affluence of households, and educational variables, cannot be included in the membership functions from the viewpoint of the information criterion; therefore, those indicators are not suitable for separating classes. This may indicate that unobserved variables, such as psychological variables, environmental concerns, and attitudes toward political parties, may be candidates for further research. Further examinations are required to verify other potential determinants.

Since this study is limited to case studies in Gorontalo Province, Indonesia, further research is pivotal for generalizing the results. As a case study, the present study is beneficial for policymakers considering the provision of job opportunities in rural areas to reduce dependency on mining and simultaneously improve rural livelihoods.

6. Conclusions

This research empirically examines the preferences for job-related attributes among rural villagers living close to ASGM in the Gorontalo Province of Indonesia. Based on hypothetical scenarios in which a private company in collaboration with the local government establishes a food processing industry in villages, the research applied BWS techniques to determine possible workers' latent demands for job opportunities. It also examined the heterogeneity of the villagers by applying an LCL model. The estimation results indicated that villagers are separated into four classes, and each class has different and unique preferences. The results suggest different policy interventions to effectively capture the heterogeneous livelihoods of rural villagers. Accounting for heterogeneous preferences for job opportunities is helpful to delink dependency on ASGM and health hazards and improve the livelihoods of rural villagers.

Author Contributions: S.K. contributed to the design of the research, data examination, statistical analysis, and writing. Y.I.D.P. conducted the field survey. M.S. provided conceptual advice and funding. T.O. provided technical advice and contributed to the writing. All authors have read and agreed to the published version of the manuscript.

Funding: This research was supported by the Ministry of Education, Culture, Sports, Science, and Technology, Japan; a Grant-in-Aid for Scientific Research, KAKENHI (No. 19K12446); and the Research Institute for Humanity and Nature "Co-Creation of Regional Innovation for Reducing Risk of Environmental Pollution" program.

Institutional Review Board Statement: The survey design, including the design of the questionnaire, was examined and passed by the ethical committee of the Research Institute for Humanity and Nature (RIHN), Kyoto, Japan (code number: 2019-3).

Informed Consent Statement: Informed consent was obtained from all questionnaire respondents.

Data Availability Statement: Since the household survey was conducted in mining areas, where security concerns and privacy restrictions exist, data are available on request.

Acknowledgments: The authors thank the editors and anonymous referees for their constructive comments and suggestions to improve the quality of an earlier version of the manuscript. The authors greatly appreciate the assistance and cooperation of the field investigators and survey respondents. The authors also convey our gratitude for members of the projects "Co-creation of Sustainable Regional Innovation for Reducing Risk of High-impact Environmental Pollution (SRIREP), Research Institute for Humanity and Nature (RIHN)" for their kind assistance.

Conflicts of Interest: The authors declare no conflict of interest in terms of financial or personal involvement that may influence the judgments expressed in this manuscript.

Appendix A

The choice probability of LCL is represented as follows:

$$P_n(ii', i \neq i') = \sum_{s=1}^{S} \pi_{ns} \prod_{t=1}^{T} \frac{\exp(\varphi(V(nit|s) - V(ni'\,t|s)))}{\sum_{\substack{j,j' \in M \\ j \neq j'}} \exp(\varphi(V(njt|s) - V(nj'\,t|s)))}$$

$$\pi_{ns} = \frac{exp(c_s + \gamma_s'\,z_n)}{\sum_{s^*=1}^{S} exp(c_{s^*} + \gamma_{s^*}'\,z_n)}$$

where t denotes the choice occasion (maximum 7), s the class into which respondents are classified, c_s is a class-s-specific constant, and z_n is a vector of covariates with parameter vector γ_s. $c_s + \gamma_s'z_n$ is a membership function, and π_{ns} is the probability of membership. The Sth parameter vector is normalized to zero to secure identification. With the LCL model, we can estimate the shares of classes to which the respondents with the same utility function belong. If we obtain the parameters of covariates in the membership function, γ_s, we can interpret why individuals are classified as they are. First, because the marginal utility parameter vector has discrete variation, LCL assures preference heterogeneity. Second, because the ratio of choice probabilities depends not only on the considered alternatives but also on all other alternatives, the IIA property is completely relaxed [45].

The IIA property of a multinomial logit model is expressed by the ratio of choice probabilities related to two alternatives, i and k ([46] (pp. 45–47)). In the context of the maxdiff model, we can arrange the ratio as follows: $P(ii', i \neq i'|M)/P(kk', k \neq k'|M)$. In the simple maxdiff model, the ratio becomes the form that is independent from unconsidered alternatives: $exp(\varphi(V(i) - V(i')))/exp(\varphi(V(k) - V(k')))$. With LCL, the ratio becomes

$$\left(\sum_{s=1}^{S} \pi_{ns} \prod_{t=1}^{T} \frac{exp(\varphi(V(nit|s) - V(ni'\,t|s)))}{\sum_{\substack{j,j' \in M \\ j \neq j'}} exp(\varphi(V(njt|s) - V(nj'\,t|s)))} \right) \Bigg/ \left(\sum_{s=1}^{S} \pi_{ns} \prod_{t=1}^{T} \frac{exp(\varphi(V(nkt|s) - V(nk'\,t|s)))}{\sum_{\substack{j,j' \in M \\ j \neq j'}} exp(\varphi(V(njt|s) - V(nj'\,t|s)))} \right)$$

where it depends on not only the considered alternatives but also all other alternatives. The only exception is when choice probability becomes as follows:

$$\forall s \in S, \sum_{s=1}^{S} \pi_{ns} \prod_{t=1}^{T} \frac{exp(\varphi(V(nit|s) - V(ni'\,t|s)))}{\sum_{\substack{j,j' \in M \\ j \neq j'}} exp(\varphi(V(njt|s) - V(nj'\,t|s)))} = \sum_{s=1}^{S} \pi_{ns} \prod_{t=1}^{T} \frac{exp(\varphi(V(nit) - V(ni'\,t)))}{\sum_{\substack{j,j' \in M \\ j \neq j'}} exp(\varphi(V(njt) - V(nj'\,t)))}$$

which represents the degeneration of LCL to the simple maxdiff model in Equation (1).

Third, correlations of parameters may exist because of behavioral phenomena—for example, employers who place high priority on the frequency of payments tend to prefer creating more job opportunities for society—and scale heterogeneity [37]. Especially in the context of scale heterogeneity, cognitive effort-related effects have been suggested, such as learning and fatigue [47,48], choice uncertainty [49], and survey engagement [50]. In the context of household surveys, especially in developing countries, these variabilities can easily occur due to differences in the degree of respondents' understanding of the scenarios provided in stated preference surveys or their involvement in the survey. Indeed, the covariance matrix of the parameter vector, β, theoretically becomes as follows [37]:

$$Cov(\beta) = \sum_{s=1}^{S} \pi_s (\beta_s - E[\beta])(\beta_s - E[\beta])'$$

where $E[\beta]$ denotes the expectation of β globally, or across classes. Therefore, any drawbacks in a multinomial logit model are relaxed by LCL.

Erdem [51] employed a scale-adjusted LCL (SALCL) to take class-specific scale parameters into account; the model was proposed by Vermunt and Magidson [52]. Although class-specific scale parameters can be estimated by SALCL and SALCL appears to be the higher model of the standard LCL, improvements in fit by SALCL are suggested as "simply as a result of increasing distributional flexibility [37] (p. 6)". Conservatively speaking, SALCL may be a model whose properties are not transparent thus far. Therefore, we decided to employ the standard model of LCL for simplicity.

References

1. UN Environment. *Global Mercury Assessment 2018*; UN Environment Programme, Chemicals and Health Branch: Geneva, Switzerland, 2019.
2. World Health Organization. *Step-By-Step Guide for Developing a Public Health Strategy for Artisanal and Small-Scale Gold Mining in the Context of the Minamata Convention on Mercury*; WHO: Geneva, Switerland, 2021.
3. Seccatore, J.; Veiga, M.; Origliasso, C.; Marin, T.; De Tomi, G. An estimation of the artisanal small-scale production of gold in the world. *Sci. Total Environ.* **2014**, *496*, 662–667. [CrossRef] [PubMed]
4. Gibb, H.; O'Leary, K. Mercury exposure and health impacts among individuals in the artisanal and small-scale gold mining community: A comprehensive review. *Environ. Health Perspect.* **2014**, *122*, 667–672. [CrossRef] [PubMed]
5. Steckling, N.; Tobollik, M.; Plass, D.; Hornberg, C.; Ericson, B.; Fuller, R.; Bose-O'Reilly, S. Global burden of disease of mercury used in artisanal small-scale gold mining. *Ann. Glob. Health* **2017**, *83*, 234–247. [CrossRef] [PubMed]
6. Veiga, M.M.; Angeloci-Santos, G.; Meech, J.A. Review of barriers to reduce mercury use in artisanal gold mining. *Extr. Ind. Soc.* **2014**, *1*, 351–361. [CrossRef]
7. LeClerq, A.T.; Gore, M.L.; Lopez, M.C.; Kerr, J.M. Local perceptions of conservation objectives in an alternative livelihoods program outside Bardia National Park, Nepal. *Conserv. Sci. Pract.* **2019**, *1*, e131. [CrossRef]
8. Novriyanto; Wibowo, J.T.; Iskandar, W.; Campbell-Smith, G.; Linkie, M. Linking coastal community livelihoods to marine conservation in Aceh, Indonesia. *Oryx* **2012**, *46*, 508–515. [CrossRef]
9. Roe, D.; Booker, F.; Day, M.; Zhou, W.; Allebone-Webb, S.; Hill, N.A.O.; Kumpel, N.; Petrokofsky, G.; Redford, K.; Russell, D.; et al. Are alternative livelihood projects effective at reducing local threats to specified elements of biodiversity and/or improving or maintaining the conservation status of those elements? *Environ. Evid.* **2015**, *4*, 22. [CrossRef]
10. Prescott, G.W.; Maung, A.C.; Aung, Z.; Carrasco, L.R.; De Alban, J.D.T.; Diment, A.N.; Ko, A.K.; Rao, M.; Schmidt-Vogt, D.; Soe, Y.M.; et al. Gold, farms, and forests: Enforcement and alternative livelihoods are unlikely to disincentivize informal gold mining. *Conserv. Sci. Pract.* **2020**, *2*, e142. [CrossRef]
11. Cartier, L.E.; Bürge, M. Agriculture and artisanal gold mining in Sierra Leone: Alternatives or complements? *J. Int. Dev.* **2011**, *23*, 1080–1099. [CrossRef]
12. Hilson, G.; Banchirigah, S.M. Are alternative livelihood projects alleviating poverty in mining communities? Experiences from Ghana. *J. Dev. Stud.* **2009**, *45*, 172–196. [CrossRef]
13. Komatsu, S.; Pongoliu, Y.I.D.; Tanaka, K.; Sakakibara, M.; Ohdoko, T. Accounting for Correlated random parameters in best-worst scaling: Case of preferences for job opportunities in an Artisanal small gold mining (ASGM) community in Indonesia. In Proceedings of the 4th International Conference of Transdisciplinary Research on Environmental Problems in Southeast Asia, Lampung, Indonesia, 18 September 2021.
14. Finn, A.; Louviere, J.J. Determining the appropriate response to evidence of public concern: The case of food safety. *J. Pub. Policy Mark.* **1992**, *11*, 12–25. [CrossRef]

15. Arifin, Y.I.; Sakakibara, M.; Takakura, S.; Jahja, M.; Lihawa, F.; Sera, K. Artisanal and small-scale gold mining activities and mercury exposure in Gorontalo Utara Regency, Indonesia. *Toxicol. Environ. Chem.* **2020**, *102*, 521–542. [CrossRef]
16. Hentschel, T.; Hruschka, F.; Priester, M. *Artisanal and Small-Scale Mining Challenges and Opprtrunities*; International Institute for Environment and Development: London, UK, 2003.
17. Langston, J.D.; Lubis, M.I.; Sayer, J.A.; Margules, C.; Boedhihartono, A.K.; Dirks, P.H.G.M. Comparative development benefits from small and large scale mines in North Sulawesi, Indonesia. *Extr. Ind. Soc.* **2015**, *2*, 434–444. [CrossRef]
18. Harianja, A.H.; Saragih, G.S.; Fauzi, R.; Hidayat, M.Y.; Syofyan, Y.; Tapriziah, E.R.; Kartiningsih, S.E. Mercury exposure in artisanal and small-scale gold mining communities in Sukabumi, Indonesia. *J. Health Pollut.* **2020**, *10*, 201209.
19. Junaidi, M.; Krisnayanti, B.D.; Juharfa; Anderson, C. Risk of mercury exposure from fish consumption at artisanal small-scale gold mining areas in West Nusa Tenggara, Indonesia. *J. Health Pollut.* **2019**, *9*, 190302. [CrossRef] [PubMed]
20. Male, Y.T.; Reichelt-Brushett, A.J.; Pocock, M.; Nanlohy, A. Recent mercury contamination from artisanal gold mining on Buru Island, Indonesia—Potential future risks to environmental health and food safety. *Mar. Pollut. Bull.* **2013**, *77*, 428–433. [CrossRef]
21. Arifin, Y.I.; Sakakibara, M.; Sera, K. Impacts of artisanal and small-scale gold mining (ASGM) on environment and human health of Gorontalo Utara Regency, Gorontalo Province, Indonesia. *Geosciences* **2015**, *5*, 160–176. [CrossRef]
22. Pateda, M.S.; Sakakibara, M. Preliminary study on human lung function of artisanal and small-scale gold miner in Gorontalo Province, Indonesia. *IOP Conf. Ser. Earth Environ. Sci.* **2020**, *536*, 12009. [CrossRef]
23. Gafur, N.A.; Sakakibara, M.; Sano, S.; Sera, K. A case study of heavy metal pollution in water of Bone river by artisanal small-scale gold mine activities in eastern part of Gorontalo, Indonesia. *Water* **2018**, *11*, 1507. [CrossRef]
24. Komatsu, S.; Tanaka, K.; Sakakibara, M.; Arifin, Y.I.; Pateda, S.M.; Manyoe, I.N. Sociodemographic attributes and dependency on artisanal and small-scale gold mining: The case of rural Gorontalo, Indonesia. *IOP Conf. Ser. Earth Environ. Sci.* **2020**, *589*, 12020. [CrossRef]
25. Katili, I.; Masia, I.; Latjompo, Z.; Lahay, A.W.; Nasrudin; Rahmayanti, D.; Otaya, A.W.; Pililie, F.; Triyono; Rahim, R. *Provincial Development Guideline of Province of Gorontalo 2013*; Provincial Working Group (A Collaboration Between Province of Gorontalo with PGSP UNDP); UNPD: New York, NY, USA, 2013.
26. Halid, A. Development of small medium enterprises of maize processed food products as a locomotive of Gorontalo District's economy. *J. Perspect. Financ. Reg. Dev.* **2019**, *6*, 729–734. [CrossRef]
27. Hasan, A.M.; Halid, A.; Ahmad, L.; Hasdiana. Developing the added value of corn chips as way of improving the community economy in Gorontalo Province of Indonesia. *Rus. J. Agric. Socio-Econ. Sci.* **2018**, *83*, 333–341. [CrossRef]
28. Street, D.J.; Burgess, L. *The Construction of Optimal Stated Choice Experiments: Theory and Methods*; John Wiley and Sons, Inc.: Hoboken, NJ, USA, 2007.
29. Hinkelmann, K.; Kempthorne, O. *Design and Analysis of Experiments Volume 2: Advanced Experimental Design*; John Wiley and Sons, Inc.: Hoboken, NJ, USA, 2005.
30. Marley, A.A.J.; Louviere, J.J. Some probabilistic models of best, worst, and best-worst choices. *J. Math. Psychol.* **2005**, *49*, 464–480. [CrossRef]
31. Louviere, J.J.; Flynn, T.N.; Marley, A.A.J. *Best-Worst Scaling: Theory, Methods and Applications*; Cambridge University Press: Cambridge, UK, 2015.
32. Thurstone, L.L. A Law of comparative judgement. *Psychol. Rev.* **1927**, *34*, 278–286. [CrossRef]
33. Manski, C.F. The structure of random utility models. *Theory Decis.* **1977**, *8*, 229–254. [CrossRef]
34. Boxall, P.C.; Adamowicz, W.L. Understanding heterogeneous preferences in random utility models: A latent class approach. *Environ. Resour. Econ.* **2002**, *23*, 421–446. [CrossRef]
35. Greene, W.H.; Hensher, D.A. A Latent Class Model for Discrete Choice Analysis: Contrasts with Mixed Logit. *Transp. Res. Part B Methodol.* **2003**, *37*, 681–698. [CrossRef]
36. Swait, J.R.A. Structural equation model of latent segmentation and product choice for cross-sectional revealed preference choice data. *J. Retail. Consum. Serv.* **1994**, *1*, 77–89. [CrossRef]
37. Hess, S.; Train, K.E. Correlation and scale in mixed logit models. *J. Choice Model.* **2017**, *23*, 1–8. [CrossRef]
38. Hynes, S.; Hanley, N.; Scarpa, R. Effects on Welfare measures of alternative means of accounting for preference heterogeneity in recreational demand models. *Am. J. Agric. Econ.* **2008**, *90*, 1011–1027. [CrossRef]
39. World Bank. *World Development Indicators, Official Exchange Rate (LCU per US$, Period Average)*; World Bank: Washington, DC, USA, 2021.
40. Li, Y.; Xiong, C.; Zhu, Z.; Lin, Q. Family migration and social integration of migrants: Evidence from Wuhan Metropolitan area, China. *Int. J. Environ. Res. Public Health* **2021**, *18*, 12983. [CrossRef] [PubMed]
41. Dhital, R.D.; Ito, T.; Kaneko, S.; Komatsu, S.; Yoshida, Y. Household access to water and education for girls: The case of villages in hilly and mountainous areas of Nepal. *Oxf. Dev. Stud.* **2021**, *in print*. [CrossRef]
42. Kanter, R.M. Some effects of proportions on group life: Skewed sex ratios and responses to token women. *Am. J. Sociol.* **1977**, *82*, 965–990. [CrossRef]
43. Dahlerup, D. From a small to a large minority: Women in Scandinavian politics. *Scand. Political Stud.* **1988**, *11*, 275–297. [CrossRef]
44. Dahlerup, D. The story of the theory of critical mass. *Polit. Gender.* **2006**, *2*, 511–522. [CrossRef]

45. Shonkwiler, J.S.; Shaw, W.D. A finite mixture approach to analyzing income effects in random utility models. In *The New Economics of Outdoor Recreation*; Hanley, N., Shaw, W.D., Wright, R.E., Eds.; Edward Elgar Publishing: Cheltenham, UK, 2003; pp. 268–278.
46. Train, K.E. *Discrete Choice Methods with Simulation*, 2nd ed.; Cambridge University Press: Cambridge, UK, 2009.
47. DeShazo, J.R.; Fermo, G. Designing choice sets for stated preference methods: The Effects of complexity on choice consistency. *J. Environ. Econ. Manag.* **2002**, *44*, 123–143. [CrossRef]
48. Balcombe, K.; Fraser, I.; McSorley, E. Visual attention and attribute attendance in multi-attribute choice experiments. *J. Appl. Econom.* **2015**, *30*, 447–467. [CrossRef]
49. Uggeldahl, K.; Jacobsen, C.; Lundhede, T.H.; Olsen, S.B. Choice certainty in discrete choice experiments: Will eye tracking provide useful measures? *J. Choice Model.* **2016**, *20*, 35–48. [CrossRef]
50. Hess, S.; Stathopoulos, A. Linking response quality to survey engagement: A combined random scale and latent variable approach. *J. Choice Model.* **2013**, *7*, 1–12. [CrossRef]
51. Erdem, S. Who do UK consumer trust for information about nanotechnology? *Food Policy* **2018**, *77*, 133–142. [CrossRef]
52. Vermunt, J.K.; Magidson, J. *Technical Guide for Latent Gold 4.0: Basic and Advanced*; Statistical Innovations Inc.: Belmont, MA, USA, 2005. Available online: https://www.statisticalinnovations.com/wp-content/uploads/LGCtechnical.pdf (accessed on 6 August 2021).

International Journal of
Environmental Research and Public Health

Article

Contamination Level in Geo-Accumulation Index of River Sediments at Artisanal and Small-Scale Gold Mining Area in Gorontalo Province, Indonesia

Basir [1,2,*], Satomi Kimijima [3], Masayuki Sakakibara [1,3,4], Sri Manovita Pateda [1,5] and Koichiro Sera [6]

1   Graduate School of Science &Engineering, Ehime University, 2-5 Bunkyo-cho, Matsuyama 790-8577, Japan; sakaki@chikyu.ac.jp (M.S.); manovita.pateda@gmail.com (S.M.P.)
2   Department of Environmental Health, Faculty of Public Health, Hasanuddin University, Jalan Perintis Kemerdekaanno No. 94, Tamalanrea, Makassar 90245, Indonesia
3   Research Institute for Humanity and Nature, 457-4 Motoyama, Kamigamo, Kita-ku, Kyoto 603-8047, Japan; kimijima@chikyu.ac.jp
4   Faculty of Collaborative Regional Innovation, Ehime University, 3 Bunkyo-cho, Matsuyama 790-8577, Japan
5   Faculty of Medicine, Universitas Negeri Gorontalo, Jenderal Sudirman Street No. 6, Gorontalo 96100, Indonesia
6   Cyclotron Research Center, Iwate Medical University, 348-58 Tomegamori, Takizawa 020-0173, Japan; ksera@iwate-med.ac.jp
*   Correspondence: baz.rasyid@gmail.com; Tel.: +62-81-2424-3295

Citation: Basir; Kimijima, S.; Sakakibara, M.; Pateda, S.M.; Sera, K. Contamination Level in Geo-Accumulation Index of River Sediments at Artisanal and Small-Scale Gold Mining Area in Gorontalo Province, Indonesia. *Int. J. Environ. Res. Public Health* **2022**, *19*, 6094. https://doi.org/10.3390/ijerph19106094

Academic Editor: Paul B. Tchounwou

Received: 26 April 2022
Accepted: 10 May 2022
Published: 17 May 2022

Publisher's Note: MDPI stays neutral with regard to jurisdictional claims in published maps and institutional affiliations.

Abstract: Substances found in watersheds and sediments in artisanal and small-scale gold mining (ASGM) areas contaminated by heavy metals are becoming tremendously critical issues in Asia. This study aimed at clarifying the pollution caused by heavy metals in sediments in river basins near ASGM sites in Gorontalo Province, North Sulawesi, Indonesia. Sediment samples collected from experimental areas were classified into nine clay samples and twenty-seven sand samples, whereas three other samples were collected from the control area. Particle-induced X-ray emission was used to analyze these samples. The Statistical Package for the Social Science and the geo-accumulation index (I_{geo}) were also used for analysis. Based on the results, Hg, Pb, As, and Zn had a concentration of 0–334 µg/g, 5.5–1930 µg/g, 0–18,900 µg/g, and 0–4923.2 µg/g, respectively, which exceeded limits recommended by the U.S. Environmental Protection Agency consensus (1991) and the Indonesian Government Regulation Number 38, 2011. Furthermore, I_{geo} showed the order of the pollution degree Hg < Zn < Pb < As and reflected an environment contaminated by heavy metals, ranging from unpolluted to extremely polluted areas. Therefore, sediments contaminated by Hg, Pb, As, and Zn could be found along the river basin of mining areas.

Keywords: heavy metal contamination; river sediments; ASGM; Gorontalo; geo-accumulation index

1. Introduction

Heavy metal pollution from artisanal and small-scale gold mining (ASGM) has been a pivotal issue in Asian countries, with heavy metals found in watersheds and sediments. Along with many volcanic areas, Gorontalo has many ASGM sites with a unique process that is influenced by hydrothermal circulation and the type of geological setting. One of the main causes is mercury (Hg) pollution due to ASGM, which uses Hg in the amalgamation process to extract gold from ore rock [1,2].

Recent investigations conducted by the United Nations Environment Program highlighted the enormity of Hg pollution in developing countries and its harmful effects on human health and ecosystems [3–7]. Over 300,000 ASGM miners work at ~1000 informal sites in Indonesia [8–12]. ASGM presents an income opportunity for those in rural communities [13–15]; therefore, farmers or fishermen with few options in terms of alternative livelihoods become miners.

In Indonesia, heavy metal pollution tends to increase because of the high use of chemical substances. Since the industrialization era, the use of Hg in this sector has polluted industrial mine minerals, and one of the causes is that tailings resulting from the amalgamation process can lead to environmental hazards. ASGM located in some areas in Indonesia, particularly in Gorontalo, is a trigger of environmental damage. Moreover, volcanoes, animal waste, and algae as well as wood from deforestation can harm the environment [8,12,14,15].

Furthermore, heavy metal pollution from natural and anthropogenic activities is frequently detected in sediments and water columns of rivers, severely contaminating a large percentage of Asia's rivers. This has become a major concern worldwide because most people lack access to clean drinking water and do not have adequate sanitation services [3,16]. In addition, heavy metals are easy to accumulate in sediments since the concentration of heavy metals is always higher in sediments than in river water [16]. Heavy metal content in sediments probably fluctuates due to the occurrence of a large undercurrent, which uplifts or displaces sediments to other locations [17–22]. Previous studies conducted using water and sediment samples from Bone River and Wubudu river of Gorontalo Province in Indonesia have revealed that these rivers have been polluted by Hg along with As and Pb due to activities at ASGM sites [19–21]. In contrast, this study particularly focused on major river sampling sites to clarify the effect of heavy metal pollution from sediments in river basins near ASGM sites in Gorontalo Province.

2. Materials and Methods

2.1. Research Areas

Gorontalo is a province in Indonesia with minerals and natural resources, particularly gold. Gold mining in Gorontalo Province has emerged as a good source of income with the potential to boost the economy and social welfare. This research was conducted in Buladu River near Hulawa Mining; Totopo River near Bumela Mining; Mopuya Daa River near Dunggilata Mining; Bone and Bula Rivers near Suwawa Mining; and Ayidu River (serving as a control area). These are parts of Gorontalo Province, Indonesia [9,12].

2.2. Samples

In this study, samples were collected based on the distance between the mining site and river, as data on Hg content in river water sediments were compared with data collected from the river closest to ASGM areas. Meanwhile, data on Hg levels in sediment samples from river water were collected from factory tailing ponds and amalgamation processing units, which are close to rivers and where mining processes occur.

Sediment samples were collected from five ASGM areas (Hulawa Mining, Bumela Mining, Dunggilata Mining, and Suwawa Mining) and one control area from November 2018 to September 2019. Nine clay samples and twenty-seven sand samples were collected (Figure 1).

Collected samples ranged from coarse sand to clay materials (approximately 2–0.002 mm). All samples were preserved in autoclaved sample bags. In addition, the samples were homogenized by grinding in an agate mortar to obtain fine particles [23–25].

2.3. Analytical Methods

In an experimental laboratory, sediment samples were dried in an oven for 48 h at 80–120 °C. Each dry sample was taken from the oven at 25% of its total volume and then filtered through a sieve to separate large-size materials, such as root, stem, or tree, which were <5 mm or 200–450 μm in size. Each sediment sample was crushed using a planetary micro mill, with jar settings of 5 min at speed 3, 3 min at speed 5, and 3 min at speed 9. In the subsequent stage, powder samples were transferred into a sample bottle using a spoon, inserted into sterile, sealed in plastic bottles, and homogenized at the facility for 60 min. Each sediment sample was measured to 50 mg and mixed with 10 mg of palladium carbon by using an agate mortar. Powder samples were added to a 3 μL collodion solution (collodion ethanol = 1:9) and then spread out (flat; homogeneous) on a thin film using a pipette tip. Concentrations of heavy metals were measured by particle-induced X-ray emission at the Cyclotron Research Centre, Iwate Medical University, Japan [24].

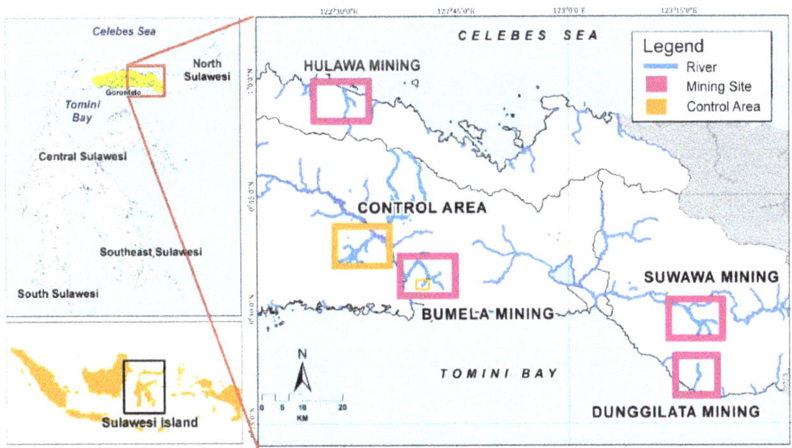

Figure 1. Map of the location of the research area.

3. Results

3.1. Concentration of Hg in River Sediments at ASGM Area and Control Area

The Hg concentration originated from mining (ASGM) sites and the control area (rivers flowing from upstream to downstream). On determination of each element in heavy metal concentration and descriptive analysis in sand and clay of sediment sample at sampling point in Gorontalo Province, Indonesia, and IBM SPSS Statistic 24 for Windows (IBM Inc., Chicago, IL, USA) was applied for the standard established that is preferred for four heavy metals (Hg, Pb, As, and Zn) because these heavy metals are related to sediment and are included in each area of the ASGM site around the river sediment.

Table 1 shows that Totopo River has the highest median Hg concentration of clay (0.15 μg/g), and the other river has the lowest concentration (0 μg/g). On the other hand, the highest median Hg concentration of sand was found in Buladu River (54.7 μg/g), and the lowest (10.1 μg/g) was found in Dunggilata River.

Table 1. Concentration of Hg (μg/g) in river sediment sample from ASGM and control area.

Location	Element	Median	Min–Max
Buladu River (Sumalata Mining)	Sand	54.7	0–147
	Clay	0	0–334
Dunggilata River (Dunggilata Mining)	Sand	10.1	0–38.1
	Clay	0	0–124
Totopo River (Bumela Mining)	Sand	53.6	0–101
	Clay	0.15	0–179
Bone River (Suwawa Mining)	Sand	33.7	86.3–9.5
	Clay	0	0–256
Ayidu River (Control area)	Sand	14.9	10.5–101
	Clay	0	0–0

In addition, some sand and clay samples were collected from the same mining sites, such as Sumalata Mining near Buladu River, Bumela Mining near Totopo River, Dunggilata Mining near Mopuya Daa River, and Suwawa Mining near Bone River, to compare the element concentration. In contrast, Hg content in the river sediment samples was very high, above the detection or threshold limit. Hg content (concentration) was above the detection limits of U.S. Environmental Protection Agency and river sediments of water quality criteria based on Indonesian Government Regulation Number 38, 2011 [26–28]. This was also in line with previous research that stated that Hg concentrations in Wubudu River sediments were far above the threshold limits stated by World Health Organization because the river is close to ASGM processing units [29,30].

3.2. Concentration of Pb in River Sediments at ASGM Area and Control Area

The Pb content originates from mining (ASGM) sites and the control area, with rivers flowing from upstream. Table 2 shows the results of the laboratory analysis in which the median concentration of Pb (µg/g) in sand sediment samples ranged from 93.6 to 237 µg/g, and clay sediment samples ranged from 36.3 to 1286.5 µg/g.

Table 2. Concentration of Pb (µg/g) in river sediment sample from ASGM and control area.

Location	Element	Median	Min–Max
Buladu River (Sumalata Mining)	Sand	230.5	30.2–589
	Clay	1286.5	0–1610
Dunggilata River (Dunggilata Mining)	Sand	124	27–412
	Clay	985.2	0–1930
Totopo River (Bumela Mining)	Sand	168.9	38.1–536
	Clay	36.3	0–308
Bone River (Suwawa Mining)	Sand	237	58.2–645
	Clay	856	0–856
Ayidu River (Control area)	Sand	93.6	5.5–187
	Clay	198	0–198

3.3. Concentrations of As in River Sediments at ASGM Area and Control Area

Table 3 shows results of the laboratory analysis in which the median concentration of As (µg/g) in sand sediment samples ranged from 0.8 to 315.1 µg/g, and that in clay sediment samples are 0 µg/g based on the interpretation of geo-accumulation index (I_{geo}) classification. For As concentration, the result indicated clay sediment samples with a grain size of 0.0025 mm had high concentration compared to sand sediment samples taken from along the river and ASGM sites, especially in Sumalata Mining and Suwawa Mining of Gorontalo Province, Indonesia. Subsequently, the concentration of As is extremely polluted if compared to others concentration of heavy metals, such as Hg, Pb, and Zn [9,21]. In addition, arsenic (As) and iron (Fe) can be associated with the environment (natural source), especially in mineral clay. Moreover, arsenic is also one of the byproducts of the processing of non-ferrous metal ores, especially gold, which has very toxic properties with damaging effects on the environment. It is easy to find in several metal ore deposits, including Cu-Zn-Pb deposits containing enargite minerals and Cu-pyrite deposits, and interestingly [31,32].

Table 3. Concentration of As (µg/g) in river sediment sample from ASGM and control area.

Location	Element	Median	Min–Max
Buladu River (Sumalata Mining)	Sand	315.1	0–587
	Clay	0	0–18,900
Dunggilata River (Dunggilata Mining)	Sand	9.2	0–11.9
	Clay	0	0–449
Totopo River (Bumela Mining)	Sand	46.3	0–88.5
	Clay	11.3	0–162
Bone River (Suwawa Mining)	Sand	108	15.9–798
	Clay	0	0–13,800
Ayidu River (Control area)	Sand	0.8	0–6.7
	Clay	0	0–729

3.4. Concentration of Zn in River Sediments at ASGM Area and Control Area

The Zn content originated from mining (ASGM) sites and the control area, with rivers flowing from upstream to downstream. Based on of results in Table 4, the median concentration of Zn (µg/g) in sand sediment samples ranged from 60.4 to 187.5 µg/g, and clay sediment samples ranged from 3.2 to 2562.1 µg/g. In addition, zinc also an element of moderate abundance in the Earth's crust. Surface sulfides liberate the soluble Zn^{2+} ion, which may form secondary carbonate and silicate minerals. Zinc sulfides are closely associated with Cd and As, and the extraction of metallic zinc is in some cases responsible for environmental increases in cadmium, copper, nickel, lead, and other heavy metals [33].

Table 4. Concentration of Zn (µg/g) in river sediment sample from ASGM and control area.

Location	Element	Median	Min–Max
Buladu River (Sumalata Mining)	Sand	187.5	3.2–673.7
	Clay	2562.1	0–4923
Dunggilata River (Dunggilata Mining)	Sand	60.4	0–138.1
	Clay	210.1	0–331
Totopo River (Bumela Mining)	Sand	114.7	1.5–234.6
	Clay	3.2	0–153.5
Bone River (Suwawa Mining)	Sand	175.1	0–282.4
	Clay	79.4	0–79.4
Ayidu River (Control area)	Sand	90.8	0–102
	Clay	549.2	0–549.2

3.5. I_{geo} in Heavy Metal Concentration

I_{geo} was originally defined by Muller (1979) to determine and define metal contamination in sediments by comparing current metal concentrations with pre-industrial levels [34]. This method was used to determine levels of contamination or accumulation of heavy metals in sediments or soil (Table 5).

$$I_{geo} = \log_2(C_i / 1.5\ B_i)$$

C_i: The measured concentration of the examined metal (n) in sediment;
B_i: Geochemical background concentration of the metal (n).

Based on the Muller scale (1981), Hg I_{geo} values in the sand in study areas (Figure 2) belonged to $I_{geo} \leq 0$ in Class 0 (46.1%), $0 < I_{geo} < 1$ in Class 1 (15.3%), $1 < I_{geo} < 2$ in Class 2 (11.5%), $2 < I_{geo} < 3$ in Class 3 (19.2%), and $3 < I_{geo} < 4$ in Class 4 (7.7%), reflecting an unpolluted to moderately polluted sand. They were classified as moderately polluted to strongly polluted sand (or I_{geo} values -0.7 to 3.1). Meanwhile, considering I_{geo}, most clays (Figure 2) were included in Class 0 ($I_{geo} \leq 0$) and Class 1 ($0 < I_{geo} \leq 1$), i.e., unpolluted to moderately polluted clay, accounting for 44.4% and 11.1% of samples, respectively, and only clay samples in Class 6 ($I_{geo} > 5$) accounting for 44.4% of the samples. I_{geo} values in the sand in study areas (Figure 3) were from $I_{geo} \leq 0$ in Class 0 (50%), $0 < I_{geo} < 1$ in Class 1 (15.3%), $1 < I_{geo} < 2$ in Class 2 (11.5%), $2 < I_{geo} < 3$ in Class 3 (15.3%), $3 < I_{geo} < 4$ in Class 4 (3.8%), and $I_{geo} > 5$ in Class 6 (3.8%), reflecting an unpolluted to moderately polluted sand, from moderately polluted to strongly polluted sand, and from strongly polluted to extremely polluted sand in the environment (or I_{geo} values -5.9 to 4.1). However, considering the I_{geo}, most clays (Figure 3) were included in Class 0 ($I_{geo} \leq 0$) and Class 1 ($0 < I_{geo} \leq 1$), i.e., unpolluted to moderately polluted, with 55.5% and 11.1% of samples, respectively, and only some clays in Class 6 ($I_{geo} > 5$) with 33.3% of the samples. Pb I_{geo} values on the sand in study areas (Figure 4) were from $I_{geo} \leq 0$ in Class 0 (26.9%), $0 < I_{geo} < 1$ in Class 1 (23%), $1 < I_{geo} < 2$ in Class 2 (23%), $2 < I_{geo} < 3$ in Class 3 (19.2%), and $3 < I_{geo} < 4$ in Class 4 (7.6%), reflecting an unpolluted to moderately polluted sand and from moderately polluted to strongly polluted sand in the environment (or I_{geo} values -3.6 to 3.1). In contrast, considering I_{geo}, most of the clay samples (Figure 4) were included in Class 0 ($I_{geo} \leq 0$) and Class 1 ($0 < I_{geo} \leq 1$), that is, unpolluted to moderately polluted clay, accounting for 44.4% and 11.1% of samples, respectively, and only some clays in Class 6 ($I_{geo} > 5$) accounted for 44.4% of the samples. Pb I_{geo} values on the sand in the study areas (Figure 4) were from $I_{geo} \leq 0$ in Class 0 (26.9%), $0 < I_{geo} < 1$ in Class 1 (23%), $1 < I_{geo} < 2$ in Class 2 (23%), $2 < I_{geo} < 3$ in Class 3 (19.2%), and $3 < I_{geo} < 4$ in Class 4 (7.6%), reflecting an unpolluted to moderately polluted sand, and from moderately polluted to strongly polluted sand in the environment (or I_{geo} values of -3.6–3.1). Subsequently, considering I_{geo}, most clay samples (Figure 4) were included in Class 0 ($I_{geo} \leq 0$) and Class 1 ($0 < I_{geo} \leq 1$), i.e., unpolluted to moderately polluted clay, accounting for 44.4% and 11.1% of samples, respectively, and only some clays in Class 6 ($I_{geo} > 5$) accounted for 44.4% of samples [35,36].

Zn I_{geo} values for the sand in study areas (Figure 5) were from $I_{geo} \leq 0$ in Class 0 (26.9%), $0 < I_{geo} < 1$ in Class 1 (7.6%), $1 < I_{geo} < 2$ in Class 2 (38.4%), $2 < I_{geo} < 3$ in Class 3 (23%), and $3 < I_{geo} < 4$ in Class 4 (3.8%), reflecting an unpolluted to moderately polluted sand, and from moderately polluted to strongly polluted sand in the environment (or I_{geo} values of -4.9–3.9). Subsequently, considering I_{geo}, most clay samples (Figure 5) were included in Class 0 ($I_{geo} \leq 0$) and Class 1 ($0 < I_{geo} \leq 1$), i.e., unpolluted to moderately polluted clay, accounting for 55.5% and 11.1% of samples, respectively, and only some clay samples in Class 6 ($I_{geo} > 5$) accounted for 33.3% of samples.

Based on the results, distribution of heavy metals in the research areas varied from upstream to downstream. Generally, content of heavy metals was relatively higher in upstream areas than downstream areas [37].

Table 5. I_{geo} by Muller's classification for geochemical index [34,38–43].

I_{geo} Value	Class	Quality of Sediment
≤ 0	0	Unpolluted
0–1	1	From unpolluted to moderately polluted
1–2	2	Moderately polluted
2–3	3	From moderately to strongly polluted
3–4	4	Strongly polluted
4–5	5	From strongly to extremely polluted
>5	6	Extremely polluted

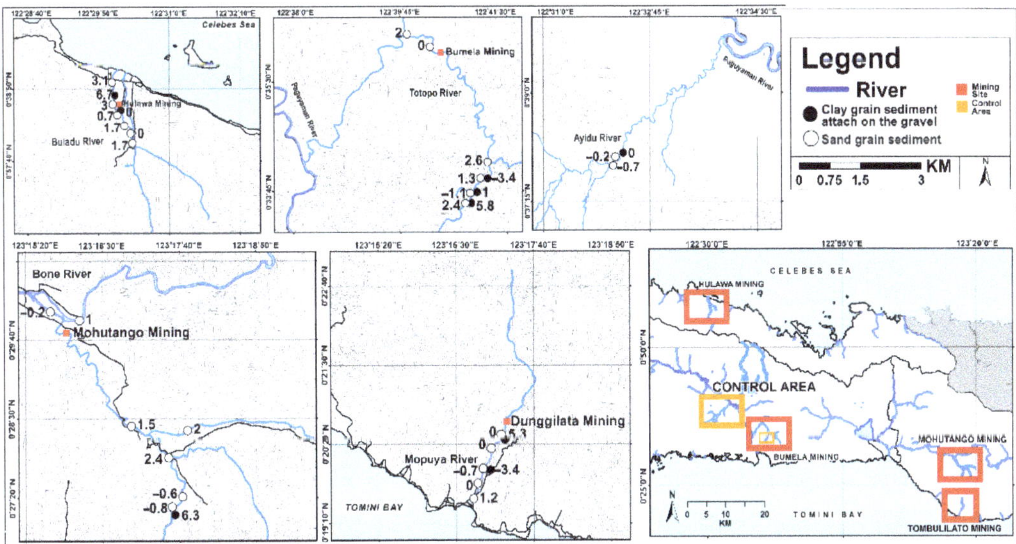

Figure 2. Plots of I$_{geo}$ and contaminant levels in Hg along the river sediments in Gorontalo Province, Indonesia (µg/g).

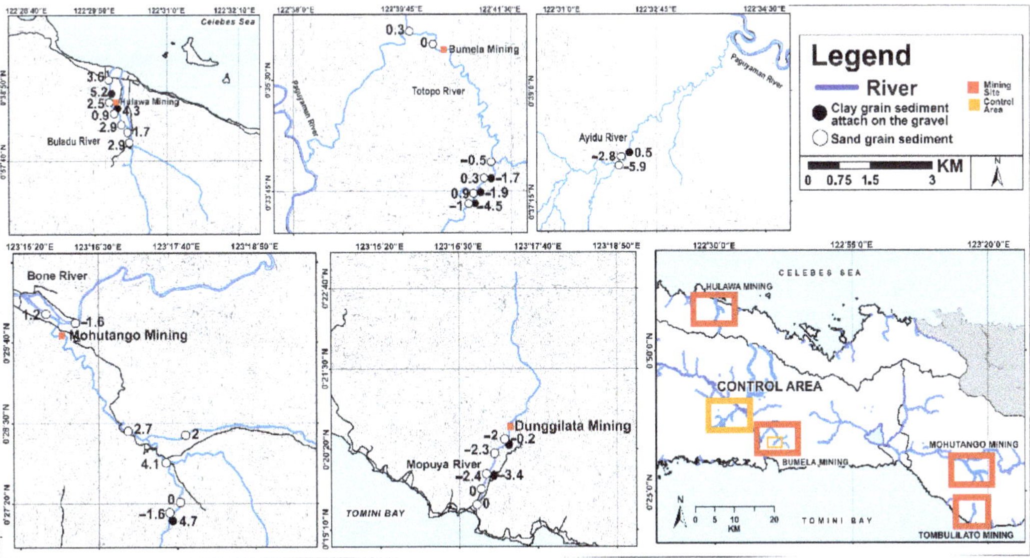

Figure 3. Plots of I$_{geo}$ and contaminant levels in As along the river sediments in Gorontalo Province, Indonesia (µg/g).

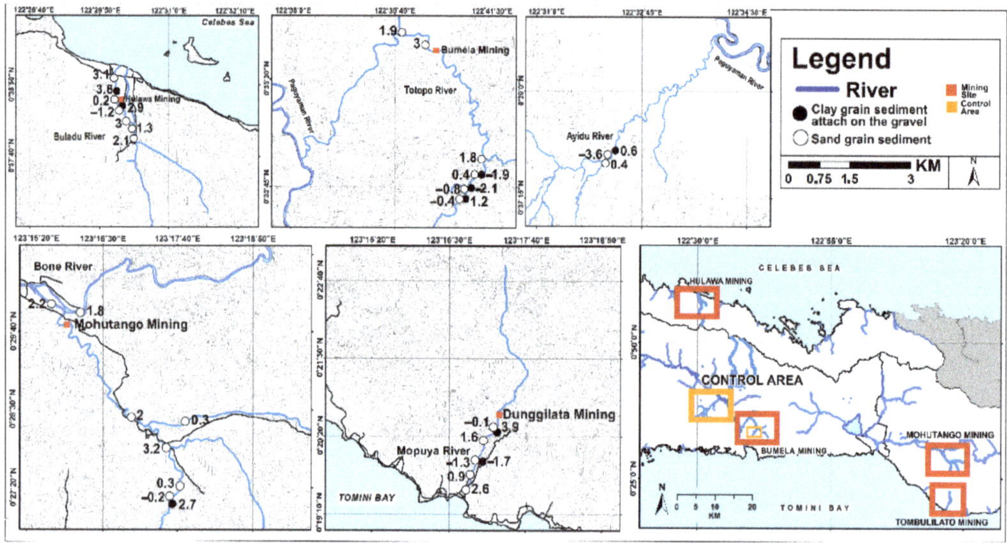

Figure 4. Plots of I_{geo} and contaminant levels in Pb along the river sediments in Gorontalo Province, Indonesia (μg/g).

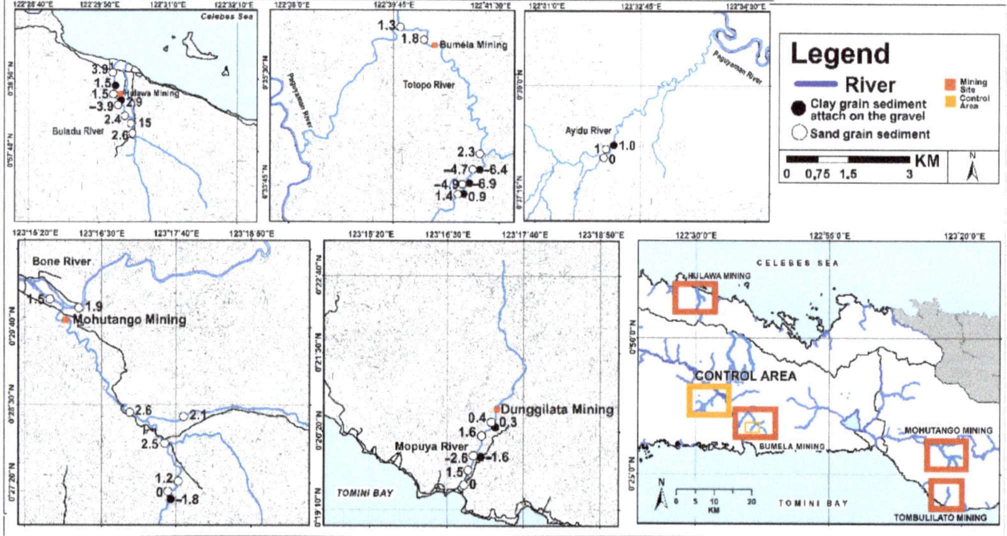

Figure 5. Plots of I_{geo} and contaminant levels in Zn along the river sediments in Gorontalo Province, Indonesia (μg/g).

3.6. Correlation Data of Each Element

Hg and As elements are positively correlated with a correlation coefficient of 0.82 (82%), represented by graph C of Figure 6, whereas the others, namely are graph A where Pb correlated with As with a correlation coefficient of 0.58 (58%) and graph B where Pb correlated with Hg with a correlation coefficient of 0.66 (66%) in Figure 6, are categorized into moderately correlated. This occurred because the community tends to add excessive amounts of Hg to strengthen the bound gold during the gold mining process. In contrast, addition of excess Hg causes tailing waste in the environment to be higher [44,45].

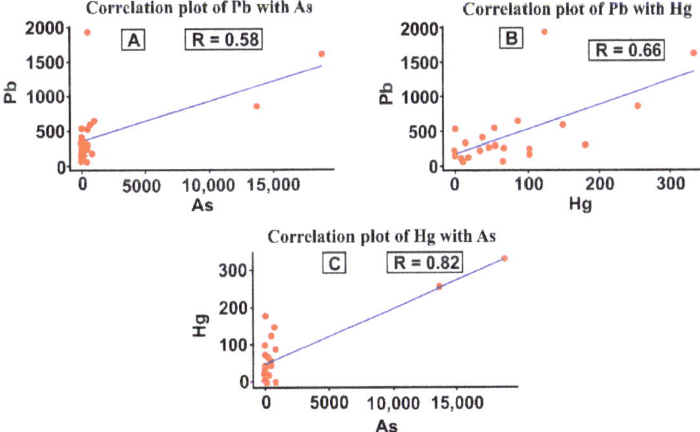

Figure 6. (**A**) Correlation between Pb and As; (**B**) Correlation between Pb, and Hg; (**C**) Correlation between Hg, and As in Gorontalo Area.

4. Discussion

4.1. Sources of Heavy Metals in the River Sediment of Gorontalo Province

Considering I_{geo}, some sediments were classified as Class 6 ($I_{geo} > 5$), indicating that they are extremely polluted, and only a few sediments were classified as Class 0 ($I_{geo} \leq 0$), Class 4 ($3 < I_{geo} < 4$), and Class 5 ($4 < I_{geo} < 5$), indicating they ranged from unpolluted to strongly polluted and from strongly polluted to extremely polluted in the environment polluted by Hg, Pb, As, and Zn.

Hg, Pb, As, and Zn elements contribute to pollution in air, water, sediments, and soils. Their contributions would negatively impact the environment on Hg contributions. It is an untypical heavy metal, as it can completely contaminate an environment, which occurs during gold mining, in which Hg can be released into the air and can also be widely distributed through tailings, especially on ASGM sites. Meanwhile, Pb and As can only be found in other heavy metals found in water, sediment, and soil [34,36,37,43,46].

4.2. The Effect of Heavy Metal Pollution on Sediments

We measured the concentration of heavy metal pollution in sediment samples collected from two field locations (ASGM area and control area) along the river in Gorontalo Province, Indonesia (Table 6). The ASGM area comprises five rivers that are still active with human activities in gold mining sites, and it also has many trommel houses and six gold mining sites. The control area is located along a highway with no mining activity. The river basin was the determining factor in changing the lower degree of chemical weathering as well as the climate from dry and cold to slightly warmer and more humid. Low chemical weathering could have influenced the lower degree of ion exchange from rock weathering to river solute in the watersheds of Gorontalo Province, Indonesia [20–22]. However, relative rivers imply that warm and humid climate conditions prevailed in their watersheds. Thus, it releases a significant number of major ions from silicate weathering into rivers [44].

Table 6. I_{geo} by Using Average Data in Control Area (Ayidu River and Totopo River) [19–21].

Classification of River Sediment Quality Standard	Element Concentration Limit			
	As	Pb	Hg	Zn
Clay Grain Sediment Attach on Gravel	342.3	88.5	2.1	184.9
Sand Grain Sediment	32	45.7	11.1	30.8
Average Data in Control Area (Ayidu River and Totopo River)				

4.3. Impact of ASGM on/for River Sediment

People in Gorontalo Province have been mining for gold since 1970. They used the traditional method, where the processing process does not use high technology and uses very simple equipment instead. Gold processing is conducted in several stages, such as rock excavation, processing unit, and waste disposal. Each stage of this process has an ecological impact that can disturb the environmental balance; hence, it must be handled carefully to reduce the risk of environmental damage. In general, almost every mining disposal area in Gorontalo Province is a land from which waste flows into the river, so metals and other materials accumulate in the waste, influencing the river ecosystem. As an ecosystem, rivers are targets for waste disposal with high levels of pollution affecting the life of the aquatic biota. It has been argued that water is often polluted by inorganic components, including dangerous heavy metals [36].

The use of these heavy metals in everyday life (directly or indirectly) intentionally, unintentionally, or intentionally but indirectly, has polluted the environment, where certain types of pollution have polluted the environment beyond the threshold for human life. Pollutants such as Hg are metals accumulated in the body of an organism and will stay in the body for a long time as a poison [31].

Traditional gold mining is one of the economic activities in the community from which miners earn a living. This activity has been a source of pollution to the surrounding environment due to the use of simple technologies such as using Hg as a binder of the gold element in the amalgamation process. Pollution occurs when Hg used as a binding agent for the gold element is discharged with washing wastewater to disposal sites both on the ground and in river water [22].

As a result, these traditional mines (ASGM) with various activities causing transportation of substances in sediments starting upstream is harmful when heavy metals are accumulated downstream because they can pollute the environment and endanger humans in the area.

Therefore, the source of mercury pollution in gold processing in Gorontalo Province for this study occurred from gold ore processing activities with the amalgamation process. In the amalgamation process, gold is separated from the binder where the gold ore that has been in the form of fine granules is an amalgamation process. Mercury will automatically bind to gold. Tailings or mining wastes from the amalgamation process, which contain a great deal of mercury, are directly discharged into the environment (rivers) without being processed first, so it is very possible to cause water pollution, especially in rivers at the gold mining location. This is what causes mercury (Hg) to have a strong relationship between leads (Pb) and arsenic (As) [33].

In addition, based on BLHRD Gorontalo Province in 2015 and 2018, the community around the river of ASGM site directly disposes of household waste and gold processing into the river. As a result, there is a decrease in river water quality because biochemical oxygen demand (BOD) levels become increased in that condition [47].

4.4. Heavy Metal Interaction (Hg, Pb, and As) Release in the Environment from Source of ASGM Site to River Sediment in this Study Area

As seen in Figure 6A–C, the element that has the strongest relationship is Hg with As, which is 0.82 (82%). Then, the second is the Hg element with Pb of 0.66 (66%), while the weakest relationship level is found in the Pb element with As of 0.58 (58%). This is based on that in the gold mining process, as the community tends to add excessive amounts of Hg with the intention to bind more to gold. On the other hand, the addition of excess Hg causes the amount of Hg that is wasted as pollution into the environment to also be higher if the gold content contained in the ore is not as large as expected.

Mercury is a heavy metal element that is widely used by humans in industrial or mining processes. In nature, mercury is a metal that can be found in two forms: organic and inorganic. Organic mercury is a mercury compound that binds to the carbon element to form methyl, ethyl, or similar functional groups, while inorganic mercury includes mercury vapor (Hg_0), mercury salt (Hg^{2+}), and mercury metal. Organic mercury is a geochemical form of mercury that has the highest level of toxicity to humans [4]. The traditional gold mining process is the largest anthropogenic contributor to mercury pollution, and one of the gold ore extraction methods widely used by traditional gold mining is amalgamation [20,21]. In the gold mining process, people tend to add excess mercury so that there are larger amounts of bound gold. The addition of excess mercury causes the mercury that is wasted as pollution in the environment to be higher. The gold content contained in the ore is often not as large as the community expects, so an experiment in adding mercury is expected with maximum results [9].

In addition, mercury pollution also occurs in the tailings process. Tailings are technically defined as fine materials that are minerals that are left after valuable minerals are extracted in an ore processing [10].

Tailings are also considered to be pulp that no longer contains valuable minerals, but given that processing costs do not reach 100%, it is still possible to have gold in the tailings [18]. Gold mining tailings contain one or more toxic hazardous materials, such as arsenic (As), cadmium (Cd), lead (Pb), mercury (Hg), and cyanide (Sn). Tailings consist of crushed rock derived from mineral rock that has been mined of minerals. Tailings can be in the form of solids such as very fine sand or slurry, namely solid tailings mixed with water to form a thin layer [1,18].

In general, tailings disposal is carried out in a terrestrial environment that is in a topographic depression or artificial reservoir, river or lake, and sea. Tailings often contain valuable mineral concentrations that do not meet the requirements for collection at the time of mining but are stored for future use. Mineralogically tailings can consist of various minerals such as silica, iron silicate, magnesium, sodium, potassium, and sulfide. Of these minerals, sulfides have chemically active properties, and when in contact with air, they will experience oxidation to form acidic salts and acidic streams containing a number of toxic metals, such as, Hg, Pb, and Cd [11].

Therefore, the source of mercury pollution in gold processing in Gorontalo Province for this study occurred from gold ore processing activities with the amalgamation process. In the amalgamation process, gold is separated from the binder where the gold ore that has been in the form of fine granules. Mercury will automatically bind to gold. Tailings or mining wastes from the amalgamation process, which contain a great deal of mercury, are directly discharged into the environment (rivers) without being processed first, so it is very possible that this causes water pollution, especially in rivers at the gold mining location. This is what causes mercury (Hg) to have a strong relationship between leads (Pb) and arsenic (As) [19–21].

5. Conclusions

In general, gold mining in Gorontalo Province is one of the potential natural resources that provide better prospects in an improved-level economy and social welfare. This economic improvement is mainly in terms of income, employment, and opportunities for new activities outside the agriculture sector plantation [12].

Despite the negative side, artisanal mining plays an essential role in developing societies. Small mines can be a major source of revenue for rural communities and can provide income for investment. Artisanal miners can exploit mineral deposits considered uneconomic by modern industry. Every USD 1 generated through artisanal mining generates about USD 3 in non-mining jobs. In Indonesia, artisanal mining is very useful as a means of livelihood for poor people and has been proven as a safety net in time of economic distress, especially during the economic crisis occurred in 1998, which lasted for about 10 years [15].

Moreover, on the basis of the interaction correlation analysis between sand and clay through a computerized analysis statistic, Pb and As, Pb and Hg, and Hg and As had a positive correlation with correlation coefficients of 0.58, 0.66, and 0.82, respectively. In addition, most ASGM sites in Gorontalo have hydrothermal processes that influence the geological setting, making them important sources of As in the environment. In addition, I_{geo} proposed by Muller in 1969 was used to evaluate the degree of heavy metal pollution in the river sediments [38,43]. It revealed that sediments along the river basin were severely polluted by heavy metals, and their estimated I_{geo} values were in the following increasing order: Hg < Zn < Pb < As. Regarding the Muller scale in 1981, I_{geo} was evaluated for all sampling sites of all five rivers in the Gorontalo Province. Of the sampling sites, 90% had I_{geo} greater than 5, which means they should be considered as extremely polluted. However, heavy metal values analyzed by using the I_{geo} indicator in Gorontalo Province were relatively high, indicating severe pollution by Hg, Pb, As, and Pb.

Furthermore, river sediments collected along river mining sites are contaminated with As, which is the highest among other heavy metals found in Gorontalo compared to any region in Indonesia or any country in the world.

Author Contributions: All authors contributed extensively to the work presented in the paper, as the principal researcher conducted the analysis and wrote the manuscript, which was undertaken in association with a Master Program; the Supervisor supervised the analyses and edited the manuscript; K.S. provided the PIXE measurement of sediment samples. All authors have read and agreed to the published version of the manuscript.

Funding: This research was supported by Research Institute for Humanity and Nature (RIHN: a constituent member of NIHU) Project No. RIHN 14200102.

Institutional Review Board Statement: Not applicable.

Informed Consent Statement: Not applicable.

Data Availability Statement: Not applicable.

Acknowledgments: The authors give special thanks to Research Institute for Humanity and Nature Gorontalo Province Government, such as Bone Bolango, Gorontalo Regencies, and North Gorontalo Government and their stakeholders. Thank you to all teams (Adhy, Rezki, and Alifia) for your support and effort to help us in the field. My sincere gratitude to the GP2KL community for your participation as a local communicator.

Conflicts of Interest: The authors declare no conflict of interest.

References

1. Miserendino, R.A.; Bergquist, B.A.; Adler, S.E.; Guimarães, J.R.D.; Lees, P.S.J.; Niquen, W.; Veiga, M.M. Challenges to measuring, monitoring, and addressing the cumulative impacts of artisanal and small-scale gold mining in Ecuador. *Resour. Policy* **2013**, *38*, 713–722. [CrossRef]
2. United Nations Environmental Protection/Global Program of Action. *Why The Marine Environment Needs Protection from Heavy Metals*; United Nations Environmental Protection/Global Program of Action: Nairobi, Kenya, 2004.
3. Unido & University. *A UNEP Global Mercury Partnership Document Produced in Conjunction with the Artisanal Gold Council and with Assistance*; Commission on Geoscience for Environmental Management: Vienna, Austria; Iugs-Gem: Wina, Austria, 2012.
4. Peplow, D. Environmental Impacts of Mining in Eastern Washington. In *Center for Water and Watershed Studies Fact Sheet*; University of Washington: Seattle, DC, USA, 1999.

5. Lenntech, K. *Water Treatment and Air Purification*; Lenntech: Rotterdamseweg, The Netherlands, 2004.
6. Habashi, F. Environmental Issues in the Metallurgical Industry–Progress and Problems. In *Environmental Issues and Waste Management in Energy and Mineral Production*; Balkama: Rotherdam, The Netherlands, 1992; Volume 2, pp. 1143–1153.
7. Graham Sustainable Institute. An Integrated Assessment of Artisanal and Small-Scale Gold Mining in Ghana: Final Report. *Ghana Mich. Gold Min. Integr. Assess.* **2015**, *12*, 5143–5176.
8. Aspinall, C. Small-scale mining in Indonesia. *Min. Miner. Sustain. Dev.* **2001**, *79*, 30. Available online: http://pubs.iied.org/pdfs/G00725.pdf (accessed on 24 May 2020).
9. Puluhulawa, F.; Harun, A.A. Policy formalization of Artisanal and Small-Scale Gold Mining (ASGM) post-ratification of Minamata Convention for Sustainability (case study of ASGM Gorontalo). *E3S Web Conf.* **2019**, *125*, 2006. [CrossRef]
10. Krisnayanti, B.D. ASGM status in West Nusa Tenggara Province, Indonesia. *J. Degrad. Min. Lands Manag.* **2018**, *5*, 1077–1084. [CrossRef]
11. Bruno, D.E.; Ruban, D.A.; Tiess, G.; Pirrone, N.; Perrotta, P.; Mikhailenko, A.V.; Yashalova, N.N. Artisanal and small-scale gold mining, meandering tropical rivers, and geological heritage: Evidence from Brazil and Indonesia. *Sci. Total Environ.* **2020**, *715*, 136907. [CrossRef] [PubMed]
12. Mahmud, M. Model Sebaran Spasial Temporal Konsentrasi Merkuri Akibat Penambangan Emas Tradisional Sebagai Dasar Monitoring Dan Evaluasi Pencemaran Di Ekosistem Sungai Tulabolo Provinsi Gorontalo. *Unspecified Thesis*. 2005. Volume 15. Available online: https://repository.ugm.ac.id (accessed on 18 August 2021).
13. Basri; Sakakibara, M.; Sera, K.; Kurniawan, I.A. Mercury contamination of cattle in artisanal and small-scale gold mining in bombana, Southeast Sulawesi, Indonesia. *Geosciences* **2017**, *7*, 133. [CrossRef]
14. Siswanto, B.; Krisnayani, B.D.; Utomo, W.H.; Anderson, C.W.N. Rehabilitation of artisanal gold mining land in West Lombok, Indonesia: Characterization of overburden and the surrounding soils. *Geology* **2012**, *4*, 1–7.
15. Limbong, D.; Kumampung, J.; Rimper, J.; Arai, T.; Miyazaki, N. Emissions and environmental implications of mercury from artisanal gold mining in north Sulawesi, Indonesia. *Sci. Total Environ.* **2003**, *302*, 227–236. [CrossRef]
16. Yu, K.C.; Tsai, L.J.; Chen, S.H.; Ho, S.T. Chemical binding of heavy metals in anoxic river sediments. *Water Res.* **2001**, *35*, 4086–4094. [CrossRef]
17. Dauvalter, V.; Rognerud, S. Heavy metal pollution in sediments of the Pasvik River drainage. *Chemosphere* **2001**, *42*, 9–18. [CrossRef]
18. Milenkovic, N.; Damjanovic, M.; Ristic, M. Study of heavy metal pollution in sediments from the Iron Gate (Danube River), Serbia and Montenegro. *Pol. J. Environ. Stud.* **2005**, *14*, 781–787.
19. Gafur, N.A.; Sakakibara, M.; Sano, S.; Sera, K. A case study of heavy metal pollution in water of Bone River by Artisanal Small-Scale Gold Mine Activities in Eastern Part of Gorontalo, Indonesia. *Water* **2018**, *10*, 1507. [CrossRef]
20. Arifin, Y.I.; Sakakibara, M.; Sera, K. Impacts of artisanal and small-scale gold mining (ASGM) on environment and human health of Gorontalo utara regency, Gorontalo Province, Indonesia. *Geosciences* **2015**, *5*, 160–176. [CrossRef]
21. Arifin, Y.I.; Sakakibara, M.; Sera, K.; Arifin, Y.I.; Sakakibara, M.; Sera, K. Arsenic, lead, and mercury concentrations of scalp hairs in ASGM miners and inhabitants of Gorontalo Utara regency, Gorontalo province, Indonesia. *Tech. Rep.* **2014**, *71*, 133–138.
22. Turekian, K.K.; Haven, N.; Hans, K.; Universitat, W.M.; Der Karl, K. Distribution of the Elements in Some Major Units of the Earth's Crust. America. *Geol. Soc. Am. Bull.* **1961**, *72*, 175–192. [CrossRef]
23. Republic of Indonesia Government Regulation 38 Regarding of River Sediment; 2011. Available online: https://jdih.sumselprov.go.id (accessed on 14 August 2020).
24. USEPA. *Sediment Quality Triad Approach*; USEPA: New York, NY, USA, 1991.
25. National Institute of Standards & Technology. Certificate of Analysis Standard Reference Material®1643f. Available online: https://www-s.nist.gov/srmors/view_detail.cfm?srm=1643F (accessed on 17 August 2020).
26. Sera, K.; Yanagisawaa, T.; Tsunoda, H.; Futatsugawa, S.; Hatakeyama, S.; Saitoh, Y.; Suzuki, S. Bio-PIXE at the Takizawa facility (bio-PIXE with a baby cyclotron). *Int. J. PIXE* **1992**, *2*, 325–330. [CrossRef]
27. SPEX Certificate. Certificate of Reference Material XSTC-13. Available online: http://seishin-syoji.co.jp/products/standard/standard004.html (accessed on 16 August 2020).
28. National Institute of Standards & Technology. Certificate of Analysis Standard Reference Material®2782. Available online: https://www-s.nist.gov/srmors/view_detail.cfm?srm=2782 (accessed on 5 August 2020).
29. World Health Organization. *Guidelines for Drinking-Water Quality First Addendum to Third Edition*; WHO: Geneva, Switzerland, 2006; Volume I, ISBN 9241546964.
30. World Health Organization. WHO's Drinking Water Standards. Available online: http://www.lenntech.com/applications/drinking/standards/who-s-drinking-water-standards.htm (accessed on 3 May 2020).
31. Uddin, M.K. A review on the adsorption of heavy metals by clay minerals, with special focus on the past decade. *Chem. Eng. J.* **2017**, *308*, 438–462. [CrossRef]
32. Rautureau, M.; Gomes, C.D.S.F.; Liewig, N.; Katouzian-Safadi, M. *Clays and Health*; Springer: Cham, Switzerland, 2017. [CrossRef]
33. Sumber, P.; Geologi, D. Tinjauan terhadap tailing mengandung unsur pencemar Arsen (As), Merkuri (Hg), Timbal (Pb), dan Kadmium (Cd) dari sisa pengolahan bijih logam. *Indones. J. Geosci.* **2006**, *1*, 31–36. [CrossRef]
34. Fardiaz. *Water and Air Pollution*; Kanisius: Yogyakarta, Indonesia, 1992; Volume 58–59, pp. 363, 739.

35. Duncan, A.E.; de Vries, N.; Nyarko, K.B. Assessment of Heavy Metal Pollution in the Sediments of the River Pra and Its Tributaries. *Water Air Soil Pollut.* **2018**, *229*, 8. [CrossRef]
36. Mortatti, J.; de Moraes, G.M.; Probst, J.-L. Open Archive TOULOUSE Archive Ouverte (OATAO) Heavy metal distribution in recent sediments along the Tietê River basin. *Geochem. J.* **2011**, *46*, 13–19. [CrossRef]
37. Abdulqaderismaeel, W.; Kusag, A. Enrichment Factor and Geo-accumulation Index for Heavy Metals at Industrial Zone in Iraq. *IOSR J. Appl. Geol. Geophys. Ver. I* **2015**, *3*, 2321–2990.
38. Zhang, N. Advance of the research on heavy metals in soil plant system. *Adv. Environ. Sci.* **1999**, *7*, 30–33.
39. Peng, J.-F.; Song, Y.-H.; Yuan, P.; Cui, X.-Y.; Qiu, G.-L. The remediation of heavy metals contaminated sediment. *J. Hazard. Mater.* **2009**, *161*, 633–640. [CrossRef] [PubMed]
40. Lacatusu, R.; Cîtu, G.; Aston, J.; Lungu, M.; Lacatusu, A.R. Heavy metals soil pollution state in relation to potential future mining activities in the Roşia Montană area. *Carpathian J. Earth Environ. Sci.* **2009**, *4*, 39–50.
41. Papastergios, G.; Georgakopoulos, A.; Fernandez-Turiel, J.L.; Gimeno, D. A correlation study of major and trace elements in sediments of River Nestos, Northern Greece and comparison with other fluvial systems. In Proceedings of the 9th International Multidicsciplinary Scientific Geoconference and EXPO-Modern Management of Mine Producing, Geology and Environmental Protection, SGEM, Albena, Bulgaria, 14–19 June 2009; pp. 431–438.
42. Namieśnik, J.; Rabajczyk, A. The speciation and physico-chemical forms of metals in surface waters and sediments. *Chem. Speciat. Bioavailab.* **2010**, *22*, 1–24. [CrossRef]
43. Zhang, J.; Liu, C.L. Riverine composition and estuarine geochemistry of particulate metals in China-weathering features, anthropogenic impact and chemical fluxes. *Estuar. Coast. Shelf Sci.* **2002**, *54*, 1051–1070. [CrossRef]
44. Miler, M.; Gosar, M. Characteristics and potential environmental influences of mine waste in the closed Mežica Pb-Zn mine (Slovenia). *J. Geochem. Explor.* **2012**, *112*, 152–160. [CrossRef]
45. Balintova, M.; Holub, M.; Singovszka, E. Study of iron, copper and zinc removal from acidic solutions by sorption. *Chem. Eng. Trans.* **2012**, *28*, 175–180.
46. Hedenquist, J.W.; Lowenstern, J.B. The role of magmas in the formation of hydrothermal ore deposits-Nature.pdf. *Nature* **1994**, *370*, 519–527. [CrossRef]
47. Badan Lingkungan Hidup dan Riset Daerah Provinsi Gorontalo Tahun. 2015. Available online: https://jurnal.poligon.ac.id/index.php/JTII/article/view/297doi.org/10.30869/jtii.v1i1 (accessed on 13 April 2022).

International Journal of
**Environmental Research
and Public Health**

Article

Environmental Survey of the Distribution and Metal Contents of *Pteris vittata* in Arsenic–Lead–Mercury-Contaminated Gold Mining Areas along the Bone River in Gorontalo Province, Indonesia

Nurfitri Abdul Gafur [1,*], Masayuki Sakakibara [2,3,*], Satoru Komatsu [4], Sakae Sano [5] and Koichiro Sera [6]

[1] Regional Planning Agency—Research and Development of Bone Bolango, Bone Bolango 96562, Indonesia
[2] Faculty of Collaborative Regional Innovation, Ehime University, Matsuyama 790-8577, Japan
[3] Research Institute for Humanity and Nature, Kyoto 603-8047, Japan
[4] School of Global Humanities and Social Sciences, Nagasaki University, Nagasaki 852-8521, Japan; skomatsu@nagasaki-u.ac.jp
[5] Faculty of Education, Ehime University, Matsuyama 790-8577, Japan; sano.sakae.mm@ehime-u.ac.jp
[6] Cyclotron Research Center, Iwate Medical University, Takizawa 020-01673, Japan; ksera@iwate-med.ac.jp
* Correspondence: vivinurv3@gmail.com (N.A.G.); sakaki@chikyu.ac.jp (M.S.)

Abstract: In this paper, we report ecological and environmental investigations on *Pteris vittata* in the As–Pb–Hg-polluted Bone River area, Gorontalo Province, Indonesia. The density distribution of *P. vittata* decreases from around the artisanal and small-scale gold mining (ASGM) site to the lower reaches of the Bone River, and it is rarely found near Gorontalo City. The maximum concentrations of As, Hg, and Pb recorded in the soil samples were 401, 36, and 159 mg kg^{-1}, respectively, with their maximum concentrations in *P. vittata* recorded as 17,700, 5.2, and 39 mg kg^{-1}, respectively. Around the ASGM sites, the concentrations of As, Pb, and Hg in *P. vittata* were highest in the study area. These data suggest that *P. vittata*, a hyperaccumulator of As, may be useful as a bioindicator for assessing environmental pollution by Pb and Hg.

Keywords: artisanal and small-scale gold mining; *Pteris vittata*; absorption; arsenic; mercury; lead; Bone River; Gorontalo Province; Indonesia

Citation: Gafur, N.A.; Sakakibara, M.; Komatsu, S.; Sano, S.; Sera, K. Environmental Survey of the Distribution and Metal Contents of *Pteris vittata* in Arsenic–Lead–Mercury-Contaminated Gold Mining Areas along the Bone River in Gorontalo Province, Indonesia. *Int. J. Environ. Res. Public Health* **2022**, *19*, 530. https://doi.org/10.3390/ijerph19010530

Academic Editor: José Ángel Fernández

Received: 29 October 2021
Accepted: 1 January 2022
Published: 4 January 2022

Publisher's Note: MDPI stays neutral with regard to jurisdictional claims in published maps and institutional affiliations.

1. Introduction

Artisanal and small-scale gold mining (ASGM) is one of the largest emitters of mercury (Hg) into the environment in the world. In ASGM, miners use elemental Hg to extract gold (Au) from finely crushed rocks or Holocene sediments, forming an Hg–Au amalgam. This amalgam is then burned, releasing Hg as vapor and leaving impure Au residue. The adverse effects on the health of miners are severe, with exposure to Hg causing neurological damage and other health issues. The people residing near the ASGM sites are also affected by the Hg contamination of water, air, and soil and by the effects of Hg on ecosystems, such as fishes and plants. Approximately 15 million people in ASGM sites in over 70 countries use Hg to refine Au, causing Hg pollution [1].

In Indonesia, over 300,000 ASGM miners work at approximately 1000 informal sites. These miners are either displaced farmers or fishermen who have very few options in terms of alternative livelihoods. These rural communities depend on the ASGM for income. The practice of using Hg to extract Au was banned by the Indonesian government in 2014 [2,3]. However, ASGM mining has released hundreds of tons of Hg into the water, soil, and air. Thus, despite the passage of laws against the use of Hg in Indonesia, it has not been effective [4].

According to the data of the Ministry of Forestry and Mining of Gorontalo Province in 2012, Gorontalo Province has a large number of ASGM sites, especially in the Bone Bolango

Regency, where the largest number of ASGM workers work [5]. According to Arifin et al. (2015), Gafur et al. (2018), and Lihawa and Mahmud (2019), in Gorontalo Province, the Hg concentrations in the hair of inhabitants are higher than those of the inhabitants of a non-ASGM area; a fish had Hg levels above the threshold limit by WHO/ICPS 1990 [6], and those of river water and sediment are also polluted by Hg [7–9].

In recent years, although many researchers have focused on Hg in environmental assessment studies at ASGM sites, little attention has been given to the other toxic metals associated with Au deposit formation. In particular, Neogene epithermal gold deposits in the Circum-Pacific Rim [10] are known to be associated with various heavy metals, such as As, Pb, Cu, and Zn. This suggests that ASGM of epithermal Au deposits is accompanied by exposure to various naturally derived harmful heavy metals in addition to anthropogenic pollution by Hg.

Generally, higher plants find it extremely difficult to survive in environments that are highly contaminated with As. However, *Pteris vittata* is a fern species with a hyperaccumulation ability for As, and it can accumulate up to approximately 28 g/kg As in its fronds [11,12]. It can also be used in the study of the bioaccumulation and biomonitoring of other heavy metals. Phytoremediation using *P. vittata* has been recently developed as an environmentally friendly and cost-effective technology to remediate heavy metal-contaminated environments [13].

Pteris. vittata is known to be an accumulator of metals in laboratory experiments. For example, Fayiga et al. demonstrated the capability of *P. vittata* to hyperaccumulate arsenic from soils in the presence of Cd, Ni, Pb, and Zn in a greenhouse study [14]. The novelty of the research is that it examines whether *P. vittata* acts as an "indicator plant" in severely contaminated soils in ASGM and examines the potential to remove Pb, As, and Hg.

The potential of *P. vittata* to remediate contaminated soils in China has also been empirically examined [15–17]. For example, Lei et al. examined the remediation efficiency of *P. vittata* in As-contaminated soils through a 2-year field experiment [15]. Since previous studies indicate the potential of *P. vittata* as an accumulator of metals, limited studies have examined the role of *P. vittata* as an accumulator of metals in the case of ASGM locations. Soils of ASGM locations are usually polluted by historical mining operations, and the potential of bioaccumulators of *P. vittata* are influenced by the level of pollutants and climatic and geographical conditions. Further empirical studies are necessary in ASGM locations to examine the potential of accumulating metals from contaminated soils. Therefore, the present study yields important policy implications for achieving sustainable livelihoods through adopting environmentally friendly technologies.

In this study, we conducted a field survey of the distribution of the population density of *P. vittata* and analyzed the metal concentrations of the samples of soils and *P. vittata* in ASGM areas polluted by multiple metals, such as Hg, As, Zn, and Pb, around the Bone River, Gorontalo Province, Indonesia. We discussed the correlations between the distributions of *P. vittata* and the metal concentrations between the soils and plants.

2. Materials and Methods

Over the past decade, mine tailings and wastewater from the two ASGM sites located upstream of the Bone River and smelted slag from the sites along the valley have been dumped along the river, and they are mixed into the river water.

The Bone River is in the eastern part of Gorontalo, Indonesia, within the Bone Bolango Regency ($0.27°–1.01°$ N by $121.23°–122.44°$ E). The regency has an area of 1985 km^2, and altitudes can reach up to 1500 m. The two ASGM sites in this region are Motomboto and Mohutango. *P. vittata* and soil samples were collected along the Bone River from downstream to upstream close to the ASGM areas (Figure 1).

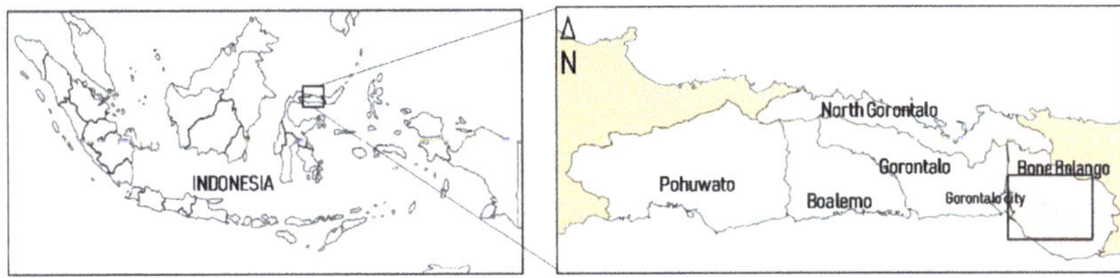

Figure 1. Location of Gorontalo in Indonesia (**left**) and map of Gorontalo (**right**) with the study area along the Bone River enclosed in a black rectangle.

2.1. Distribution of the Population Density of P. vittata and Sampling of the Plants and Soils

We mapped the distribution of the population density of *P. vittata* along the Bone River from the ASGM area to Gorontalo City. Plant population density was determined using the quadrat counts method [18] by counting the plants on the slopes along the Bone River; during this process, the locations were recorded by the global positioning system and mapped using ArcMap v. 10.3 (Esri Company, Redlands, CA, USA).

In all, 37 plant samples were collected from the field from upstream to downstream along the Bone River at appropriate intervals, and one whole plant was taken at each point. The plant tissue we used for analysis was the fronds. Soil samples ($n = 20$) were collected from depths of 10–20 cm from the *P. vittata* collection points. The soil samples were dried (120 °C for 48 h) with a ventilated oven ISUZU drying oven Model 2-2045 and stored in sterilized plastic bags prior to analysis. The dried soil samples were sieved using a 2 mm sieve for the removal of larger organic components, and the smaller components were removed using tweezers. The samples were ground to a powder in a planetary ball mill (Varian PM-2005 m, Osaka Chemical Co., Ltd., Osaka, Japan) and homogenized before analysis by particle-induced X-ray emission (PIXE) at the Cyclotron Research Center, Iwate Medical University, Japan. All the samples were imported into Japan under a permit issued by the Ministry of Agriculture, Forestry, and Fisheries, Japan, in accordance with the Plant Protection Law.

Our sampling procedures precisely estimated the correlation between contamination in the soil samples and uptake by *P. vittata* because the soil samples and *P. vittata* were simultaneously collected at the same location.

2.2. Chemical Analysis

For the PIXE analysis, the homogenized soil samples were mixed with Pd–C as an internal standard at a ratio of 3:1, set up onto a thin-film holder by using a collodion solution, and bombarded with a 2.9 MeV proton beam energy from a cyclotron [19]. PIXE is a valuable ion-beam analyzer that has proton beams of 1–4 MeV energy for the analysis of geological and biological samples.

The plant samples were washed with deionized water and dried in a ventilated oven at 80 °C for 48 h. The dried leaves were ground to a powder in a ball mill and prepared for analysis as follows: the sample powder (20 mg) was digested with 1 mL HNO_3 on a hotplate at 80–100 °C for 2 h and then cooled to room temperature; subsequently, deionized water was added to obtain the final volume required for analysis. The As, Hg, and Pb concentrations in the plant samples were determined by inductively coupled plasma–mass spectrometry (ICP–MS) performed using a Varian 820-MS instrument (Agilent Technologies, Santa Clara, CA, USA) at the Integrated Center for Sciences, Ehime University, Japan. The flowchart of the methodology is illustrated in Figure 2. Due to the characteristics of PIXE, analytical errors are potentially larger if the concentration of metals is less than the minimum threshold (less than 10 ppm). However, PIXE is suitable because sample preparation is easier, and a large number of samples can be examined for a relatively short duration.

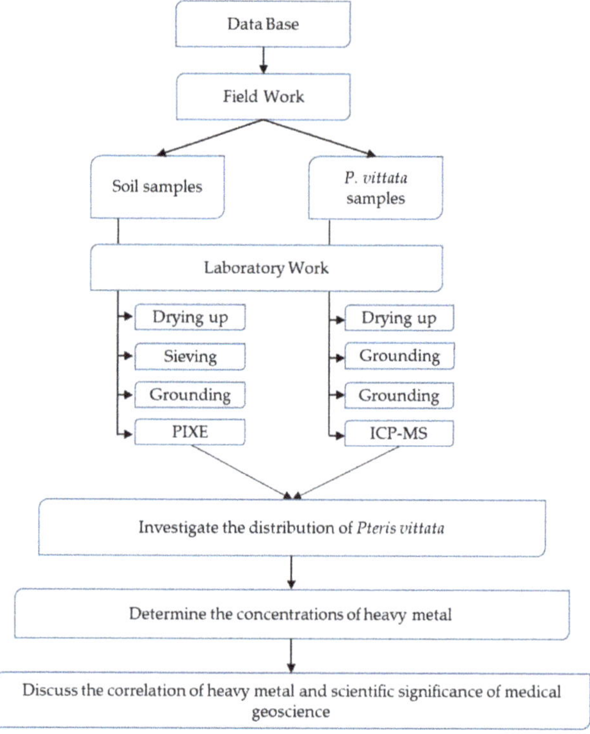

Figure 2. Flowchart of methodology.

2.3. Quality Control and Statistical Analysis

The US National Institute of Standard and Technology (NIST) CRM 1643f [20] and multielement standard XSTC-13 from SPEX CertiPrep (Metuchen, NJ, USA) [21] were used as the calibration standards, and the certified reference material NIST SRM 2782 [22] was used as the check standard for the As, Hg, and Pb analyses of the soil samples. The analytical accuracy and precision of the plant analyses were verified using National Institute for Environmental Studies (NIES) No.9 Sargasso [23].

The significant differences among the sampling sites were determined using Stata 16.1 (StataCorp, College Station, TX, USA) and OriginLab v. 9.60 program for Windows.

3. Results and Discussion

The distribution of the population density of *P. vittata* in the study area (shown in Figure 1) is shown in Figure 3. The densest overgrowth of *P. vittata* is observed along the sunny slopes and valleys around the Motomboto ASGM site (Figure 4). The plant reaches a maximum height of approximately 80 cm. The density of *P. vittata* decreases from the Motomboto and Mohutango ASGM sites to Gorontalo City. It is rarely found in the western Bone Bolango Regency. The plants form communities along rivers and on slopes in slightly humid environments. The distributions, each more than 50 stocks/m^2, are the highest around the Motomboto and Mohutango ASGM sites (Figure 4). The distribution becomes roughly lower toward the downstream side and becomes zero on the eastern side near Gorontalo City.

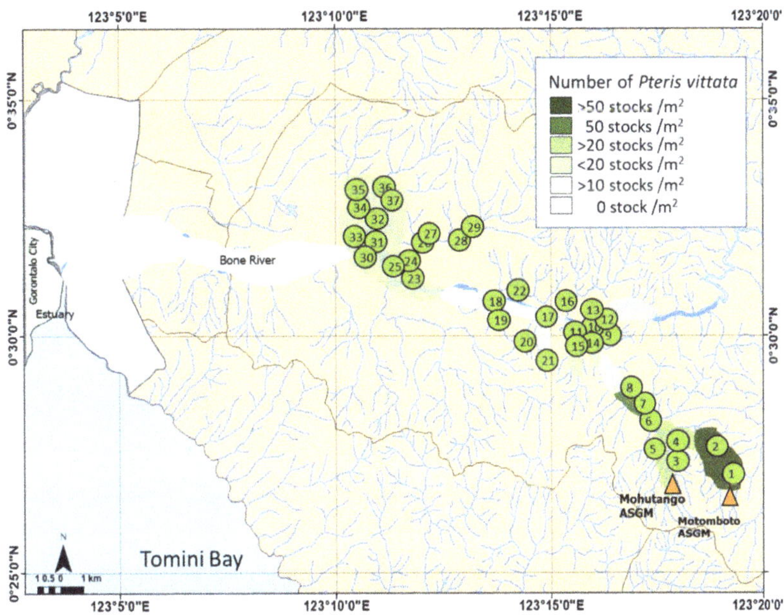

Figure 3. *P. vittata* distribution map and sampling point.

Figure 4. Mode of occurrence of *Pteris vittaae* around the Motomboto ASGM sites in Bone Bolango Regency, Gorontalo Province. (**a**) *P. vittata* growing in mine trommels; (**b**) *P. vittata* growing at a site contaminated with As–Pb at the Motomboto ASGM site.

3.1. As, Hg, and Pb Concentrations in Soil and P. vittata

The soil concentrations in this study and regulatory guidelines for As, Hg, and Pb as per the US Environmental Protection Agency (EPA) (1993) [24] are listed in Tables 1 and 2. Large differences were found in the As and Pb concentrations of the soil samples. These differences indicate that soil samples of the study area were relatively heterogeneous in terms of metal concentrations. The As and Pb concentrations were in the ranges of not detected (ND)–36,500 and 8–11,400 mg kg^{-1}, respectively. As and Pb concentrations exceed the criteria set by the US EPA 1993 (Table 2). The overall tendency is that the concentration of each metal decreases from the ASGM site toward the downstream side.

Table 1. Concentrations of As, Hg, and Pb in the soil samples along the Bone River.

No.	Elements					
	Concentration (µg/g)		Concentration (µg/g)		Concentration (µg/g)	
	As	Error	Hg	Error	Pb	Error
1	15.0	24.0	11.0	17.0	46.0	18.0
2	1.80	4.20	ND	ND	70.0	16.0
3	ND	ND	ND	ND	90.0	31.0
4	4060	149	ND	ND	1700	178
5	92.0	8.80	10.3	17.0	60.0	32.0
6	40.0	29.0	ND	ND	8.20	18.0
7	401	21.0	33.0	19.0	29.0	59.0
8	6.30	5.50	ND	ND	92.0	21.0
9	199	14.0	ND	ND	47.0	48.0
10	110	11.0	18.0	22.0	58.0	39.0
11	11.0	7.50	ND	ND	74.0	29.0
12	3.20	11.0	ND	ND	159	42.0
14	61.0	49.0	ND	ND	25.0	33.0
15	6600	298	ND	ND	75.0	340
16	25.0	7.42	ND	ND	87.0	29.0
17	112	14.0	26.0	22.0	106	51.0
18	56.0	14.0	ND	ND	69.0	51.0
19	1.30	4.20	16.0	15.0	63.0	16.0
20	295	13.0	ND	ND	32.0	36.0
21	16.0	4.00	ND	ND	40.0	15.0
22	12.0	11.0	ND	ND	146	44.0
23	29.0	35.0	ND	ND	130	26.0
24	6.90	6.60	36.0	21.0	86.0	25.0
25	6.90	8.20	ND	ND	63.0	32.0
26	221	19.0	ND	ND	31.0	64.0
27	36,500	1376	ND	ND	11,400	735
28	49.0	29.0	ND	ND	73.0	19.0
29	218	19.0	20.0	27.0	170	65.0

ND, not detected; error, analytical error. Investigations were performed three times at each location.

Table 2. As, Hg, and Pb concentrations in the soil samples taken along the Bone River in Gorontalo Province and the regulatory guidelines for As, Hg, and Pb according to the US EPA (1993).

Metal/Metalloid	Soil Concentration Range mg kg^{-1}	Soil Regulatory Limits * mg kg^{-1}
As	0–36,500	20
Hg	0–36	270
Pb	8–11,400	600

* US EPA 1993.

The concentrations of As, Pb, and Hg in the samples of *P. vittata* and soil samples are summarized in Tables 1 and 3, and their geographical variations are shown in Figure 5. Table 1 shows that there were 17 sampling points that were over the regulatory limit (Table 3) for the concentration of arsenic and two sampling points for the Pb concentration. This result indicates that the soils in this area were naturally polluted by As and Pb because of the weathering of the hydrothermally altered mother rocks. On the other hand, the maximum Hg concentration in the soil was 36 mg kg^{-1}, which was relatively low but tended to increase toward the ASGM sites.

Table 3. Concentrations of As, Hg, and Pb in the *P. vittata* samples taken from the Bone River area.

Sample Number	As Concentrations mg kg^{-1}	RSD	Hg Concentrations mg kg^{-1}	RSD	Pb Concentrations mg kg^{-1}	RSD
1	15,600	1.30	3.20	3.00	39.0	4.20
2	17,700	1.20	5.20	7.70	14.0	3.30
3	220	1.10	0.20	20.0	2.20	3.80
4	2810	5.50	0.90	5.60	22.0	4.80
5	570	1.60	0.40	4.10	0.30	1.20
6	5800	2.60	0.20	5.10	0.60	7.80
7	220	1.70	0.04	12.0	0.002	3.10
8	5800	0.50	0.30	21.0	4.10	2.20
9	14.0	0.90	0.01	10.0	0.002	1.40
10	32.0	18.0	0.20	11.0	0.90	1.50
11	2.50	2.10	0.60	5.00	0.10	3.00
12	570	1.60	0.40	4.00	3.60	1.20
13	77.0	0.40	0.45	7.00	0.50	1.60
14	32.0	0.70	0.02	24.0	0.03	5.00
15	690	1.50	0.80	3.00	0.90	1.10
16	160	0.90	0.20	4.00	0.53	1.50
17	2.80	1.10	0.01	17.0	0.02	1.20
18	140	1.30	0.04	7.10	ND	ND
19	39.0	0.90	0.14	19.0	0.90	1.70
20	69.0	0.80	ND	ND	0.70	2.50
21	39.0	2.00	ND	ND	0.04	1.10
22	71.0	3.10	0.10	8.20	0.30	2.40
23	3.10	0.70	0.01	7.20	0.004	3.10
24	8.30	2.00	0.01	6.50	ND	ND
25	47.0	1.90	0.00	0.79	4.80	26.0
26	1.30	1.33	0.02	14.2	0.03	1.30
27	130	0.90	ND	ND	0.70	0.90
28	95.0	3.10	0.20	5.89	3.80	1.80
29	8.80	0.80	0.10	6.87	0.20	8.00
30	5.40	0.90	0.01	15.5	ND	2.20
31	7.00	0.70	0.01	18.3	0.01	3.90
32	8.20	4.00	0.70	26.3	0.10	6.10
33	6.40	1.30	0.02	5.60	ND	2.40
34	95.0	0.90	0.20	3.90	3.82	2.00
35	34.0	1.20	ND	ND	0.10	17.0
36	153	0.80	ND	ND	0.90	1.40
37	2.10	1.20	0.01	15.0	ND	6.40

ND, not detected; RSD, relative standard deviation. Investigations were performed three times at each location.

The concentrations of As, Pb, and Hg in *P. vittata* were in the ranges of ND–17,700, ND–39 mg kg^{-1}, and ND–5.2, respectively. The ND data correspond to a situation in which the concentration is lower than the detection limit of the analytical instrument.

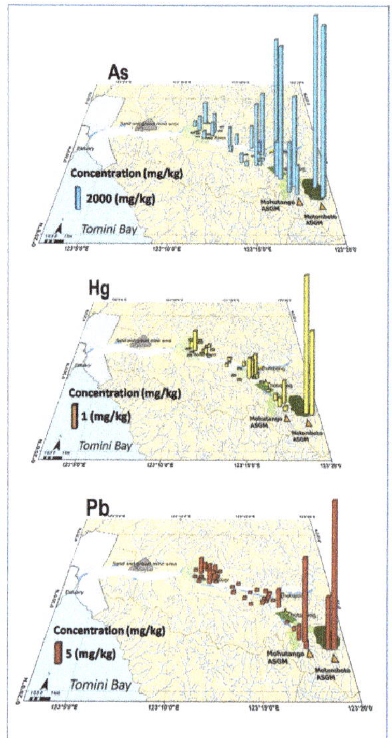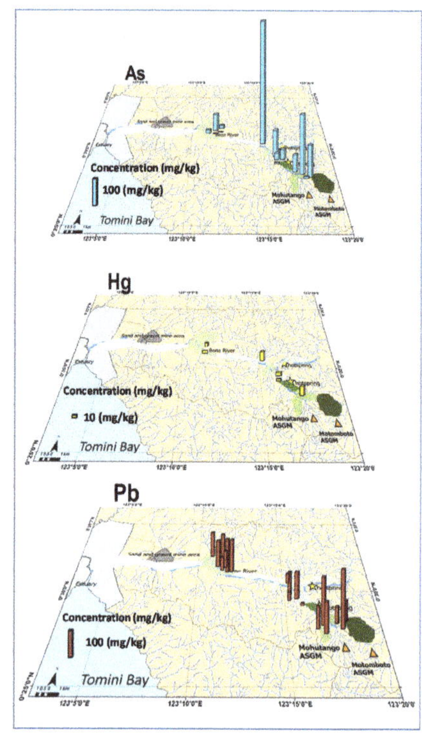

Figure 5. As, Hg, and Pb concentrations in *P. vittata* (**left**) and soil (**right**) samples were taken along the Bone River (mg kg^{-1}).

3.2. Relationship between the Distributions of the Population Density of P. vittata and the As Concentration in Soil

There is generally a positive correlation between the density distribution of *P. vittata* and the As concentration in soil. This correlation is also consistent with *P. vittata* being found in As-contaminated areas [25]. *P. vittata* grows actively in a neutral or slightly alkaline environment [11,25]. Although the pH of the soil samples was not measured in this study, it is likely that the soil pH is neutral or weakly alkaline at many sampling points.

3.3. Metal Pollution and Uptake by P. vittata along the Bone River

We found that *P. vittata* absorbed large amounts of As, reaching a maximum concentration of 17,700 mg kg^{-1} (Table 3). The rocks around the ASGM sites in Gorontalo are Pliocene altered andesites containing sulfide minerals, mainly pyrite. At present, most of the sulfide minerals in the altered andesitic rocks on the ground surface have been decomposed by weathering under tropical weather conditions. The As concentration in pyrite is not clearly known, but since weathered soil contains a certain amount of As, it is presumed that pyrite also contains a similar amount of As.

The results of this study show dispersion and no significant correlation between *P. vittata* and As concentration in the rhizosphere soil (Figure 6). The results of previous laboratory and field studies on *P. vittata* indicate that the removal efficiency of *P. vittata* depends on the ratio of soluble As and the pH range of the soil [12,13,25]. The results of our study suggest that the ratio of soluble As and pH of the soil in the study area may be diverse.

(a) Concentration of As and Hg in the *P. vittata* sample

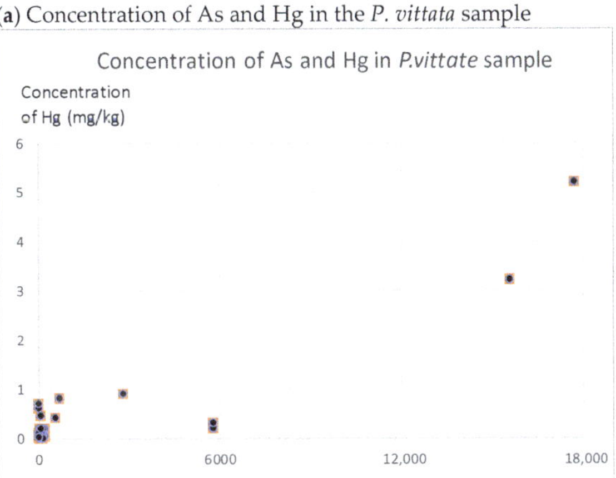

(b) Concentration of As and Pb in the *P. vittata* sample

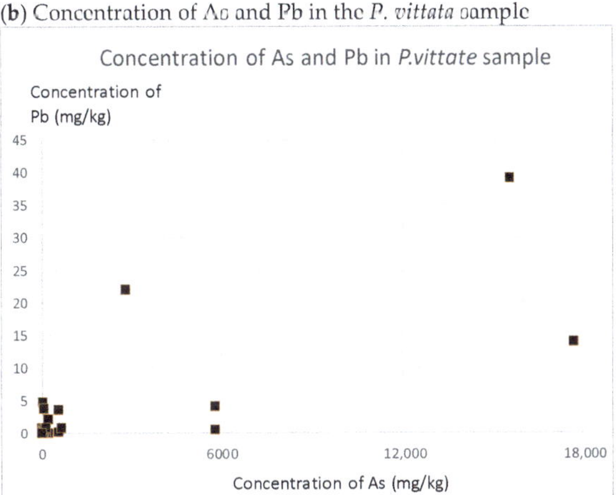

(c) Concentration of Hg and Pb in the *P. vittata* sample

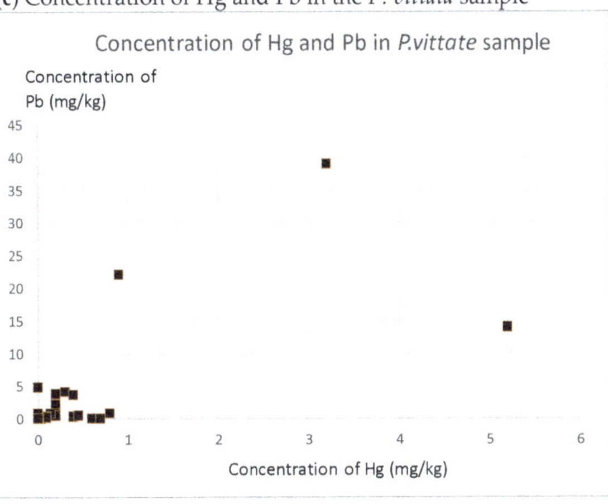

Figure 6. Scatter plots depicting the relationships between As, Hg, and Pb concentrations in *P. vittata* samples.

Generally, Hg accumulation in the soil is related to deposition caused by anthropogenic activities through the biosphere or atmosphere [26]. The ASGM activity includes techniques, such as crushing and mixing the raw materials with Hg, and the waste produced after amalgamation also contains a large amount of Hg [27–30]. We found that the Hg uptake by *P. vittata* in the mining areas was higher than that in the downstream area. The source of the high concentration of Hg in the ASGM area was mainly Au extraction activities. Hg could be transferred in forms such as methylmercury in the ecosystem [24] and was then absorbed by *P. vittata*. Based on the ability to absorb Hg, *P. vittata* is considered to have the potential for use as a bioindicator for Hg [31,32].

The maximum soil concentrations of Pb in this study area are higher than the limit prescribed by the US EPA for agricultural soil [33]. *P. vittata* had a maximum Pb concentration of 39 mg kg^{-1} that tended to be considerably higher around the ASGM sites. These results indicate the possibility of using this plant as a bioindicator for Pb in contaminated areas.

The correlations among As, Hg, and Pb concentrations in *P. vittata* are shown in Figure 6. Hg is of anthropogenic origin, whereas As and Pb are of natural origin. Despite their different origins, they generally show a positive correlation. According to the results of Spearman's rank correlation tests, a positive and statistically significant relationship was identified among As and Hg (10% level), As and Pb (1% level), and Hg and Pb (10% level). This finding suggests that Hg is methylated over time and exposed to the ecosystem.

The correlation between As, Hg, and Pb in *P. vittata* and soil samples is shown in Figure 7. With regard to the Hg concentration, the results of Spearman's rank correlation tests between the plant and soil samples indicate a negative and statistically significant relationship at the 5% level. This result suggests that the locations where the Hg contamination of soils is higher show lower accumulation in the plant samples. The results are uncertain because of the limited number of samples, and further investigation is needed to determine if Hg absorption saturation has been attained at the study sites.

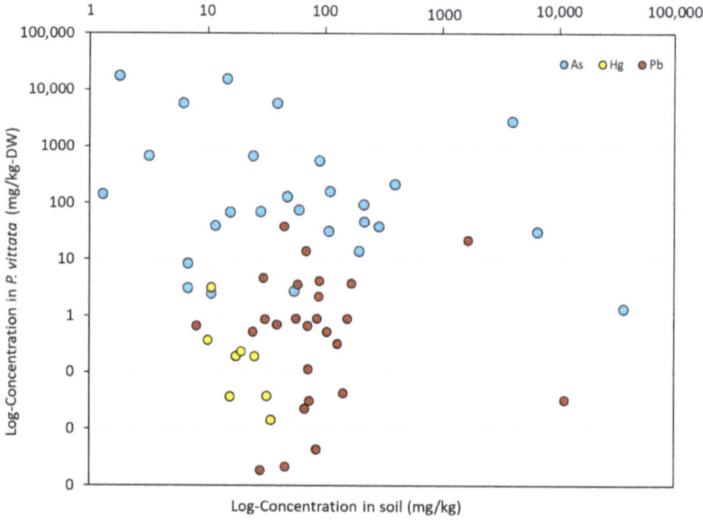

Figure 7. As, Hg, and Pb uptake intensity from the soil to shoot of *P. vittata*; DW: dry weight.

The findings raise concern for biodiversity in the local environment. Since *P. vittata* can adopt and survive in metal-contaminated soils, there is a concern that it could be an invasive species, which thereby influences local biodiversity. Rathinasabapathi discusses the competitive ability and invasiveness of several ferns [34]. As there is no empirical evidence about the invasive nature of *P. vittata* in the surveyed areas, further studies are required to understand the influences of expanding *P. vittata* for other local species.

The study did not examine the mobile/bioavailable fraction of metals that could potentially correlate with uptake by *P. vittata*. Examining the composition of mobile fractions extracted from environmental solid samples may provide a clearer explanation of the uptake. There are two reasons why the study did not examine the composition of mobile/bioavailable fractions. First, the current study provides a relationship between uptake and total metal contamination, even though mobile/bioavailable composition is not fully investigated. More total metal contamination is identified, and more plant uptake is empirically identified. Second, even mobile/bioavailable and solid fractions are mixed in the field, providing evidence for total metal contamination and plant uptake yields important implications for policy-makers, especially in developing countries, regarding whether contamination is prevalent. At this point, the research is considered a preliminary investigation, and further research is necessary. Examining the composition of mobile fractions from environmental solid samples to show clear characteristics between metal compositions and plant uptake is beneficial to elaborate on the findings of the research.

4. Conclusions

We conducted ecological and environmental investigations on *P. vittata* in the As–Pb–Hg-polluted Bone River area, Gorontalo Province, Indonesia. The density distribution of *P. vittata* decreases from around the ASGM sites to the lower reaches of the Bone River, and the plant is rarely found near Gorontalo City. The maximum concentrations of As, Pb, and Hg in *P. vittata* were 17,700, 5.2, and 39 mg kg^{-1}, respectively. Around the ASGM sites, the metal concentrations in *P. vittata* were the highest in the study area. These data suggest that *P. vittata*, which is a hyperaccumulator of As, may also be useful as a bioindicator for assessing the environmental pollution caused by Pb and Hg.

Since this research was limited to case studies in the ASGM regions of Gorontalo Province, Indonesia, further research is pivotal for generalizing the results. For example, a high level of mercury contamination has been reported in ASGM communities in various provinces, such as in West Java [35], West Nusa Tenggara [36], South Sulawesi [37], and Buru Island, Mollucas [38]. Further studies are necessary to determine whether *P. vittata* acts as an indicator plant in other areas. Identifying the potential and limitations of *P. vittata* as an indicator plant by extensive investigations will be helpful for policy-makers to assess the environmental pollution of respective ASGM areas.

Author Contributions: All authors contributed to the work presented in the manuscript. N.A.G. undertook the study and wrote the paper. Supervisor M.S. provided advice, revised the paper, and made recommendations on analytical methods. S.K. provided comments and revised the paper. S.S. provided advice concerning ICP–MS analyses. K.S. provided advice concerning PIXE analysis. All authors have read and agreed to the published version of the manuscript.

Funding: This research was funded by the Japanese Government through JSPS KAKENHI Grant Number 16H02706 to M. Sakakibara for graduate study at Ehime University. It was also supported by the feasibility study "Social Acceptance of Regional Innovation for Reducing High-impact Environmental Pollution," Research Institute for Humanity and Nature (RIHN), Project Number 14200102.

Institutional Review Board Statement: Not applicable.

Informed Consent Statement: Not applicable.

Data Availability Statement: The data were generated during the study.

Acknowledgments: The authors thank the government of Bone Bolango Regency, Gorontalo Province, Indonesia, for allowing the study. We gratefully acknowledge the funding received by M. Sakakibara and thank the Gorontalo Province government and their stakeholders as well as anonymous reviewers for recommendations. This research was supported by the Research Institute for Humanity and Nature (RIHN; a constituent member of NIHU), Project Number 14200102.

Conflicts of Interest: The authors declare no conflict of interest.

References

1. United Nations Environment Programme-UN Environmental Program. The Global Mercury Assessment. 2013. Available online: http://www.unep.org/PDF/PressReleases/GlobalMercuryAssessment2013.pdf (accessed on 14 August 2019).
2. McGrew, L. Asia Pacific Foundation of Canada Home Page. Available online: https://www.asiapacific.ca/blog/artisanal-and-small-scale-gold-mining-sector-problems-and (accessed on 29 October 2021).
3. GOLD-ISMIA Team. United Nation Development Programme Indonesia Home Page. Available online: https://www.id.undp.org/content/indonesia/en/home/presscenter/articles/2021/090621.html (accessed on 29 October 2021).
4. Bose-O'Reilly, S.; Drasch, G.; Beinhoff, C.; Rodrigues-Filho, S.; Roider, G.; Lettmeier, B.; Maydl, A.; Maydl, S.; Siebert, U. Health Assessment of Artisanal Gold Miners in Indonesia. *Sci. Total Environ.* **2010**, *408*, 713–725. [CrossRef]
5. Puluhulawa, F.; Junus, N. *A Final Report: Perlindungan Hukum Terhadap Usaha Pertambangan Rakyat di Provinsi Gorontalo*, Gorontalo State University: Gorontalo, Indonesia, 2013; *unpublished work*.
6. WHO/IPCS. Methylmercury—Environmental Health Criteria 101. Available online: https://wedocs.unep.org/handle/20.500.11822/29413 (accessed on 3 September 2021).
7. Arifin, Y.; Sakakibara, M.; Sera, K. Impacts of Artisanal and Small-Scale Gold Mining (ASGM) on Environment and Human Health of Gorontalo Utara Regency, Gorontalo Province, Indonesia. *Geosciences* **2015**, *5*, 160–176. [CrossRef]
8. Gafur, N.A.; Sakakibara, M.; Sano, S.; Sera, K. A Case Study of Heavy Metal Pollution in Water of Bone River by Artisanal Small-Scale Gold Mine Activities in Eastern Part of Gorontalo, Indonesia. *Water* **2018**, *10*, 1507. [CrossRef]
9. Lihawa, F.; Mahmud, M. The Content of Mercury in Sediments Around Artisanal Smallscale Gold Mining (ASGM) Bumela District, Gorontalo Regency, Gorontalo Province, Indonesia. *IOP Conf. Ser. Earth Environ. Sci.* **2019**, *314*, 012016. [CrossRef]
10. Hedenquist, J.W. Mineralization Associated with Volcanic-Related Hydrothermal Systems in the Circum-Pacific Basin. In Proceedings of the 4th Circum Pacific Energy and Mineral Resources Conference, Singapore, 17–22 August 1986; pp. 513–524.
11. Ma, L.Q.; Komar, K.M.; Tu, C.; Zhang, W.; Cai, Y.; Kennelley, E.D. A Fern That Hyperaccumulates Arsenic. *Nature* **2001**, *409*, 579. [CrossRef] [PubMed]
12. Natarajan, S.; Stamps, R.H.; Saha, U.K.; Ma, L.Q. Phytofltration of Arsenic-Contaminated Groundwater Using *Pteris vittata* L.: Effect of Plant Density and Nitrogen and Phosphorus Levels. *Int. J. Phytoremediat.* **2008**, *10*, 220–233. [CrossRef]
13. Sakakibara, M.; Watanabe, A.; Inoue, M.; Sano, S.; Kaise, T. Phytoextraction and Phytovolatilization of Arsenic from as-Contaminated Soils by *Pteris vittata*. In Proceedings of the Annual International Conference on Soils, Sediments, Water and Energy, Amherst, MA, USA, 18–21 October 2010; Volume 12, pp. 264–272.
14. Fayiga, A.O.; Ma, L.Q.; Cao, X. Rathinasabapathi, B. Effects of heavy metals on growth and arsenic accumulation in the arsenic hyperaccumulator *Pteris vittata* L. *Environ Pollut.* **2004**, *132*, 289–296. [CrossRef]
15. Lei, M.; Wan, X.; Guo, G.; Yang, J.; Chen, T. Phytoextraction of arsenic-contaminated soil with *Pteris vittata* in Henan Province, China: Comprehensive evaluation of remediation efficiency correcting for atmospheric depositions. *Environ. Sci. Pollut. Res. Int.* **2018**, *25*, 124–131. [CrossRef]
16. Zeng, P.; Guo, Z.; Xiao, X.; Peng, C.; Feng, W.; Xin, L.; Xu, Z. Phytoextraction potential of *Pteris vittata* L. co-planted with woody species for As, Cd, Pb and Zn in contaminated soil. *Sci. Total Environ.* **2019**, *650 Pt 1*, 594–603. [CrossRef]
17. Ma, J.; Lei, E.; Lei, M.; Liu, Y.; Chen, T. Remediation of Arsenic contaminated soil using malposed intercropping of *Pteris vittata* L. and maize. *Chemosphere* **2018**, *194*, 737–744. [CrossRef] [PubMed]
18. Raymond, R.C.; Madan, K.O.; John, B.W.; Joan, E.B.; William, D.M. *Analysis of Gopher Tortoise Population Estimation Techniques*; Engineer Research and Development Center: Gainesville, FL, USA, 2005. Available online: https://citeseerx.ist.psu.edu/viewdoc/download?doi=10.1.1.665.3013&rep=rep1&type=pdf (accessed on 21 November 2016).
19. Sera, K.; Yanagisawa, T.; Tsunoda, H.; Futatsugawa, S.; Hatakeyama, S.; Saitoh, Y.; Suzuki, S.; Orihara, H. Bio-PIXE at the Takizawa Facility (Bio-PIXE with a Baby Cyclotron). *Int. J. Pixe* **1992**, *2*, 325–330. [CrossRef]
20. Carlos, A.G.; Robert, J. National institute of Standards & Technology Certificate of Analysis. Standard Reference Materials® 1643f. 2015. Available online: https://www-s.nist.gov/srmors/certificates/1643F.pdf (accessed on 5 October 2016).
21. SPEX Certificate. Certificate of Reference Material XSTC-13. Available online: http://www.seishin-syoji.co.jp/files/libs/567/201604271251467620.pdf (accessed on 16 October 2018).
22. Carlos, A.G.; Robert, J.; National institute of Standards & Technology Certificate of Analysis. Standard Reference Materials® 2782. 2015. Available online: https://www-s.nist.gov/m-srmors/certificates/2782.pdf (accessed on 5 October 2016).
23. NIES. *NIES No. 9 Sargasso. Certificate of Analysis*; National Institute of Environmental Studies: Tsukuba, Japan, 1988.
24. USEPA (U.S. Environmental Protection Agency). *Clean Water Act*; Section 503; US EPA: Washington, DC, USA, 1993; Volume 58.
25. Anh, B.T.K.; Minh, N.N.; Ha, N.T.H.; Kim, D.D.; Kien, N.T.; Trung, N.Q.; Cuong, T.T.; Danh, L.T. Field Survey and Comparative Study of *Pteris vittata* and Pityrogramma Calomelanos Grown on Arsenic Contaminated Lands with Different Soil pH. *Bull. Environ. Contam. Toxicol.* **2018**, *100*, 720–726. [CrossRef]

26. Hindersah, R.; Risamasu, R.; Kalay, A.M.; Dewi, T.; Makatita, I. Mercury Contamination in Soil, Tailing and Plants on Agricultural Fields near Closed Gold Mine in Buru Island, Maluku. *J. Degrad. Min. Lands Manag.* **2018**, *5*, 1027–1034. [CrossRef]
27. Basu, N.; Clarke, E.; Green, A.; Calys-Tagoe, B.; Chan, L.; Dzodzomenyo, M.; Fobil, J.; Long, R.N.; Neitzel, R.L.; Obiri, S.; et al. Integrated Assessment of Artisanal and Small-Scale Gold Mining in Ghana—Part 1: Human Health Review. *Int. J. Environ. Res. Public Health* **2015**, *12*, 5143–5176. [CrossRef]
28. Fashola, M.O.; Ngole-Jeme, V.M.; Babalola, O.O. Heavy Metal Pollution from Gold Mines: Environmental Effects and Bacterial Strategies for Resistance. *Int. J. Environ. Res. Public Health* **2016**, *13*, 1047. [CrossRef]
29. Rajaee, M.; Obiri, S.; Green, A.; Long, R.; Cobbina, S.J.; Nartey, V.; Buck, D.; Antwi, E.; Basu, N. Integrated Assessment of Artisanal and Small-Scale Gold Mining in Ghana—Part 2: Natural Sciences Review. *Int. J. Environ. Res. Public Health* **2015**, *12*, 8971–9011. [CrossRef]
30. Wilson, M.L.; Renne, E.; Roncoli, C.; Agyei-Baffour, P.; Tenkorang, E.Y. Integrated Assessment of Artisanal and Small-Scale Gold Mining in Ghana—Part 3: Social Sciences and Economics. *Int. J. Environ. Res. Public Health* **2015**, *12*, 8133–8156. [CrossRef] [PubMed]
31. Naveed, N.H.; Batool, A.I.; Rehman, F.; Hameed, U. Leaves of Roadside Plants as Bioindicator of Traffic Related Lead Pollution During Different Seasons In. *Afr. J. Environ. Sci. Technol.* **2010**, *4*, 770–774. [CrossRef]
32. Peer, W.A.; Baxter, I.R.; Richards, E.L.; Freeman, J.L.; Murphy, A.S. Phytoremediation and Hyperaccumulator Plants. *Top. Curr. Genet.* **2005**, *14*, 299–340. [CrossRef]
33. Statescu, F.; Cotiusca-Zauca, D. Heavy Metal Soil Contamination. *Environ. Eng. Manag. J.* **2018**, *5*, 1205–1213. [CrossRef]
34. Rathinasabapathi, B. Ferns represent an untapped biodiversity for improving crops for environmental stress tolerance. *New Phytol.* **2006**, *172*, 385–390. [CrossRef]
35. Harianja, A.H.; Saragih, G.S.; Fauzi, R.; Hidayat, M.Y.; Syofyan, Y.; Tapriziah, E.R.; Kartiningsih, S.E. Mercury Exposure in Artisanal and Small-Scale Gold Mining Communities in Sukabumi, Indonesia. *J. Health Pollut.* **2020**, *10*, 201209. [CrossRef]
36. Junaidi, M.; Krisnayanti, B.D.; Juharfa; Anderson, C. Risk of Mercury Exposure from Fish Consumption at Artisanal Small-Scale Gold Mining Areas in West Nusa Tenggara, Indonesia. *J. Health Pollut.* **2019**, *9*, 190302. [CrossRef]
37. Abbas, H.H.; Sakakibara, M.; Sera, K.; Arma, L.H. Mercury Exposure and Health Problems in Urban Artisanal Gold Mining (UAGM) in Makassar, South Sulawesi, Indonesia. *Geosciences* **2017**, *7*, 44. [CrossRef]
38. Male, Y.T.; Reichelt-Brushett, A.J.; Pocock, M.; Nanlohy, A. Recent mercury contamination from artisanal gold mining on Buru Island, Indonesia—Potential future risks to environmental health and food safety. *Mar. Pollut. Bull.* **2013**, *77*, 428–433. [CrossRef]

International Journal of
Environmental Research and Public Health

MDPI

Article

Evaluation of the Total Mercury Weight Exposure Distribution Using Tree Bark Analysis in an Artisanal and Small-Scale Gold Mining Area, North Gorontalo Regency, Gorontalo Province, Indonesia

Hendra Prasetia [1,2,*]**, Masayuki Sakakibara** [3,4]**, Koichiro Sera** [5] **and Jamie Stuart Laird** [6]

[1] Graduate School of Science and Engineering, Ehime University, Matsuyama 790-8577, Japan
[2] Department of Forestry, Faculty of Agriculture, University of Lampung, Bandar Lampung 35145, Indonesia
[3] Research Institute for Humanity and Nature, Kyoto 603-8047, Japan; sakaki@chikyu.ac.jp
[4] Faculty of Collaborative Regional Innovation, Ehime University, Matsuyama 790-8577, Japan
[5] Cyclotron Research Center, Iwate Medical University, Tomegamori 348-58 Tomegamori, Takizawa 020-0173, Japan; ksera@iwate-med.ac.jp
[6] School of Chemistry, University of Melbourne, Parkville, VIC 3010, Australia; jslaird@unimelb.edu.au
* Correspondence: hendra.prasetia@fp.unila.ac.id

Abstract: It is well known that atmospheric mercury (Hg) contaminates air, water, soil, and living organisms, including trees. Therefore, tree bark can be used for the environmental assessment of atmospheric contamination because it absorbs heavy metals. This study aimed to establish a new biomonitoring for the assessment of atmospheric Hg pollution. Reporting on atmospheric Hg contamination in an artisanal and small-scale gold mining (ASGM) area in North Gorontalo, Indonesia, we calculated the total weight of Hg (THg) and quantitatively measured the concentrations of Hg in the tree bark of *Mangifera indica*, *Syzygium aromaticum*, *Terminalia catappa*, and *Lansium domesticum*. The THg of Hg in the *M. indica* tree bark samples ranged from not detected (ND) to 74.6 µg dry weight (DW) per sample. The total Hg in the tree bark of *S. aromaticum*, *T. catappa*, and *L. domesticum* ranged from ND to 156.8, ND to 180, and ND to 63.4 µg DW, respectively. We concluded that topography significantly influences the accumulation of Hg together with local weather conditions. A mapped distribution of the THg suggested that the distribution of THg in the tree bark was not affected by the distance to the amalgamation site. Therefore, tree bark can be used as biomonitoring of atmospheric Hg contamination for the assessment of ASGM areas.

Keywords: atmospheric; mercury; ASGM; amalgamation; accumulation; tree bark

Citation: Prasetia, H.; Sakakibara, M.; Sera, K.; Laird, J.S. Evaluation of the Total Mercury Weight Exposure Distribution Using Tree Bark Analysis in an Artisanal and Small-Scale Gold Mining Area, North Gorontalo Regency, Gorontalo Province, Indonesia. *Int. J. Environ. Res. Public Health* **2022**, *19*, 33. https://doi.org/10.3390/ijerph19010033

Academic Editor: Richard A. Lord

Received: 17 November 2021
Accepted: 18 December 2021
Published: 21 December 2021

Publisher's Note: MDPI stays neutral with regard to jurisdictional claims in published maps and institutional affiliations.

1. Introduction

Artisanal and small-scale gold mining (ASGM), which provides income to many poor communities in developing countries, such as Indonesia, uses several gold extraction methods that use mercury (Hg). The International Labor Organization estimates that there are currently around 13 million artisanal miners in 55 countries [1]. During the processes of panning and amalgamation when amalgam is burned in a small charcoal fire, ASGM releases Hg into the atmosphere [2,3]. ASGM is a widely recognized major source of Hg contamination, and its activities cause serious Hg pollution.

Mercury is extremely dangerous and contaminates air, water, soil, and living organisms. The health of miners and people living within or outside ASGM areas is affected by the inhalation of atmospheric Hg [3]. Anthropogenic Hg emissions to the atmosphere significantly interfere with the natural Hg cycle [4]; however, estimates of natural global Hg emissions vary by orders of magnitude [4,5]. The increasing total weight of Hg (THg) in the soil is likely due to the deposition of Hg released into the atmosphere [6].

Plants are sensitive to their environmental conditions, and their elemental compositions actively reflect changes in these conditions [7–9]. Tree bark, in particular, can be used to assess the status of the environment, especially the level of Hg contamination, and sources of pollution can be traced by the enrichment of trace elements in tree bark [10]. Airborne particles, trapped within the structure of tree bark, accumulate over several years [11]. The uptake of trace elements by plants involves both root uptake and foliar absorption, including from the deposition of particulate matter on leaves [12]. Different plant uptake patterns are based on three factors: plant species, element species, and conditions at specific sites [13]. Canopy crops act to trap gaseous and particulate Hg, which can then be trapped by tree bark depending on its roughness and porosity [14].

Although the use of tree bark has been studied for environmental pollution assessments, corresponding atmospheric contamination has not been comprehensively discussed in relation to the distance from the source of the contamination and the transport of Hg in the atmosphere. Consequently, the practical application of tree bark as a biomonitoring for the atmosphere has not been previously proposed. Therefore, several tropical species, including *Mangifera indica*, *Syzygium aromaticum*, *Terminalia catappa*, and *Lansium domesticum*, were comprehensively studied to establish a new biomonitoring for the assessment of atmospheric Hg pollution in an ASGM area in North Gorontalo Regency, Gorontalo Province, Indonesia.

2. Materials and Methods

2.1. Sampling Plots

We performed a field survey and laboratory analyses to determine heavy metal concentrations (particularly Hg) in tree bark to assess the environmental contamination in the study area. This study research was conducted during August and September 2016, entering the rainy season in Indonesia, in an ASGM area of north Gorontalo Regency. In this regency, there are three ASGM sites, which are located in different districts, shown in Figure 1. The study area of this research was located in east Sumalata District, shown in Figure 1. This study was obtained tree bark about 65 samples in total with details 21 samples of *M. indica*, 20 samples of *S. aromaticum*, 15 samples of *T. catappa*, and 9 samples of *L. domesticum* from the study area, as shown in Figure 2, and sampled randomly selected that were around settlements area. The mercury is very harmful to humans and, therefore, the tree barks samples, mostly found in the inhabitant's yard, were selected in this study due to its capability as biomonitoring of atmospheric Hg contamination. These sampling species were grown naturally in this area, which dominated in the lower topography of this area.

The tree bark samples were collected from 1.3 m above the ground, being the diameter at breast height standard height. The bark was collected as 10×10 cm fragments to ensure homogeneous sampling. About 12–18 cm^2 of 100 cm^2 bark samples were analyzed to indicate the Hg concentrations.

2.2. Analytical Methods

The tree bark samples were dried at ~80 °C for 2 days in a ventilated oven. About 12–18 cm^2 of each sample was crushed to a fine powder with a powder mill (Varian PM-2005 m, Osaka Chemical Co., Ltd., Osaka, Japan) to produce homogeneous samples for particle-induced X-ray emission (PIXE) analysis. The tree bark powders (30 mg) were then digested by a mixture of indium (In) and HNO_3 in a ratio of 3:100 before the heavy metal concentrations, such as Pb, Zn, Fe, Hg, and As, were determined by PIXE [15–17] at Iwate Medical University (Iwate, Japan). The dimensions of the tree bark samples were calculated by ImageJ, Version 1.48 software. The analytical conditions followed [15]. A small cyclotron provided a 2.9 MeV-proton beam on the target after passing through a beam collimator of graphite. The maximum beam intensity on the target was approximately 40 nA for a beam spot diameter of 2 mm and 80 nA for a diameter of 6 mm. Elements from Na to

U were detected by two ORTEC Si (Li) detectors. The elements heavier than Ca were detected by the first detector, which had a 0.025 mm-thick Be window and a 6 mm active diameter, with X-rays with an energy resolution of 154 eV at 5.9 keV and a 300 to 500 μm thick Mylar absorber inserted between the target and the detector. The other low atomic number elements were detected by the second detector, which had a 0.008 mm Be window and a 4 mm active diameter, a resolution of 157 eV, and a small graphite aperture without an absorber.

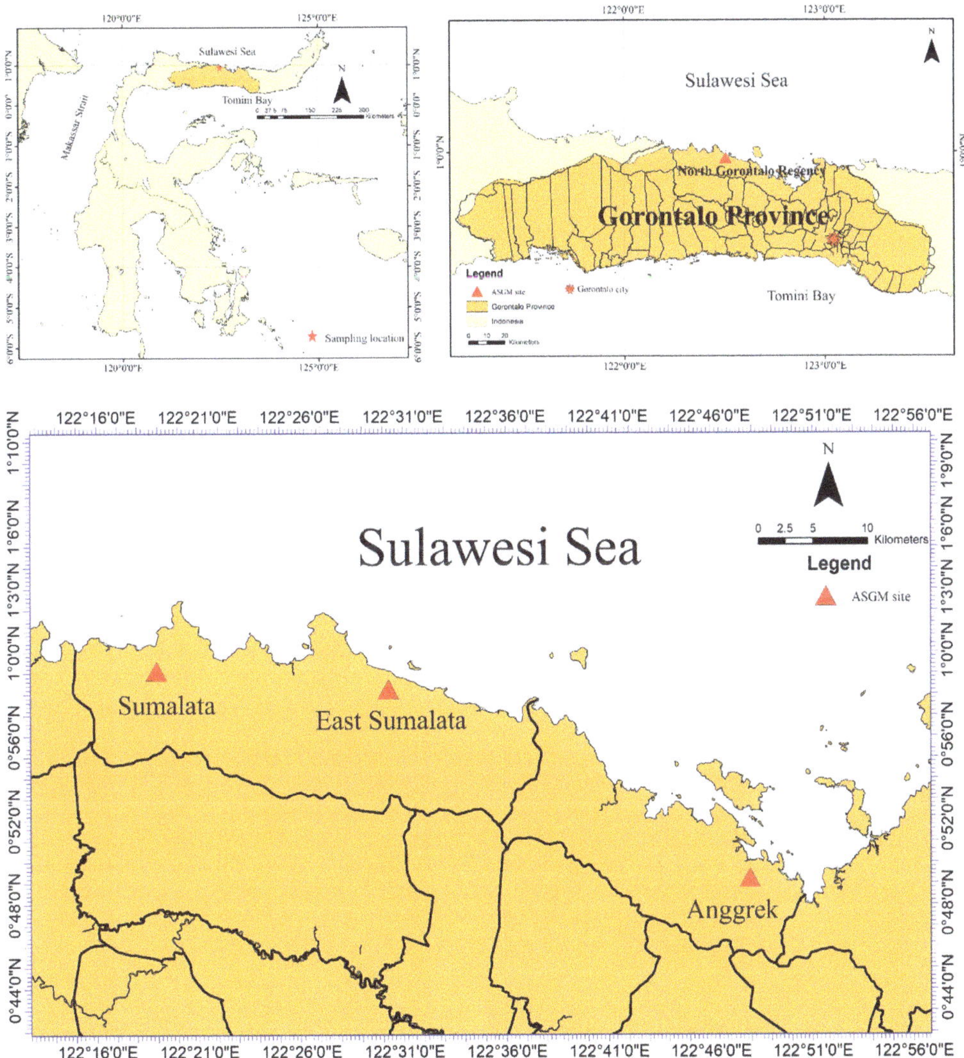

Figure 1. Artisanal and small-scale gold mining (ASGM) sites in North Gorontalo Regency, Gorontalo Province, Indonesia.

2.3. Calculation of THg

The bioaccumulation of Hg was estimated using the THg, defined as the dry weight of the sample multiplied by the Hg concentration determined by the PIXE analysis in 100 cm^2 of sample [17].

$$\text{THg} = (\text{DW} \times \text{C}_{\text{Hg}}) \times (\text{FD}/\text{real square}) \tag{1}$$

where DW is the dry weight of the sample, C_{Hg} is the Hg concentration, FD is fragment dimensions (100 cm^2), and real square is the sample dimension, as measured by ImageJ.

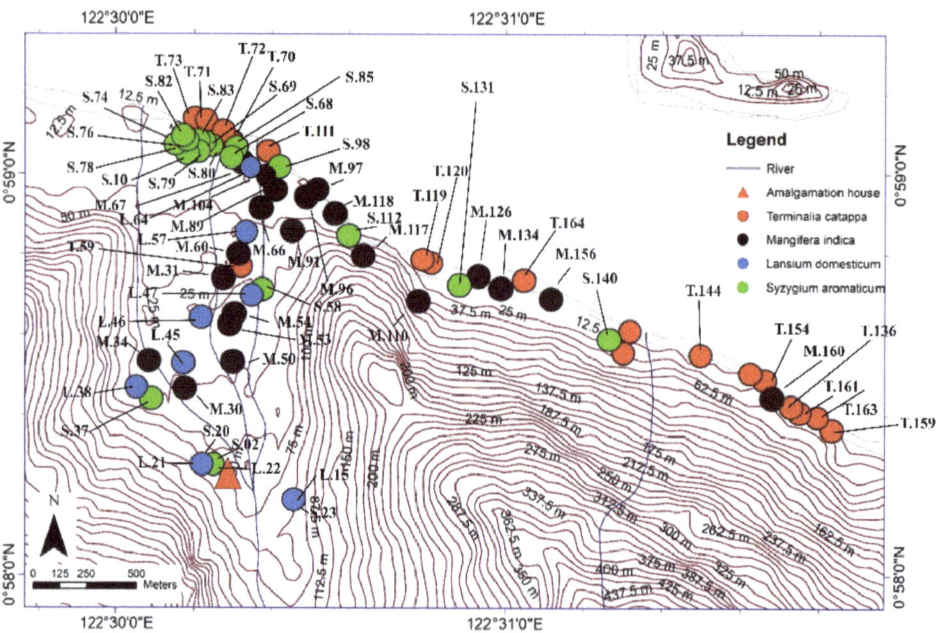

Figure 2. *Terminalia catappa, Mangifera indica, Lansium domesticum*, and *Syzygium aromaticum* sampling points in East Sumalata District, Gorontalo Province, Indonesia (N = 65). The sampling point coordinates was measured using a GPS (Oregon 650 TCJ; Garmin), and the map contours were created by ArcGIS 10.3 and Global Mapper 10 software.

2.4. Statistical Analysis

Statistical analyses were performed using IBM SPSS Statistic 21 for Microsoft Windows. The Shapiro–Wilk test was used to check the normality of the Hg concentrations. The data were log-normally distributed, so the Kruskal–Wallis ANOVA test was used to test for significant differences, with $p < 0.05$ considered statistically significant.

3. Results

3.1. Estimation of THg

The THg in the *M. indica* bark samples ranged from not detected (ND) to 74.6 μg DW (see Table 1). The ND results were probably attributable to the absorption of Hg due to different weather conditions under different topographic conditions. The THg in the bark samples of *S. aromaticum, T. catappa,* and *L. domesticum* ranged from ND to 156.8, ND to 180, and ND to 63.4 μg DW, respectively (see Tables 2 to 4). The average results of THg showed that the *T. catappa* has highest mean of THg about 66.2 μg DW. Then, it was followed by *S. aromaticum, M. indica,* and *L. domesticum* with THg concentrations about 42.4, 26.1, and 15.4 μg DW, respectively. The highest THg was located in areas of lower topography, as shown in Figure 3. This suggested that atmospheric Hg contamination is most dominant in estuaries. A plant is categorized as toxic if the concentration of Hg exceeds 1 ppm [9], so based on this study results, the bark of *M. indica, S. aromaticum, T. catappa* and *L. domesticum* could be used as a biomonitoring of atmospheric Hg contamination in the environmental assessments of ASGM areas.

Table 1. Total weight of Hg (THg) in the bark and diameter of *Mangifera indica*.

No	Sample	T(Hg) (µg—DW) ± SD	Diameter (cm)
1	*M. indica* 30	49.8 ± 27.7	42.7
2	*M. indica* 117	15.9 ± 15.3	40.1
3	*M. indica* 118	64.5 ± 20.1	46.8
4	*M. indica* 156	13.0 ± 14.3	25.5
5	*M. indica* 126	63.5 ± 49.1	30.3
6	*M. indica* 160	ND	19.1
7	*M. indica* 134	22.2 ± 31.4	27.4
8	*M. indica* 53	8.70 ± 16.4	20.7
9	*M. indica* 66	47.9 ± 31.1	69.7
10	*M. indica* 96	ND	23.9
11	*M. indica* 31	35.5 ± 34.6	72.6
12	*M. indica* 89	ND	36.3
13	*M. indica* 97	24.3 ± 12.3	89.2
14	*M. indica* 104	54.1 ± 27.3	47.8
15	*M. indica* 34	4.10 ± 7.10	41.4
16	*M. indica* 67	6.60 ± 13.7	11.8
17	*M. indica* 54	ND	29.9
18	*M. indica* 91	30.9 ± 15.0	24.2
19	*M. indica* 110	74.6 ± 27.6	43.9
20	*M. indica* 60	32.9 ± 23.4	58.3
21	*M. indica* 50	ND	41.1
	Mean	26.1 ± 17.4	40.1

T: Total; DW: Dry Weight; SD: Standard Deviation; ND: Not Detected.

Table 2. Total weight of Hg (THg) in the bark and diameter of *Syzygium aromaticum*.

No	Sample	T(Hg) (µg—DW) ± SD	Diameter (cm)
1	*S. aromaticum* 23	20.9 ± 28.9	11.1
2	*S. aromaticum* 79	9.10 ± 18.2	39.8
3	*S. aromaticum* 140	51.0 ± 33.7	19.1
4	*S. aromaticum* 68	24.2 ± 28.5	16.2
5	*S. aromaticum* 76	16.0 ± 20.6	33.4
6	*S. aromaticum* 69	31.2 ± 48.5	10.8
7	*S. aromaticum* 82	138 ± 68.7	18.5
8	*S. aromaticum* 112	156.8 ± 79.6	18.5
9	*S. aromaticum* 85	23.0 ± 17.7	13.7
10	*S. aromaticum* 20	58.8 ± 33.3	23.9
11	*S. aromaticum* 98	54.3 ± 26.5	25.5
12	*S. aromaticum* 10	ND	28.4
13	*S. aromaticum* 02	22.6 ± 10.2	29.8
14	*S. aromaticum* 83	39.5 ± 27.4	26.1
15	*S. aromaticum* 58	47.8 ± 19.6	10.5
16	*S. aromaticum* 78	ND	16.9
17	*S. aromaticum* 74	42.6 ± 16.3	17.2
18	*S. aromaticum* 37	41.3 ± 23.5	11.1
19	*S. aromaticum* 80	40.6 ± 24.4	30.6
20	*S. aromaticum* 131	29.6 ± 15.3	11.5
	Mean	42.4 ± 27.0	20.3

T: Total; DW: Dry Weight; SD: Standard Deviation; ND: Not Detected.

Table 3. Total weight of Hg (THg) in the bark and diameter of *Terminalia catappa*.

No	Sample	T(Hg) (µg—DW) ± SD	Diameter (cm)
1	*T. catappa* 71	8.70 ± 22.0	17.5
2	*T. catappa* 120	35.9 ± 17.9	20.7
3	*T. catappa* 136	68.4 ± 26.3	55.1
4	*T. catappa* 144	ND	17.8
5	*T. catappa* 163	150 ± 42.4	43.6
6	*T. catappa* 73	35.4 ± 38.7	32.8
7	*T. catappa* 164	113 ± 69.7	50.3
8	*T. catappa* 59	180 ± 105	44.9
9	*T. catappa* 72	26.4 ± 43.5	41.7
10	*T. catappa* 70	92.7 ± 43.6	36.6
11	*T. catappa* 159	72.5 ± 56.4	58.9
12	*T. catappa* 119	40.1 ± 69.4	53.8
13	*T. catappa* 154	ND	89.5
14	*T. catappa* 161	16.8 ± 19.2	53.8
15	*T. catappa* 111	152 ± 63.3	31.5
	Mean	66.2 ± 41.2	43.2

T: Total; DW: Dry Weight; SD: Standard Deviation; ND: Not Detected.

Table 4. Total weight of Hg (THg) in the bark and diameter of *Lansium domesticum*.

No	Sample	T(Hg) (µg—DW) ± SD	Diameter (cm)
1	*L. domesticum* 26	25.0 ± 18.8	38.5
2	*L. domesticum* 57	ND	34.1
3	*L. domesticum* 64	10.7 ± 14.7	37.6
4	*L. domesticum* 38	2.50 ± 15.5	37.9
5	*L. domesticum* 45	14.1 ± 25.1	38.5
6	*L. domesticum* 15	ND	30.7
7	*L. domesticum* 47	63.4 ± 26.8	27.4
8	*L. domesticum* 21	13.0 ± 13.6	10.4
9	*L. domesticum* 46	9.50 ± 25.0	10.5
	Mean	15.4 ± 15.5	29.5

T: Total; DW: Dry Weight; SD: Standard Deviation; ND: Not Detected.

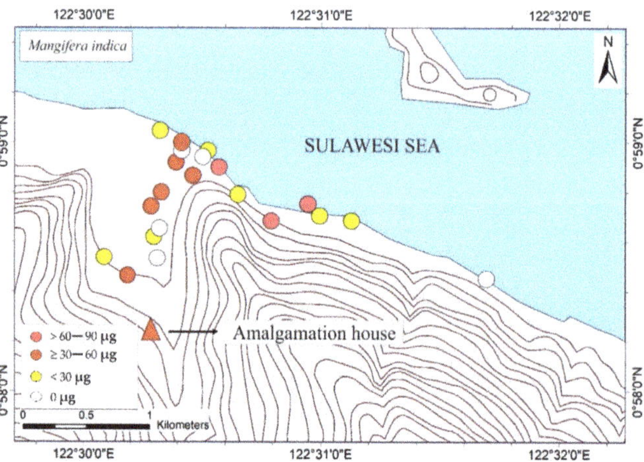

Figure 3. *Cont.*

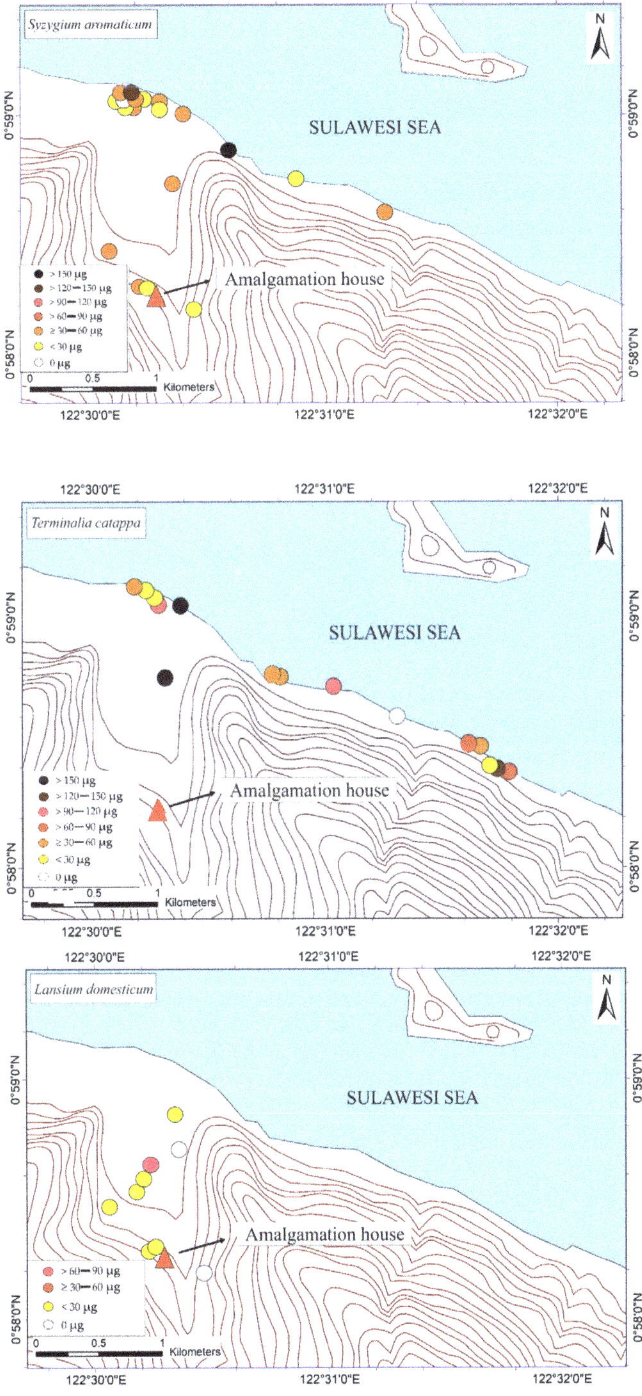

Figure 3. Mapped distribution of total weight of Hg (THg) in the bark of *Mangifera indica*, *Syzygium aromaticum*, *Terminalia catappa*, and *Lansium domesticum* (*N* = 65).

4. Discussion

4.1. Mapping Distribution of THg in ASGM Area

The mapped distribution of THg in the bark of *M. indica, S. aromaticum, T. catappa* and *L. domesticum* (Figure 3) suggested that topography significantly influences the accumulation of Hg in the atmosphere together with local weather conditions. However, the distribution was not affected by the distance to the amalgamation site, as shown in Figures 3 and 4. This is probably attributable to the wind direction, which transports and deposits the atmospheric Hg in the estuary area. The concentrations on the tree barks were affected by atmospheric attachment to the barks, not from the root absorption in soil and/or water [17].

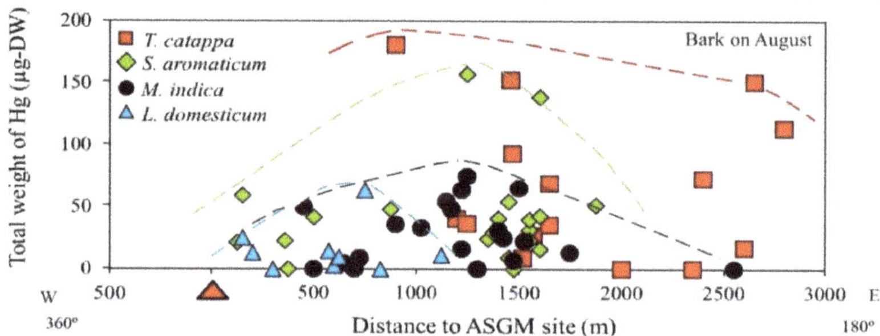

Figure 4. Total weight of Hg (THg) in the bark of various tree species.

Zang et al.'s [18] study showed that Hg transport and dispersion in the atmosphere may not always be explained by so-called prevailing winds. In their study area, there was no prevailing wind strong enough to control the direction of the Hg diffusion in the atmosphere, so the evenly distributed Hg in their bark samples was attributed to precipitation rather than dry deposition or contamination from a reservoir. Local influences can diminish the relative importance of the general atmospheric transport of Hg [18]. According to Barnes et al. [19], tree bark accumulates metals as a function of its proximity to the pollution source.

Elemental Hg is the predominant (95%) form in the atmosphere, with reactive and particulate forms also being important [20]. Plants have been shown to accumulate atmospheric elemental Hg (Hg°) in foliage over time as a function of exposure to concentrations of Hg in the air and soil [21,22]. Elemental Hg is thought to be directly taken up via stomata and/or cuticles and possibly transformed into a water-soluble Hg compound and absorbed via leaves [23,24]. In addition to the concentration of atmospheric Hg, environmental factors, such as solar irradiation, air temperature, altitude, and biological factors, such as plant species, leaf age, and leaf placement, also significantly influence the uptake of Hg by foliage [22,25–27].

In addition, climate change has the potential to alter the sequestration of Hg from forest soils via direct pressures (meteorological) or indirect pressures (vegetation changes) [28]. This could have indirect consequences for forests that may also affect Hg cycling. According to Richardson and Friedland [28], vegetation type can affect many aspects of Hg cycling in forest soils. The variable foliar morphology and biomass characteristics of different vegetation types can affect Hg levels in litterfall. The physical attributes of the canopy structure of each species can also directly affect the accumulation of Hg in foliage [29,30].

4.2. Distribution of THg Based on Distance to Source and Elevation

The distance of the samples from the ASGM site did not influence the THg attachment; however, the total weight of Hg in the bark taken from *T. catappa* was higher compared to *S. aromaticum, M. indica,* and *L. domesticum,* as shown in Figure 4. In the study area, *T. catappa* and *M. indica* grow naturally, whereas *S. aromaticum* and *L. domesticum* are cultivated for economic purposes.

Living at lower topographic levels along the coastline, *T. catappa* is mainly an estuary plant, and some of the *S. aromaticum* sampled in this study was also cultivated at lower topographic levels along the coastline. As shown in Section 3.1, the highest THg was found in the *S. aromaticum* and *T. catappa*, located in the coastline area.

A wide range of total Hg concentrations in the air has been reported in the literature. The latitudinal distribution of total gaseous Hg indicates a background level of about 2 ng m^{-3} in the lower troposphere of the northern hemisphere and just over 1 ng m^{-3} in the southern hemisphere, at least in an oceanic environment [31]. In general, elemental Hg seems to be the dominant form [31]. The Hg associated with aerosol particles normally makes up only a small fraction of the total airborne Hg; however, the role of particulate Hg in the atmospheric is important. The atmospheric cycle retains Hg in the atmosphere for long periods, and, consequently, transports it over very long distances [32]. Mercury vapor, which comprises 95–99% of total Hg in the atmosphere, has an atmospheric residence time of 1 year [4], allowing for global dispersal and the contamination of ecosystems through both wet and dry deposition [33,34].

4.3. Distribution of THg Based on Tree Species

A THg boxplot for the various tree species showed that the mean values for *T. catappa*, *S. aromaticum*, *M. indica*, and *L. domesticum* were 66.2, 42.4, 26.1, and 15.4 μg-DW, respectively, but there were no significant differences ($p < 0.05$), as shown in Figure 5. This indicated that tree species significantly influenced the attachment of Hg to the bark. We assumed that *T. catappa bark* has a high porosity, so it retains Hg better than the other species.

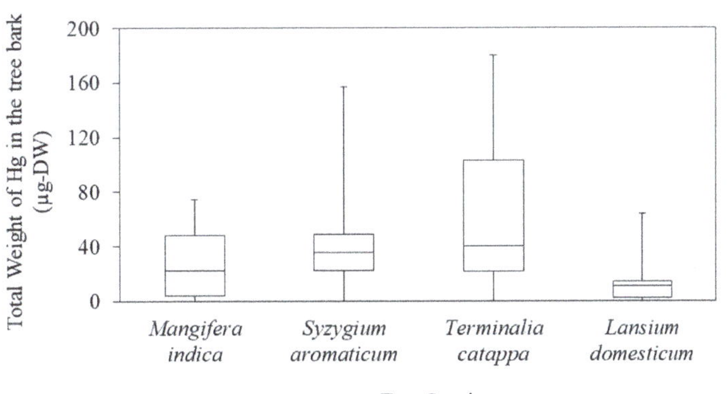

Figure 5. Boxplot of the total weight of Hg (THg) in the bark of the various species.

5. Conclusions

Our results showed that there was a high level of heterogeneity in the THg in the bark of both the naturally grown and cultivated tropical tree species we studied. The sampled *M. indica* and *T. catappa* were located in areas of lower topography on the coastline of the study area. *M. indica* is a native plant mostly cultivated close to houses as a garden plant. *S. aromaticum* and *L. domesticum* are native plants cultivated for economic purposes in the study area. This variability suggested that topography significantly influenced the accumulation of Hg together with local weather conditions, but was not affected by distance from the amalgamation site. This study indicated that tree species significantly influenced the attachment of Hg to the bark. We assumed that *T. catappa bark* has a high porosity, so it retains Hg better than the other species. We found that the tree bark of *M. indica*, *S. aromaticum*, *T. catappa* and *L. domesticum* could be used as a biomonitoring of atmospheric contamination assessment in the ASGM area.

Author Contributions: All authors contributed to the work presented in this manuscript. H.P., as principal researcher, undertook this work in association with his PhD program at Ehime University. M.S. is his PhD supervisor. J.S.L. provided the micro-PIXE measurement of the tree bark samples, and K.S. provided the PIXE measurement of the tree bark samples. All authors have read and agreed to the published version of the manuscript.

Funding: This research was funded by Research Institute for Humanity and Nature Project No. 14200102. This work was also supported by JSPS KAKENHI Grant Number 16H02706.

Institutional Review Board Statement: Not applicable.

Informed Consent Statement: Not applicable.

Acknowledgments: Hendra Prasetia wishes to thank the Japanese Government for providing a Monbukagakusho Scholarship for graduate studies at Ehime University, RIHN for funding this research and the North Gorontalo Regency Government of Gorontalo Province that allowed the authors to conduct the research activity.

Conflicts of Interest: The authors declare no conflict of interest.

References

1. Hinton, J.J.; Veiga, M.M.; Veiga, A.T.C. Clean Artisanal Gold Mining: A Utopian Approach? *J. Clean. Prod.* **2003**, *11*, 99–115. [CrossRef]
2. Veiga, M.M.; Maxson, P.A.; Hylander, L.D. Origin and Consumption of Mercury in Small-Scale Gold Mining. *J. Clean. Prod.* **2006**, *14*, 436–447. [CrossRef]
3. Taylor, H.; Appleton, J.D.; Lister, R.; Smith, B.; Chitamweba, D.; Mkumbo, O.; Machiwa, J.F.; Tesha, A.L.; Beinhoff, C. Environmental Assessment of Mercury Contamination from the Rwamagasa Artisanal Gold Mining Centre, Geita District, Tanzania. *Sci. Total Environ.* **2005**, *343*, 111–133. [CrossRef] [PubMed]
4. Fitzgerald, W.F.; Engstrom, D.R.; Mason, R.P.; Nater, E.A. The Case for Atmospheric Mercury Contamination in Remote Areas. *Environ. Sci. Technol.* **1998**, *32*, 1–7. [CrossRef]
5. Rasmussen, P.E. Temporal Variation of Mercury in Vegetation. In *Mercury as a Global Pollutant*; Springer: Berlin/Heidelberg, Germany, 1995; Volume 80, pp. 1039–1042. [CrossRef]
6. Tomiyasu, T.; Kono, Y.; Kodamatani, H.; Hidayati, N.; Rahajoe, J.S. The Distribution of Mercury Around the Small-Scale Gold Mining Area Along the Cikaniki River, Bogor, Indonesia. *Environ. Res.* **2013**, *125*, 12–19. [CrossRef] [PubMed]
7. Vtorova, V.N. Substantiation of Methods and Objects of Observations over Chemical Composition of Plants During Monitoring of Forest Ecosystem. *Inf. Bull. Probl. III Counc. Mutual Econ. Help* **1987**, *1*, 1–2.
8. Vtorova, V.N. Quantitative Evaluation of the Chemical Similarity of Needles of Picea Schrenkiana with Other Spruce Species in Natural and Artificial Growth Conditions. *Biol. Bull. Acad. Sci. USSR* **1991**, *17*, 245–253.
9. Kabata-Pendias, A.; Pendias, H. *Trace Elements in Soils and Plants*; CRC Press: Boca Raton, FL, USA, 1984; Volume 315.
10. Geagea, M.L.; Stille, P.; Millet, M.; Perrone, T. REE Characteristics and Pb, Sr and Nd Isotopic Compositions of Steel Plant Emissions. *Sci. Total Environ.* **2007**, *373*, 404–419. [CrossRef]
11. Catinon, M.; Ayrault, S.; Boudouma, O.; Asta, J.; Tissut, M.; Ravanel, P. The Inclusion of Atmospheric Particles into the Bark Suber of Ash Trees. *Chemosphere* **2009**, *77*, 1313–1320. [CrossRef] [PubMed]
12. Olajire, A.A.; Ayodele, E.T. Study of Atmospheric Pollution Levels by Trace Elements Analysis of Tree Bark and Leaves. *Bull. Chem. Soc. Ethiop.* **2003**, *17*, 11–17. [CrossRef]
13. Markert, B. Presence and Significance of Naturally Occurring Chemical Elements of the Periodic System in the Plant Organism and Consequences for Future Investigations on Inorganic Environmental Chemistry in Ecosystems. *Plant Ecol.* **1992**, *103*, 1–30.
14. Fostier, A.H.; Santos, C.; Carpi, A.; Windm, C.C.; Melendez-Perez, J.J. Soil and biomass mercury emissions during a prescribed fire in the Amazonian rain forest. *Environ. Res.* **2014**, *96*, 415–422.
15. Sera, K.; Yanagisawa, T.; Tsunoda, H.; Futatsugawa, S.; Hatakeyama, S.; Saitoh, Y.; Suzuki, S.; Orihara, H. Bio-PIXE at the Takizawa Facility (Bio-PIXE with a Baby Cyclotron). *Int. J. PIXE* **1992**, *02*, 325–330. [CrossRef]
16. Prasetia, H.; Sakakibara, M.; Sueoka, Y.; Sera, K. Pteris cretica as a Potential Biomarker and Hyperaccumulator in an Abandoned Mine Site, Southwest Japan. *Environments* **2016**, *3*, 15. [CrossRef]
17. Prasetia, H.; Sakakibara, M.; Omori, K.; Laird, J.S.; Sera, K.; Kurniawan, I.A. Mangifera indica as Bioindicator of Mercury Atmospheric Contamination in an ASGM Area in North Gorontalo Regency, Indonesia. *Geosciences* **2018**, *8*, 31. [CrossRef]

18. Zhang, L.; Qian, J.-L.; Planas, D. Mercury Concentration in Tree Rings of Black Spruce (Picea mariana Mill. BSP) in Boreal Quebec, Canada. *Water Air Soil Pollut.* **1995**, *81*, 163–173. [CrossRef]
19. Barnes, D.; Hamadah, M.A.; Ottaway, J.M. The Lead, Copper and Zinc Content of Tree Rings and Bark A Measurement of Local Metallic Pollution. *Sci. Total Environ.* **1976**, *5*, 63–77. [CrossRef]
20. Millhollen, A.G.; Gustin, M.S.; Obrist, D. Foliar Mercury Accumulation and Exchange for Three Tree Species. *Environ. Sci. Technol.* **2006**, *40*, 6001–6006. [CrossRef] [PubMed]
21. Frescholtz, T.E.; Gustin, M.S.; Schorran, D.E.; Fernandez, G.C.J. Assessing the Source of Mercury in Foliar Tissue of Quaking Aspen. *Environ. Toxicol. Chem.* **2003**, *22*, 2114–2119. [CrossRef] [PubMed]
22. Ericksen, J.A.; Gustin, M.S.; Schorran, D.E.; Johnson, D.W.; Lindberg, S.E.; Coleman, J.S. Accumulation of Atmospheric Mercury in Forest Foliage. *Atmos. Environ.* **2003**, *37*, 1613–1622. [CrossRef]
23. Browne, C.L.; Fang, S.C. Uptake of Mercury Vapor by Wheat: An Assimilation Model. *Plant Physiol.* **1978**, *61*, 430–433. [CrossRef]
24. Lindberg, S.E.; Stratton, W.J. Atmospheric Mercury Speciation: Concentrations and Behavior of Reactive Gaseous Mercury in Ambient Air. *Environ. Sci. Technol.* **1998**, *32*, 49–57. [CrossRef]
25. Zhu, W.; Lin, C.-J.; Wang, X.; Sommar, J.; Fu, X.; Feng, X. Global Observations and Modeling of Atmosphere—Surface Exchange of Elemental Mercury: A Critical Review. *Atmos. Chem. Phys.* **2016**, *16*, 4451–4480. [CrossRef]
26. Kos, G.; Ryzhkov, A.; Dastoor, A.; Narayan, J.; Steffen, A.; Ariya, P.A.; Zhang, L. Evaluation of Discrepancy Between Measured and Modelled Oxidized Mercury Species. *Atmos. Chem. Phys.* **2013**, *13*, 4839–4863. [CrossRef]
27. Laacouri, A.; Nater, E.A.; Kolka, R.K. Distribution and Uptake Dynamics of Mercury in Leaves of Common Deciduous Tree Species in Minnesota, USA. *Environ. Sci. Technol.* **2013**, *47*, 10462–10470. [CrossRef]
28. Richardson, J.B.; Friedland, A.J. Mercury in Coniferous and Deciduous Upland Forests in Northern New England, USA: Implications of Climate Change. *Biogeosciences* **2015**, *12*, 6737–6749. [CrossRef]
29. Obrist, D.; Johnson, D.W.; Lindberg, S.E.; Luo, Y.; Hararuk, O.; Bracho, R.; Battles, J.J.; Dail, D.B.; Edmonds, R.L.; Monson, R.K.; et al. Mercury Distribution Across 14 US Forests. Part I: Spatial Patterns of Concentrations in Biomass, Litter, and Soils. *Environ. Sci. Technol.* **2011**, *45*, 3974–3981. [CrossRef] [PubMed]
30. Blackwell, B.D.; Driscoll, C.T. Deposition of Mercury in Forests Along a Montane Elevation Gradient. *Environ. Sci. Technol.* **2015**, *49*, 5363–5370. [CrossRef]
31. Lindqvist, O.; Rodhe, H. Atmospheric Mercury, a Review. *Tellus B* **1985**, *37*, 136–156. [CrossRef]
32. Friedli, H.R.; Radke, L.F.; Payne, N.J.; McRae, D.J.; Lynham, T.J.; Blake, T.W. Mercury in Vegetation and Organic Soil at an Upland Boreal Forest Site in Prince Albert National Park, Saskatchewan, Canada. *J. Geophys. Res.* **2007**, *112*. [CrossRef]
33. Vandal, G.M.; Mason, R.P.; Fitzgerald, W.F. Cycling of Volatile Mercury in Temperate Lakes. *Water Air Soil Pollut.* **1991**, *56*, 791–803. [CrossRef]
34. Zillioux, E.J.; Porcella, D.B.; Benoit, J.M. Mercury Cycling and Effects in Freshwater Wetland Ecosystems. *Environ. Toxicol. Chem.* **1993**, *12*, 2245–2264. [CrossRef]

International Journal of
***Environmental Research
and Public Health***

Article

Element Rich Area Associated with Human Health Disorders: A Geomedical Science Approach to Potentially Toxic Elements Contamination

Sri Manovita Pateda [1,*], Masayuki Sakakibara [2,3] and Koichiro Sera [4]

[1] Medical Faculty, Universitas Negeri Gorontalo, Jenderal Sudirman Street 6, Gorontalo City 96100, Indonesia
[2] Research Institute for Humanity and Nature, Kyoto 603-8047, Japan; sakaki@chikyu.ac.jp
[3] Graduate School of Science and Engineering, Ehime University, Matsuyama City 790-8577, Japan
[4] Cyclotron Research Center, Iwate Medical University, Takizawa 020-01673, Japan; ksera@iwate-med.ac.jp
* Correspondence: manovitapateda@ung.ac.id; Tel.: +62-822-3967-8077

Abstract: (1) Background: Geomedical science focuses on the relationship between environmental impact and human health. The abundance of elements in a geographic area is reflected accumulation of these elements in humans. This study aims to describe the relationship between concentrations of geologic elements and accumulations in the human body as well as element-related symptoms. (2) Methods: Geogenic sampling was conducted in an Artisanal and Small-Scale Gold Mining (ASGM) area and around residential areas in Indonesia, and samples were analyzed using particle-induced X-ray Emission (PIXE). Head hair was sampled, and health assessments were performed to determine heavy metal exposure, especially to copper and mercury. (3) Results: Results show that potentially toxic elements' accumulation in the human body follows the abundance of these elements in the geographic area, which then affect health and manifest with specific signs and symptoms. East Tulabolo is an area rich in copper (hazard quotient (HQ) in dust = 152.8), and most of the population shows the sign of Kayser–Fleischer rings. Likewise, the Dunggilata area has the highest concentration of mercury, especially in the dust (HQ = 11.1), related to ASGM activity in residential areas. (4) Conclusions: This study concludes that the geogenic concentration of elements parallels the accumulation of human tissue and manifests with element-related signs and symptoms.

Keywords: copper; mercury; geomedical science; Gorontalo

Citation: Pateda, S.M.; Sakakibara, M.; Sera, K. Element Rich Area Associated with Human Health Disorders: A Geomedical Science Approach to Potentially Toxic Elements Contamination. *Int. J. Environ. Res. Public Health* **2021**, *18*, 12202. https://doi.org/10.3390/ijerph182212202

Academic Editor: Paul B. Tchounwou

Received: 13 October 2021
Accepted: 17 November 2021
Published: 20 November 2021

Publisher's Note: MDPI stays neutral with regard to jurisdictional claims in published maps and institutional affiliations.

1. Introduction

1.1. Background

Human health is constantly influenced by exposure to potentially toxic elements in the environment [1–8]. Since ancient times, human interaction with potentially toxic elements has been quite close, through exposure to medicine, equipment, and other sources. Heavy metal concentrations in the environment are quite high, with contaminants in food or geogenic materials. In this industrial age, large-scale mining has caused occupational diseases in the form of poisoning by various toxic metals [1]. Potentially toxic elements do not undergo metabolism in the human body; instead, they accumulate there and combine with ligands that are biochemical, resulting in toxic effects [1,9–11].

Heavy metal contamination can be viewed from the perspective of geomedical science usually known as medical geology, a science that studies health problems related to "location", based on the abundance of potentially toxic elements in the area that affects human health both directly and indirectly [2,12]. The journey of heavy metal elements into the body's metabolic system is a complex process, influenced by many heterogeneous determinants. The essential threat of heavy metal exposure is access to the elements and their fate after entering human cell system. The basic principle of exposure to an agent is that it must enter the human body, which acts as a host, by first passing through a medium.

The media of toxic elements in the geosphere and sociosphere are air, dust, food, water, and soil. Exposure to potentially toxic elements in humans occurs through inhalation, ingestion, and skin absorption [13–15].

The complexity of metabolic processes in the human body depends on many factors, both internal and external. Toxicokinetics and toxicodynamics of potentially toxic elements in the body vary depending on the character and history of human health. The elimination process is through two routes, primarily excretion through urine or feces and through short- or long-term accumulation pathways. Long-term retention manifests as signs of heavy metal exposure; for example, Kayser–Fleischer rings are as a sign of copper accumulation in the iris, and analysis of heavy metal concentrations in hair reveal it as a place of accumulation.

Analysis of long-term heavy metal accumulation in the human body shows that it manifests in specific signs and symptoms. Previous studies have focused on the analysis of mercury as the main pollutant in gold mining. This study focuses on two elements based on geologic character studies and sources of heavy metal pollution: copper (Cu) and mercury (Hg). The overall aim of this study is to provide information that can prove the predictability of exposure to these elements leading to accumulation in the human body.

1.2. Geologic Character of Gorontalo Province, Indonesia

Gorontalo Province consists of two geologic sheets: Tilamuta and Kotamobagu. The north side of Sulawesi Island, where Gorontalo Province is located, is a volcano-plutonic arc [16]. Outcrop observations show a malachite layer in the porphyry Cu oxidation zone. In the process of dismantling Cu ore from sulfide minerals, the Cu is carried away by rainwater, which is then deposited in the soil and water [16].

1.3. Artisanal and Small-Scale Gold Mining

Artisanal and Small-Scale Gold Mining (ASGM), as defined by the United Nations Environment Program, is gold mining conducted by individual miners or small enterprises with limited capital investment and production costs [3,17–19]. Processing of ore bearing gold in the ASGM area uses mercury as a gold metal binder. Initially, the effects of mine pollution affected workers and communities around the mine; however, the impact of Hg contamination can extend to the communities who live far from the mine area, and Hg contamination from mining is slowly but surely expanding.

Pateda et al., in their research on Gorontalo Province, reported a tendency for chronic Hg accumulation in human hair samples, and the chronic effect of Hg vapor on respiratory health [20]. This detrimental effect is not only experienced by the miners, but also by the people who live around the mining area even extending to a radius of a longer distance [21–23].

Generally, ASGM activity usually comprises the following steps: (a) Extraction, miners exploit alluvial deposits (river sediments) or hard rock deposits; (b) Processing, the gold is liberated from other minerals. Trommels are widely used and added with mercury in this process; (c) Concentration, in Gorontalo mining, the ore is then concentrated and washed with the help of sluices and pans; (d) Amalgamation, elemental mercury is used to obtain a mercury–gold alloy called an "amalgam"; (e) Burning, the amalgam is heated to vaporize the mercury and separate the gold; and (f) Refining, sponge gold is further heated to remove residual mercury and other impurities [24–27].

2. Materials and Methods

The research data for this study include geologic information and results of human health collected from 2018 to 2019. The geologic data comprise two types of samples: soil and dust. The health assessments were carried out on 192 respondents classified into two groups based on residence in a polluted or a non-polluted (control) area. Of the 192 participants, 108 were from polluted areas (villages of East Tulabolo, Dunggilata, Hulawa, and Bumela) and 84 were from control areas (villages of Bongo and Longalo).

The polluted area was defined as an area that is close to ASGM, whereas the control area is an area where no mining activity is present. Geologic samples were taken from several points, both in the mining area and around residential areas.

2.1. Soil and Dust Sample Analysis

Soil samples taken from 4 mining areas (East Tulabolo, Dunggilata, Hulawa, and Bumela) and from 2 control areas (Bongo and Longalo). Ground soil was sampled from a depth of 20–30 using a shovel and then placed into a plastic bag.

Dust sampling was conducting using a soft toothbrush, and the samples were stored in a paper bag. Each dust sample taken at a point close to the location of the ground soil sample. All soil and dust samples were analyzed using particle-induced X-ray emission (PIXE) conducted at the Cyclotron Research Center, Iwate Medical University, Japan. Sample preparation was required based on a standard method for PIXE. First, the soil and dust samples were ground to powder using a ball mill powdering system. Furthermore, the samples were mixed with the internal standard using the palladium on carbon powder method [28]. The normal concentration range of the soil and dust samples were calculated from the average values in the world [29]. The threshold value for Hg and Cu were 52 and 20 mg/kg, respectively.

2.2. Hair Sample Analysis

A grouping of scalp hair approximately 3 cm long was obtained from the root in the occipital region of each respondent. This approach meant that long hairs were sampled and then cut to 3 cm and only the 3 cm on the root side was analyzed. Hair samples went through several processes before being ready to send for inductively coupled plasma mass spectrometry (ICP-MS) analysis. This process was divided into three stages: stage (1) pre-washing; stage (2) washing; and stage (3) post-washing. The washing process by the ultrapure water (MilliQ®) aimed to remove contaminants from the hair, such as dust, dirt, bacteria, and other possible elements.

The scalp hair samples were analyzed by ICP-MS in the Research Institute for Humanity and Nature, Kyoto, Japan. Several steps for the sample and standard solution preparation were carried out. Analytical accuracy and precision were verified using Certified Reference Material No.13 (PerkinElmer, Waltham, MA, USA). Rhodium solution (Wako Pure Chemical Industries, Osaka, Japan) was used as internal standard solution.

2.3. Hazard Quotient Quantification

An appropriate assessment method depends on the known toxic effect of the chemicals and comprises the chemical mixture, availability of toxicity data, and quality of available exposure data [13]. The potential health hazard from exposure to each chemical is estimated by calculating its individual hazard quotient (HQ) with the following formula [13]:

$$HQ = \frac{Chemical\ Exposure}{Standard\ exposure}$$

2.4. Neurologic Assessment

Evaluation of the neurologic system was performed by examining several clinical signs, for which the most dominant neurologic sign is tremor. A tremor is a rhythmic shaking movement in one or more parts of the body. According to the Protocols for Environmental and Health Assessment of Mercury Released by Artisanal and Small-Scale Gold Miners by United Nations Industrial Development Organization, this study implemented the finger to nose test. The procedure for this test is for the participant to stand still, with their legs together, arms outstretched, and eyes closed, and then to touch their fingertip to their nose. The rating scale score range is: 0, no tremor; 1, slight to moderate (amplitude 0.5–1 cm), may be intermittent; 2, marked amplitude (1–2 cm); and 3, severe amplitude (>2 cm) [30].

2.5. Eye Assessment

Examination of Cu accumulation used the presence of a sign called Kayser–Fleischer rings, which are dark rings that appear to encircle the iris of the eye. Using a medical flashlight, the research team checked the participants' eyes and could see the rings with the naked eye on the inner surface of the cornea in the Descemet membrane. The rings typically appear as a golden, brown ring in the peripheral cornea.

3. Results

3.1. Element-Rich Area

3.1.1. Distribution of Elements per Area

Table 1 shows the ratio of the lowest and highest values of Cu and Hg concentrations in soil and dust samples from four polluted areas and two control areas. Most of the highest Cu and Hg concentrations were present in the dust samples with the maximum values of 3055 mg/kg in the East Tulabolo area and 577 mg/kg in the Dunggilata area. Compared with the dust concentration, the ground soil sample showed a concentration of 50% of the value of the dust sample.

Table 1. Ratio of Cu and Hg concentrations in soil and dust samples per area.

Heavy Metal in Soil	Limit Regulation (mg/kg)	Range of Potentially Toxic Elements Concentration (mg/kg) per Area (Min–Max)					
		East Tulabolo	Dunggilata	Hulawa	Bumela	Longalo	Bongo
		In Soil					
Cu	20	86–**1470**	96–183	37–536	86–122	52–66	37–52
Hg	52	32–131	10–**294**	2–128	44.7	DL	33.3
		In dust					
Cu	20	399–**3055**	63–329	-	-	1–24	39–50
Hg	52	15–91	63–**577**	-	-	56–98	48–65

Cu = copper; Hg = mercury; and DL = detection limit; The bold number show the highest concentration of element.

The distribution of Cu and Hg concentrations compared with the limit regulation value is depicted in Figure 1 for polluted and control areas. Dust samples are unavailable for two areas, Hulawa and Bumela. Most data show that the Cu concentration is above the limit value of 20 mg/kg, except in the control region. Concentration variations were found in the Hg samples, where the highest polluted area was the Dunggilata area, as one of the active ASGMs in Gorontalo Province. A high Hg value in the control area was found in the dust sample with a median value above the limit value.

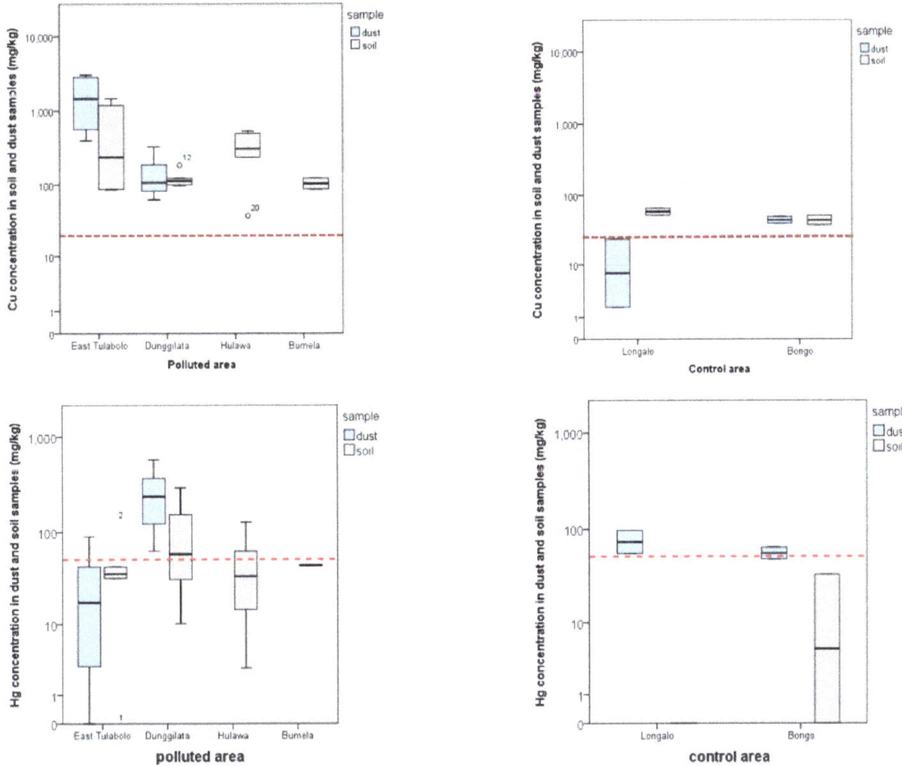

Figure 1. Distribution of copper (Cu) and mercury (Hg) concentrations per area based on polluted and control areas. The mean concentration value uses a logarithmic scale as there is a large difference in the lowest and highest values. The red dotted line indicates the threshold limit according to the regulation. The confidence interval of this study is 95%.

3.1.2. Hazard Quotient Quantification

Hazard quotient is a comparison parameter between the exposure value with the standard allowable level. Table 2 describes the HQ of each metal element by area. The concentration of Cu in the dust showed an extreme increase compared with the increase in Hg, either in the dust or in the ground soil. In fact, the Cu concentration was markedly increased in the control region. The Hg value in the control area was also quite high in the dust sample.

Table 2. Hazard quotient of element based on area.

Area	Cu		Hg	
	Soil	**Dust**	**Soil**	**Dust**
East Tulabolo	**73.5**	**152.8**	2.5	1.8
Dunggilata	9.2	16.4	**5.7**	**11.1**
Hulawa	26.8	NS	2.5	NS
Bumela	6.1	NS	0.9	NS
Longalo (control)	3.3	1.2	-	1.9
Bongo (control)	2.6	2.5	0.6	1.3

Cu = copper; Hg = mercury; and NS = not sampled; The bold number show the highest HQ of element.

3.1.3. Mapping of Elements

Mapping was used to show areas based on HQs parametrically due to the large gap between the lowest and the highest values. The concentration of Cu is quite high throughout the Gorontalo area, which is represented by the research sampling areas, both in the polluted and control areas. The largest HQ for Cu is depicted in the East Tulabolo area, whereas the largest Hg HQ is in the Dunggilata area, and the smallest is in the Bumela area, an average of more than 5 of HQ. These are shown in Figure 2.

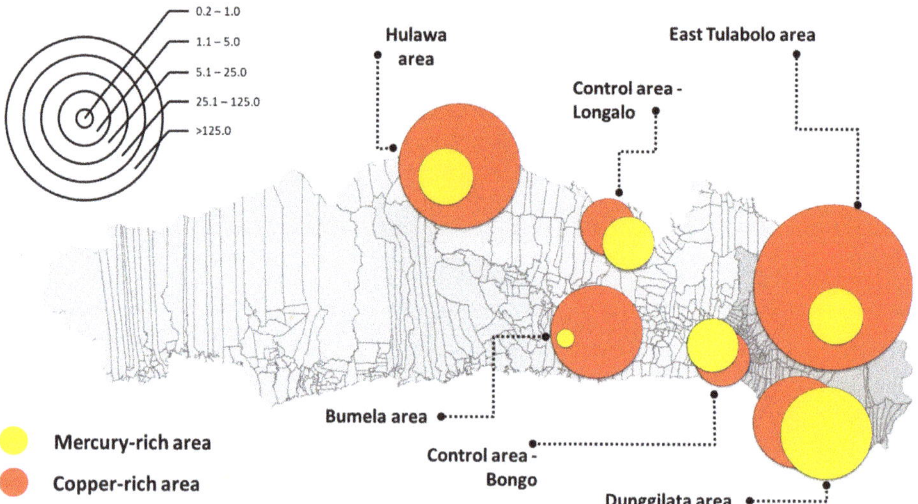

Figure 2. Mapping of copper (Cu) and mercury (Hg) hazard quotient per area. The scale in the left corner provides a parametric picture of the magnitude of the hazard quotient. The size of the circle is adjusted according to the hazard quotient ratio of each element, red for copper and yellow for mercury.

3.2. Sign and Symptom Distribution per Area

Public health assessments related to exposure to potentially toxic elements, especially Cu and Hg, provide data for the sign and symptoms observed and reported from the general health check-up. The presence of Kayser–Fleischer rings is a sign of chronic accumulation of Cu exposure in humans, whereas exposure to Hg is indicated by tremor as a symptom of neurologic disorders related to Hg contamination. Figure 3 describe it well.

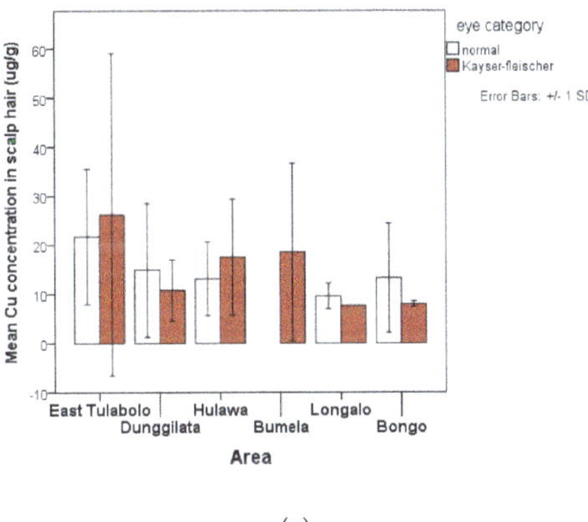

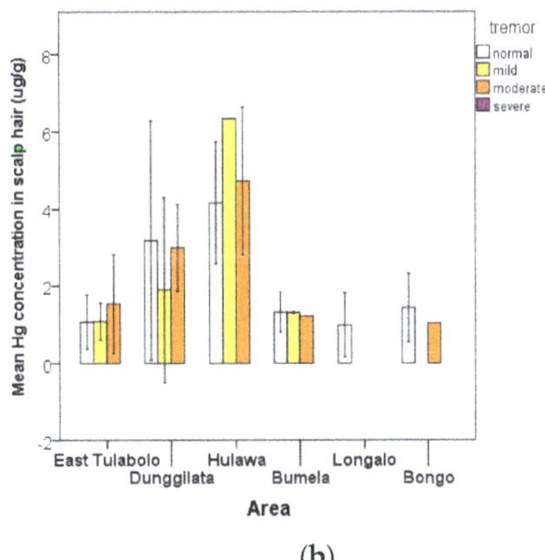

(a) (b)

Figure 3. Graph depicting the relationship between the presence of health signs and symptoms related to heavy metal exposure and the accumulation of these potentially toxic elements in scalp hair: (**a**) average accumulation of copper (Cu) in human scalp hair per area shows a high number in the East Tulabolo area, and is almost as common in all areas; (**b**) average accumulation of mercury (Hg) in scalp hair per area shows a high number in the Hulawa and Dunggilata areas.

4. Discussion

The uptake of elements is a process that may vary considerably depending on the complexity of the living system. Moreover, the bioavailability of elements is the main factor that influences the impact of an environmental chemical species in the human body. The bioaccumulation of elements in geologic samples is used as a consideration in risk assessment, whereas biological samples are used as bioindicators of exposure to an element including potentially toxic elements such as Cu and Hg.

A potentially toxic element that has made its way into a human body has travelled through an environmental medium [2]. The media are air, water, and soil/food, which enter the human body through three pathways: inhalation, skin absorption, and digestion. After undergoing the process of cellular metabolism, the element is then eliminated in either of two ways: excreted or accumulated. The process of excretion is through urine and feces, whereas the accumulation process occurs in several organs, including scalp hair. This accumulation is used as a bioindicator to assess the effect of exposure to a heavy metal. Once accumulated, potentially toxic elements will be trapped in organ tissues for a long time.

In the context of environmental monitoring studies, bioindicators reflect organisms— or parts of organisms or communities of organisms—that contain information on the quality of the environment—or a part of the environment [31,32]. The abundance of an element in a geographic area, be it the result of natural or anthropogenic processes, characterizes the accumulation of elements in the tissues of the human body. In line with the research conducted, the abundance of Cu levels in the Gorontalo area is due to its geologic character, which is the porphyry Cu oxidation zone.

The high concentration of Hg in the environment is from ASGM activity in polluted areas. The dust samples, representative of the presence of Hg in the air, provide evidence that Hg can be carried by the wind to more distant locations. This effect is reflected in the study results showing that the Hg concentration was also high in the control area where no ASGM activity was present. The high levels of Hg in the environment in the Dunggilata area are caused by mining work, especially mining practices that use Hg in production, such as concentration and burning of amalgam, which is carried out in residential settings that cover a wider area. Therefore, this exposure leads to quite a lot of neurologic disorder effects in this area.

The passage of elements from the environment into the human body and give effect, through complex stages. There is an intermediate analysis that must be studied previously to explain the relationship between the abundance of geological elements and their concentration in the human body. This is the limitation in this study that needs to be clarified with other more in-depth studies.

5. Conclusions

Accumulated elements from geogenic activities in human biological tissue can result from exposure from environmental sources. Humans take potentially toxic elements into their bodies, both consciously and unconsciously, from both natural and anthropogenic sources. This incorporation leads to the accumulation of potentially toxic elements in the tissues of the human body which then manifest as the most likely signs and symptoms to appear associated with the toxicity.

Author Contributions: All authors contributed extensively to the work presented in the paper. S.M.P. as principal researcher conducted the analysis and wrote the manuscript which was undertaken in association with the project theme. M.S. as the head of project team supervised the analyses and edited the manuscript. K.S. as the analyst of PIXE measurement of soil and dust. All authors have read and agreed to the published version of the manuscript.

Funding: This research was funded by The Research Institute for Humanity and Nature (RIHN: a constituent member of NIHU), Project No.14200102.

Institutional Review Board Statement: The study was conducted according to the guidelines of the Declaration of Helsinki, and approved by the medical ethical committee of RIHN, Japan (protocol code 2018-6, approved on 2018).

Informed Consent Statement: Informed consent was obtained from all subjects involved in the study.

Data Availability Statement: Not applicable.

Acknowledgments: Sincere appreciation goes to The Research Institute for Humanity and Nature (RIHN: a constituent member of NIHU) for the financial support under project No.14200102. Special thanks to Gorontalo Province government. My indeed thank you for my family and also for all colleagues in Universitas Negeri Gorontalo.

Conflicts of Interest: The authors declare no conflict of interest.

References

1. Syarif, A.; Estuningtyas, A.; Setiawati, A. *Farmakologi dan Terapi FK UI Ed 5 (Cetak Ulang dengan Tambahan, 2012)*; UI Press: Jakarta, Indonesia, 2012.
2. Selinus, O.; Davies, B.E.; Bowman, C.; Davies, P.C. *Essentials of Medical Geology: Revised Edition*; Elsevier Academic Press: London, UK, 2013.
3. UNEP. Reducing Mercury Use in Artisanal and Small-Scale. In *United Nations Environ. Policy*; UNEP: Nairobi, Kenya, 2012; p. 76.
4. Rajaee, M.; Obiri, S.; Green, A.; Long, R.; Cobbina, S.J.; Nartey, V.; Buck, D.; Antwi, E.; Basu, N. Integrated Assessment of Artisanal and Small-Scale Gold Mining in Ghana—Part 2: Natural Sciences Review. *Int. J. Environ. Res. Public Health* **2015**, *12*, 8971–9011. [CrossRef] [PubMed]
5. Eisler, R. Health risk of gold miners: A synoptic review. *Env. Geochem Heal.* **2003**, *25*, 325–345. [CrossRef] [PubMed]
6. Baldwin, D.R.; Marshall, W.J. Heavy metal poisoning and its laboratory investigation. *Ann. Clin. Biochem.* **1999**, *36*, 267–300. [CrossRef] [PubMed]
7. Adal, A. Heavy Metal Toxicity Background, Pathophysiology, Epidemiology. *Medscape* **2016**, *Volume 7*.
8. Harada, M. Minamata disease:methylomercury poisoning in Japan causes by environmental pollution. *Crit. Rev. Toxicol.* **1995**, *25*, 1–24. [CrossRef] [PubMed]
9. Tchounwou, P.B.; Yedjou, C.G.; Patlolla, A.K.; Sutton, D.J. *Molecular, Clinical and Environmental Toxicology*; Springer: Berlin/Heidelberg, Germany, 2012; Volume 101, pp. 1–30.
10. Bensefa-Colas, L.; Andujar, P.; Descatha, A. Mercury poisoning. *Rev. Med. Interne* **2011**, *32*, 416–424. [CrossRef] [PubMed]
11. Zalups, R.K. Molecular interactions with mercury in the kidney. *Pharmacol. Rev.* **2000**, *52*, 113–143. [PubMed]
12. Selinus, O. *Medical Geology: A Regional Synthesis*; Springer: Berlin/Heidelberg, Germany, 2010.
13. Friberg, L.; Nordberg, G.F.; Vouk, V.B. *Handbook on the Toxicology of Metals*; Academic Press: Cambridge, MA, USA, 2015.

14. Pavilonis, B.; Grassman, J.; Johnson, G.; Diaz, Y.; Caravanos, J. Characterization and risk of exposure to elements from artisanal gold mining operations in the Bolivian Andes. *Environ. Res.* **2017**, *154*, 1–9. [CrossRef] [PubMed]

15. Wu, M.; Deng, J.; Lin, K.; Tsai, W. Lead, Mercury, and Arsenic Poisoning Due to Topical Use of traditional Chinese medicines. *Am. J. Med.* **2013**, *126*, 451–454. [CrossRef]

16. Yamamoto, M.; Maulana, A.; Yonezu, K.; Watanabe, K.; Subehan, A. Copper-Gold Mineralization Characteristics of the Sungai Mak Deposit in Gorontalo, Northen Sulawesi, Indonesia. *Int. J. Eng. Sci. Appl.* **2016**, *2*, 38–42.

17. Telmer, K.H.; Veiga, M.M. World emissions of mercury from artisanal and small scale gold mining. In *Mercury Fate and Transport in the Global Atmosphere: Emissions, Measurements and Models*; Springer: Berlin/Heidelberg, Germany, 2009; pp. 131–172.

18. Futsaeter, G.; Wilson, S. The UNEP Global Mercury Assessment: Sources, Emissions and Transport. In *E3S Web of Conferences*; EDP Sciences: Les Ulis, France, 2013; p. 36001.

19. Keane, S.E.; Artisanal and Small Scale Gold Mining Area. UNEP Global Mercury Partnership. 2019. Available online: https://wedocs.unep.org/handle/20.500.11822/30811 (accessed on 12 October 2021).

20. Pateda, S.; Sakakibara, M.; Sera, K.; Pateda, S.M.; Sakakibara, M.; Sera, K. Lung Function Assessment as an Early Biomonitor of Mercury-Induced Health Disorders in Artisanal and Small-Scale Gold Mining Areas in Indonesia. *Int. J. Environ. Res. Public Health* **2018**, *15*, 2480. [CrossRef]

21. Nakazawa, K.; Nagafuchi, O.; Kawakami, T.; Inoue, T.; Yokota, K.; Serikawa, Y.; Cyio, B.; Elvince, R. Human health risk assessment of mercury vapor around artisanal small-scale gold mining area, Palu city, Central Sulawesi, Indonesia. *Ecotoxicol. Environ. Saf.* **2016**, *124*, 155–162. [CrossRef] [PubMed]

22. Gafur, N.A.; Sakakibara, M.; Sano, S.; Sera, K. A case study of heavy metal pollution in water of Bone River by Artisanal Small-Scale Gold Mine Activities in Eastern Part of Gorontalo, Indonesia. *Water* **2018**, *10*, 1507. [CrossRef]

23. Arifin, Y.; Sakakibara, M.; Sera, K. Impacts of Artisanal and Small-Scale Gold Mining (ASGM) on Environment and Human Health of Gorontalo Utara Regency, Gorontalo Province, Indonesia. *Geosciences* **2015**, *5*, 160–176. [CrossRef]

24. Telmer, K.; Stapper, D. *Reducing Mercury Use in Artisanal and Small-Scale Gold Mining—A Practical Guide*; United Nations Environment Programme; Artisanal Gold Council: Geneva, Switzerland, 2012.

25. Esdaile, L.J.; Chalker, J.M. The Mercury Problem in Artisanal and Small-Scale Gold Mining. *Chem. A Eur. J.* **2018**, *24*, 6905–6916. [CrossRef] [PubMed]

26. World Health Organization. *Artisanal and Small-Scale Gold Mining and Health—Environmental and Occupational Health Hazards Associated with Artisinal and Small-scale Gold Mining*; Technical Paper; WHO Press: Geneva, Switzerland, 2016.

27. Richard, M.; Moher, P.; Telmer, K. *Health Issues in Artisanal and Small-Scale Gold Mining: Training for Health Professionals. Version 1.0*; Artisanal Gold Council: Victoria, BC, Canada, 2014.

28. Goto, S.; Hosokawa, T.; Saitoh, Y.; Sera, K. Soil sample preparation for PIXE analysis. *Int. J. PIXE* **2014**, *24*, 77–83. [CrossRef]

29. Nieder, R.; Benbi, D.K.; Reichl, F.X. *Soil Components and Human Health*; Springer: Berlin/Heidelberg, Germany, 2018.

30. Veiga, M.M.; Baker, R.; Baker, R.F.; Fried, M.B.; Withers, D. *Protocols for Environment and Health Assessment of Mercury Released by Artisanal and Small-Scale Gold Miners*; GEF/UNDP/UNIDO: Vienna, Austria, 2005.

31. Knezović, Z.; Trgo, M.; Sutlović, D. Monitoring mercury environment pollution through bioaccumulation in meconium. *Process. Saf. Environ. Prot.* **2016**, *101*, 2–8. [CrossRef]

32. Markert, B. Definitions and principles for bioindication and biomonitoring of trace metals in the environment. *J. Trace Elem. Med. Biol.* **2007**, *21*, 77–82. [CrossRef] [PubMed]

International Journal of
*Environmental Research
and Public Health*

MDPI

Case Report

Transdisciplinary Online Health Assessment of an Artisanal and Small-Scale Gold Mining Community during the COVID-19 Pandemic in the Mandalay Region of Myanmar

Win Thiri Kyaw [1,*], **Yee Mon Myint** [2], **Xiaoxu Kuang** [1] **and Masayuki Sakakibara** [1,3]

[1] Research Institute for Humanity and Nature, Kyoto 603-8047, Japan; xxkuang@chikyu.ac.jp (X.K.); sakaki@chikyu.ac.jp (M.S.)
[2] Magway General Hospital, Magway 04011, Myanmar; yeemonmyint@gmail.com
[3] Graduate School of Science and Engineering, Ehime University, Matsuyama 790-8577, Japan
[*] Correspondence: thiri@chikyu.ac.jp; Tel.: +81-075-707-2443

Abstract: Artisanal and small-scale gold mining (ASGM) has a known negative effect on the community's health; therefore, assessment to monitor community health is essential to detect any issues and enable early treatment. Because ASGM-related health issues are complex and cannot be addressed effectively with a traditional one-time health assessment alone, both long-term and regular health assessments using a transdisciplinary approach should be considered. In response to this need, we designed an online health assessment tool as a reference for a future long-term health assessment system. An online video interview was conducted with 54 respondents living in the ASGM area of Chaung Gyi Village, Thabeikkyin Township, Mandalay Region, Myanmar, via a social networking service application. The tool was used to evaluate community health during the coronavirus 2019 pandemic, including mercury intoxication symptoms, mining-related diseases, and other diseases. Results show that persons working in mining versus non-mining occupations had a greater prevalence of pulmonary diseases, such as pulmonary tuberculosis, silicosis, and bronchial asthma, in addition to malaria. Based on these findings, online health assessment using a transdisciplinary approach can be recommended as an effective tool for sustainable and long-term health assessment of ASGM-related disease and should be performed regularly following physical health surveys.

Keywords: artisanal and small-scale gold mining; online health assessment; Myanmar; coronavirus disease; mining community

Citation: Kyaw, W.T.; Myint, Y.M.; Kuang, X.; Sakakibara, M. Transdisciplinary Online Health Assessment of an Artisanal and Small-Scale Gold Mining Community during the COVID-19 Pandemic in the Mandalay Region of Myanmar. *Int. J. Environ. Res. Public Health* **2021**, *18*, 11206. https://doi.org/10.3390/ijerph182111206

Academic Editor: Paul B. Tchounwou

Received: 22 September 2021
Accepted: 21 October 2021
Published: 25 October 2021

Publisher's Note: MDPI stays neutral with regard to jurisdictional claims in published maps and institutional affiliations.

1. Introduction

Artisanal and small-scale gold mining (ASGM) is a major livelihood of rural communities in developing countries; however, ASGM is also widely known as a major source of atmospheric pollution by mercury (Hg) [1] and other heavy metals, as well as the cause of serious health problems in mining communities. Elementary Hg is used in the ASGM process to extract gold (Au) from ore, and the Au-Hg amalgam is further burned to obtain Au, which releases Hg into the atmosphere. Because some ASGM activities take place in residential communities without a proper containment system, the Hg vapor released during the amalgam-burning phase is widely dispersed, leading to Hg infiltration into the bodies of persons working in mining (hereafter, miners) and other local community residents through inhalation.

Several countries are focusing on reducing the use of Hg in the ASGM process, but Hg amalgamation remains a favored method to obtain Au because of its accessibility, affordability, and applicability in local settings. According to previous studies in ASGM areas internationally, Hg is a serious health hazard in ASGM workers, regardless of the origin [2–11]. In addition to Hg-related health issues, ASGM can also cause infectious diseases, pulmonary diseases, and accidents resulting from dust inhalation [12]. All these studies point out that regular health assessments of ASGM communities are essential in monitoring their long-term health.

However, several major limitations exist for regular health assessment in ASGM communities because of the nature of the industry. Typically, ASGM is performed in remote areas situated far away from the medical facilities available, and thus, the communities cannot easily receive critical health care. It is also challenging to conduct regular assessments of ASGM community health. In addition, health surveys in ASGM areas are usually performed either only once or in limited frequency, which cannot sufficiently contribute to the understanding of the health status of the community, especially because of the severe nature of the ASGM-related problems. Moreover, this dire situation has escalated during the global pandemic of coronavirus disease (COVID-19) because onsite health surveys and in-person interactions are not possible.

Therefore, it is essential to develop effective and long-term sustainable health assessments of ASGM areas to monitor the community health status and solve related issues with a transdisciplinary approach. In this study, an online health survey was conducted in the ASGM community of Chaung Gyi Village, Thabeikkyin Township, Mandalay Region, Myanmar, during the COVID-19 pandemic. A video health interview was conducted using a transdisciplinary approach with collaboration among researchers, the local physician, the government, and local stakeholders as a case study for designing a future, long-term, sustainable health assessment. According to the findings of our previous preliminary research in the study area, a small number of miners had the signs and symptoms of chronic Hg intoxication [13]. This case study reports the health status of the ASGM community and discusses the effectiveness of the collaborative online health assessment tool.

2. Materials and Methods

2.1. Study Area

The location of the Chaung Gyi Village is depicted in Figure 1. According to the March 2019 data of the administrative office, Chaung Gyi Village has 1772 households and a population of 8375 people. The ASGM activity of Chaung Gyi Village can be classified into formal and informal types [13].

Figure 1. Study area. Map of Chaung Gyi Village. (Map is provided by Network Activities Group (NAG), a local NGO of Myanmar).

2.2. Transdisciplinary Approach Research Design

Between December 2020 and February 2021, 54 individuals from Chaung Gyi Village, Chaung Gyi Village Tract, Thabeikkyin Township, Mandalay Region, Myanmar, were recruited for this study. The mean age of the participants was 36.8 ± 13.4 years; among them, 32 were male, and 22 were female. As shown in Figure 2, Myanmar government entities, including the Environmental Conservation Department (ECD) of the Ministry of Natural Resources and Environmental Conservation (MONREC), and the local stakeholders of Chaung Gyi Village, researchers, and local physicians worked together and contributed to the implementation of the online health assessment. The ECD and MONREC shared ASGM information for the study area with researchers through collaboration with local stakeholders, and the local stakeholders contributed their support to connect the ASGM community of Chaung Gyi Village with the researchers and the local physician. Then, the researchers and local physicians contributed to the design and development of the online health assessment tool to make it suitable for the study area through collaboration with the government and local stakeholders, including the mining community. Then, the researchers and local physicians engaged the community (hereinafter referred to as "respondents") in a conversation by video interviews of the respondents' social networking service application to explain and conduct the study. Verbal informed consent was obtained from the respondents because the nature of the study made it impossible to receive their written consent.

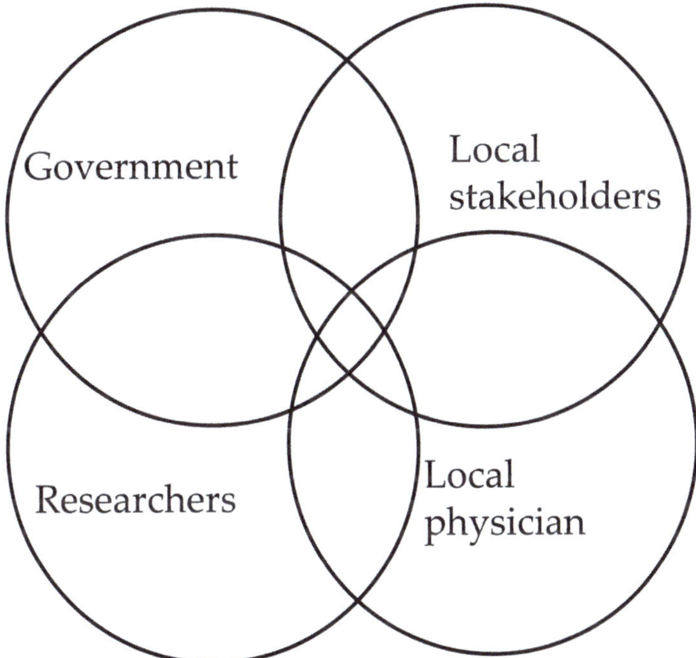

Figure 2. Transdisciplinary approach research design.

2.3. Interview Questions

Respondents used the questionnaire form to provide the following data: (1) general information, including living status as a native to the study area or migrant, education, occupation, and income; (2) risk of exposure to Hg and other heavy metals, such as the distance of their homes to ASGM sites, use of Hg and cyanide in the ASGM process, and use of personal protection for miners; (3) health complaint; (4) present and past medical histories; (5) symptoms of Hg intoxication, such as metallic taste, excessive salivation, tremors, sleep disturbances, tiring easily, feeling sleepy or drowsy, lack of energy, concentration problems, forgetfulness, feeling nervous or sad, palpitations, headaches, nausea, loss of appetite, weight loss, numbness, prickling, and aching sensations. Then, during the video interview, respondents were evaluated for signs of Hg intoxication, gait ataxia, tremor on the finger-to-nose test, and alternate movements on the twist-hand test. Based on their type of occupation at the time of health assessment, respondents were divided into two groups: miners and non-miners.

2.4. Statistical Analysis

An unpaired *t*-test was employed for comparison of the mean values, and Fisher's exact test was performed for the analysis of comparison among the variables of miners and non-miners, respectively. The level of significance was set at $p < 0.05$.

3. Results

3.1. Socioeconomic Assessment

Table 1 describes the results of the demographic and socioeconomic assessments. The number of miners was 24 (male; 18 and female 6), and that of non-miners was 30 (male 14 and female 16). The percentage of male and female participants and age were equally distributed. Most participants were aged 20–29 years. No difference was noted between the two groups in terms of living status as native or migrant and education status. Most respondents had completed primary education. The income of the miners ranged from \$6–\$31 USD/day, which was considered to meet the threshold of monthly family expenses, whereas those in local non-mining occupations earn less ($p = 0.01$). Some of the miners of this study work in the underground mines, 20–30 m below the surface of the ground, whereas some work to retrieve gold using the sluicing methods in the rivers. Miners who work for the excavation of ores underground mostly work long continuous hours for one to two days at the mining site. The underground mining area is situated far from the village, from where the excavated Au-containing ore is usually transported to the village for further processing, such as crushing, screening, and refining at the homes of miners, local gold shops, or in open spaces on the streets of the village. Crushing ore is done using simple equipment in front of their houses, and then screening ore is made in the reservoir dug by the miners in the yard. The Au-amalgam is prepared by mixing the Au-containing ore with the Hg purchased from local convenience stores, after which the amalgam is burned to evaporate the Hg into the atmosphere and retrieve the Au. This process is known as refining.

The categories of occupants in the non-miner group were agriculture, bricklaying, vehicle driving, buying and selling goods, miscellaneous jobs, and household jobs. Three men from the non-miner groups worked previously at random in excavation sites and in crushing ores of mining temporally, but not as the occupation. The respondents engaged in agriculture were the owners, or the owners' family members, of the lands where they were working, and their working routine varied according to the seasons and types of crops and vegetables they were growing. The agricultural products produced were consumed by the producers as well as sold in the local wet markets and other distant cities. The bricklaying category of occupation was irregular and dependent on the availability of the job offer. The earnings from buying and selling of goods also depended on the demand of the consumers, which is similar to the nature of miscellaneous jobs and household jobs in the aspect of income being unstable. Compared with these non-mining jobs, mining work contributed to a more stable income in the study area.

Table 1. Demographic and socioeconomic assessments.

Variable	Category	54 Respondents (%)	Miners	Non-Miners
Sex	Male vs. Female	59.3	18 vs. 6	14 vs. 16
Age (years)	Mean $\pm$ SD		37.3 $\pm$ 11.7	36.5 $\pm$ 14.2
Age (years)	20–29	38.9	9 (n)	12 (n)
	30–39	24.1	4 (n)	9 (n)
	40–49	16.7	6 (n)	3 (n)
	50–69	20.3	5 (n)	6 (n)
Living status	Native vs. migrant	59.3	13 vs. 11	19 vs. 11
Education status	Monastic School Education	7.4	1 (n)	3 (n)
	Primary School	51.9	8 (n)	20 (n)
	Middle School	16.7	7 (n)	2 (n)
	High School Completed	13	4 (n)	3 (n)
* Income per day (USD)	Mean $\pm$ SD		14.7 $\pm$ 10.8	2.5 $\pm$ 1.5
Income per day (USD)	Range (minimum–maximum)		$6–$31	$1–$8

* Statistically significant ($p = 0.01$) in unpaired t-test. Monastic school education contributes to the basic education need of the people with poverty who could not afford to attend regular schools and provide the curriculum education and ethics to the students.

3.2. Risk of Exposure to Hg and Other Heavy Metals

The factors influencing exposure to Hg and other heavy metals are detailed in Table 2. Of the 54 respondents, 25.9% ($n = 14$) reported living at a distance less than 1 km from ASGM activity. Most miners used Hg in their ASGM activities. Moreover, more than two-thirds of the miner group did not use personal protection, such as gloves and masks. These factors represent a high possibility of exposure to Hg and other heavy metals. Regarding the types of mining activities, most miners worked in excavation and crushing ores. In addition, most miners had worked in the ASGM occupation for 10–20 years.

Table 2. Risk of exposure to Hg and other heavy metals.

Risk Factor	Numbers (n)
Living distance from ASGM activity of all respondents	
<1 km	14
1–5 km	17
5–10 km	23
>10 km	0
Miner-Specific Risk Factor of miners	
Hg use	12
Cyanide use	3
Personal protection use during mining activity of miners	
Yes	9
No	15
Mining activities of miners	
Working across all areas of mining	6
Excavation and crushing ores primarily	10
Panning and amalgamation only	4
Refining (burning amalgam) only	2
Carrying gold ores only	2
Mining experience of miners	
<10 years	8
10–20 years	14
>20 years	2

3.3. Disease Prevalence by Occupation

Health complaints, including present and past medical histories of the respondents, were summarized as the disease prevalence, as shown in Figure 3. According to the interview results of the respondents, disease prevalence was classified into "normal" representing no health issues; "lung diseases" with pulmonary health issues; "malaria;" "minor illness" with minor health complaints; "musculoskeletal diseases," problems in joints, bones, muscles, and spine; "cardiovascular diseases" having typical symptoms of cardiovascular diseases including pain or pressure in the chest, pain or discomfort in the arms, left shoulder, elbows, jaw, or back, shortness of breath, nausea and fatigue, lightheadedness or dizziness and cold sweats; and "liver disease" which includes jaundice, abdominal pain and swelling, nausea, vomiting, and melaena.

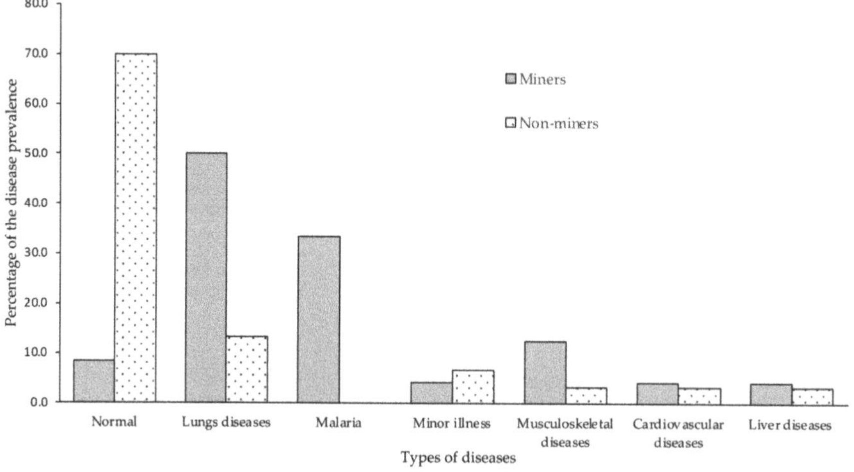

Figure 3. Disease prevalence in miners and non-miners.

On comparison of miners versus non-miners, 50% ($n = 12$) of miners, especially those who engaged in excavation and crushing ores, had greater pulmonary issues, including pulmonary tuberculosis, silicosis, and bronchial asthma, versus 13.3% ($n = 4$) of the non-miner group, which had pulmonary tuberculosis and bronchial asthma ($p = 0.01$). Miners with mining experience greater than 10 years had pulmonary disease, and they had to reduce their workload due to their health problems. History of malaria infection was also only detected in the miner group ($n = 8$, 33.3%). The prevalence of musculoskeletal diseases was 12.5% ($n = 3$) for miners and 3.3% ($n = 1$) for non-miners. The prevalence of cardiovascular diseases and liver diseases for miners were 4.2% for each category ($n = 1$) and non-miners were 3.3% for each category ($n = 1$) respectively.

3.4. Chronic Hg Intoxication Signs and Symptoms

Of the miners, 19% reported symptoms of Hg intoxication, such as sleep disturbance, loss of appetite and weight, depression, and numbness of the digits. Non-miners did not have any symptoms. However, no abnormal findings on the examination for the signs of Hg intoxication were noted for either group.

4. Discussion

4.1. Factors in the Choice of ASGM as the Main Livelihood in the Study Area

According to the nature of the socioeconomic characteristics, it can be clearly noted that ASGM can contribute to a stable and higher income for individuals within the community who have the same educational background compared with those in the non-mining occupations available in the study area, such as agriculture and other job options. Because ASGM activities can be conducted within the study area at a close distance to the community dwelling places, the accessibility to the ASGM areas is high. Therefore, ASGM may be the preferred occupation in the study area for the aforementioned factors, combined with the context of regional poverty.

4.2. Community Health Status

The study area community had a significant incidence of pulmonary disease, including pulmonary tuberculosis, silicosis, and bronchial asthma, in addition to malaria and musculoskeletal, cardiovascular, and liver diseases. Compared with the non-miner group, the miner group had an increased incidence of pulmonary diseases and malaria. Symptoms of pulmonary diseases, such as silicosis and pulmonary tuberculosis, have been reported by ASGM miners who worked in excavating and crushing ores as a major ASGM activity; however, these symptoms have not been detected in miners who performed only panning, amalgamation, or burning amalgam. The risk for silicosis is caused by inhalation of silica crystalline, which is a common health hazard in mining and increases the risk for silicosis, malignancy, and other diseases [14]. Silica exposure is related to the development of various diseases, including pulmonary tuberculosis, chronic obstructive pulmonary disease, and rheumatoid arthritis [15]. Therefore, the cause of the high incidence of pulmonary tuberculosis in miners in this study may be due to the presence of silicosis, which can also cause secondary infection in the non-mining community, such as in the families of miners. Because Myanmar is a developing country with poverty in remote areas, pulmonary tuberculosis is common. Therefore, it is necessary to decrease the incidence of ASGM-related silicosis to relieve the national burden of pulmonary tuberculosis. Malaria is prevalent in the Thabeikkyin Township [16], and the miners' nature of working in the mines and sleeping without sleeping nets could have led to increased exposure to malaria, in the present study, relative to that among the non-miners.

Regarding the health hazard of Hg exposure, only 19% of the miners had the symptoms of chronic Hg intoxication, although most of the miners were exposed to Hg through the ASGM process. Similarly, in our previous preliminary study, we detected the signs and symptoms of chronic Hg intoxication in only a minority of the miners, ref. [13] and Hg concentration in their hair was not high compared with that in ASGM areas of other nations. Considering these factors together, it can be assumed that the major health problem in the study area mining community is silicosis and its associated complications rather than chronic Hg intoxication. The reason for the absence of chronic Hg intoxication in the present study may be the lower Hg usage in the study area compared to that in other ASGM areas worldwide, which may have caused the decreased formation of methylmercury in the environment. However, it is difficult to draw conclusions only on this basis owing to the small sample size; thus, the continuous assessment remains necessary to evaluate the status of chronic Hg intoxication.

4.3. Effectiveness of Online Health Assessment in ASGM Context

To our knowledge, this study is the first to present an online health assessment conducted for the ASGM community during the COVID-19 pandemic. The researchers, local physicians, local government, and local stakeholders contributed equally to the study design during this difficult time of COVID-19, which limited access to the respondents. Being far from the health facilities, the unavailability of laboratory services to detect Hg concentrations in hair as biomonitoring data, and the limited ability to install the remote health monitoring systems in the study area results in the community remaining vulnerable to mining-related health hazards to date. However, applying this remote health assessment in the study area through video interviews using the individuals' mobile phones enables preliminary health assessment, which can help facilitate the diagnosis of diseases in the mining community to decide on further necessary treatment and management.

When compared to the past onsite preliminary study of the study area conducted during the non-COVID-19 pandemic period [13], the present study offers the following merits (1) shorter processing time for the survey investigators, (2) ability to adjust the interviewing time as per the availability of both the survey investigators and respondents,

(3) ease of contacting the investigators by the researcher, local physician, and respondents, (4) greater chance to learn the lifestyles of the mining community that is important in disease prevalence, and (5) the ability to monitor the health of the mining community in a consecutive manner, which is a challenging aspect for onsite surveys. Despite these advantages, there are some disadvantages, including the lack of clinical examination and the lack of analysis of the biomarkers of Hg, such as in hair and blood Hg contents. Therefore, it is recommended that onsite health assessment surveys be conducted to evaluate clinical examinations and the biomarkers once every year, followed by regular multiple online health assessments to effectively manage the health of the mining community.

Our study also had some limiting factors, such as the small sample size and the aforementioned fact of being unable to conduct the physical examinations of the respondents in person. However, the design of the present study allows the health assessment of the study area through communication among the researchers, local physician, and community by means of an individual online connection with community members, through which we hope that we can also implement regular and long-term assessment of the ASGM community in the future, conceived as an "online assessment system for the ASGM community." Furthermore, because the ASGM context reflects the severity of the associated health and environmental hazards, the traditional assessment using biomonitoring, such as the evaluation of Hg content in hair, is solely insufficient for the ASGM community health assessment. In the future, a trained local health assessment team, including the nurses and local stakeholders, for physical monitoring of the health of the local ASGM community should be formed with the support of physicians and researchers. As an adjunct to that team, the "online assessment system for the ASGM community" by health professionals and researchers in various fields is recommended for an effective management strategy to address the health-related problems of ASGM.

5. Conclusions

We present a case study of the effectiveness of the first online health assessment in an ASGM community in Myanmar, conducted with a transdisciplinary approach. The main health problems of the community of the study area were pulmonary disease and health-related complications due to the ASGM process. This assessment can be conducted serially and regularly in the future to monitor the health status of the ASGM community from various aspects in collaboration with local stakeholders, health professionals, and researchers.

Author Contributions: Authors contributed to the work equally. The research method was constructed by W.T.K. and M.S. Investigations were conducted by W.T.K., Y.M.M. and X.K. Original draft was written by W.T.K. Editing of the manuscript was performed by M.S. All authors have read and agreed to the published version of the manuscript.

Funding: This research was funded by the Research Institute for Humanity and Nature (RIHN; a constituent member of NIHU), Project No. 14200102 and a research grant from the Health Care Science Institute https://www.iken.org (accessed on 20 September 2021).

Institutional Review Board Statement: The study was conducted according to the guidelines of the Declaration of Helsinki and approved by the Ethics Committee of Research Institute for Humanity and Nature (protocol code; 2020-2 and date of approval; 26 August 2020).

Informed Consent Statement: Informed consent was obtained from all subjects involved in the study.

Data Availability Statement: The data that support the findings of the study are available from the corresponding author (Kyaw, W.T.) upon reasonable request.

Acknowledgments: This research was supported by the Research Institute for Humanity and Nature (RIHN: a constituent member of NIHU), Project No. 14200102. The authors would like to thank ECD, MONREC, ECD-Mandalay, and the headman of the Chaung Gyi Village for their important contributions to this study.

Conflicts of Interest: The authors declare no conflict of interest. The funders had no role in the design of the study; in the collection, analyses, or interpretation of data; in the writing of the manuscript, or in the decision to publish the results.

References

1. UNEP. *A Practical Guide to Reducing Mercury Use in Artisanal and Small-Scale Gold Mining*; United Nations Environment Policy: Nairobi, Kenya, 2012; 76p.
2. Steckling, N.; Bose-O'Reilly, S.; Pinheiro, P.; Plass, D.; Shoko, D.; Drasch, G.; Bernaudat, L.; Siebert, U.; Hornberg, C. The burden of chronic mercury intoxication in artisanal small-scale gold mining in Zimbabwe: Data availability and preliminary estimates. *Environ. Health A Glob. Access Sci. Source* **2014**, *13*. [CrossRef] [PubMed]
3. Yard, E.E.; Horton, J.; Schier, J.G.; Caldwell, K.; Sanchez, C.; Lewis, L.; Gastañaga, C. Mercury Exposure Among Artisanal Gold Miners in Madre de Dios, Peru: A Cross-sectional Study. *J. Med. Toxicol.* **2012**, *8*, 441–448. [CrossRef] [PubMed]
4. Harari, R.; Harari, F.; Gerhardsson, L.; Lundh, T.; Skerfving, S.; Strömberg, U.; Broberg, K. Exposure and toxic effects of elemental mercury in gold-mining activities in Ecuador. *Toxicol. Lett.* **2012**, *213*, 75–82. [CrossRef] [PubMed]
5. Tomicic, C.; Vernez, D.; Belem, T.; Berode, M. Human mercury exposure associated with small-scale gold mining in Burkina Faso. *Int. Arch. Occup. Environ. Health* **2011**, *84*, 539–546. [CrossRef] [PubMed]
6. Bose-O'Reilly, S.; Drasch, G.; Beinhoff, C.; Tesha, A.; Drasch, K.; Roider, G.; Taylor, H.; Appleton, D.; Siebert, U. Health assessment of artisanal gold miners in Tanzania. *Sci. Total Environ.* **2010**, *408*, 796–805. [CrossRef] [PubMed]
7. Gardner, R.M.; Nyland, J.F.; Silva, I.A.; Ventura, A.M.; de Souza, J.M.; Silbergeld, E.K. Mercury exposure, serum antinuclear/antinucleolar antibodies, and serum cytokine levels in mining populations in Amazonian Brazil: A cross-sectional study. *Environ. Res.* **2010**, *110*, 345–354. [CrossRef] [PubMed]
8. Bose-O'Reilly, S.; Lettmeier, B.; Gothe, R.M.; Beinhoff, C.; Siebert, U.; Drasch, G. Mercury as a serious health hazard for children in gold mining areas. *Environ. Res.* **2008**, *107*, 89–97. [CrossRef] [PubMed]
9. Silva, I.A.; Nyland, J.F.; Gorman, A.; Perisse, A.; Ventura, A.M.; Santos, E.C.O.; De Souza, J.M.; Burek, C.L.; Rose, N.R.; Silbergeld, E.K. Mercury exposure, malaria, and serum antinuclear/antinucleolar antibodies in amazon populations in Brazil: A cross-sectional study. *Environ. Health A Glob. Access Sci. Source* **2004**, *3*, 11. [CrossRef] [PubMed]
10. Drake, P.L.; Rojas, M.; Reh, C.M.; Mueller, C.A.; Jenkins, F.M. Occupational exposure to airborne mercury during gold mining operations near EL Callo, Venezuela. *Int. Arch. Occup. Environ. Health* **2001**, *74*, 206–212. [CrossRef]
11. Drasch, G.; Böse-O'Reilly, S.; Beinhoff, C.; Roider, G.; Maydl, S. The Mt. Diwata study on the Philippines 1999—Assessing mercury intoxication of the population by small scale gold mining. *Sci. Total Environ.* **2001**, *267*, 151–168. [CrossRef]
12. Eisler, R. Health risks of gold miners: A synoptic review. *Environ. Geochem. Health* **2003**, *25*, 325–345. [CrossRef] [PubMed]
13. Kyaw, W.T.; Kuang, X.; Sakakibara, M. Health Impact Assessment of Artisanal and Small-Scale Gold Mining Area in Myanmar, Mandalay Region: Preliminary Research. *Int. J. Environ. Res. Public Health* **2020**, *17*, 6757. [CrossRef] [PubMed]
14. NIOSH. *Health Effects of Occupational Exposure to Respirable Crystalline Silica*; NIOSH: Washington, DC, USA, 2002.
15. Calvert, G.M.; Rice, F.L.; Boiano, J.M.; Sheehy, J.W. Occupational silica exposure and risk of various diseases: An analysis using death certificates from 27 states of the United States. *Occup. Environ. Med.* **2003**, *60*, 122–129. [CrossRef] [PubMed]
16. Liu, Z.; Soe, T.N.; Zhao, Y.; Than, A.; Cho, C.; Aung, P.L.; Li, Y.; Wang, L.; Yang, H.; Li, X.; et al. Geographical heterogeneity in prevalence of subclinical malaria infections at sentinel endemic sites of Myanmar. *Parasites Vectors* **2019**, *12*. [CrossRef]

International Journal of
**Environmental Research
and Public Health**

Review

Mercury Pollution from Artisanal and Small-Scale Gold Mining in Myanmar and Other Southeast Asian Countries

Pyae Sone Soe [1,2,*], Win Thiri Kyaw [3], Koji Arizono [4], Yasuhiro Ishibashi [2] and Tetsuro Agusa [2]

[1]   Graduate School of Environmental and Symbiotic Sciences, Prefectural University of Kumamoto, Kumamoto 862-8502, Japan

[2]   Faculty of Environmental and Symbiotic Sciences, Prefectural University of Kumamoto, Kumamoto 862-8502, Japan; yisibasi@pu-kumamoto.ac.jp (Y.I.); te-agusa@pu-kumamoto.ac.jp (T.A.)

[3]   Research Institute for Humanity and Nature, Kyoto 603-8047, Japan; thiri@chikyu.ac.jp

[4]   Graduate School of Pharmaceutical Sciences, Kumamoto University, Kumamoto 862 0973, Japan; arizono@kumamoto-u.ac.jp

*   Correspondence: g2075002@pu-kumamoto.ac.jp

Citation: Soe, P.S.; Kyaw, W.T.; Arizono, K.; Ishibashi, Y.; Agusa, T. Mercury Pollution from Artisanal and Small-Scale Gold Mining in Myanmar and Other Southeast Asian Countries. *Int. J. Environ. Res. Public Health* **2022**, *19*, 6290. https://doi.org/10.3390/ijerph19106290

Academic Editor: José Ángel Fernández

Received: 31 March 2022
Accepted: 16 May 2022
Published: 22 May 2022

Publisher's Note: MDPI stays neutral with regard to jurisdictional claims in published maps and institutional affiliations.

Abstract: Mercury (Hg) is one of the most harmful metals and has been a public health concern according to the World Health Organization (WHO). Artisanal and small-scale gold mining (ASGM) is the world's fastest-growing source of Hg and can release Hg into the atmosphere, hydrosphere, and geosphere. Hg has been widely used in ASGM industries throughout Southeast Asia countries, including Cambodia, Indonesia, Laos, Malaysia, Myanmar, the Philippines, and Thailand. Here, 16 relevant studies were systematically searched by performing the PRISMA flow, combining the keywords of "Hg", "ASGM", and relevant study areas. Mercury concentrations exceeding the WHO and United States Environmental Protection Agency guideline values were reported in environmental (i.e., air, water, and soil) and biomonitoring samples (i.e., plants, fish, and human hair). ASGM-related health risks to miners and nonminers, specifically in Indonesia, the Philippines, and Myanmar, were also assessed. The findings indicated severe Hg contamination around the ASGM process, specifically the gold-amalgamation stage, was significantly high. To one point, Hg atmospheric concentrations from all observed studies was shown to be extremely high in the vicinity of gold operating areas. Attentions should be given regarding the public health concern, specifically for the vulnerable groups such as adults, pregnant women, and children who live near the ASGM activity. This review summarizes the effects of Hg in Myanmar and other Southeast Asian countries. In the future, more research and assessment will be required to investigate the current and evolving situation in ASGM communities.

Keywords: Hg; artisanal and small-scale gold mining; air; water; soil; plant; fish; human hair; health risk; Myanmar; Southeast Asia

1. Introduction

1.1. Mercury

Mercury (Hg) is listed among the top 10 most harmful metals by the World Health Organization (WHO), and its chemical forms are considered a public health concern [1]. All its common forms, including elemental (metallic), inorganic, and organic are highly toxic. In particular, methylmercury (MeHg) is the most dangerous form because it can bioaccumulate in microorganisms and biomagnify or enhance the trophic levels in aquatic food webs [2]. Meanwhile, elemental Hg can be converted to MeHg in aquatic sediments. The use of elemental Hg in artisanal and small-scale gold-mining (ASGM) sector can be hazardous because of the inhalation of Hg vapor, which easily penetrates the blood–brain barrier and induces neurotoxicity [3]. A famous catastrophic Hg outbreak occurred in Minamata Bay, Japan, in the 1950s, when factory wastewater containing MeHg from a factory was discharged into the Shiranui Sea, poisoning the people who ingested the contaminated seafood [3]. This became one of the first and the most critical incidents of Hg poisoning due to an industrial site.

Different forms of Hg can be released into the atmosphere, water, and across land as a result of human activities such as burning of fossil fuels (e.g., coal and petroleum), industrial effluents, product waste (e.g., electronic) from intentional use, dental amalgamation, agricultural practices, and ASGM and natural processes, including volcanic eruptions, rock weathering, and forest fires. Thus, Hg is discharged worldwide into the environments. ASGM is the world's fastest-growing source of Hg and can discharge Hg into both the aquatic environment and the terrestrial ecosystem. The emission of Hg into the atmosphere via ASGM account for 37.7% of global outputs among the other Hg emission sources, with South America, Asia, and Sub-Saharan Africa as the primary sources [4]. Meanwhile, ASGM occurs in more than 70 nations worldwide, with an estimated 16 million miners who afford up to 20% of the annual global gold production [4]. According to the global inventory of the United Nations Environmental Program (UNEP) in 2015 [4], 2220 tons of Hg was emitted to the atmosphere from all anthropogenic sources. Notably, 38% (838 tons) of Hg was emitted from the ASGM sources (Figure 1).

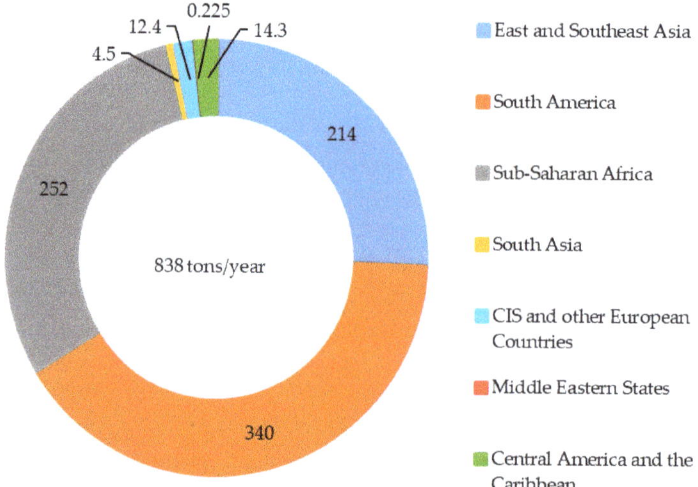

Figure 1. Regional results of global Hg emissions into air from the ASGM sources. Each value represents tons/year [4].

1.2. Hg Use in ASGM Communities

The use of Hg is quite common in ASGM sectors worldwide, including Southeast Asian countries such as Cambodia, Indonesia, Laos, Myanmar, the Philippines, and Thailand. The earliest records of Hg use in alchemy and amalgamation were from Egypt and China more than 3000 years ago [5]. Hg has been used in inexpensive, easy, and rapid approaches for extracting gold from its ore and soil [6]. Notably, owing to weak legislation, poor engagement, contribution of artisanal miners, and easily accessible of robust black market for Hg usage in the ASGM will continue to persist [7–9].

Myanmar, a developing country in Southeast Asia, has various mineral and natural resources, such as for jade, gold, ruby, and copper. Myanmar has been facing overexploitation of natural resources for more than two decades because its people seek to extract its natural resources illegally [10]. Additionally, 70% of the population of Myanmar is in rural areas and rely on natural resources. Meanwhile, gold production in Myanmar has increased slightly in recent years, reaching 1700 kg in 2016 [11]. ASGM activities in this country have been conducted throughout the states and regions of the country, namely Bago, Kachin, Mandalay, Mon, Sagaing, Shan, and Tanintaryi (Figure 2) [12].

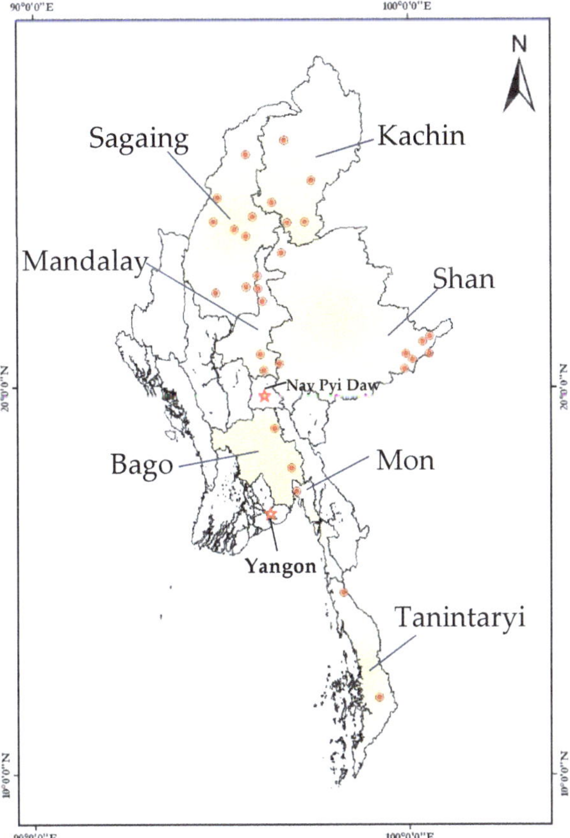

Figure 2. Distribution of artisanal and small-scale gold mining (ASGM) activities in the states and regions of Myanmar by township.

Various ASGM processes can be categorized on the basis of region and depend on the gold deposit types. ASGM can be classified using several methods such as panning, river mining with bucket dredges, suction dredging, and hydraulic mining. Alluvial deposits (river sediments) and hard-rock deposits (typically gold in quartz veins) have been exploited the gold ore by local and migrant miners. In the most common method for ASGM in Myanmar, gold ore is excavated via underground or open-pit mining, after which it is dried and ground using a powdering machine. The resulting powder is placed in a pan containing water to separate gold particles via gravity settling. Gold particles are collected at the bottom of the pan together with sand. A piece of Hg (like a finger-tip) is added to the pan to extract gold by forming a gold–Hg amalgam, which is squeezed by hand through a fabric cloth. Subsequently, a mine operator vaporizes the Hg in the gold-amalgam using a burner to obtain pure gold. Hg vapors are consequently released into the atmosphere and deposited into aquatic and terrestrial ecosystems. The ASGM practices in the considered study areas are summarized in Figures 3 and 4.

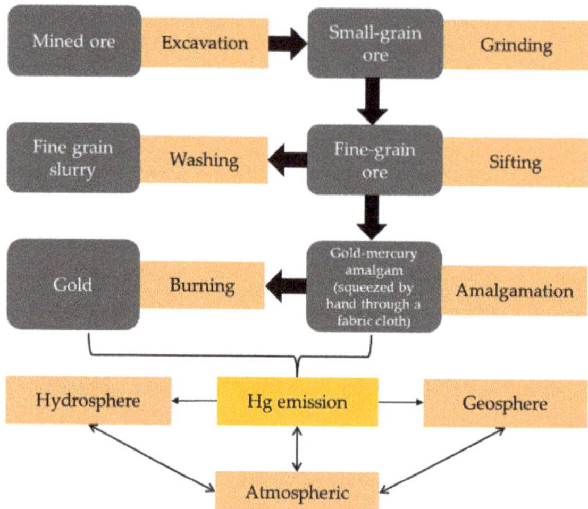

Figure 3. Flow chart showing general ASGM processes.

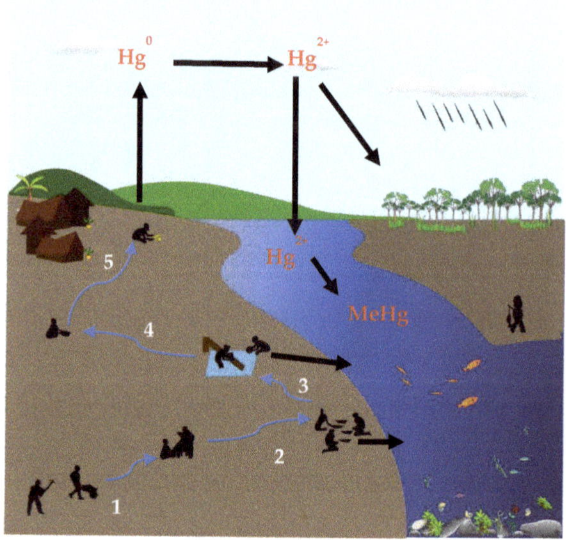

Figure 4. General ASGM processes involving (1) excavation, (2) grinding and sifting, (3) washing, (4) panning, and (5) gold–Hg-amalgam burning.

1.3. Minamata Convention on Hg

The Minamata Convention on Hg became one of the first worldwide environmental agreements in the 21st century. The convention was adopted in 2013, and to date, 123 countries have signed the agreement. The convention aims "to protect human health and the environment from anthropogenic emissions and releases of Hg and the compounds and it sets out a range of measures to meet the objective" [13]. According to Article 7 of the Minamata Convention, each party from an ASGM shall develop a national action plan regarding Annex C, which indicates implementing national objectives and reducing targets and actions to eliminate the Hg and related compounds. In the national action plan (NAP) for ASGM sectors, the informal ASGM sector must be regulated to accomplish the requirements for reducing the Hg in the country. Key to an NAP is the development of Hg inventories and baselines in the ASGM sector to monitor improvements and establish regulatory standards for Hg emission reduction. The pertinent parties must cooperate with the relevant stakeholders from governments, industry, NGOs, and academia. Subsequently, the parties need to build awareness regarding all Hg compounds in an ASGM process, promote non-Hg alternative practices, and provide technical and financial support. The countries that have not ratified the Minamata Convention in the Association of Southeast Asian Nations (ASEAN) include Brunei, Laos, and Myanmar. However, in Myanmar, the NAP for Minamata Initial Assessment, which is funded by a global environmental facility, has started a national Hg inventory. The status of the Minamata Convention in ASEAN countries is summarized in Table 1.

Table 1. Minamata Convention status in ASEAN countries [14].

Participants	Signature Date	Status	Date (Ratification/Accession/Approval)
Indonesia	10/10/2013	Ratification	09/22/2017
The Philippines	10/10/2013	Ratification	8/7/2020
Cambodia	10/10/2013	Ratification	8/4/2021
Vietnam	11/10/2013	Approval	06/26/2017
Malaysia	24/09/2014	Signature	
Laos		Accession	09/21/2017
Thailand		Accession	06/22/2017
Brunei	N.A.	N.A.	N.A.
Myanmar	N.A.	N.A.	N.A.

N.A.; not available.

1.4. Objective

Mercury pollution is a worldwide problem especially in ASGM countries. The number of artisanal miners has increased over the years and now totals to ~45 million people [15], with at least half of them engaged in gold mining, extracting up to 450 tons of gold per year in at least 70 countries [16]. ASGM activities produce increasing amounts of gold from countries in Africa (e.g., Ghana, Mali, Sudan, Tanzania, and Zimbabwe), Latin America (e.g., Brazil, Colombia, and Peru), and Asia (e.g., China, Indonesia, Mongolia, Myanmar, and the Philippines).

This study emphasized Myanmar and other ASEAN countries that are practicing ASGM. Similar to other developing countries, environmental challenges in Myanmar have been given strong consideration since natural resources have been extracted by illegal measures. Achieving an environmental balance has become a crucial role for such challenges. Unfortunately, Myanmar has insufficient professional labors with acquired skills; an ineffective governing mechanism; no transparency for trading by-products of gold, specifically Hg; and minimal research activities. Therefore, there are only six publications on Hg pollution in Myanmar up to now [17–22].

This review has identified and assessed the critical Hg pollution issues in Myanmar and other Southeast Asian countries. This study outlines the Hg problems that are of crucial concern to the citizens of these nations. We evaluated the Hg contamination in environmental media and the risks of Hg exposure on human health and propose a relevant policy framework regarding Hg issues.

2. Materials and Methods

2.1. Study Selection

The study identified the relevant literature published between 2000 and 2021 using databases including *PubMed*, *Web of Science*, *Springer*, *Science Direct*, and *Google Scholar*. The keywords used during the search were "Hg", "ASGM", "Myanmar", "Indonesia", "the Philippines", and "Malaysia". Research materials on international regulations, laws, and procedures related to Hg problems were also considered. The study focused on research articles from Myanmar and other ASEAN countries. To conduct the screening study, two reviewers (P.S.S. and T.A.) compared the titles and abstracts of the studies according to the inclusion and exclusion criteria presented in Table 2.

Table 2. Criteria for inclusion and exclusion of a study.

Inclusion	Exclusion
Studies related to ASGM communities.	Studies in other industries, such as coal-fired power plant.
Studies on Hg compounds, total Hg, inorganic Hg, and MeHg.	Unrelated compounds, such as ethylmercury.
Emphasis on Hg concentration in environmental media (e.g., air, water, and soil)	Measurement of Hg in other environmental media.
Emphasis on Hg concentrations in biomonitors of plants, fish, and human hair.	Measurement of Hg in other biological indicators.
Reports relating to human-health risk assessment in ASGM communities.	Reports relating to human-health risk assessment in other industries.

2.2. Quality Assessment

In selecting the literature, the study focused on the Preferred Reporting Items for Systematic Reviews and Meta-Analyses (PRISMA) statements [23] on identification, and screening, and included research as shown in Figure 5. The search method considered mainly published literature on Hg-related studies in environmental science, social science, and public health. The study focused on original research articles and systematic reviews. To ensure the quality of the evaluation, duplications were evaluated and checked rigorously. In the exclusion criteria, we considered the publication year (2000–2021) the language used in the research articles. Only research published in English from the study areas in Southeast Asian countries were included. Third-party tools (e.g., Microsoft Excel and Mendeley) were used for importing website data and data screening.

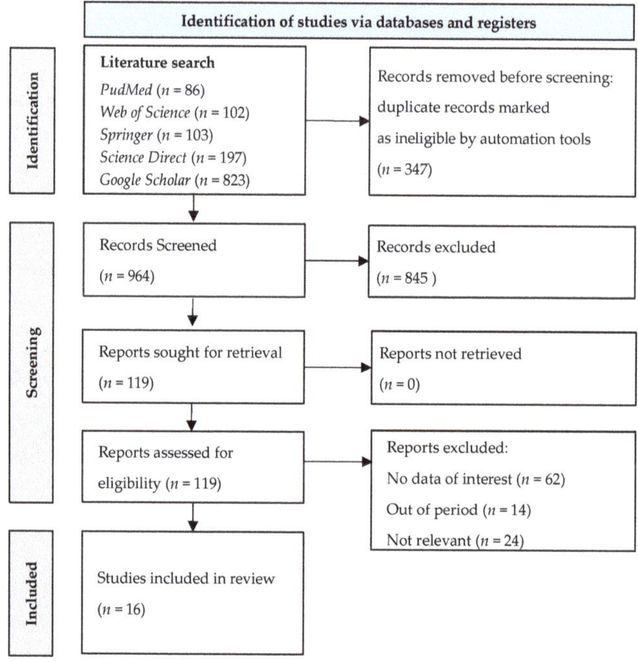

Figure 5. PRISMA flow diagram showing the search and selection process.

3. Results

3.1. Hg Concentrations in Air

Hg in the atmosphere occurs primarily in three forms, namely the gaseous state of elemental Hg (Hg(0)), reactive gaseous Hg (Hg(II)), and total particulate Hg (Hg(p)) [24]. Hg(0) emission from ASGM activities is the highest Hg emission source. Hg vapor, (mainly in the chemical form of elemental Hg(0)) can travel vast distances in the air and be deposited or captured in forest treetops and leaves [25]. According to the 2018 global Hg assessment report, the global emission of Hg into the air in 2015 from ASGM sources was 838 tons, with East and Southeast Asian countries accounting for ~214 tons [4]. Recent studies have showed very high Hg concentrations in the atmosphere resulting from ASGM activities in Central Sulawesi, Indonesia [26], Camarines Norte, the Philippines [27], and Mandalay region, Myanmar [21]. The Hg concentrations in air from ASGM activities in Indonesia, Myanmar, and the Philippines are summarized in Table 3.

Table 3. Summary of Hg concentrations in air from Indonesia, Myanmar, and the Philippines.

Location(s)	Sample Sources	n	Hg Concentration (ng/m^3)	Reference
Palu city, Sulawesi, Indonesia	Gold-processing area	21	9172 ± 16,422 (mean ± SD)	[26]
	Northern area of city		514 ± 420 (mean ± SD)	
	Central area of city		141 ± 141 (mean ± SD)	
	Western area of city		22 ± 15 (mean ± SD)	
	Southern area of city		116 ± 135 (mean ± SD)	
Mandalay region, Myanmar	ASGM site	13	0–10,900	[21]
		19	0.66–74,000	
Province of Camarines Norte, The Philippines	ASGM site	4	7.8–314,000	[27]

N.A.; not available, S.D.; standard deviation.

In Central Sulawesi, Indonesia, the highest average concentrations of 24 h ambient Hg(0) of 9172 ng/m^3 were found in the gold-processing areas that refined gold (including the stages of Hg amalgam burning) [26]. This total value was nine times higher than the WHO guideline limit of 1000 ng/m^3 [28]. Further, this study also considered indoor and outdoor air Hg(0) concentration in the Palu city area and the village of Mangkahui. The highest indoor and outdoor air concentrations of Hg(0) in the Palu city were 450 and 2250 ng/m^3, respectively. In the village of Mangkahui, the Hg(0) concentrations in the indoor and outdoor air were 196 and 103 ng/m^3, respectively, at site A. Meanwhile, the values were 238 and 279 ng/m^3, respectively, at site B.

A study investigated the atmospheric Hg concentration at an ASGM site in the Mandalay region of central Myanmar via two surveys [21]. In the first and second survey, the highest Hg concentrations of 10,900 and 74,000 ng/m^3, respectively, were noted in an amalgamation-burning area of an ASGM site. These values were several times higher than the Hg limit value in the WHO guidelines [28]. In addition, the study suggested that Hg was dispersed not only in the ASGM areas but also in nearby residential areas.

Atmospheric Hg pollution has been identified in the Province of Camarines Norte, the Philippines, by Murao et al. [27]. The authors focused on an area of a rod-mill station in the ASGM that recently burned gold amalgamation. The highest Hg concentration in air was 314,000 ng/m^3, which was considerably higher than WHO guideline (1000 ng/m^3) [28] at the rod-mill station in Benit, the Philippines. Meanwhile, the lowest concentration was 7.8 ng/m^3 at the same place 4 weeks after the burning.

In comparison with other air constituents, gaseous Hg(0) is relatively inert. Hg(0) from both anthropogenic and natural emissions can be transported over large distances by air and stay in the atmosphere for a year; therefore, Hg(0) can deposit in terrestrial and aquatic ecosystems [29,30]. The studies from Indonesia, Myanmar, and the Philippines have revealed that the atmospheric concentration of Hg were much higher in the ASGM areas that burn gold amalgamation. The concentrations were also higher than the WHO guidelines [28]. Therefore, a Hg recovery method should be considered in the ASGM industries.

3.2. Hg Concentrations in Water Bodies

Water resources can be impacted by various ASGM operation steps, such as mined-ore sifting and washing. An amalgamation process used in ASGM sectors typically discharges wastewater into water bodies. Subsequently, aquatic organisms are exposed to elevated Hg levels. Furthermore, inorganic Hg can be transformed into toxic MeHg [31]. MeHg in aquatic organisms is biomagnified through the food chain. Notably, fish intake is the primary source of Hg exposure in humans [32]. Mercury concentrations in waterbodies of Indonesia, Malaysia, Myanmar, the Philippines, and Thailand are presented in Figure 6.

A study in the Cikaniki River, Bogor, Indonesia, reported Hg concentrations in the river water ranging from 0.4 to 9.4 μg/L. The highest concentrations were found near an ASGM village. In this study, significant correlations were observed between Hg(0) and MeHg since MeHg concentration was considerably lower than Hg(0). The fact assumes that mining wastes was not a direct source of MeHg in the Cikaniki River [33]. Otherwise, Hg(0) deposited in river water can be subjected to methylation, suggesting that the change in chemical forms of Hg in water systems should be conducted in the future. An earlier study of the river conducted in 2009 reported Hg concentrations of 0.09–9.1 μg/L [34] and showed similar results, indicating the continuous pollution of the Cikanaki River.

In a study conducted in the Mandalay region, Myanmar, two groundwater samples were collected from five ASGM areas [21]. Irrawaddy River is one of the main rivers in Myanmar, which is located near an ASGM area. Surface water samples from upstream and downstream areas of this river were collected. Hg concentrations were also determined in groundwater at the nearby residential areas. The Hg concentrations in the sampled groundwater were in the range of 0–0.04 μg/L. The area closest to gold-mining activities typically showed Hg contamination [35]. The samples obtained in the Irrawaddy River, which was downstream from the ASGM area, showed a Hg concentration of 0.005 μg/L Hg, and samples taken from the upper stream of the river contained 0.004 μg/L of Hg. The reported Hg concentrations in the Irrawaddy river were slightly higher than the typical concentrations of Hg in lakes and rivers (0.001–0.003) [36] but not exceptional.

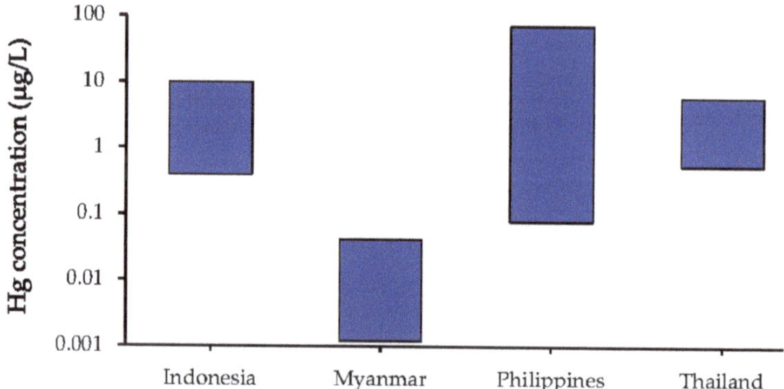

Figure 6. Hg concentrations in waterbodies from Indonesia [33], Myanmar [21], the Philippines [37], and Thailand [31]. Each piece of data represents a Hg concentration of range value, which was from a single study.

Interestingly, the Hijo River in the Philippines supports food security and is a means of living for local people who participate in it for gold processing. Wastewater containing Hg and cyanide from mine tailings is discharged into the river without treatment [36]. Additionally, the Naboc River in the Philippines receives effluent water from mining operations. The Hg concentrations found in the Hijo, Naboc, and Kingking Rivers were 78.4, 72.8, and 75.2 µg/L, respectively [37]. These levels are much higher than the national standard limit (i.e., 1 µg/L) of the Philippines [38,39].

A study in Phichit Province, Thailand, focused on the surface water of the Klong Dai Nam Khun and Klong Sa Luang canals, which were connected with a mining area [31]. Thirteen locations with aquatic habitats, including upstream, downstream, reservoir, and other water bodies near a gold-processing area and separatory ditch that were used for gold-ore separation processes, were considered. The study emphasized evaluation of the Hg concentrations level-related Hg-contaminated sites and their distances from ASGM sites, showing Hg concentrations in the range of 0.6–5.4 µg/L. The workplace area showed a higher Hg concentration than areas at greater distances upstream and downstream because the amalgam processes were conducted near the sampling locations.

In the considered studies, the studied samples from Indonesia [33], the Philippines [37], and Thailand [31] exceeded WHO guideline limit of 0.5 µg/L of Hg [40]. ASGM areas could affect Hg concentrations in surface water because Hg can be derived from atmospheric sources [41]. Further, Hg can transfer to the food chain in aquatic environments. Therefore, the effluent from improperly treated wastewater can be detrimental to marine life and people who consume seafood.

3.3. Hg Concentrations in Soil

Soil is a key indicator for monitoring environmental Hg concentration because Hg entering the atmosphere from amalgamation burning can deposit in the top layers of soil. It is necessary to understand the level of Hg concentration to prevent Hg pollution of soil. For the topsoil profile, Hg concentrations have been found to decline from the top soil to the deeper horizon [42,43]. In addition, Hg sorption from the air may contribute to Hg accumulation in topsoil horizons through litter accumulation and decomposition. The sources of Hg contamination of soil are fertilizers, lime, sludges, and manures [44]. A summary of the Hg levels in soils from ASGM sites in Indonesia, Myanmar, the Philippines, and Thailand is shown in Figure 7.

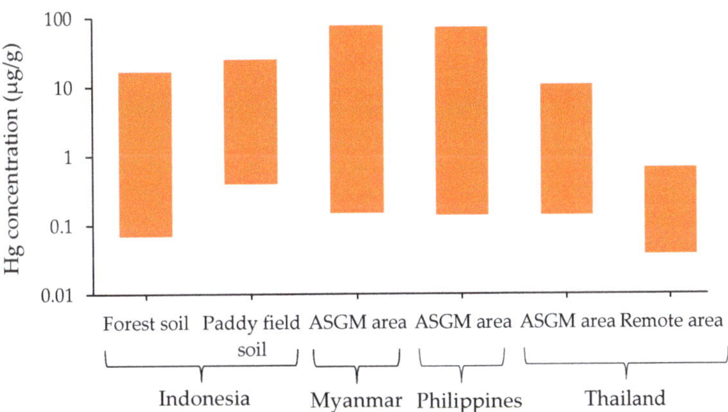

Figure 7. Hg concentrations in soil from Indonesia [33], Myanmar [17], the Philippines [27], and Thailand [31]. Each piece of data represents the Hg concentration of range value, which was from a single study.

An Indonesian study classified forest and paddy field soils impacted by ASGM activities [33]. The Hg concentrations analyzed in forest soil and paddy field soils were 0.07–16.7 and 0.4–24.9 μg/g, respectively. Additionally, the concentrations of MeHg were in the range of 0.07–2 μg/kg in forest soils and 0.07–56.3 μg/kg in paddy field soils. These data demonstrated that paddy field soil is particularly affected by ASGM activities [33].

ASGM activities are widely performed in the upper part of Banmauk Township, Sagaing Region, Myanmar. A study investigated soil samples from the placer gold-deposition area and identified soil matrices based on ASGM operation processes, such as ore processing, sluicing, panning, and amalgamation [17]. The soil matrix from the amalgamation process exhibited the highest Hg concentration of 77.44 μg/g, whereas Hg concentrations during soil ore processing, sluicing, and panning stages (gold-amalgamation stage) were 0.68, 0.51, and 4.86 μg/g, respectively [17].

A study conducted in the Philippines determined Hg concentrations in soils obtained from potentially contaminated hotspots and the areas distant from such spots. The highest Hg concentrations observed were 71.75 μg/g in the sample from a rod-mill station in the amalgamation-burning workplace. By contrast, the lowest concentration observed was 0.15 μg/g in the sample from a nonmining area, showing that the higher Hg concentrations had contaminated the vicinity of the ASGM area [27].

A study of an ASGM operation in the Phichit region, Thailand, considered surface soil (0–5 cm depth) from mining and remote areas. The Hg concentrations in the mining and remote areas were in the range of 0.14–10.56 and 0.038–0.632 μg/g, respectively. The higher Hg concentration observed in the mining area indicated that Hg vapor emitted into the atmosphere was likely deposited on soil surfaces near the burning stoves. This was because of the 7.8 h/day amalgamation process takes place for extraction of 60–150 g of gold [31].

A study reported that Hg concentrations in soil do not typically exceed 0.1 μg/g, and normal levels in soil were reported. Moreover, the normal Hg levels in soil were 0.05–0.08 μg/g [36]. The data from Myanmar, the Philippines, and Thailand showed higher Hg levels than United States Environmental Protection Agency (US EPA) generic soil guidelines value (0–0.2 μg/g) [45].The study found that amalgamation process in ASGM areas contributed considerably to the Hg concentrations found. Therefore, people residing near the ASGM area could be impacted by Hg exposure during the amalgamation process.

3.4. Hg Concentrations in Plants

Plants are widely used as biomonitors for monitoring environmental Hg [46]. Rasmussen et al. found that among vegetative structures, the leaves contained the highest Hg concentrations [47]. In plants that absorb Hg primarily from the soil, Hg contents were found to be higher in the roots. Conversely, for the plants that adsorb Hg primarily from the air, Hg contents were found to be higher in the shoots and leaf tissues [48]. Some studies have reported that crops such as vegetables are the sources of Hg exposure for people living in Hg-mining areas [49]. The Hg contents found in plant samples from Indonesia, Myanmar, the Philippines, and Thailand are presented in Figure 8.

Some recent studies from Indonesia have reported high Hg concentrations of 1.4 μg/g dry weight (d.w.) in leaves from plants that grew near ASGM locations [50]. Similarly, contaminated forage plants (an edible animal feedstock) were found at a gold-mining site in Southeast Sulawesi Province, Indonesia [51]. Fresh forage plant samples from the Rarowatu and North Rarowatu Districts of Bombana were studied [51]. The sampling locations were divided into reference, mining commercial, and ASGM. The highest Hg content of 9.9 ± 14 μg/g d.w. was found in the ASGM area. The values in the commercial mining and reference areas were 3.20 ± 3.50 and 2.70 ± 2.80 μg/g d.w., respectively. According to the critical limits for Hg related to ecotoxicological effects on plants, the Hg levels in forage plants can be divided into three categories: high (>3 μg/g), low–moderate (0.1–3.0 μg/g), and low (0.1 μg/g) [51].

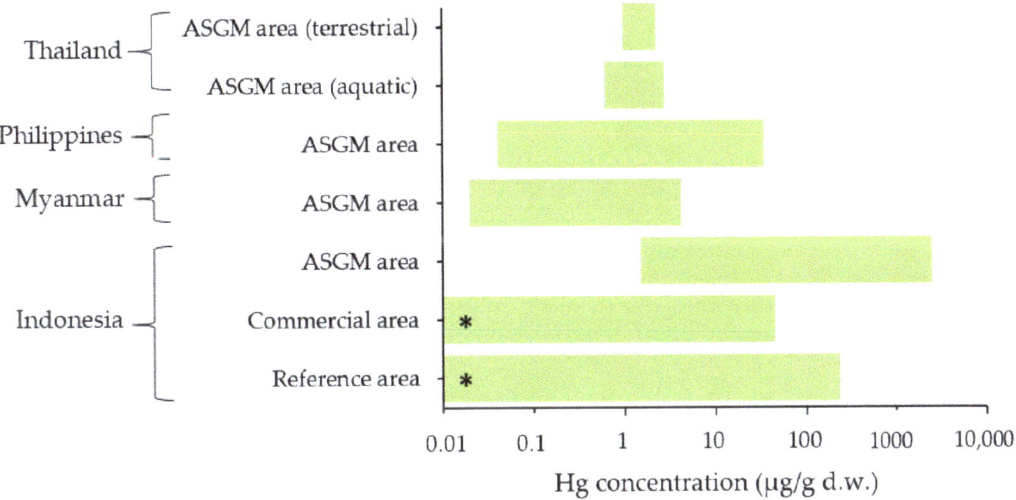

Figure 8. Hg concentrations in plants from Indonesia [51], Myanmar [18], the Philippines [27], and Thailand [31]. Each datum represents Hg concentration of range value, which was from a single study. *. minimum value of "0".

A study from the Mandalay region, Myanmar, included a preliminary survey that assessed the Hg air pollution in advance of future studies by examining tree bark, leaves, and blades of grass from *Typha latifolia* L. (leaf) species, *Azadirachta indica* (bark), *Terminalia catappa* L. (bark), *Manifera indica* L. (leaf), and *Naringi crenulata* (Thanaka (leaf)) [18]. Hg from the atmosphere and the soil can be deposited in plants [52]. The highest Hg concentration found was 4.17 µg/g d.w. in a Thanaka leaf near a gold shop, whereas the lowest concentration found was 0.02 µg/g d.w. in *Typha latifolia* L. leaf sample obtained some distance away from ASGM areas. The highest Hg concentrations found were in samples obtained near the gold refinery area.

Plants can serve as an indicator to regulate the uptake and transport of pollutants to the air because their internal pollutant concentrations are generally identical to the pollutant concentrations detected in the parent soil [53]. A study from the Philippines considered Hg concentrations in plant species including *Dadvalia* sp., *Alugbatging puti*, *Citrus* sp., cacao, and *Dilang aso* [27]. The study analyzed plant samples from an ASGM area and a few meters away from it. The Hg concentration in plants ranged between 0.04 and 34 µg/g d.w. [27]. The highest Hg concentrations was found in *Dadvalia*, which grew near a rod-mill station in the ASGM area [27].

In a study from Thailand, neem leaves and flowers from aquatic and terrestrial environments in the ASGM workplace in Phichit Province were investigated [31]. Neem flowers from aquatic sites showed Hg levels of 0.62–2.151 µg/g d.w. [31]. Similarly, neem leaves from terrestrial sites exhibited Hg levels in the range of 0.967 and 1.30 µg/g d.w.. The neem flowers were purposely collected from the aquatic tract, and the results showed that the highest Hg concentrations was found near the Hg emission source. Further, the study suggested that Hg concentration in neem flowers growing along the aquatic sampling site was related to the concentration of Hg in sediment at the same location [31]. The Hg levels were higher than the maximum permissible limit of Hg content (0.5 µg/g w.w.) for biota tissue [30]. Moreover, a study highlighted that the concentration of Hg can be deposited in legume species, such as *Indigofera enneaphylla* and *Desmodium triflorum* [31]. Therefore, the study suggested avoiding eating the plants near potentially Hg-contaminated areas.

Plants obtained in Indonesia, Myanmar, the Philippines, and Thailand have been found to contain Hg concentrations that were above the FAO/WHO guideline values (0.5 µg/g w.w.) [54,55].

3.5. Hg Concentration in Fish

Fish consumption is one of the most important factors contributing to Hg uptake in humans. Hg concentrations in fish tissues can be affected by the age, length, and weight of the fish [56]. Freshwater biota accumulate Hg from both natural and anthropogenic sources. Most fish have natural Hg levels of 0.02–0.3 µg/g wet weight (w.w.); however, small and short-lived herbivorous fish species have been found with a Hg level of 0.01 µg/g w.w. [57]. According to the recommendation of the FAO/WHO, the Hg content in a fish should not be more than 0.5 µg/g w.w. [58]. Hg concentration in fish from Indonesia and the Philippines based on w.w. are presented in Figure 9.

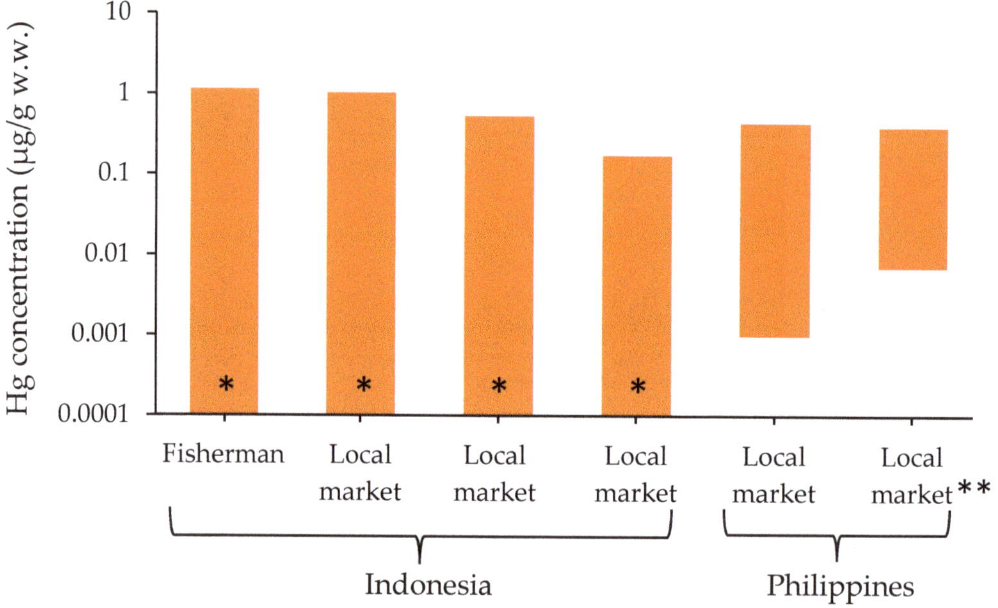

Figure 9. Hg concentrations in fish from Indonesia [59] and the Philippines [37]. Each datum represents Hg concentration of range value, which was from a single study. *; minimum value of "0", **; MeHg concentration.

A study in Cambodia involved the collection of 82 fish species from local fishermen/fisherwomen in Kampi pool near Kraite, which is located near the O Tron gold-mining area [60]. The Hg concentration in the collected fish samples (n = 160) ranged between 0.008 and 0.64 µg/g. Additionally, the study grouped the size of fish as "small size" and "big size." The big-sized fish showed an average Hg content of 0.128 µg/g, which was considerably higher than the average Hg of 0.086 µg/g in the small-sized fish [60]. However, the considered study had not determined the wet weight or dry weight for fish sample analysis. According to Baran et al. [61], Cambodians consume an average of 1.26 kg of fish each week; thus, Cambodians who consume more fish than the normal quantity intake higher Hg levels and are at a higher health risk [60].

Recent studies from Indonesia conducted fish sampling during 2007 to 2011 in Ratatotok Subdistrict, North Sulawesi, Indonesia, which was near the Mesel gold-mining area [59]. Local people from the studied area, which has an active fishing economy, have faced health issues during the period of active mining. The study involved the collection of fish samples from fishermen/fisherwomen and local market [59]. The fish samples from a Buyat Pantai fisherman showed Hg levels of 0.00–1.13 μg/g w.w. The samples from Buyat, Ratatotok, and Manado fish markets showed levels of 0.00–1.03, 0.00–0.53, and 0.00–0.17 μg/g w.w., respectively [59]. Thus, except for the Manado fish market, the fish samples from other sources exceeded the WHO standard guideline [58]. Nonetheless, the reported mean Hg concentrations in fish were within the standard limit for consumption.

Surveys were conducted to determine Hg concentration in marine samples in Davao del Norte, south of Manila, the Philippines, and near a gold-processing area. At the local market of Apokon, Tagum, seventeen specimens of fish and one seaweed sample were examined to determine the Hg and MeHg concentrations, which ranged between 0.001 and 0.44 μg/g w.w. and 0.007 and 0.38 μg/g w.w., respectively [37].

The considered studies revealed that the maximum concentration of Hg found in fish from Cambodia, Indonesia, and the Philippines were 1.13 μg/g, 0.44 μg/g w.w., and 0.64 μg/g w.w., respectively. Compared with the studies from Latin America, such as Brazil (1.04–2.84 μg/g w.w.), Colombia (1.60–4.50 μg/g w.w.), Bolivia (1.08–2.86 μg/g w.w.), and Ecuador (1.39–1.6 μg/g w.w.) [62], the reported Hg concentrations in Southeast Asia were relatively lower.

3.6. Hg Concentrations in Human Hair

Hair is a common biomarker for characterizing MeHg exposures [63]. Low Hg concentrations have been considered risks for neurosis (50 μg/g) and health issues (11 μg/g) in unborn fetuses [64]. Moreover, a low Hg level in hair has also been associated with a low susceptibility of hair for Hg vapor. Hg concentrations measured in human hair from Cambodia, Indonesia, Myanmar, the Philippines, and Thailand are summarized in Table 4.

A study in Cambodia study involved the collection of human hair samples around the Mekong River, which is one of the world's major rivers [60]. Hair samples were taken from people including mine workers living in the area of the O Tron gold mines and upstream and downstream along the Mekong River. These results revealed that the mean Hg concentration (5.21 μg/g) in the hair samples from men ($n = 32$) was higher than that of women (3.08 μg/g) ($n = 46$). When the female hair samples were sorted by sample area, the women from Ratanakiri province, near mine-affected areas, had a substantially higher Hg concentration of (3.47 μg/g) ($n = 23$) than a control group (2.7 μg/g) ($n = 23$) [60]. Hg levels in the Cambodian hair sample exceeded those observed near gold mines in the Philippines, where association of impaired human health with Hg concentration was observed [37].

A study involved the collection of human hair samples from an active ASGM area operating for more than 20 years in Lebaksitu, Indonesia [65]. The Hg hotspot village (Lebak-1) and downstream village (Lebak-2) were considered as high and low-risk areas, respectively. Human hair samples from both villages showed a mean Hg content of 3.2 μg/g, with a range of 0.847 to 9.015 μg/g [65]. The samples from Lebak-1 residents showed a considerably higher mean MeHg value (2.12 μg/g) than other residents, indicating that Lebak-1 residents were more exposed to MeHg than Lebak-2 residents. After comparison with other research on Hg-affected areas in Colombia [66], MeHg accumulation in hair from Indonesia was primarily caused due to consumption of food, such as fish and rice [65].

In the Mandalay region of Myanmar, human hair surveys were conducted of miners and nonminers living around the ASGM areas. The maximum Hg concentration in the hair samples of miners and nonminers were 5.7 and 2.9 μg/g, respectively [21]. The fact indicates that Hg concentration in human hair from the considered study was not at a level that would adversely affect human health because the approximate lowest levels of Hg that can cause neurosis and health problems in unborn fetuses are 50 and 11 μg/g, respectively [64,67].

Table 4. Summary of Hg concentrations in human hair from Cambodia, Indonesia, Myanmar, the Philippines, and Thailand.

Location	Sample Source	Number of Samples	THg Concentration (µg/g)	MeHg Concentration (µg/g)	Reference
Mekong River, Cambodia (near O Tron gold mine)	Tonle Srepok	25	4.54 *	N.A.	[60]
	Tonle Kong	17	4.22 *		
	Mekong N. Stung Treng	16	3.36 *		
	Mekong Kratie	20	3.47 *		
	All males	32	5.21 *		
	All females	46	3.08 *		
	All adults	59	4.01 *		
	All children (aged < 13 y)	19	3.38 *		
	Women Ratanakirri (mine impacted)	23	3.47 *		
	Women Mekong	23	2.7 *		
Lebaksitu, Lebak regency Java Island, Indonesia	ASGM area	41	0.847–9.015	0.37–4.33	[65]
Mandalay Region, Myanmar	ASGM area (miners and nonminers)	50	0.4–5.7	N.A.	[21]
Acupan region, Benguet, the Philippines	ASGM area	70	0–26.6	N.A.	[68]
Nong Pra subdistrict, Wang Sai Poon district, Phichit Province, Thailand	Gold miners	79	1.17 ± 0.05 (mean ± SD)	N.A.	[69]
	School children	59	0.93 ± 0.01 (mean ± SD)		

N.A.; not analyzed, S.D; standard deviation, *; mean concentration.

Human hair samples from 70 inhabitants of Acupan, Benguet, in the northern regions of the Philippines, were obtained [68]. In the large studied ASGM community, the age of the participants was in the range of 8–66 years. The results showed that the average Hg content in the inhabitants was 3.47 µg/g. Hg concentrations in nine interviewees were higher than the human biomonitor limit of 5 µg/g [68]. Additionally, the highest Hg concentration was 26.6 µg/g, which was found in a 46-year-old male participant who was involved actively in amalgamation burning and lived only 5 m away from the ASGM location.

Hair samples were obtained from miners, schoolchildren, and a control group from the Phanom Pha gold-mining area located in Nong Pra subdistrict, Wang Sai Poon District, Phichit Province, Thailand [69]. The study considered the miners involved in the amalgamation process and working the ore-preparation area as groups I and II, respectively. The schoolchildren belonged to the group involved in gold-mining activities. The hair samples from miners showed an average Hg concentration of 1.17 (µg/g), which was within the reference group's Hg concentration range [69]. The average Hg content in hair samples from schoolchildren in group I and II were 0.95 µg/g and 0.90 µg/g, respectively. Both schoolchildren groups showed Hg concentrations that were within the range of the control [69]. The fact suggests that lower Hg concentrations are expected in the hair because exposure to Hg is primarily due to inorganic Hg (i.e., Hg vapor) [69].

3.7. Health Risk Assessment of ASGM Communities

A study in Central Sulawesi, Indonesia, examined the health risks Hg exposure caused by the Poboya ASGM sites at the residential areas of Palu city [26]. The study focused on miners and other residents to estimate their health risk exposure. In each of the five locations studied, the frequencies of each hazard quotient (HQ) ratio (HQ ratio ≥ 1) from gaseous Hg (0) inhalation risk were determined [26]. Based on daytime Hg (0) concentrations, only 1.5% in the gold-processing area showed HQ ratios of <1, suggesting no risk. However, 93% of the sample population was found to be at risk. There are high chances of inhaling Hg released via ASGM activities in studied area. The human health risk from Hg exposure is particularly high in the Poboya gold-processing area and the areas close to Palu city. Moreover, 93% of the sample population in the Poboya area exceeded the no-risk values with HQ ratio > 1. These findings suggest that people who work in the gold-processing industry and nonminers in Palu city are at risk of adverse health impacts due to inhalation of Hg vapor.

A preliminary health survey conducted in ASGM area of Thabeikkyhin Township, Mandalay region, Myanmar, involved the health inspection of men ($n = 18$) and women ($n = 11$) [19] to determine the health of the neurological system and respiratory functions. Based on the neurological assessment, three female miners who participated in ASGM panning and amalgamation processes for more than 5 years were diagnosed with mild tremors and ataxia. The respiratory assessment by spirometry on miners showed 38.9% normal, 27.8% mild, 27.8% moderate, and 5.6 % severe conditions. Meanwhile, the nonminer group exhibited 27.3% normal, 27.3% mild, and 45.5% moderate influence conditions [19]. Furthermore, the study found that with an increased duration of mining activities, the FEC and FEV1 values declined, indicating a chronic damage of the respiratory function in the enrolled miners. Therefore, health inspections of the ASGM community in Myanmar should be conducted intensively.

The health impacts of Hg on miners and children in the vicinity of ASGM mining in Apokon, Tagum, Davao del Norte, the Philippines, were studied [37]. The neurological effects found were mainly located on the cranial nerves (17.1%), reflexes (5.1%), sensory (5.1%), cerebellar (3.89%), and motor nerves (1.2%). The neurological effects were characterized as follows: cranial nerve VIII abnormalities (6.87%), distally decreased vibratory sense (2.69%), palmomental reflex deficiency (2.4%), cranial nerve I (2.40%), visual acuity (2.10%), and Babinski (1.50%). Based on physical examination, abnormalities were found in all 163 children enrolled in the study with the following five predominant abnormalities: below-average height, gingival discoloration, below-average weight, adenopathy, and dermatologic irregularities (Figure 10).

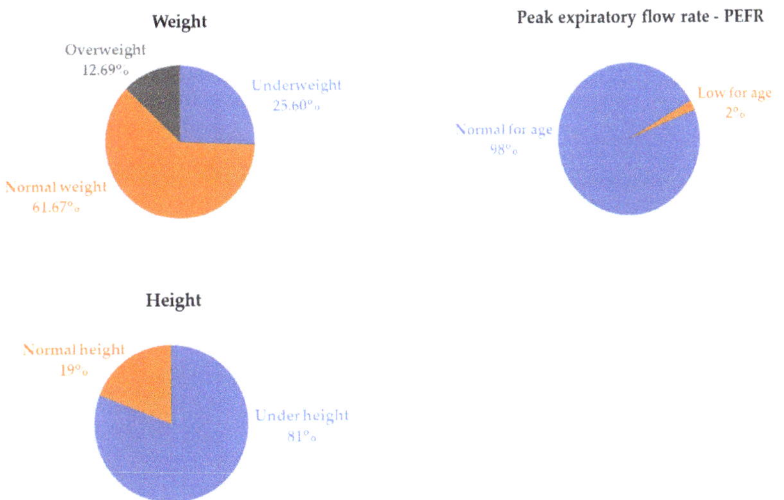

Figure 10. Predominant abnormalities found in schoolchildren of Apokon, Tagum, Davao del Norte, the Philippines [36].

A study was conducted in the Phanom Pha gold-mining area of Thailand [69]. Two groups of miners and schoolchildren were divided into groups I (involvement in mining activities) and II (no involvement in mining activities) to estimate individual health risks. According to the U.S. EPA, the reference dosage is 0.0003 mg/kg/day [70], and the HQ ratio represents the estimated exposure intake. Low exposure to Hg vapor in the group of miners was evidenced by the range of Hg in the air of 0.005–0.021 mg/m^3. The HQ ratios of group II indicated no risk [69]. However, the HQ ratios ranged from 16 to 218 in group I, which were much higher than in the HQ values of group II. Regarding the group of schoolchildren, group I exhibited a low HQ value of 0.02–0.23, while group II showed an even lower HQ value of 0.01–0.02 [69]. The higher HQ values of group I could be attributed to the Hg exposure from gold mining in the vicinity of amalgamation open burning. This suggests that the miners who work in the amalgamation process are at the greatest risk of inhaling Hg vapor. Therefore, miners who work in the amalgamation process are at the greatest risk of Hg vapor inhalation. Mitigation strategies to lower Hg contamination in the workplace must be considered.

4. Discussion

This paper reviewed the Hg pollution from ASGM areas in Myanmar and other Southeast Asian countries. Environmental indicators (e.g., air, water, and soil) and biomonitors (e.g., plants, fish, and human hair) were used in the considered studies. The concentrations of Hg in the air found at various areas in Indonesia, Myanmar, and the Philippines were higher than the standard limit values indicated in WHO guidelines [28]. The high Hg concentrations in the air were mainly due to the burning of gold amalgamation in the studies areas. By contrast, the reported concentrations of Hg in the air around ASGM areas in Idrija, Slovenia (<10 ng/m^3) [71], and Guizhou, China (17.8 ng/m^3) [72], were low. Moreover, in Almadén, Spain, where cinnabar was melted to produce Hg, the Hg levels reported were in the range of 100–14,000 ng/m^3 [73], which were lower than the Hg levels in the ASGM areas in Myanmar and the Philippines.

In ASGM areas, water is essential for drinking and the domestic purposes of local people. Additionally, water purification is critical in the ASGM area because mine wastewater can be discharged directly into water bodies. Thus, the reviewed studies considered Hg concentrations in river water and groundwater around the ASGM areas. Hg concentrations in the water samples from Indonesia, the Philippines, and Thailand exceeded the WHO standard (0.5 µg/L) [40]. Hg concentrations in water samples from Myanmar was relatively lower compared with those from Indonesia, the Philippines, and Thailand.

Atmospheric deposition is the primary source of Hg in remote environments. Additionally, soil is another primary receiver of atmospheric Hg deposition in terrestrial ecosystems. Moreover, Hg can be retained by soil over long periods because of its elemental impurities [17]. Hg contents found in samples from Myanmar and the Philippines exceeded the standard limits of 1 µg/g in the U.S. (California), 6.6 µg/g in Canada, and 0.83 µg/g in the European Union (the Netherlands) [74].

Plants use their radicle system to absorb organic and inorganic Hg forms, which are then delivered to the leaves [75]. Temmerman et al. [76] found that Hg absorption also occurs through plant roots depending on soil exposure levels to Hg. Another theory is that Hg from the atmosphere can accumulate in most plants [77]. In this considered studies, Hg content in plants sampled from the Philippines showed the highest values. This was followed by Indonesia, Myanmar, and Thailand. Based on the findings of this review, the Hg levels found in the studied areas were higher than those reported in the Lanmuchang Hg-mining area, Guizhou Province, China (0.175 µg/g w.w.) [78]. Meanwhile, the reported Hg concentrations in vegetable samples collected at the Idrija Hg-mining area in Solvenia were <0.215 µg/g w.w. [71]. Compared with a study in the Alacran mine, Colombia [79], where the maximum value of Hg found in a leaf was 2.78 µg/g d.w., the values reported in the reviewed studies were higher. In addition, the Hg levels found in plant samples from the Almadén mining district, Spain, showed extremely high values in leaves, in the range of 0.16–1278 µg/g [80].

Fish in polluted water bodies are potentially contaminated by Hg. The reviewed studies investigated fish species obtained from local markets and fishermen within ASGM areas. Although the levels of Hg in some fish samples were below the WHO standard limits (0.5 µg/g w.w.) [58], the Hg levels in fish from fishermen sources in Indonesia and Cambodia were very high. Meanwhile, fish samples from the Philippines exceeded the U.S. EPA standard in fish tissue of 0.3 µg/g [62]. Generally, more than 75% of the Hg accumulated in the muscle tissues of freshwater fish is in the organic form of MeHg [81]. Moreover, seasonal variation such as precipitation should be considered, as there is a wide variety of aquatic habitats in the studied regions, which are affected by seasonal variations. For example, floods can temporarily modify the biogeochemical components (e.g., oxygen content, pH, and prey availability) of a system. Thus, the fish-sampling condition with respect to season (e.g., during dry or wet season) is important [82,83]. We therefore suggest that people living near ASGM areas should practice caution when consuming fish.

Mercury concentrations in human hair have been associated with both the endogenous Hg contamination through consumption of food that were contaminated by Hg species and the Hg concentrations in the air because elemental Hg can adhere to human hair [71,84]. The Hg concentrations in human hair samples from the studied ASGM areas in Cambodia, Myanmar, and Thailand were lower than those of residents of the Wuchuan Hg-mining area, China, (mean value and range of 34 and 7.6–93.1 µg/g, respectively) [78]. The lower Hg level may be attributed to lower susceptibility of human hair for Hg vapor [21]. In addition, Hg-mining areas in Guizhou Province and valley in the southern part of Shaanxi Province in central China showed Hg levels with mean values of 4.3 µg/g (1.6–12.6 µg/g) [85]. However, the Hg content found in the hair samples from the ASGM area in Lebaksitu, Indonesia, and the Acupan region, the Philippines, showed concentrations of 0.84–9.015 µg/g, and 0–26.6 µg/g, respectively [65,68]. Those value were above the allowable limit as per the WHO guidelines [67]. Meanwhile, the Hg concentrations found in Cambodia, Myanmar, and Thailand were within the recommended limit. In addition, a study reported links between high fish consumption and burning gold-amalgam exposure to high levels of Hg in human hair [86].

In an ASGM process, the final stage is the most critical in Hg inhalation because miners are exposed to Hg vapors during amalgamation burning. The average Hg concentration in the air of Palu city, Indonesia, was 12,782 ng/m^3 [26]. The study indicated that 93% of the population was above the no-risk HQ ratio Therefore, both miners and nearby residents were at the risk of adverse health effects resulting from inhalation Hg vapor [26].

A study in Myanmar conducted a health inspection around an ASGM area in the Mandalay region. Based on the Human Biomonitoring Commission Standard, seven miners were in the range of warning status. Furthermore, the study highlighted that 16% of miners showed signs of Hg poisoning, such as nervous system damage, whereas nonminers did not demonstrate aberrant symptoms [19].

A study in the Philippines conducted physical examinations on 163 children [37]. All children showed the following common abnormalities: lower-than-average height, gingival discoloration, lower-than-average weight, adenopathy, and dermatologic abnormalities. According to the WHO, an adult who had an intake of 200 mg/day of Hg (e.g., from fish) has a 0.3% and 8% chance of experiencing paresthesia symptoms, respectively [87].

In the Phanom Pha gold-mining area, Thailand, the Hg exposure of miners and schoolchildren after ASGM activities was investigated. The HQ to a reference dosage (0.0003 mg/kg/day) was below the level at which unfavorable health effects on miners should be predicted [69]. The high-exposure miner group and schoolchildren showed the HQ ratios of 16–218 and 0.02–0.23, respectively. Inhalation of gold-amalgamation vapor can accumulate in the brain and kidneys [69]. Indeed, a study reported that miners in Brazil who used the open-burning method without using a Hg retort showed Hg levels that were higher than the normal Hg concentration in urine with using a Hg retort [69].

5. Conclusions

This review assessed and identified the Hg pollution in the ASGM areas of Myanmar and some other relevant Southeast Asian countries. Research should continue to focus on the current situation of ASGM activities in the studied countries and other parts of the world that allow ASGM activities because Hg is released by ASGM areas, which is a persistent and toxic global pollutant that can be transported through the atmosphere and deposited in terrestrial and aquatic ecosystems. The study aims to contribute to further research activities, such as health inspection and Hg management from ASGM areas because Hg is still used in ASGM activities. For example, Myanmar has not recognized the national Hg inventory, and its research activities are still limited. Additionally, Myanmar is still not a part of the Minamata Convention. Because gold prices remain high, despite countries such as Indonesia ratifying the Minamata Convention, Hg demand is still high in ASGM activities.

According to these reviewed studies, it is evident that Hg continues to cause contamination in the vicinity of ASGM areas, including nearby residential areas. Human epidemiological assessments on Hg-related diseases should be undertaken on a regular basis in the ASGM areas. Hg-contaminated areas should be controlled using reasonable regulations, policies, and frameworks. In addition, innovative Hg-free processing technologies and alternative economies should be introduced to support ASGM communities to reduce Hg emissions. Consequently, awareness of Hg problems can effectively reduce Hg pollution in ASGM communities.

Author Contributions: K.A. and Y.I. supervised the research. T.A. supervised and edited the manuscript. W.T.K., supported and funded the manuscript. P.S.S., wrote this manuscript. All authors have read and agreed to the published version of the manuscript.

Funding: This research was funded by the Research Institute for Humanity and Nature (RIHN; a constituent member of NIHU), Project No. RIHN 14200102.

Institutional Review Board Statement: Not applicable.

Informed Consent Statement: Not applicable.

Data Availability Statement: Not applicable.

Acknowledgments: The author would like to sincerely thank Tetsuro Agusa, who gave kind support, comments, and suggestions to complete this manuscript. This research was supported by Research Institute for Humanity and Nature (RIHN: a constituent member of NIHU) Project No. RIHN 14200102 and by the Bilateral Partnership Joint Research Project (JPJSBP 120209934) of Japan Society for the Promotion of Science (JSPS).

Conflicts of Interest: The authors declare no conflict of interest.

References

1. McNutt, M. Mercury and Health. *Science* **2013**, *341*, 14–30. Available online: https://www.who.int/news-room/fact-sheets/detail/mercury-and-health (accessed on 30 March 2022). [CrossRef] [PubMed]
2. Moreno-Brush, M.; McLagan, D.S.; Biester, H. Fate of mercury from artisanal and small-scale gold mining in tropical rivers: Hydrological and biogeochemical controls. A critical review. *Crit. Rev. Environ. Sci. Technol.* **2020**, *50*, 437–475. [CrossRef]
3. Harada, M. Minamata disease: Methylmercury poisoning in Japan caused by environmental pollution. *Crit. Rev. Toxicol.* **1995**, *25*, 1–24. [CrossRef] [PubMed]
4. United Nations Environment Programme (UNEP). *Global Mercury Assessment 2018*; United Nations: New York, NY, USA, 2019.
5. Hylander, L.D.; Meili, M. 500 Years of mercury production: Global annual inventory by region until 2000 and associated emissions. *Sci. Total Environ.* **2003**, *304*, 13–27. [CrossRef]
6. UNEP. *Reducing Mercury Use in Artisanal and Small-Scale*; United Nations Environment Programme, United Nations: New York, NY, USA, 2012; p. 76.
7. Hilson, G. Abatement of mercury pollution in the small-scale gold mining industry: Restructuring the policy and research agendas. *Sci. Total Environ.* **2006**, *362*, 1–14. [CrossRef]
8. Clifford, M.J. Future strategies for tackling mercury pollution in the artisanal gold mining sector: Making the Minamata Convention work. *Futures* **2014**, *62*, 106–112. [CrossRef]

9.   Sousa, R.; Veiga, M.; van Zyl, D.; Telmer, K.; Spiegel, S.; Selder, J. Policies and regulations for Brazil's artisanal gold mining sector: Analysis and recommendations. *J. Clean. Prod.* **2011**, *19*, 742–750. [CrossRef]

10. Htun, K. Sustainable Mining in Myanmar. *Appl. Environ. Res.* **2014**, *36*, 25–35. [CrossRef]

11. McFarlane, D.; Villalobos, R. The State of Artisanal Mining in Myanmar. 2019. Available online: https://delvedatabase.org/uploads/resources/The-State-of-Artisanal-Mining-in-Myanmar-Report.-PACT.-May28.2019.pdf (accessed on 4 November 2021).

12. Environmental Conservation Department (ECD), Ministry of Natural Resources and Environmental Conservation—MONREC. *Distribution of ASGM Activities in Myanmar*; MONREC: Naypyidaw, Myanmar, 2018.

13. United Nations Environnent Programme (UNEP). Minamata Convention on Mercury. Available online: www.unep.org; www.mercuryconvention.org (accessed on 1 September 2019).

14. UNEP. Status of Signature, and Ratification, Acceptance, Approval or Accession in Minamata Convention. Available online: https://www.mercuryconvention.org/en/parties (accessed on 30 March 2022).

15. Delve. A Global Platform for Artisanal and Small Scale Mining Data. 2020. Available online: https://stateofthesector.delvedatabase.org (accessed on 30 March 2022).

16. Seccatore, J.; Veiga, M.; Origliasso, C.; Marin, T.; de Tomi, G. An estimation of the artisanal small-scale production of gold in the world. *Sci. Total Environ.* **2014**, *496*, 662–667. [CrossRef]

17. Tun, A.Z.; Wongsasuluk, P.; Siriwong, W. Heavy Metals in the Soils of Placer Small-Scale Gold Mining Sites in Myanmar. Available online: https://meridian.allenpress.com/jhp/article-abstract/10/27/200911/445447 (accessed on 1 November 2021).

18. Kuang, X.; Kyaw, W.T.; Soe, P.S.; Thandar, A.M.; Khin, H.E. A Preliminary Study on Mercury Contamination in Artisanal and Small-Scale Gold Mining Area in Mandalay Region, Myanmar by using Plant Samples. *Pollution* **2022**, *8*, 225–238. [CrossRef]

19. Kyaw, W.T.; Kuang, X.; Sakakibara, M. Health impact assessment of artisanal and smallscale gold mining area in Myanmar, Mandalay Region: Preliminary research. *Int. J. Environ. Res. Public Health* **2020**, *17*, 6757. [CrossRef] [PubMed]

20. Osawa, T.; Hatsukawa, Y. Artisanal and small-scale gold mining in Myanmar: Preliminary research for environmental mercury contamination. *Int. J. Hum. Cult. Stud.* **2015**, *25*, 221–230. [CrossRef]

21. Kawakami, T.; Konishi, M.; Imai, Y.; Soe, P.S. Diffusion of mercury from artisanal small-scale gold mining (ASGM) sites in Myanmar. *Int. J. GEOMATE* **2019**, *17*, 228–235. [CrossRef]

22. Wongsasuluk, P.; Tun, A.Z.; Chotpantarat, S.; Siriwong, W. Related health risk assessment of exposure to arsenic and some heavy metals in gold mines in Banmauk Township, Myanmar. *Sci. Rep.* **2021**, *11*, 22843. [CrossRef] [PubMed]

23. Yepes-Nuñez, J.J.; Urrútia, G.; Romero-García, M.; Alonso-Fernández, S. The PRISMA 2020 statement: An updated guideline for reporting systematic reviews. *Rev. Esp. Cardiol.* **2021**, *74*, 790–799. [CrossRef]

24. Munthe, J.; Wängberg, I.; Pirrone, N.; Iverfeldt, A.; Ferrera, R.; Ebinghaus, R.; Feng, X.; Gardfeldt, K.; Keeler, G.; Lanzillotta, E.; et al. Intercomparison of methods for sampling and analysis of atmospheric mercury species. *Atmos. Environ.* **2001**, *35*, 3007–3017. [CrossRef]

25. Crespo-Lopez, M.E.; Augusto-Oliveira, M.; Lopez-Araujo, A.; Santos-Sacramento, L.; Takeda, P.Y.; Macchi, B.d.M.; Martins do Nascimento, J.L.; Maia, C.S.F.; Lima, R.R.; Arrifano, G.P. Mercury: What can we learn from the Amazon? *Environ. Int.* **2021**, *146*, 106223. [CrossRef]

26. Nakazawa, K.; Nagafuchi, O.; Kawakami, T.; Inoue, T.; Yokota, K.; Serikawa, Y.; Cyio, B.; Elvince, R. Human health risk assessment of mercury vapor around artisanal small-scale gold mining area, Palu city, Central Sulawesi, Indonesia. *Ecotoxicol. Environ. Saf.* **2016**, *124*, 155–162. [CrossRef]

27. Murao, S.; Tomiyasu, T.; Ono, K.; Shibata, H.; Narisawa, N.; Takenaka, C. Mercury Distribution in Artisanal and Small-Scale Gold Mining Area: A Case Study of Hot Spots in Camarines Norte, Philippines. Available online: http://www.ijesd.org/vol10/1160-L0008.pdf (accessed on 2 November 2021).

28. WHO/Europe, Chapter 6.9 Mercury, 2000. Available online: http://www.euro.who.int/_data/assests/pdf_file/0004/123079/AQG2ndEd_6_9Mercury.PDF (accessed on 2 November 2021).

29. Schroeder, H. Atmospheric Mercury-an Overview. *Atmos. Environ.* **1998**, *32*, 809–822. [CrossRef]

30. Sommar, J.; Osterwalder, S.; Zhu, W. Recent advances in understanding and measurement of Hg in the environment: Surface-atmosphere exchange of gaseous elemental mercury (Hg0). *Sci. Total Environ.* **2020**, *721*, 137648. [CrossRef]

31. Pataranawat, P.; Parkpian, P.; Polprasert, C.; Delaune, R.D.; Jugsujinda, A. Mercury emission and distribution: Potential environmental risks at a small-scale gold mining operation, Phichit Province, Thailand. *J. Environ. Sci. Health Part A Toxic Hazard. Subst. Environ. Eng.* **2007**, *42*, 1081–1093. [CrossRef] [PubMed]

32. Nevado, J.J.B.; Bermejo, L.F.G.; Martín-Doimeadios, R.C.R. Distribution of mercury in the aquatic environment at Almadén, Spain. *Environ. Pollut.* **2003**, *122*, 261–271. [CrossRef]

33. Tomiyasu, T.; Kodamatani, H.; Hamada, Y.K.; Matsuyama, A.; Imura, R.; Taniguchi, Y.; Hidayati, N.; Rahajoe, J.S. Distribution of total mercury and methylmercury around the small-scale gold mining area along the Cikaniki River, Bogor, Indonesia. *Environ. Sci. Pollut. Res.* **2017**, *24*, 2643–2652. [CrossRef]

34. Tomiyasu, T.; Kono, Y.; Kodamatani, H.; Hidayati, N.; Rahajoe, J.S. The distribution of mercury around the small-scale gold mining area along the Cikaniki river, Bogor, Indonesia. *Environ. Res.* **2013**, *125*, 12–19. [CrossRef] [PubMed]

35. Elvince, R.; Inoue, T.; Tsushima, K.; Takayanagi, R. Assessment of Mercury Contamination in the Kahayan River, Central Kalimantan, Indonesia. *J. Water Environ. Technol.* **2008**, *6*, 103–112. [CrossRef]

36. World Bank Group. Sources and Uses. In *Pollution Preventon and Abatement Handbook*; World Bank Group: Washington, DC, USA, 1998; pp. 219–222.
37. Akagi, H.; Castillo, E.S.; Cortes-Maramba, N.; Francisco-Rivera, A.T.; Timbang, T.D. Health assessment for mercury exposure among schoolchildren residing near a gold processing and refining plant in Apokon, Tagum, Davao del Norte, Philippines. *Sci. Total Environ.* **2000**, *259*, 31–43. [CrossRef]
38. Department of Environment and Natural Resources. Philippine National Standard for Drinking-Water. 1993. Available online: http://www.wepa-db.net/policies/law/philippines/1993std_drinking.htm (accessed on 2 November 2021).
39. Department of Environment and Natural Resources. DENR Administrative Order No. 2016-08: Water Quality Guidelines and General Effluent Standards. 2016. Available online: https://server2.denr.gov.ph/uploads/rmdd/dao-2019-12.pdf (accessed on 2 November 2021).
40. WHO. *Guidelines for Drinking-Water Quality*, 4th ed.; WHO: Geneva, Switzerland; Available online: https://www.who.int/publications/i/item/9789241549950 (accessed on 2 November 2021).
41. Vandal, G.M.; Mason, R.P.; Fitzgerald, W.F. Cycling of Volatile Mercury in Temperate Lakes. *Water Air Soil Pollut.* **1991**, *56*, 791–803. [CrossRef]
42. Wang, S.; Zhong, T.; Chen, D.; Zhang, X. Spatial Distribution of Mercury (Hg) Concentration in Agricultural Soil and Its Risk Assessment on Food Safety in China. *Sustainability* **2016**, *8*, 795. [CrossRef]
43. Santos-Francés, F.; García-Sánchez, A.; Alonso-Rojo, P.; Contreras, F.; Adams, M. Distribution and mobility of mercury in soils of a gold mining region, Cuyuni river basin, Venezuela. *J. Environ. Manag.* **2011**, *92*, 1268–1276. [CrossRef]
44. Azevedo, R.; Rodriguez, E. Phytotoxicity of Mercury in Plants: A Review. *J. Bot.* **2012**, *2012*, 848614. [CrossRef]
45. EPA. *Soil Screening Guidance: Technical Background Document Soil*; EPA: Frankfurt, Germany, 1996.
46. Lodenius, M. Use of plants for biomonitoring of airborne mercury in contaminated areas. *Environ. Res.* **2013**, *125*, 113–123. [CrossRef] [PubMed]
47. Rasmussen, P.E.; Mierle, G.; Nriagu, J.O. The analysis of vegetation for total mercury. *Water Air Soil Pollut.* **1991**, *56*, 379–390. [CrossRef]
48. Li, R.; Wu, H.; Ing, J.D.; Fu, W.; Gan, L.; Li, Y. Mercury pollution in vegetables, grains and soils from areas surrounding coal-fired power plants. *Sci. Rep.* **2017**, *7*, 46545. [CrossRef] [PubMed]
49. Zhang, H.; Feng, X.; Larssen, T.; Qiu, G.; Vogt, R.D. In inland China, rice, rather than fish, is the major pathway for methylmercury exposure. *Environ. Health Perspect.* **2010**, *118*, 1183–1188. [CrossRef]
50. Mahmud, M.; Lihawa, F.; Saleh, Y.; Desei, F.; Banteng, B. Study of mercury concentration in plants in Traditional Buladu Gold Mining. *IOP Conf. Ser. Earth Environ. Sci.* **2019**, *314*, 12018. [CrossRef]
51. Sakakibara, M.; Sera, K. Mercury in soil and forage plants from artisanal and small-scale gold mining in the bombana area, Indonesia. *Toxics* **2020**, *8*, 15. [CrossRef]
52. Browne, C.L.; Fang, S.C. Uptake of Mercury Vapor by Wheat. *Plant Physiol.* **1978**, *61*, 430–433. [CrossRef]
53. Peralta-Videa, J.R.; Lopez, M.L.; Narayan, M.; Saupe, G.; Gardea-Torresdey, J. The biochemistry of environmental heavy metal uptake by plants: Implications for the food chain. *Int. J. Biochem. Cell Biol.* **2009**, *41*, 1665–1677. [CrossRef]
54. Rajaee, M.; Obiri, S.; Green, A.; Long, R.; Cobbina, S.J.; Nartey, V.; Buck, D.; Antwi, E.; Basu, N. Integrated Assessment of Artisanal and Small-Scale Gold Mining in Ghana—Part 2: Natural Sciences Review. *Int. J. Environ. Res. Public Health* **2015**, *12*, 8971–9011. [CrossRef]
55. FAO; WHO. *Codex General Standard for Contaminants and Toxins in Food and Feed*; WHO: Geneva, Switzerland, 2013; pp. 1–48.
56. Castilhos, Z.C.; Rodriguez-Filho, S.; Rodriguez, A.P.C.; Villas-Boas, R.C.; Siegel, S.; Veiga, M.M.; Beinhoff, C. Mercury contamination in fish from gold mining areas in Indonesia and human health risk assessment. *Sci. Total Environ.* **2006**, *368*, 320–325. [CrossRef]
57. Suckcharoen, S.; Nuorteva, P.; Hasanen, E. Alarming signs of mercury pollution in a freshwater area of Thailand. *Ambio* **1978**, *7*, 113–116.
58. FAO. *Report of the Joint FAO/WHO Expert Consultation on the Risks and Benefits of Fish Consumption*; FAO: Rome, Italy, 2015; pp. 25–29. Available online: http://www.fao.org/docrep/014/ba0136e/ba0136e00.pdf (accessed on 2 November 2021).
59. Bentley, K.; Soebandrio, A. Arsenic and mercury concentrations in marine fish sourced from local fishermen and fish markets in mine-impacted communities in Ratatotok Sub-district, North Sulawesi, Indonesia. *Mar. Pollut. Bull.* **2017**, *120*, 75–81. [CrossRef] [PubMed]
60. Murphy, T.P.; Irvine, K.N.; Sampson, M.; Guo, J.; Parr, T. Mercury Contamination along the Mekong River, Cambodia. *Asian J. Water Environ. Pollut.* **2008**, *6*, 1–9.
61. Baran, E.; Jantunen, T.; Kieok, C.C. *Values of Inland Fisheries in the Mekong River Basin*; WorldFish Center: Phnom Penh, Cambodia, 2007; p. 76. Available online: https://digitalarchive.worldfishcenter.org/bitstream/handle/20.500.12348/1671/WF_895.pdf?sequence=1&isAllowed=y (accessed on 2 November 2021).
62. Canham, R.; González-Prieto, A.M.; Elliott, J.E. Mercury Exposure and Toxicological Consequences in Fish and Fish-Eating Wildlife from Anthropogenic Activity in Latin America. *Integr. Environ. Assess. Manag.* **2021**, *17*, 13–26. [CrossRef]
63. Environmental Health Perspectives. Guidance for Identifying Populations at Risk from Mercury Exposure. *Exposure* **2008**, *113*, 1381–1385. [CrossRef]

64. Loi, V.D.; Gian, T.; Anh, L.; Duc, T.; Huy, T.; Mon, P.; Ha, T.; Mineshi, S. NIMD Forum 2006 II-Current Issue on Mercury Pollution in the Asia-Pacific Region—Mercury Exposure to Workers at Gold Mining and Battery Plants in Vietnam. 2006. Available online: http://nimd.env.go.jp/english/kenkyu/nimd_forum/nimd_forum_2006_II.html (accessed on 4 November 2021).

65. Novirsa, R.; Dinh, Q.P.; Jeong, H.; Addai-Arhin, S.; Nugraha, W.C.; Hirota, N.; Wispriyono, B.; Ishibashi, Y.; Arizono, K. The Dietary Intake of Mercury from Rice and Human Health Risk in Artisanal Small-Scale Gold Mining Area, Indonesia. Available online: https://www.jstage.jst.go.jp/article/fts/7/5/7_215/_article/-char/ja/ (accessed on 4 November 2021).

66. Salazar-Camacho, C.; Salas-Moreno, M.; Marrugo-Madrid, S.; Marrugo-Negrete, J.; Díez, S. Dietary human exposure to mercury in two artisanal small-scale gold mining communities of northwestern Colombia. *Environ. Int.* **2017**, *107*, 47–54. [CrossRef]

67. National Institute for Minamata Disease (NMID), Ministry of Environment, Japan. *Mercury and Health*; 2013. Available online: http://nimd.env.go.jp/english/kenkyu/docs/Mercury_and_health.pdf (accessed on 2 November 2021).

68. Clemente, E.; Sera, K.; Futatsugawa, S.; Murao, S. PIXE analysis of hair samples from artisanal mining communities in the Acupan region, Benguet, Philippines. *Nucl. Instrum. Methods Phys. Res. Sect. B Beam Interact. Mater. Atoms* **2004**, *219–220*, 161–165. [CrossRef]

69. Umbangtalad, S.; Parkpian, P.; Visvanathan, C.; Delaune, R.D.; Jugsujinda, A. Assessment of Hg contamination and exposure to miners and schoolchildren at a small-scale gold mining and recovery operation in Thailand. *J. Environ. Sci. Health Part A Toxic/Hazard. Subst. Environ. Eng.* **2007**, *42*, 2071–2079. [CrossRef]

70. United States Environmental Protection Agency. *Integrated Risk Information System (IRIS)*; National Center for Environmental Assessment, Office of Research and Development: Washington, DC, USA, 1999.

71. Kobal, A.B.; Tratnik, J.S.; Mazej, D.; Fajon, V.; Gibicar, D.; Miklavcic, A.; Kocman, D.; Kotnik, J.; Briski, A.S.; Osredkar, J.; et al. Exposure to mercury in susceptible population groups living in the former mercury mining town of Idrija, Slovenia. *Environ. Res.* **2017**, *152*, 434–445. [CrossRef]

72. Wang, S.; Feng, X.; Qiu, G.; Fu, X.; Wei, Z. Characteristics of mercury exchange flux between soil and air in the heavily air-polluted area, eastern Guizhou, China. *Atmos. Environ.* **2007**, *41*, 5584–5594. [CrossRef]

73. Higueras, P.; Oyarzun, R.; Kotnik, J.; Esbri, J.M.; Martinez-Coronado, A.; Horvat, M.; Lopez-Berdonces, M.A.; Llanos, W.; Vaselli, O.; Nisi, B.; et al. A compilation of field surveys on gaseous elemental mercury (GEM) from contrasting environmental settings in Europe, South America, South Africa and China. *Int. J. Environ. Res. Public Health* **2016**, *13*, 160. [CrossRef]

74. Guney, M.; Akimzhanova, Z.; Kumisbek, A.; Beisova, K.; Kismelyeva, S.; Satayeva, A.; Inglezakis, V.; Karaca, F. Mercury (HG) contaminated sites in kazakhstan: Review of current cases and site remediation responses. *Int. J. Environ. Res. Public Health* **2020**, *17*, 8936. [CrossRef] [PubMed]

75. Hanson, P.J.; Lindberg, S.E.; Tabberer, T.A.; Owens, J.G.; Kim, K.H. Foliar exchange of mercury vapor: Evidence for a compensation point. *Water Air Soil Pollut.* **1995**, *1*, 373–382. [CrossRef]

76. De Temmerman, L.; Waegeneers, N.; Claeys, N.; Roekens, E. Comparison of concentrations of mercury in ambient air to its accumulation by leafy vegetables: An important step in terrestrial food chain analysis. *Environ. Pollut.* **2009**, *157*, 1337–1341. [CrossRef] [PubMed]

77. Patra, M.; Sharma, A. Mercury toxicity in plants. *Bot. Rev.* **2000**, *66*, 379–422. [CrossRef]

78. Feng, X.; Qiu, G. Mercury pollution in Guizhou, Southwestern China—An overview. *Sci. Total Environ.* **2008**, *400*, 227–237. [CrossRef]

79. Marrugo-Negrete, J.; Marrugo-Madrid, S.; Pinedo-Hernández, J.; Durango-Hernández, J.; Díez, S. Screening of native plant species for phytoremediation potential at a Hg-contaminated mining site. *Sci. Total Environ.* **2016**, *542*, 809–816. [CrossRef]

80. Molina, J.A.; Oyarzun, R.; Esbrí, J.M.; Higueras, P. Mercury accumulation in soils and plants in the Almadén mining district, Spain: One of the most contaminated sites on Earth. *Environ. Geochem. Health* **2006**, *28*, 487–498. [CrossRef]

81. Zhang, L.; Wong, M.H. Environmental mercury contamination in China: Sources and impacts. *Environ. Int.* **2007**, *33*, 108–121. [CrossRef]

82. Guédron, S.; Point, D.; Acha, D.; Bouchet, S.; Baya, P.A.; Tessier, E.; Monperrus, M.; Molina, C.I.; Groleau, A.; Chauvaud, L.; et al. Mercury contamination level and speciation inventory in Lakes Titicaca & Uru-Uru (Bolivia): Current status and future trends. *Environ. Pollut.* **2017**, *231*, 262–270. [CrossRef]

83. Obrist, D.; Kirk, J.L.; Zhang, L.; Sunderland, E.M.; Jiskra, M.; Selin, N.E. A review of global environmental mercury processes in response to human and natural perturbations: Changes of emissions, climate, and land use. *Ambio* **2018**, *47*, 116–140. [CrossRef] [PubMed]

84. Li, P.; Feng, X.; Qiu, G.; Shang, L.; Wang, S. Mercury exposure in the population from Wuchuan mercury mining area, Guizhou, China. *Sci. Total Environ.* **2008**, *395*, 72–79. [CrossRef] [PubMed]

85. Jia, Q.; Zhu, X.; Hao, Y.; Yang, Z.; Wang, Q.; Fu, H.; Yu, H. Mercury in soil, vegetable and human hair in a typical mining area in China: Implication for human exposure. *J. Environ. Sci.* **2017**, *8*, 73–82. [CrossRef] [PubMed]

86. Langeland, A.L.; Hardin, R.D.; Neitzel, R.L. Mercury levels in human hair and farmed fish near artisanal and small-scale gold mining communities in the madre de dios River Basin, Peru. *Int. J. Environ. Res. Public Health* **2017**, *14*, 302. [CrossRef] [PubMed]

87. WHO. Inorganic Mercury (Environmental Health Criteria), International Program on Chemistry Safety, Vol. 118. WHO. 1991. Available online: https://apps.who.int/iris/handle/10665/40626 (accessed on 4 November 2021).

International Journal of
*Environmental Research
and Public Health*

MDPI

Review

Transdisciplinary Communities of Practice to Resolve Health Problems in Southeast Asian Artisanal and Small-Scale Gold Mining Communities

Win Thiri Kyaw [1],* and Masayuki Sakakibara [1,2]

[1] Research Institute for Humanity and Nature, Kyoto 603-8047, Japan; sakaki@chikyu.ac.jp
[2] Graduate School of Science and Engineering, Ehime University, Matsuyama 790-8577, Japan
* Correspondence: thiri@chikyu.ac.jp; Tel.: +81-075-707-2443

Abstract: Artisanal and small-scale gold mining (ASGM) has been a major part of people's livelihood in the rural areas of many developing countries, including those in Southeast Asia (SEA). Nevertheless, because of the use of mercury, ASGM activities have significant local and global adverse impacts on the environment and ASGM community health. Although there have been many monodisciplinary projects by academic researchers and governments to solve the environmental and health problems in SEA ASGM communities, they have not been sufficient to solve the complex socioeconomic problems. This review first outlines the nature of the SEA ASGM activities and the consequent environmental, community health, and socioeconomic problems and then introduces an approach using transdisciplinary communities of practice that involves both academic and nonacademic participants to relieve these wicked ASGM problems and to improve the environmental governance and community health in ASGM communities in SEA.

Keywords: artisanal and small-scale gold mining; ASGM; TDCoPs; transdisciplinary; mercury; Southeast Asia

Citation: Kyaw, W.T.; Sakakibara, M. Transdisciplinary Communities of Practice to Resolve Health Problems in Southeast Asian Artisanal and Small-Scale Gold Mining Communities. *Int. J. Environ. Res. Public Health* **2022**, *19*, 5422. https://doi.org/10.3390/ijerph19095422

Academic Editor: Paul B. Tchounwou

Received: 26 March 2022
Accepted: 27 April 2022
Published: 29 April 2022

Publisher's Note: MDPI stays neutral with regard to jurisdictional claims in published maps and institutional affiliations.

1. Introduction

Artisanal and small-scale gold mining (ASGM) is an informal occupational sector in many rural parts of developing nations. It relies on unskilled labor to mine and process gold, particularly in areas where agricultural income alone is unable to support community livelihoods. The International Labor Organization (ILO) reported that ASGM activities are labor-intensive, involve a small number of people, and rely on basic equipment. Globally, over 100 million people are directly or indirectly involved in ASGM for their livelihoods. Although ASGM contributes to poverty alleviation and generates national income, it has also been negatively associated with social, environmental, and health issues [1,2]. On a social level, catastrophic devastation, gender inequality, and child labor issues occur as a result of the discovery of deposits and the consequent influx of small-scale miners into formerly native or rural communities [2]. Nevertheless, the social issues differ between continents; for example, the primary challenges in Africa are AIDS and sustainable community development; in Asia/Pacific, the challenges are related to intercultural elements and cultural rights; and in Latin American/Caribbean, the most important issues are the environment, indigenous peoples, and legal aspects [2].

Many developing countries in Southeast Asia (SEA) practice ASGM because of poverty and tradition. Most ASGM is practiced in Indonesia, the Philippines, and Myanmar, with a significantly smaller percentage of people working in ASGM in Thailand, Cambodia, and Laos. The more specific ASGM profiles in these countries are discussed later in this paper.

Similar to other countries involved in ASGM activities, mercury (Hg) is commonly applied as a cheap, rapid method for extracting gold, although it is highly toxic and can have serious health consequences for miners, their families, and the neighboring communities. ASGM sectors in over 70 countries have been found to have the largest global Hg emissions at 640 to 1350 tons annually, and a recent United Nations Environment Program (UNEP) global mercury assessment found that global ASGM activities were increasing [3]. Although Hg is illegal in some countries, as it is affordable and easy to use, it continues to be widely used. The Hg used in SEA is directly sourced from Spain and China or from less transparent transit routes through Singapore [4]. In Indonesia, approximately 280 tons of illegal Hg was imported in 2010 [5]; however, UN Comtrade (2020), an international UN branch that reports goods that were officially exported and imported by countries [6], reported that Indonesia had shifted from importing to exporting Hg in 2015 because of its development of national cinnabar (HgS) mines [7]. A previous report in 2002 on the estimation of annual Hg emissions from ASGM activities found that Indonesia emitted the most of all SEA countries at 145 tons, followed by the Philippines with 25.0 tons, Cambodia and Vietnam with 7.5 tons each, Myanmar with 6.5 tons, Malaysia with 3.5 tons, Thailand with 1.5 tons, and Laos with 1.3 tons [8].

While Hg is useful for extracting gold, it causes significant environmental pollution and is a health hazard for the miners and the community. Hg (elemental Hg) is used to bind the gold in the ore, which is then known as a gold–Hg amalgam. The amalgam is then smelted to extract the gold [9], which emits an Hg vapor into the atmosphere, where it can be oxidized into ionic Hg. It can either fall into water bodies directly or be deposited on the land's surface and washed into an aquatic system by run off. When ionic mercury enters water, it undergoes three main transformations. First, it may be volatilized back into elemental Hg. The second channel involves adsorption into sediments in lakes, rivers, and reservoirs. Finally, in anaerobic conditions at the bottom of the water bodies and the sediment–water interface, the sulfate-reducing bacteria may methylate ionic Hg to become methylmercury (MeHg), the most toxic form of Hg, which can be ingested by planktons, after which it enters the food chain through bioaccumulation in long-lived predatory species such as sharks and shellfish [10]. Thus, fish can collect significant levels of MeHg, which can be consumed by humans or wildlife. MeHg is found in fish protein and is not degraded by cooking or washing [11]. Aquatic changes in Hg occur in both directions. In certain cases, these processes can be reversed; thus, several types of Hg can be found in aquatic systems. However, increased Hg contents were discovered not only in marine food but also in rice samples of ASGM areas in Indonesia [12,13], which may occur through the absorption of MeHg by the paddy roots, and where it is stored in the rice grain until it reaches ripening [14]. This may also occur in other agricultural lands due to the waste discharged from the process of ASGM.

In addition, Hg vapor can be inhaled by miners during the smelting process, which is their main Hg exposure route [15]. Thus, regardless of whether they are directly engaging in ASGM activities, ASGM community health is doubly affected by the direct inhalation of Hg vapor and by the consumption of contaminated fish and shellfish. Children are more susceptible to the adverse health impacts of Hg, and even before they are born, they can suffer neurological disorders from prenatal Hg dietary exposure from the consumption of marine fish by their mothers.

Despite these serious issues, continuing poverty in SEA means that ASGM activities are expected to increase, which will escalate these "wicked problems"; therefore, a sound, sustainable solution is needed. It is hypothesized that the major contributors to these wicked problems are (1) the lack or insufficient awareness in SEA ASGM communities because of poor information and (2) ineffective approaches by academic researchers, local experts, governments, and other stakeholders in tackling these problems. Thus, solutions to these wicked ASGM problems in SEA are needed in order to address the ASGM contributing factors, the causative ASGM hazards, and the health hazards in mining communities.

2. Methods

To identify relevant studies, we conducted searches related to information on the ASGM of SEA countries and their socioeconomic and community health hazards in search engines such as PubMed, Google Scholar and Scopus. Terms such as "mercury", "ASGM", "socio-economic", "health", "health impact", "health assessment", and "mercury intoxication" were included in the search. The reports were then filtered by the name of SEA countries, reports with the clinical findings, and the usage of English. The reports with clinical assessments were considered relevant for the summary of health assessments.

3. ASGM Profiles in SEA Countries

3.1. General Overview of Hg Usage in ASGM

Although Hg has serious environmental and health impacts, it is widely used as part of the ASGM process because it is cheap, easily available, and easy to handle. Developing countries such as those in SEA use Hg in both whole-ore-amalgamation and concentration methods. The whole-ore-amalgamation method results in higher Hg emissions from the tailings than the concentration method [16]. With the whole-ore-amalgamation method, the Hg come into contact with 100% of the ore, with four parts of Hg being used for every one part of gold; that is, the gold-to-Hg ratio is 4:1, and it can be even higher, such as 20:1 or 50:1 [16]. However, with the concentration method, the gold is initially reduced to a smaller quantity before the amalgamation is performed, which is commonly done using gravity. Then, only the concentrations that have the heaviest minerals and gold are treated with the Hg, which means that the Hg-to-gold ratio is substantially lower than in the whole ore amalgamation method (usually 1:1 vs. 1.3:1), and significantly smaller quantities of Hg remain in the tailings [16]. Thus, the nature and the degree of environmental destruction and damage to human health depend on the type of minerals in the local areas and the type of ASGM processes being used.

3.2. Indonesia

ASGM activity in Indonesia is illegal, and Indonesian health and environmental regulations have banned ASGM. Nonetheless, Indonesia is well known for having the most ASGM activities in SEA, with more than 850 mining sites and 200,000 miners [12,17,18]. ASGM involves panning, dredging, or using high-pressure pumps on river banks, open pits, or vertical excavations to expose the secondary or tertiary alluvial ores or to physically extract the ore from the neighboring hills or mountains using traditional instruments, such as wide hoes, bars, and basic pulleys. Hammers and other similar tools are used to hand-smash the ore, and homemade mechanical crushers are used in some processing industries. The ore is then physically loaded into sacks and hauled to the processing facilities, where the crushed ore (30–40 kg) is put into a trommel, a steel mill grinder, where the gold is extracted from the ore using water and decimeter-sized boulders and milled for 3–4 h until the material is fine enough to release the gold. Then, roughly 500 g of Hg is fed into the rotating trommel for approximately 30 min to complete the amalgamation [19], after which the material is discharged, separated, and washed with bare hands to separate the gold from the nonbinding Hg while keeping the sediment aside for a second use. The resulting amalgam is burned inside or outside dwellings and gold shops, which emits Hg vapor into the surrounding area. Approximately 20–30% of gold is obtained from the amalgamation method.

In Indonesia, similar to many other nations in Latin America and Africa, the processing facilities in Indonesia offer free or low-cost ore amalgamation to miners in exchange for the tailings [20]. The processing facility keeps the Hg-contaminated tailings to recover the residual gold using cyanide (CN), which is significantly more effective than Hg at extracting fine gold particles. Usually, 80–90% of the gold in the tailings is retrieved in this way. Nevertheless, CN dissolves the remaining Hg in the tailings, which generates the mercury cyanide complexes, $Hg(CN)_{(2+N)}{}^{(-N)}$ (aq) compound, and causes significantly greater aquatic biota pollution than the elemental Hg [21]. In rivers, the sediment is washed and panned by women and children, and the gold is extracted from the sediment using Hg [19]. In some areas, the muddy water that is tainted by Hg-contaminated tailings is released directly into the local waterways, which then drains into the rice paddy fields and fish ponds [12]. Thus, Hg contaminates the soil, the sediment, the water and biota, the plants, the tree bark, and even cattle [20,22–25]. As gold mining activities involve materials extraction, other heavy metals are also exposed, such as arsenic (As), cadmium (Cd), lead (Pb), cobalt (Co), manganese (Mn), and zinc (Zn) [26,27]. High levels of As and Pb that exceed the safe drinking water limit defined by the WHO have been detected in water samples from the Bone River, which is close to an ASGM site [28].

Soil remediation solutions for Hg-polluted soils have been developed, such as phytoremediation using green plants and their microbiome, to remediate polluted areas [29]. However, the expense of traditional phytoremediation increases each year, and the owners of polluted lands lose money throughout the cleanup process, limiting its practical implementation. Therefore, recently, high-biomass crops such as Indian mustard, maize, sunflower, and sweet sorghum have been grown to be able to remediate heavy metals for practical and economic reasons [30]. A field study in Indonesia suggested that KCS105 sweet sorghum might be a promising energy crop for phytoremediation of mercury-polluted soil because it grew well on mercury-contaminated soil and accumulated mercury in its root and shoot. Furthermore, *Agrobacterium tumefaciens* inoculation increased the phytoremediation efficacy of Hg to 934 mg/ha [31].

3.3. The Philippines

The Philippines ranks 20th in global gold production, with 70% of this coming directly from the estimated 500,000 ASGM sites operating in over 40 of its 81 provinces. The ILO estimated that around 19,000 children work in ASGM in the Philippines [32]; however, after the Philippines committed to legalizing ASGM activities, several measures were introduced between 2016 and 2020 to end ASGM child labor, such as the CARING Gold Mining Project in the Philippines operated by the ILO and executed with support from BanToxics and the United States Department of Labor [33]. Gold ore is mined mostly around Diwalwal in Mindanao, which is one of the Philippines' major islands. Dominated by Mount Diwata, Diwalwal is a prominent gold town of 15,000 inhabitants, in which gold mining has been conducted for over a century [25]. The Mount Diwata ASGM area comprises small industrial complexes, such as ball mill and cyanidation facilities, as well as stores and housing, all of which are scattered around the site, with their wastes, including human waste, being thrown into the rivers or discarded. Small communities in the Diwalwal community living area crush the mining ore into powder, and Hg is added to make the gold–Hg amalgam, which is later burned by small local companies or in the miners' homes. Thus, Hg tailings are found throughout the region [34]. Tuberculosis is the leading cause of death in Diwalwal, and the local health clinic, which has had only midwives and "helots" and no doctor for years, is therefore underequipped and unable to diagnose or treat Hg toxicity.

3.4. Myanmar

ASGM is mainly practiced in 300 officially recorded areas in Kachin State, Sagaing Region, Mandalay Region, and Bago Region, in which an estimated 730,000 people work [35]. The main ASGM activities are panning, river mining with bucket dredges, suction dredging, hydraulic mining, open-pit mining for alluvial and colluvial gold deposits [36], and underground mining for hard rock deposits [37–39]. CN is often used in underground mining to crush rocks and dissolve gold [35]. The final gold is recovered using Hg, which is either collected for a second use or discharged into the waterways. Similar to the ASGM activities in Indonesia, to retrieve the gold, the Hg amalgam is burned in local gold shops, inside the houses, or in the open air [37]. The Hg emitted from the ASGM activities has polluted the environment [36,37] and has contaminated groundwater and the atmosphere [39]. High concentrations of heavy metals, such as As, Cd, Pb, and Hg, have also been found in ASGM site soil samples [40].

3.5. Other SEA Countries

Vietnam has approximately 63,000 ASGM workers, Cambodia has approximately 6000, and Laos has approximately 3000 [8]. The ASGM activities in the Lao PDR are relatively unknown; however, the Department of Gold Mining has claimed that there are ASM sites in Borikhamxay, Saravanh, Vientiane, and Luang Prabang provinces, although the extent and usage of Hg are unclear. A baseline study by Earth Systems Lao in Luan Prabang Province, Lao PDR, found that the ASGM activities were primarily conducted by families, with the ore/alluvium extraction normally performed by men using shovels and chisels, with the women and children carrying the ore to bowls and sluice boards, panning the ore, and extracting the gold at home by applying the Hg at the panning stage [41]. Similarly, ASGM in Cambodia is practiced at the family level, with most gold extraction activities occurring in the northeast part of the country. The estimated annual usage of Hg is from 34.5 to 1182 kg [42], and CN is also used [43].

4. Socioeconomic Hazards and Community Health Hazards Due to ASGM Activities

4.1. Socioeconomic Hazards Due to ASGM Activities

Due to the lack of infrastructure, mining communities have many socioeconomic problems, such as (1) conflicts between the native miners and the migrant miners; (2) gender inequality, which means that females have less income and less opportunity; and (3) poor or no educational support for children or the adults [44].

4.2. Health Hazards from Exposure to Hg and Other Heavy Metals

As discussed above, ASGM workers are exposed to elemental Hg mainly by inhaling the Hg vapor during the amalgamation and amalgam-burning processes. There are two ways in which elemental Hg in the atmosphere can be changed into two other forms of Hg: oxidizing into inorganic mercury salts (Hg^+ and $Hg2^+$) and methylating into methylmercury (MeHg), which has been found in fish [34]. The burning of the amalgam in homes or in open air exposes ASGM workers, their families, and their neighbors to the Hg vapor, which passes through the alveolar membrane, is absorbed into the blood, and travels to tissues, primarily affecting the respiratory system [45]. The major elemental Hg absorption takes place in the lungs (80%), after which it rapidly travels into the blood and other organs [46]. The inhaled Hg vapor can also cross the blood–brain barrier and blood–placenta barrier and accumulates in the central nervous system. Elemental Hg is primarily deposited in the brain and kidneys in its oxidized form, with the kidney being the organ that can be most damaged from repeated exposure [46].

Thus, ASGM communities can be exposed to Hg and other heavy metals, such as Pb, Cd, Co, and Mn. The International Agency for Research on Cancer has classified As as a Category 1 carcinogen that can cause bladder, skin, and lung cancers. Previous ecological research has linked high As concentrations in the soil to an increased risk of cancer. Lead is also a carcinogen that can cause kidney damage, hypertension, and a decrease in mental abilities. Furthermore, chronic low-level Mn exposure has been linked to an increase in Parkinsonism in exposed populations, and cobalt exposure can cause lung cancer, cardiomyopathy, and hearing and vision loss [47].

4.2.1. Acute Effects from Exposure to Elemental Hg

Dermatitis can result from acute Hg exposure. After inhaling elemental Hg vapors, people may suffer from coughs, chills, fevers, shortness of breath, and gastrointestinal symptoms such as nausea, vomiting, and diarrhea, which can be followed by a metallic taste, dysphagia, salivation, weakness, headaches, and visual disorders [48]. Lungs can also be severely damaged from acute high-level exposure, and in severe cases, the resulting hypoxia can result in death. Reportedly, two children died, and their parents suffered from severe respiratory distress because of gold processing in a kitchen with poor ventilation [49].

4.2.2. Chronic Effects from Exposure to Elemental Hg through Inhalation and MeHg through the Food Chain

When elemental Hg accumulates in the central nervous and renal systems, people are mainly affected by chronic intoxication, the major clinical presentations for which are unintentional or intentional tremors, psychological disturbances or erethism, proteinuria, and gingivitis [50]. Erethism can cause behavioral changes, such as irritability, low self-confidence, depression, apathy and shyness, and proteinuria resulting in tubular damage [46]. Other disorders include allergies or autoimmunity because of the reduced resistance to infection and cancers [51]. Children are particularly sensitive to exposure from eating MeHg-contaminated seafood. During pregnancy, MeHg bioaccumulates in fish, causing neurodevelopmental issues in the unborn child. The fetal brain is particularly vulnerable to transplacental exposure. Mental retardation, seizures, visual and hearing loss, developmental delays, language difficulties, and memory loss are all neurological signs. Chronic Hg exposure in children causes acrodynia, a condition marked by red and aching limbs [52,53].

4.3. Other Health Hazards

Most studies have focused on Hg-related health impacts in ASGM miners and communities; however, there are also other community health hazards [54], such as respiratory damage from the silica dust from drilling the ores (silicosis) [55], which increases the risk of tuberculosis from silicosis, and malaria, noise exposure, and injury, all of which are aggravated by the lack of infrastructure and the crowded living conditions.

4.4. ASGM Health Assessments in SEA Countries

There is no universal diagnosis for chronic Hg intoxication. Many of the health surveys in ASGM communities in Indonesia, the Philippines, and the Laos PDR have been conducted with the support of the United Nations Industrial Development Organization (UNIDO) Global Mercury Project. Clinical examinations and biomonitoring of Hg levels in the hair, blood, and urine have found instances of chronic elemental Hg intoxication and chronic MeHg intoxication from the food chain that has caused the loss of peripheral vision; ataxia; pins and needles in the hands, feet, and around the mouth; speech and hearing impairments; and muscle weakness. Because they are easily accessible, hair Hg levels are commonly analyzed to assess the Hg exposure in ASGM miners and communities; however, hair and blood Hg levels generally indicate the presence of MeHg through contamination of the food chain, and urine Hg levels indicate the occupational elemental Hg exposure from the ASGM activities.

Table 1 summarizes the findings of surveys conducted in SEA countries, which include clinical chronic Hg intoxication analyses in miners and ASGM communities and, as suggested by the German Human Biomonitoring Commission, the collection of biomonitoring samples that assessed the Hg content in hair ($\mu g/g$). Normal levels are below 1.0 $\mu g/g$, alert levels are 1.0–5.0 $\mu g/g$, and levels over 5.0 $\mu g/g$ indicate a substantial health risk [56]. Most studies have found that the miner and community participants were suffering from Hg health impacts.

Overall, however, there have been few ASGM health impact evaluation projects in SEA. To the best of our knowledge, there was only one preliminary health survey conducted in Myanmar in 2020, which included clinical examinations and hair sample analyses [57] and only one online health survey [58] of the ASGM communities in the Mandalay Region, Thabeikkyin Township, Chaung Gyi Village. However, no clinical health assessments have been conducted in ASGM communities in Vietnam or Cambodia, and the one survey conducted in Thailand only included biomonitoring of human samples but no clinical examinations. The survey was conducted in the Phanom Pha gold mining area of Thailand,

and it found that the environmental Hg contamination and open amalgam burning were the likely sources of the miners' health problems, the Hg exposure at work surpassed permissible values, and the urine Hg levels indicated that the miners were being exposed to inorganic Hg [59]. Although no health assessments have been conducted in Cambodia, miners there have reported skin rashes, and animal deaths have been reported near the mining operations [43]. The atmospheric Hg level of these countries are also mentioned in Table 1 in terms of the level of Hg contamination. The gold-production region in central Sulawesi, Indonesia has the highest average 24-h ambient Hg values at 9172 ng/m^3, which was nine times the WHO's limit of 1000 ng/m^3. The amalgam burning sites of both the Philippines and Myanmar also showed the high atmospheric Hg values at 314,000 ng/m^3 and 74,000 ng/m^3, respectively.

Table 1. Health assessment findings in SEA countries and Hg contamination levels.

Country	Year	Sample Size	Health Assessment Findings	References	Atmospheric Hg Level by Countries (ng/m^3)
Indonesia (Galangan in Central Kalimantan and Talawaan in Northern Sulawesi)	2010	281	Ataxia, tremor, dysdiadochokinesia, etc. -Mean value of Hg-blood (μg/L); control group (A) (4.92), only living group in Kalimantan (B) (12.86) and Sulawesi (C) (7.05), panning workers in Kalimantan (D) (20.35) and Sulawesi (E) (15.18), smelting workers in Kalimantan (F) (38.92) and Sulawesi (G) (27.43). -Mean value of Hg-urine (μg/L); A (0.90), B (21.47), C (4.48), D (37.45), E (13.37), F (177.69), G (54.46). -Mean value of Hg-urine (μg/g creatinine); A (0.43), B (10.44), C (2.70), D (15.65), E (5.58), F (69.35), G (31.89). -Mean value of Hg-hair (μg/g); A (1.64), B (7.14), C (2.30), D (42.56), E (5.73), F (17.09), G (13.14	Bose-O'Reilly, S. et al., 2010a [60]	9172 $\pm$ 16,422 (mean $\pm$ SD) [61]
Indonesia (Banten/Cisitu Village)	2015	28 children	Tremors in adults, and neurological deficits in children and teenagers such as developmental delays, hydrocephalus, deafness, vision disorders, and other congenital deformities	Ismawati, Y., 2015 [17]	
Indonesia (Banten/Cisitu Village)	2016	18	Typical signs and symptoms of chronic Hg intoxication (excessive salivation, sleep disturbances, tremors, ataxia, dysdiadochokinesia, pathological coordination tests, gray to bluish discoloration of the oral cavities, and proteinuria). The mean values of Hg-urine (μg/L) were increased in eight patients (>7 μg). All 18 people had increased hair levels (>1 μg Hg/g hair)	Bose-O'Reilly, S. et al., 2016 [12]	
Indonesia (Gorontalo)	2015	44	Bluish gums, Babinski reflex, labia reflex, tremor, rigidity, ataxia, alternating movements, and nystagmus in ASGM miners and inhabitants of Angrrek and Sumalata. -Hg-hair (μg/g); 14.2 μg	Arifin Y. I. et al., 2015 [62]	
Indonesia (Sulawesi, Makassar)	2016	40 gold workers and 17 residents as control	Tremors in the tongue, eyelid, finger, nose, pouring, posture holding, and Romberg test, unbalanced rigidity and ataxia, pathology reflex, sensory disturbance, constricted field of vision, and slow knee jerk and bicep reflexes. Hg-hair (μg/g); directly exposed group (10.8), indirectly exposed group (6.5), and control group (2.8)	Abbas H.H. et al., 2017 [63]	
Philippines (Mindanao/ Monkayo and Diwalwal) Davao as Control area	2000	323 (workers from Diwalwal, local families from Monkayo including children and a control group in Davao)	Fatigue, tremor, memory problems, restlessness, loss of weight, metallic taste, sleeping disturbances (reported symptoms), and intentional tremors, mainly fine tremors of the eyelids, lips, and fingers, ataxia, hyperreflexia, sensory disturbances, and bluish discoloration of the gums (symptoms) were observed in approximately 65% of the population in the Mt. Diwata area, and 85% in the ball mill and amalgam smelter workers. To a lesser but still not acceptable extent (approximately 33%) Hg intoxication (headache, vision problems, and nausea) was found in the nonoccupationally exposed population at Mt. Diwata and downstream in the Monkayo plain (38%). No Hg intoxication was found in the control area of Davao. -Hg-blood (μg/L); span < 0.25–107.6, median 8.2, arithmetic mean 11.48 -Hg-urine (μg/L); span < 0.25–294, median 2.5, arithmetic mean 11.08 -Hg-urine (μg/g creatinine); span < 0.1–196.3, median 2.4, arithmetic mean 8.40 -Hg-hair (μg/g); span 0.03–37.76, median 2.72, arithmetic mean 4. 14	Bose-O'Reilly, S. et al., 2000 [25]	314,000 [64]
Philippines (Apokon, Tagum, Davao del Norte)	2000	162 (school children aged 5–17 years)	Under-height, gingival discoloration, adenopathy, underweight and dermatologic abnormalities. -Total Hg-blood (μg/L); 0.757–56.88 -MeHg-blood (μg/L); 1.36–46.73 -Total Hg-hair (μg/g); 0.278–20.393 -MeHg-hair (μg/g); 1.36–46.73	Akagi, H. et al., 2000 [65]	
Myanmar (Mandalay Region/Chaung Gyi Village)	2020	29	Tremor, Ataxia, decreased lung function in miners. -Hg-hair (μg/g); miners 0.93 (0.72–1.44) (median–interquartile range) Nonminers 0.63 (0.53–0.67) (median–interquartile range)	Kyaw WT, 2020 [57]	74,000 [39]
Thailand (Phanom Pha)	2007	79 miners 59 school children	No clinical assessments were included. Hg-urine level; miners (μg/g creatinine) (22.85 $\pm$ 0.04 μg/g), School children (μg/g creatinine) (13.93 $\pm$ 0.33). Hg-hair level; miners (1.17 $\pm$ 0.05 μg/g), School children (0.93 $\pm$ 0.01)	UMBANGTALAD S, 2007 [59]	
Laos PDR -District of Chomphet (Houay Gno Village and Houay Koh), -District of Pak Ou (Latthahai Village and Pak Ou Village)	2004	191	The study observed neurological abnormalities in 56% of men (47 out of 83) and 41% of women (44 out of 107); however, only 16 % of men and 71% of women were using Hg. The author suggested considering other environmental and genetic factors as possible causes of the neurological abnormalities. Maximum level; Hg-blood level μg/L (12.2), Hg-urine level μg/L (15), and Hg-hair level μg/g (18.6)	Bose-O'Reilly, S. et al., 2004 [66]	

5. ASGM Problems in SEA Countries and a Sustainable Solution

As discussed above, ASGM community health assessments have generally been conducted by global projects and institutional researchers. The "Minamata Convention on Mercury" was adopted in 2013 at a diplomatic conference of the UNEP to resolve the problems resulting from anthropogenic Hg exposure. As of 2022, there were 128 signatories to the treaty out of 137 parties, including most SEA countries [67]. Projects such as the UNIDO global mercury project, which is focused on Hg environmental and health impact assessments in ASGM areas, are also being conducted. Nevertheless, despite these projects, the ASGM communities are still being exposed to these health hazards primarily because of poverty, a lack of information, and the ineffective monodisciplinary approach of academic researchers and governments. For example, researchers only conduct bottom-up ASGM community evaluations, whereas governments only focus on top-down ASGM rules and regulations, neither of which tackle the wicked nature of these issues. Furthermore, local medical professionals and health care service personnel are often unaware of chronic Hg intoxication symptoms or the other ASGM-related health issues, which results in misdiagnoses and a failure to provide prevention measures or early treatment.

Thus, the key to the sustainable resolution of these problems is to encourage transdisciplinary collaboration and active community participation to effectively share information, increase awareness, and add value to ASGM-related issues, which we refer to as transdisciplinary communities of practice (TDCoPs). The authors' project, which is known as the Sustainable Regional Innovations for Reducing Environmental Pollution project, is a five-year project funded by the Institute for Humanity and Nature that runs from 2019 and 2023 to elucidate pathways to alleviate ASGM-related Hg contamination in Indonesia using a transdisciplinary approach. In collaboration with local stakeholders, several TDCoPs groups have been successfully formed in the study area to encourage sustainable community livelihoods and establish new industries to reduce the reliance on ASGM, promote environmental conservation, stop Hg vapor emissions into the atmosphere from amalgam burning, and promote community health [68]. Based on our experiences and following a review on the role of transdisciplinary research (TDR) and communities of practice (CoPs) in public health, TDCoPs are introduced in this article to demonstrate the environmental governance and health improvement possibilities for ASGM communities in SEA countries.

6. Environmental Governance and Community Health Improvements

6.1. General View of TDR and Its Role in Public Health

As the activities in the context of ASGM in SEA countries are similar to those in Ghana, Figure 1 summarizes the ASGM community health hazards in SEA countries based on previously published human health issues in Ghana [69] and proposes an environmental governance process to solve these hazards and improve community health. As summarized in Figure 1, the ASGM communities in SEA countries are affected by both socioeconomic hazards and community health hazards due to the nature of ASGM activities. These problems are defined as wicked problems due to the multifaceted, complicated, and interrelated nature of the challenges; thus, transdisciplinary (TD) rather than monodisciplinary approaches are needed that involve TDR for which interdisciplinary researchers work together with stakeholders to develop new knowledge to tackle the above-mentioned ASGM issues. Recent research has emphasized the value of interdisciplinary and transdisciplinary methods to advance scientific knowledge and solve important societal issues [70–73].

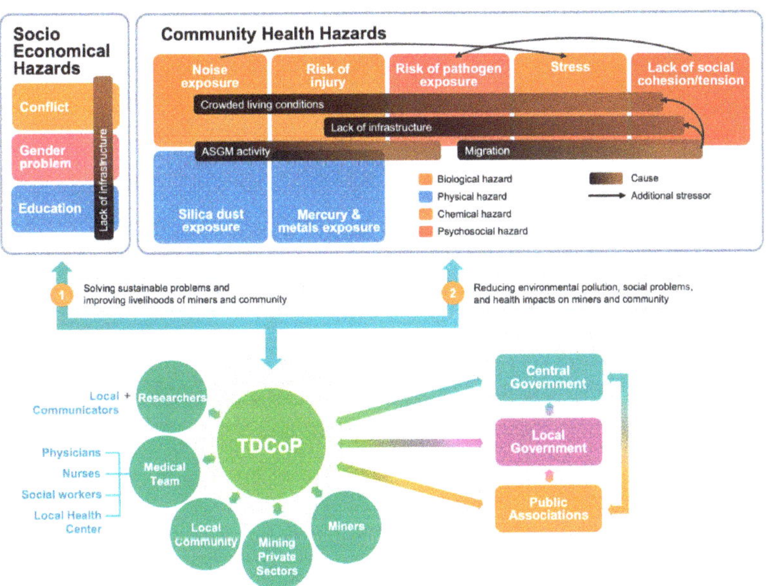

Figure 1. Transdisciplinary communities of practice (TDCoPs): a process to solve the wicked ASGM problems in SEA. Reproduced with permission from [69]. 2015, Basu N et al. The credit for the summary on the community health hazards goes to Basu N et al., 2015 "Integrated Assessment of Artisanal and Small-Scale Gold Mining in Ghana—Part 1: Human health Review".

The term TDR initially appeared in the 1970s [74,75] and has various definitions. TDR is a research concept that involves a paradigm shift in research practice to solve complex societal challenges [76]. Although the TDR approach is similar to interdisciplinary and multidisciplinary approaches, it differs in the way the information is used and shared. In multidisciplinary approaches, everyone works on the same subject inside their own discipline's limits, makes their own assumptions, and develops their own methodologies and frames of reference. In interdisciplinary approaches, the disciplinary boundaries are merged so that the assumptions, limitations, and ideologies are blended. Nevertheless, TDR emphasizes the integration of academic and nonacademic information to coproduce knowledge that transcends disciplinary and sectoral borders, which is achieved by increasing the participative cooperation between the various academic and experiential stakeholders, that is, the people with lay or "lived" knowledge [77–79]. For TDR activities to be effective, team members must be open to new ideas and must respect the views of other disciplines, and they must be willing to engage in the exchange of ideas, mutual debates, problem solving, and conflict resolution when there are opposing opinions or ideals [71,80]. A major advantage of TDR approaches is that they allow for the integration of concepts from multiple disciplines into a new, common, conceptual framework [81] and therefore can help establish new conceptual models that better elucidate the complex processes involved in producing and sustaining public health challenges and provide evidence for formulating solutions and public policy [80].

A recent review on the TD approach in public health stated that there had been transdisciplinary initiatives by scholars in the field of environmental health [82] and provided evidence that despite the obstacles and demands on researchers, the collaborations between social science and environmental health had the greatest potential to improve public and environmental health [83]. Another study introduced an approach that supported science-based dialog among stakeholders in cumulative burden assessments (CuBA) and conducted two workshops focused on this approach to examine its utility and feasibility. The results revealed that the stakeholders were able to conduct joint discourse on the cumulative burdens, learn about the technical and social issues related to CuBA, and coproduce knowledge. The study also suggested exploring ways the CuBA methodology could promote or hinder social learning and knowledge coproduction in diverse institutional, social, and political multiple environmental hazard situations [84].

Despite these efforts, however, TDR has several barriers of implementation between academic researchers and the nonacademic participants [85], the most challenging of which are encouraging collaboration between the community residents and the other nonacademic participants [86] and dealing with trust issues between the participants and the community. An evaluation of the members of a Canadian aging and technology research network was conducted to assess their perceptions and experiences with TDR, and its results found that TDR was effective in encouraging mutual learning and understanding and in attempting to address the complicated challenges in the targeted sector. Nevertheless, individual, systemic, and cultural impediments to the implementation of TDR were also found. The primary individual impediments were related to technology-based communication, the unfamiliar terminologies and goals, and the conflicting priorities when the business-oriented goals and the demands of the industrial partners did not line up with other priorities of academics and people from other fields. Many also found it hard to work with researchers who were more traditional and focused on a single discipline because this research approach tends to have a top-down decision-making structure, which does not work well with TDR collaborative team approach. Participants also said that a major systemic problem was that there was not enough time to meet the needs and interests of all stakeholders and to complete the research in a way that met the deadlines set by the funding agencies [77].

As TDR expects people from different backgrounds and interests to work together to solve problems and produce new ideas, there has been a significant interest in social theories on how people learn, such as CoPs, which could assist TDR participants and better facilitate their work [87].

6.2. CoPs in Public Health

Jean Lave and Etienne Wenger were the first to use the term CoPs in 1991 as part of their theory of situated learning to explain how knowledge and learning occurred in specific places, showing that the social relationships in these settings were more important for professional development than in the classroom. Within a group of CoPs, contextual learning occurs spontaneously as individuals become more proficient in the area of knowledge [88]. CoPs, which are defined as groups of people who learn together to resolve mutual concerns and make improvements through regular interactions, have three fundamental elements: domain, community, and practice [89]. In a CoPs learning community, there is a more equal relationship between the experts and the students; that is, the teachers and students work together. Although CoPs were initially formed to tackle specific problems using a team approach, this does not necessarily mean that CoPs can be artificially created [87]. Although CoPs can overcome barriers, foster new knowledge, and facilitate the coming together of normally segregated experts and individuals concerned about issues [89], participants remain within their formal organizations as they participate in CoPs. As they come and go from the CoPs, they learn about the local work practices.

CoPs have been used for TDR and have had great results. CoPs have been used in Canada to share and learn about climate change and health issues [90] and to develop professional and organizational public-health-development strategies in context. They have also been used as knowledge-to-action mobilizers for health practitioners in the Senior Health Research Transfer Network (later Seniors Health Knowledge Network (SHKN)) to improve the health of Ontario seniors [91]. However, although there have been significant benefits found from CoPs, some barriers have also been observed. A study that analyzed an SHKN case to examine the value of CoPs in facilitating system change found that the CoPs successfully served as an initiator to focus on best practice, research, and experience by providing a reflective learning cycle that motivated participants to work together; however, organizational and sector-level cultural norms that were driven by structural goals impeded the CoPs' initiatives to modify behaviors in the long-term-care system [92]. Thus, it was concluded that unless provincial CoPs are properly supported in their attempts to transform systems, any improvements in elder health care would be lost. The authors also stated that the literature on CoPs lacked recommendations on how to utilize CoPs to encourage system change and suggested that more primary research on the CoPs' functions and effects on system-level transformations was needed [92].

6.3. TDCoPs and Their Role in Securing Environmental Governance and Resolving ASGM Socioeconomic and Community Health Hazards in SEA Countries

6.3.1. The Importance of TDCoPs

Matsumoto et al. (2022) claimed that there were three barriers faced by CoPs to overcome in TDR: "barriers of indifference", where stakeholders abstain from participating due to disinterest in the problems; "barriers of position", where stakeholders do not collaborate within the CoPs due to internal differences in position; and "barriers of continuity", where the activities cease once researchers leave the CoPs [93]. Barriers of indifference occur when the stakeholder participation in CoPs is not enthusiastic; however, collaboration and knowledge development with stakeholders such as community inhabitants are critical when seeking to resolve environmental and sustainability issues [94]. Barriers of position occur when social class inequalities contribute to power imbalances [95] and negatively impact participation [86]. Although it is challenging to bring diverse stakeholders together because of their differing perspectives and attitudes [96,97], even if there is successful stakeholder collaboration, it is necessary to address the issue of sustainability, and it is here that the final barriers of continuity arise. Academic participants often leave when they have finished their research tasks [33]; thus, academic participants must commit themselves to being the CoPs' coordinators in the early stages so that they can stabilize the CoPs before handing over the management to the local stakeholders [33]. If the academic scholars always lead the cooperation and knowledge generation and the stakeholders only "cooperate," the activities would be more likely to cease when the researchers leave, leaving the challenges unaddressed [33].

These barriers to CoPs can be overcome when the CoPs take a transdisciplinary approach, as TDCoPs use transformative boundary objects (TBOs). Compared with CoPs, TDCoPs have more diversified stakeholder groups, which may also include "indifferent" participants, and they encourage boundary crossing and engagement between the members and the others around them. The stakeholder engagement then evolves into a collaboration, which then evolves into autonomy via learning and practice [33]. The TDCoP approach has been used to examine health, the environment, and socioeconomic and cultural aspects. Our ongoing TDCoPs experience of collaborating with ASGM workers, the community, and stakeholders has made the ASGM workers and their communities aware of the health impacts of their ASGM activities, which has made them more willing to explore and implement alternative sustainable livelihoods by adding value to their existing local and traditional knowledge, which they were made aware of through discussions with the stakeholders and researchers. One of the ongoing focuses of TDCoPs has been the Healthy and Resilient Village, which seeks to improve the socioeconomic condition and community health in the ASGM area in East Suwawa, Bone Bolango Regency in Gorontalo Province, Indonesia. The local government figures, such as heads of the villages, actively participate and collaborate with local researchers, local medical professionals, the local mining community, and academic researchers [68].

6.3.2. The Process and Role of TDCoPs in Securing Environmental Governance and Resolving Socioeconomic and Community Health Hazards

The TDCoPs that will work towards resolving SEA's ASGM issues could comprise (1) academic researchers from host countries and joint academic institutes with scientific knowledge who would be supported by local communicators; (2) medical teams, such as physicians, nurses, social workers, staff from local health centers, and specialists to deal with the long-term ASGM health problems; (3) local community members; (4) private mining sector members; (5) local miners to share their issues and livelihood goals; (6) public associations; and (7) authorities such as local governments and central governments (Figure 1). The transdisciplinary approach has been broadly applied to a range of long-term health care services in the community, with most published works being focused on rehabilitation [98,99], dementia care [100], the oral health care of stroke patients [100], palliative care in acute care settings [101], bladder health, and preventing lower urinary tract symptoms in women [102].

The starting point for the TDCoPs would be the local government, such as the village, district, or provincial departments concerned with ASGM-related environmental and health issues in the area, who would then initiate the development of a collaborative network of professionals to be responsible for taking practical actions. TDCoPs go through five stages: potential, coalescing, maturing, activity, and transformation [89]. During the potential stage, the participants share the purpose of the TDCoPs, collect the basic information from the diverse disciplines and build trust through dialog [93], which requires researchers to initially take the leading role and then gradually take a supporting role as the TDCoPs move into the transformation stage [93]. For instance, in the TDCoPs formed to deal with the ASGM issues, through dialog, (1) the miners and local community can discuss the health and socioeconomic problems they are facing and the traditional, cultural, and local knowledge unique to their living area; (2) the medical teams can share basic lay information on the general health of the community, the kind of support they can offer, the barriers they are facing, and knowledge on Hg-related health issues; (3) the private mining sectors can provide information on the issues between the ASGM miners and government officials, and especially those related to local laws, such as disputes over legislation and the permissions needed by government officials, miners, and mining companies; and (4) government officials can share knowledge about the official processes important to the ASGM profile. Several TDCoPs can be formed across multiple layers depending on the situation and the cultures, which are known as multilayer TDCoPs. The TDCoPs' members determine the problems during this stage, and TBOs are identified, which leads to the formation of TDCoPs in the "coalescing" stage. Examples of TBOs are chronic Hg intoxication and respiratory health.

After the formation of TDCoPs, basic research is conducted. The local government take the roles of planning and action by collaborating with the upper and central governments; the local medical teams conduct surveys; and the researchers contribute research techniques, such as the methodologies for the environmental, health, and socioeconomic surveys in the study area. The local communicators and medical teams perform the surveys and work in collaboration with the community under the guidance of the local government. The key stakeholders oversee the surveys and research execution and support these activities by offering contact points for resident collaboration and information sharing. During the "maturing" and "activity" stages, the TDCoPs members achieve transformation by reflecting on the basic research findings and solving their own problems, at which time the researchers move into supportive roles by sharing their knowledge. Finally, the communities transform and make plans for practical activities that can evolve independently of the TDCoPs. The instigation of TDCoPs can reduce and/or solve sustainability problems and improve ASGM community livelihoods by reducing environmental pollution, resolving the social and health problems of both the miners and the mining community, and providing the community with an autonomous community-based team.

6.3.3. A Brief Report of the Current TDCoPs Developed in Authors' Project

Among several ongoing TDCoPs of the authors' project, an example of a TDCoPs for securing environmental governance and resolving ASGM socioeconomic and health issues, called "Healthy and Resilient Village" (comprising several TDCoPs), by ASGM miners and owners, was formed in 2021 in East Suwawa, Bone Bolango Regency in Gorontalo Province, Indonesia by local and central stakeholders, the local medical team, and academic researchers after discussions since 2018. Healthy and Resilient Village is attempting to promote the public understanding of proper mining operations, the consequences of using Hg in ASGM, environmental issues, and community safety and health, and in the future, it will be transformed into an independent community that will tackle the sustainable environmental problems and improve the community livelihood [103].

7. Conclusions

This study presented a country-specific ASGM profile for SEA countries and outlined the community health hazards experienced because of the wicked problems in the areas of ASGM. Most monodisciplinary studies have focused on Hg-related health problem assessments, in which academic researchers have taken on the main roles. Although these studies were able to identify Hg-related issues in the ASGM communities, they have been unable to resolve the community health problems because of their monodisciplinary focus. Governments have also been trying to tackle the problems through the implementation of rules and regulations; however, the problems remain unsolved because there is no community involvement.

TD healthcare approaches have been found to have a positive effect on community health management in the long-run; moreover, the authors' project in ASGM areas in terms of developing TDCoPs, which addresses ASGM-induced, complex, socioeconomical and health problems, have made significant progress in making the community sustainable. Therefore, based on these data, a TDCoPs in ASGM areas in SEA countries was proposed and would involve active participation by researchers, the medical profession, the local community, and other invested stakeholders to resolve the wicked ASGM problems and to provide environmental governance and community health improvement guidance in ASGM communities in SEA countries.

Author Contributions: W.T.K., a physician and medical researcher, wrote the manuscript, and M.S., as the project leader, supervised and edited the manuscript. All authors have read and agreed to the published version of the manuscript.

Funding: This research was funded by Research Institute for Humanity and Nature (RIHN; a constituent member of NIHU), Project No. RIHN 14200102.

Institutional Review Board Statement: Not applicable.

Informed Consent Statement: Not applicable.

Data Availability Statement: Not applicable.

Acknowledgments: This research was supported by Research Institute for Humanity and Nature (RIHN: a constituent member of NIHU), Project No. RIHN 14200102.

Conflicts of Interest: The authors declare no conflict of interest.

References

1. Ofosu-Mensah, A.E. Mining As a Factor of Social Conflict in Ghana. *Glob. J. Hist. Cult.* **2012**, *1*, 7–21.
2. Hentschel, T.; Priester, M.; Hruschka, F. Global Report on Artisanal & Small-Scale Mining. *Min. Miner. Sustain. Dev.* **2002**, *70*, 67.
3. UN Environment. *Global Mercury Assessment 2018 | UN Environment Programme*; Chemicals and Health Branch: Geneva, Switzerland, 2019.
4. Veiga, M.M.; Maxson, P.A.; Hylander, L.D. Origin and Consumption of Mercury in Small-Scale Gold Mining. *J. Clean. Prod.* **2006**, *14*, 436–447. [CrossRef]

5. Drwiega; Ismawati; Yuyun. Illegal and Illicit Trade of Mercury in Indonesia | PlanetGOLD. 2018. Available online: https://www.planetgold.org/illegal-and-illicit-trade-mercury-indonesia (accessed on 20 March 2022).

6. UN Comtrade. International Trade Statistics Database. Mercury. 2020. Available online: https://comtrade.un.org/ (accessed on 20 March 2022).

7. Spiegel, S.J.; Agrawal, S.; Mikha, D.; Vitamerry, K.; Le Billon, P.; Veiga, M.; Konolius, K.; Paul, B. Phasing Out Mercury? Ecological Economics and Indonesia's Small-Scale Gold Mining Sector. *Ecol. Econ.* **2018**, *144*, 1–11. [CrossRef]

8. Telmer, K.H.; Veiga, M.M. World Emissions of Mercury from Artisanal and Small Scale Gold Mining and the Knowledge Gaps about Them. In *Mercury Fate and Transport in the Global Atmosphere*; Springer: Boston, MA, USA, 2002; p. 43.

9. United Nations Environment Programme (UNEP). *United Nations Environment Programme Global Mercury Assessment 2013 Sources, Emissions, Releases and Environmental Transport*; UNEP Division of Technology, Industry and Economics, Chemical Branch International Environmental House: Geneva, Switzerland, 2015; p. 38.

10. Krabbenhoft, D.P.; Rickert, D.A. USGS FS 216-95—Mercury Contamination of Aquatic Ecosystems. Available online: https://pubs.usgs.gov/fs/1995/fs216-95/ (accessed on 19 April 2022).

11. Li, P.; Feng, X.; Qiu, G. Methylmercury Exposure and Health Effects from Rice and Fish Consumption: A Review. *Int. J. Environ. Res. Public Health* **2010**, *7*, 2666–2691. [CrossRef]

12. Böse-O'reilly, S.; Schierl, R.; Nowak, D.; Siebert, U.; William, J.F.; Owi, F.T.; Ismawati, Y. A Preliminary Study on Health Effects in Villagers Exposed to Mercury in a Small-Scale Artisanal Gold Mining Area in Indonesia. *Environ. Res.* **2016**, *149*, 274–281. [CrossRef]

13. Wispriyono, B.; Quang, P.; Jeong, H.; Fukushima, S.; Ishibashi, Y.; Arizono, K.; Novirsa, R. The Evaluation of Mercury Contamination in Upland Rice Paddy Field around Artisanal Small-Scale Gold Mining Area, Lebaksitu, Indonesia. *J. Environ. Saf.* **2019**, *10*, 119–125.

14. Qiu, G.; Feng, X.; Li, P.; Wang, S.; Li, G.; Shang, L.; Fu, X. Methylmercury Accumulation in Rice (*Oryza Sativa* L.) Grown at Abandoned Mercury Mines in Guizhou, China. *J. Agric. Food Chem.* **2008**, *56*, 2465–2468. [CrossRef]

15. UNEP DTIE Chemicals Branch and WHO Department of Food Safety, Zoonoses and Foodborne Diseases. *Guidance for Identifying Populations at Risk from Mercury Exposure*; IOMC (Inter-Organization Program. Sound Manag. Chem. Acooperative Agreem. among UNEP, ILO, FAO, WHO, UNIDO, UNITAR OECD); World Health Organization: Geneva, Switzerland, 2008; p. 176.

16. United Nations Environment Programme (UNEP). *A Practical Guide in Reducing Mercury Use in Artisanal and Small-Scale Gold Mining*; United Nations Environment Policy: Nairobi, Kenya, 2012; p. 76.

17. Ismawati, Y. *Preliminary Report on Suspected Mercury Poisoning in 3 ASGM Hotspots of Indonesia: Case Reports Bombana-Southeast Sulawesi, Sekotong-West Lombok, and Cisitu-Lebak*; BaliFokus & Medicuss Foundations: Bali, Indonesia, 2015.

18. Ismawati, Y.; Petrlik, J.; DiGangi, J. Mercury Hotspots in Indonesia. ASGM Sites: Poboya and Sekotong in Indonesia. IPEN Mercury-Free Campaign Report. 2013. Available online: http://www.ipen.org/hgmonitoring/pdfs/indonesia-report-en.pdf (accessed on 20 March 2022).

19. Arifin, Y.I.; Sakakibara, M.; Takakura, S.; Jahja, M.; Lihawa, F.; Sera, K. Artisanal and Small-Scale Gold Mining Activities and Mercury Exposure in Gorontalo Utara Regency, Indonesia. *Toxicol. Environ. Chem.* **2020**, *102*, 521–542. [CrossRef]

20. Veiga, M.M. A Critical Review of Suitable Methods to Eliminate Mercury in Indonesia's Artisanal Gold Mining: Co-Existence Is the Solution. Report to UNDP Indonesia. 2020. Available online: https://www.goldismia.org/sites/default/files/2020-06/VEIGA%20-%20FINAL%20Report.pdf (accessed on 16 March 2022).

21. Drace, K.; Kiefer, A.M.; Veiga, M.M. Cyanidation of Mercury-Contaminated Tailings: Potential Health Effects and Environmental Justice. *Curr. Environ. Healthy Rep.* **2016**, *3*, 443–449. [CrossRef]

22. Prasetia, H.; Sakakibara, M.; Omori, K.; Laird, J.S.; Sera, K.; Kurniawan, I.A. Mangifera Indica as Bioindicator of Mercury Atmospheric Contamination in an ASGM Area in North Gorontalo Regency, Indonesia. *Geosciences* **2018**, *8*, 31. [CrossRef]

23. Krisnayanti, B.D. ASGM Status in West Nusa Tenggara Province, Indonesia. *J. Degrad. Min. Lands Manag.* **2018**, *5*, 1077–1084. [CrossRef]

24. Basri; Sakakibara, M.; Sera, K.; Kurniawan, I.A. Mercury Contamination of Cattle in Artisanal and Small-Scale Gold Mining in Bombana, Southeast Sulawesi, Indonesia. *Geosciences* **2017**, *7*, 133. [CrossRef]

25. Böse-O'reilly, S.; Maydl, S.; Drasch, G.; Roider, G. *Mercury as a Health Hazard Due to Gold Mining and Mineral Processing Activities in Mindanao/Philippines*; Final Report, UNIDO Project DP/PHI/98/005; Institute of Forensic Medicine, Ludwig-Maximilians University: Munich, Germany, 2000.

26. Acosta, J.A.; Arocena, J.M.; Faz, A. Speciation of Arsenic in Bulk and Rhizosphere Soils from Artisanal Cooperative Mines in Bolivia. *Chemosphere* **2015**, *138*, 1014–1020. [CrossRef]

27. Appleton, J.D.; Williams, T.M.; Orbea, H.; Carrasco, M. Fluvial Contamination Associated with Artisanal Gold Mining in the Ponce Enríquez, Portovelo-Zaruma and Nambija Areas, Ecuador. *Water Air Soil Pollut.* **2001**, *131*, 19–39. [CrossRef]

28. Gafur, N.A.; Sakakibara, M.; Sano, S.; Sera, K. A Case Study of Heavy Metal Pollution in Water of Bone River by Artisanal Small-Scale Gold Mine Activities in Eastern Part of Gorontalo, Indonesia. *Water* **2018**, *10*, 1507. [CrossRef]

29. Ahmadpour, P.; Ahmadpour, F.; Mahmud TM, M.; Abdu, A.; Soleimani, M.; Tayefeh, F.H. Phytoremediation of Heavy Metals: A Green Technology. *Afr. J. Biotechnol.* **2012**, *11*, 14036–14043. [CrossRef]

30. Utami, D.; Takahi, S.; Prijambada, I.D. Mercury Accumulation in Gold Mine Tailing by Sweet Sorghum Inoculated with Chromium Uptake Enhancing Rhizobacteria. *Int. J. Biosci. Biotechnol.* **2013**, *1*, 86–90.
31. Oh, K.; Prijambada, I.D. Phytoremediation of Mercury Contaminated Soils in a Small Scale Artisanal Gold Mining Region of Indonesia. *Int. J. Biosci. Biotechnol.* **2015**, *3*, 14–21.
32. ILO (International Labour Organization). *Handbook on Hazardous Child Labour*; ILO (International Labour Organization): Geneva, Switzerland, 2011.
33. World Bank. *2019 State of the Artisanal and Small-Scale Mining Sector*; World Bank: Washington, DC, USA, 2019, p. 85.
34. Drasch, G.; Böse-O'Reilly, S.; Beinhoff, C.; Roider, G.; Maydl, S. The Mt. Diwata Study on the Philippines 1999—Assessing Mercury Intoxication of the Population by Small Scale Gold Mining. *Sci. Total Environ.* **2001**, *267*, 151–168. [CrossRef]
35. McFarlane, D.; Villalobos, R. The State of Artisanal Mining in Myanmar (Pact, 2018). 2019. Available online: https://delvedatabase.org/uploads/resources/The-State-of-Artisanal-Mining-in-Myanmar-Report.-PACT.-May28.2019.pdf (accessed on 16 March 2022).
36. Osawa, T.; Hatsukawa, Y. Artisanal and Small-Scale Gold Mining in Myanmar: Preliminary Research for Environmental Mercury Contamination. *Int. J. Hum. Cult. Stud.* **2015**, *2015*, 221–230. [CrossRef]
37. Kuang, X.; Kyaw, W.T.; Soe, P.S.; Thandar, A.M.; Khin, H.E. A Preliminary Study on Mercury Contamination in Artisanal and Small-Scale Gold Mining Area in Mandalay Region, Myanmar by Using Plant Samples. *Pollution* **2022**, *8*, 225–238. [CrossRef]
38. Oo, K.T.; Kyi, H. Assessment on Environmental Impacts of Gold Mining in Wetthe-Phatshe Area of Thabeikkyin Township. *Yadanabon Univ. Res. J.* **2019**, *10*, 1.
39. Kawakami, T.; Konishi, M.; Imai, Y.; Soe, P.S. Diffusion of Mercury from Artisanal Small-Scale Gold Mining (ASGM) Sites in Myanmar. *Int. J. Geomate* **2019**, *17*, 228–235. [CrossRef]
40. Tun, A.Z.; Wongsasuluk, P.; Siriwong, W. Heavy Metals in the Soils of Placer Small-Scale Gold Mining Sites in Myanmar. *J. Healthy Pollut.* **2020**, *10*, 200911. [CrossRef] [PubMed]
41. Earth Systems Lao. Luang Prabang Artisanal Gold Mining and Sociological Survey, Lao PDR. UNIDO Global Mercury Report. 2003. Available online: https://iwlearn.net/resolveuid/f4a6bed798cf80bc493568d01f1fcc66 (accessed on 20 March 2022).
42. Ministry of Environment Cambodia. National Assessment Report on Mercury in Cambodia. 2016. Available online: https://www.mercuryconvention.org/sites/default/files/documents/submission_from_government/Cambodia%2520MIA%2520report.pdf (accessed on 26 April 2022).
43. Penh, P. *Artisanal Small-Scale Gold Mining (ASGM) in Battambang and Preah Vihear Provinces Investigation Report 2012*; Development and Partnership in Action: Phnom Penh, Cambodia, 2012.
44. Long, R.N.; Renne, E.P.; Basu, N. Understanding the Social Context of the ASGM Sector in Ghana: A Qualitative Description of the Demographic, Health, and Nutritional Characteristics of a Small-Scale Gold Mining Community in Ghana. *Int. J. Environ. Res. Public Health* **2015**, *12*, 12679. [CrossRef] [PubMed]
45. Glezos, J.D.; Albrecht, J.E.; Gair Bsc, R.D. Pneumonitis after Inhalation of Mercury Vapours. *Can. Respir. J.* **2006**, *13*, 150–152. [CrossRef]
46. Park, J.-D.; Zheng, W. Human Exposure and Health Effects of Inorganic and Elemental Mercury. *J. Prev. Med. Public Healthy* **2012**, *45*, 344. [CrossRef]
47. Pavilonis, B.; Grassman, J.; Johnson, G.; Diaz, Y.; Caravanos, J. Characterization and Risk of Exposure to Elements from Artisanal Gold Mining Operations in the Bolivian Andes. *Environ. Res.* **2017**, *154*, 1–9. [CrossRef]
48. Nelson, L.S.; Lewin, N.A.; Howland, M.A.; Hoffman, R.S.; Goldfrank, L.R.F.N. *Goldfrank's Toxicologic Emergencies*; McGrawHill: New York, NY, USA, 2011.
49. Solis, M.T.; Yuen, E.; Cortez, P.S.; Goebel, P.J. Family Poisoned by Mercury Vapor Inhalation. *Am. J. Emerg. Med.* **2000**, *18*, 599–602. [CrossRef]
50. Clarkson, T.W.; Magos, L. The Toxicology of Mercury and Its Chemical Compounds. *Crit. Rev. Toxicol.* **2006**, *36*, 609–662. [CrossRef] [PubMed]
51. Moszczyński, P. Immunological Disorders in Men Exposed to Metallic Mercury Vapour—A Review. Available online: https://pubmed.ncbi.nlm.nih.gov/10084014/ (accessed on 26 January 2022).
52. WHO (World Health Organization). Exposure to Mercury: A Major Public Health Concern. In *Preventing Disease through Healthy Environments*; WHO: Geneva, Switzerland, 2006.
53. WHO (World Health Organization). *Mercury: Assessing the Environmental Burden of Disease at National and Local Levels*; (WHO Environmental Burden of Disease Series No. 16); Annette, P.-Ü., Ed.; WHO: Geneva, Switzerland, 2008.
54. Cossa, H.; Scheidegger, R.; Leuenberger, A.; Ammann, P.; Munguambe, K.; Utzinger, J.; Macete, E.; Winkler, M.S. Health Studies in the Context of Artisanal and Small-Scale Mining: A Scoping Review. *Int. J. Environ. Res. Public Health* **2021**, *18*, 1555. [CrossRef] [PubMed]
55. Green, A.; Jones, A.D.; Sun, K.; Neitzel, R.L.; Basu, N.; Keane, S.; Black Moher, P. The Association between Noise, Cortisol and Heart Rate in a Small-Scale Gold Mining Community—A Pilot Study. *Int. J. Environ. Res. Public Healthy* **2015**, *12*, 9952–9966. [CrossRef] [PubMed]
56. Schulz, C.; Angerer, J.; Ewers, U.; Kolossa-Gehring, M. The German Human Biomonitorng Commission. *Int. J. Hyg. Environ. Health* **2007**, *210*, 373–382. [CrossRef]

57. Kyaw, W.T.; Kuang, X.; Sakakibara, M. Health Impact Assessment of Artisanal and Small-Scale Gold Mining Area in Myanmar, Mandalay Region: Preliminary Research. *Int. J. Environ. Res. Public Health* **2020**, *17*, 6757. [CrossRef]
58. Kyaw, W.T.; Myint, Y.M.; Kuang, X.; Sakakibara, M. Transdisciplinary Online Health Assessment of an Artisanal and Small-Scale Gold Mining Community during the COVID-19 Pandemic in the Mandalay Region of Myanmar. *Int. J. Environ. Res. Public Health* **2021**, *18*, 11206. [CrossRef]
59. Umbangtalad, S.; Parkpian, P.; Visvanathan, C.; Delaune, R.D.; Jugsujinda, A. Assessment of Hg Contamination and Exposure to Miners and Schoolchildren at a Small-Scale Gold Mining and Recovery Operation in Thailand. *J. Environ. Sci. Healthy Part A Toxic Hazard. Substain. Environ. Eng.* **2007**, *42*, 2071–2079. [CrossRef]
60. Bose-O'Reilly, S.; Drasch, G.; Beinhoff, C.; Rodrigues-Filho, S.; Roider, G.; Lettmeier, B.; Maydl, A.; Maydl, S.; Siebert, U. Health Assessment of Artisanal Gold Miners in Indonesia. *Sci. Total Environ.* **2010**, *408*, 713–725. [CrossRef]
61. Nakazawa, K.; Nagafuchi, O.; Kawakami, T.; Inoue, T.; Yokota, K.; Serikawa, Y.; Basir-Cyio, M.; Elvince, R. Human Health Risk Assessment of Mercury Vapor around Artisanal Small-Scale Gold Mining Area, Palu City, Central Sulawesi, Indonesia. *Ecotoxicol. Environ. Saf.* **2016**, *124*, 155–162. [CrossRef]
62. Arifin, Y.I.; Sakakibara, M.; Sera, K. Impacts of Artisanal and Small-Scale Gold Mining (ASGM) on Environment and Human Health of Gorontalo Utara Regency, Gorontalo Province, Indonesia. *Geosciences* **2015**, *5*, 160–176. [CrossRef]
63. Abbas, H.H.; Sakakibara, M.; Sera, K.; Arma, L.H. Mercury Exposure and Health Problems in Urban Artisanal Gold Mining (UAGM) in Makassar, South Sulawesi, Indonesia. *Geosciences* **2017**, *7*, 44. [CrossRef]
64. Murao, S.; Tomiyasu, T.; Ono, K.; Shibata, H.; Narisawa, N.; Takenaka, C. Mercury Distribution in Artisanal and Small-Scale Gold Mining Area: A Case Study of Hot Spots in Camarines Norte, Philippines. *Int. J. Environ. Sci. Dev.* **2019**, *10*, 122–129. [CrossRef]
65. Akagi, H.; Castillo, E.S.; Cortes-Maramba, N.; Francisco-Rivera, A.T.; Timbang, T.D. Health Assessment for Mercury Exposure among Schoolchildren Residing near a Gold Processing and Refining Plant in Apokon, Tagum, Davao Del Norte, Philippines. *Sci. Total Environ.* **2000**, *259*, 31–43. [CrossRef]
66. Boese-O'Reilly, S.; Dahlmann, F.; Lettmeier, B.; Drasch, G. Removal of Barriers to the Introduction of Cleaner Artisanal Gold Mining and Extraction Technologie in Kadoma, Zimbabwe. Final Report, 130 p., 28 Figures, 25 Pictures, 22 Tables, 3 Appendix. 2004. Available online: https://iwlearn.net/resolveuid/886be3abb7f723d8653bf38e9e678e7b (accessed on 26 April 2022).
67. UN Environment Programme. Parties and Signatories Minamata Convention on Mercury. Available online: https://www.mercuryconvention.org/en/parties (accessed on 20 March 2022).
68. SRIREP. TDCOPs. Available online: https://srirep.org/ (accessed on 14 March 2022).
69. Basu, N.; Clarke, E.; Green, A.; Calys-Tagoe, B.; Chan, L.; Dzodzomenyo, M.; Fobil, J.; Long, R.N.; Neitzel, R.L.; Obiri, S.; et al. Integrated Assessment of Artisanal and Small-Scale Gold Mining in Ghana-Part 1: Human Health Review. *Int. J. Environ. Res. Public Health* **2015**, *12*, 5143–5176. [CrossRef]
70. Pohl, C. From Transdisciplinarity to Transdisciplinary Research. *Transdiscipl. J. Eng. Sci.* **2010**, *1*, 65–73. [CrossRef]
71. Stokols, D. Toward a Science of Transdisciplinary Action Research. *Am. J. Community Psychol.* **2006**, *38*, 63–77. [CrossRef]
72. Annerstedt, M. Transdisciplinarity as an Inference Technique to Achieve a Better Understanding in the Health and Environmental Sciences. *Int. J. Environ. Res. Public Health* **2010**, *7*, 2692–2707. [CrossRef]
73. Lawrence, M.G.; Williams, S.; Nanz, P.; Renn, O. Characteristics, Potentials, and Challenges of Transdisciplinary Research. *One Earth* **2022**, *5*, 44–61. [CrossRef]
74. Jantsch, E. Inter- and Transdisciplinary University: A Systems Approach to Education and Innovation. *Policy Sci.* **1970**, *1*, 403–428. [CrossRef]
75. Jantsch, E. *Interdisciplinarity: Problems of Teaching and Research in Universities*; Organization for Economic Cooperation and Development (OECD) and Center for Educational Reserch and Innovation (CERI): Paris, France, 1972.
76. OECD. Addressing Societal Challenges Using Transdiciplinary Research. *Policy Pap.* **2020**, *88*, 39–51.
77. Wada, M.; Grigorovich, A.; Fang, M.L.; Sixsmith, J.; Kontos, P. An Exploration of Experiences of Transdisciplinary Research in Aging and Technology. In *Forum: Qualitative Social Research*; Institut für Qualitative Forschung: Berlin, Germany, 2020; Volume 21. [CrossRef]
78. Boger, J.; Jackson, P.; Mulvenna, M.; Sixsmith, J.; Sixsmith, A.; Mihailidis, A.; Kontos, P.; Miller Polgar, J.; Grigorovich, A.; Martin, S. Principles for Fostering the Transdisciplinary Development of Assistive Technologies. *Disabil. Rehabil. Assist. Technol.* **2016**, *12*, 480–490. [CrossRef] [PubMed]
79. Choi, B.C.K.; Pak, A.W.P. Multidisciplinarity, Interdisciplinarity, and Transdisciplinarity in Health Research, Services, Education and Policy: 2. Promotors, Barriers, and Strategies of Enhancement. *Clin. Investig. Med.* **2007**, *30*, E224–E232. [CrossRef] [PubMed]
80. Abrams, D.B. Applying Transdisciplinary Research Strategies to Understanding and Eliminating Health Disparities. *Healthy Educ. Behav.* **2006**, *33*, 515–531. [CrossRef]
81. Rosenfield, P.L. The Potential of Transdisciplinary Research for Sustaining and Extending Linkages between the Health and Social Sciences. *Soc. Sci. Med.* **1992**, *35*, 1343–1357. [CrossRef]
82. Cannon, C.E.B. Towards Convergence: How to Do Transdisciplinary Environmental Health Disparities Research. *Int. J. Environ. Res. Public Health* **2020**, *17*, 2303. [CrossRef] [PubMed]
83. Cordner, A.; Poudrier, G.; Divalli, J.; Brown, P. Combining Social Science and Environmental Health. *Int. J. Environ. Res. Public Health* **2019**, *16*, 3483. [CrossRef]

84. Shrestha, R.; Flacke, J.; Martinez, J.; van Maarseveen, M. Interactive Cumulative Burden Assessment: Engaging Stakeholders in an Adaptive, Participatory and Transdisciplinary Approach. *Int. J. Environ. Res. Public Health* **2018**, *15*, 260. [CrossRef]
85. Tress, B.; Tress, G.; Fry, G. Defining Concepts and Process of Knowledge Production in Integrative Research. *Landsc. Res. Landsc. Plan* **2007**, *12*, 13–26. [CrossRef]
86. Mobjörk, M. Consulting versus Participatory Transdisciplinarity: A Refined Classification of Transdisciplinary Research. *Futures* **2010**, *42*, 866–873. [CrossRef]
87. Regeer, B.J.; Bunders, J.F.G. The Epistemology of Transdisciplinary Research: From Knowledge Integration to Communities of Practice. *Interdiscip. Environ. Rev.* **2003**, *5*, 98. [CrossRef]
88. Lave, J.; Wenger, E. *Situated Learning: Legitimate Peripheral Participation*; Cambridge University Press: Cambridge, UK, 1991.
89. Wenger, E.; McDermott, R.A.; Snyder, W. *Cultivating Communities of Practice: A Guide to Managing Knowledge*; Harvard Business Review Press: Boston, MA, USA, 2002.
90. El Amiri, N.; Abernethy, P.; Spence, N.; Zakus, D.; Kara, T.A.; Schuster-Wallace, C. Community of Practice: An Effective Mechanism to Strengthen Capacity in Climate Change and Health. *Can. J. Public Healthy* **2020**, *111*, 862–868. [CrossRef] [PubMed]
91. Conklin, J.; Kothari, A.; Stolee, P.; Chambers, L.; Forbes, D.; Le Clair, K. Knowledge-to-Action Processes in SHRTN Collaborative Communities of Practice: A Study Protocol. *Implement. Sci.* **2011**, *6*, 12. [CrossRef] [PubMed]
92. Kothari, A.; Boyko, J.A.; Conklin, J.; Stolee, P.; Sibbald, S.L. Communities of Practice for Supporting Health Systems Change: A Missed Opportunity. *Healthy Res. Policy Syst.* **2015**, *13*, 33. [CrossRef] [PubMed]
93. Matsumoto, Y.; Kasamatsu, H.; Sakakibara, M. Challenges in Forming Transdisciplinary Communities of Practice for Solving Environmental Problems in Developing Countries. *World Futures* **2021**, *0*, 1–20. [CrossRef]
94. McGregor, S.L.T.; Donnelly, G. Transleadership for Transdisciplinary Initiatives. *World Futures* **2014**, *70*, 164–185. [CrossRef]
95. Jahn, T.; Bergmann, M.; Keil, F. Transdisciplinarity: Between Mainstreaming and Marginalization. *Ecol. Econ.* **2012**, *79*, 1–10. [CrossRef]
96. Crona, B.I.; Parker, J.N. Network Determinants of Knowledge Utilization: Preliminary Lessons From a Boundary Organization. *Sci. Commun.* **2011**, *33*, 448–471. [CrossRef]
97. Cundill, G.; Roux, D.J.; Parker, J.N. Nurturing Communities of Practice for Transdisciplinary Research. *Ecol. Soc.* **2015**, *20*, 22. [CrossRef]
98. Reilly, C. Transdisciplinary Approach: An Atypical Strategy for Improving Outcomes in Rehabilitative and Long-Term Acute Care Settings. *Rehabil. Nurs.* **2001**, *26*, 216–244. [CrossRef]
99. Aubin, T.; Mortenson, P. Experiences of Early Transdisciplinary Teams in Pediatric Community Rehabilitation. *Infants Young Child.* **2015**, *28*, 165–181. [CrossRef]
100. Galvin, J.E.; Valois, L.; Zweig, Y. Collaborative Transdisciplinary Team Approach for Dementia Care. *Neurodegener. Dis. Manag.* **2014**, *4*, 455–469. [CrossRef] [PubMed]
101. Daly, D.; Matzel, S.C. Building a Transdisciplinary Approach to Palliative Care in an Acute Care Setting. *Omega* **2013**, *67*, 43–51. [CrossRef] [PubMed]
102. Harlow, B.L.; Bavendam, T.G.; Palmer, M.H.; Brubaker, L.; Burgio, K.L.; Lukacz, E.S.; Miller, J.M.; Mueller, E.R.; Di Newman, K.; Rickey, L.M.; et al. The Prevention of Lower Urinary Tract Symptoms (PLUS) Research Consortium: A Transdisciplinary Approach Toward Promoting Bladder Health and Preventing Lower Urinary Tract Symptoms in Women Across the Life Course. *J. Womens Healthy* **2018**, *27*, 283–289. [CrossRef] [PubMed]
103. KTK (Kampung Tangguh Kesehatan) Healthy and Resilient Village-SRIREP. Available online: https://srirep.org/ktk-kampung-tangguh-kesehatan-healthy-and-resilient-village/ (accessed on 20 April 2022).

International Journal of
***Environmental Research
and Public Health***

MDPI

Article

Artisanal and Small Gold Mining and Petroleum Production as Potential Sources of Heavy Metal Contamination in Ecuador: A Call to Action

José Luis Rivera-Parra [1,2,*], Bernardo Beate [3], Ximena Diaz [4] and María Belén Ochoa [1]

[1] Departamento de Petróleos, Escuela Politécnica Nacional, Ladrón de Guevara E11-253, 170525 Quito, Ecuador; mabe_ochoa@hotmail.com
[2] Departamento de Biología, Escuela Politécnica Nacional, Ladrón de Guevara E11-253, 170525 Quito, Ecuador
[3] Departamento de Geología, Escuela Politécnica Nacional, Ladrón de Guevara E11-253, 170525 Quito, Ecuador; bernardo.beate@epn.edu.ec
[4] Departamento de Metalurgia Extractiva, Escuela Politécnica Nacional, Ladrón de Guevara E11-253, 170525 Quito, Ecuador; ximena.diaz@epn.edu.ec
* Correspondence: jose.riverap@epn.edu.ec

Citation: Rivera-Parra, J.L.; Beate, B.; Diaz, X.; Ochoa, M.B. Artisanal and Small Gold Mining and Petroleum Production as Potential Sources of Heavy Metal Contamination in Ecuador: A Call to Action. *Int. J. Environ. Res. Public Health* **2021**, *18*, 2794. https://doi.org/10.3390/ijerph18062794

Academic Editors: Fayuan Wang and Pedro Brito

Received: 6 January 2021
Accepted: 3 March 2021
Published: 10 March 2021

Abstract: Mining and petroleum production are the source of many elements and base materials fundamental for our modern way of life. The flip side of these keystone industries is the environmental degradation they can cause if not properly managed. Metallic mining and petroleum production can contaminate the local ecosystem with sediments, chemicals used in the industrial processes and heavy metals, part of the metallic ore or oil reservoir. The objective of this project was to analyze the spatial distribution of the presence of different potentially hazardous elements that make up the metallic deposits and oil reservoirs in Ecuador, focused mainly on artisanal and small-scale gold mining (ASGM) districts. Additionally, we were interested in analyzing this information under the local political and administrative contexts which are key to determining how likely it is that mismanagement of the local mineral deposits and petroleum exploitation projects will end up causing environmental degradation. An extensive and intensive literature search was conducted for information on the presence and concentration of 19 potentially harmful elements. We analyzed data on 11 metallic deposits throughout Ecuador and a major oilfield in the Ecuadorian Amazon basin. We used geographic information systems to analyze the spatial distribution of these reservoirs and their mineral compositions. The results indicated a widespread distribution and high concentration of elements potentially harmful for human health, such as mercury, cadmium and arsenic, throughout the metallic deposits in Ecuador. This is particularly true for long-exploited ASGM districts, such as Ponce-Enríquez, Portovelo-Zaruma and Nambija. This study highlights the importance of understanding geological diversity and its potential risks to better protect the biological diversity and public health of its inhabitants. Furthermore, we consider our work not as a call to stop ASGM mining nor petroleum production, but on the contrary as a strong call to plan every mining and petroleum production project considering these risks. Moreover, our work is a call to action by the local government and authorities to stop corruption and fulfill their duties overseeing the activities of mining and petroleum companies, stopping illegal mining, helping ASGM communities to improve their environmental standards, finding alternative income sources and protecting the local environment.

Keywords: artisanal and small gold mining; contaminant movement; Ecuador; environmental risks; extractive industries

1. Introduction

Mining and petroleum production provide the basic resources that sustain a diverse array of industries which are keystones of our modern way of life. Precious metals such as gold or platinum are fundamental for electronics [1] and even for cutting edge medical therapies such as gene editing [2]. The same applies for the products derived from petroleum and the petrochemical industry, which create ubiquitous plastics and base materials for countless industries. The flip side of mining and petroleum production lies in the environmental impacts they can generate, such as water contamination, people displacement and large-scale disasters such as tailings dam breakage or large oil spills, e.g., [3–9]. Cases such as the Brumadinho dam collapse, which occurred in 2019, are extreme examples of the impacts large-scale mining can cause [10]. In the petroleum industry, oil spills such as Exxon Valdez [11] or Deepwater Horizon [12] are well-known cases of major environmental disasters which affected kilometers of coastline and countless species at sea. Even when examples like these are the ones that come to mind when thinking about environmental impacts from mining and petroleum production, usually, the environmental impacts are more discrete. Usually, small-scale impacts contaminate local rivers and ravines that are later used by local communities and contaminate local ecosystems. These impacts have two main origins: problems handling the different chemicals and petrous material products of mining and petroleum production [13], and risks directly related to the chemistry of the exploited deposits and reservoirs and its surrounding rock, including the formation water that lies in the oil reservoirs [14].

The geochemistry of the mineral deposit itself and its associated surrounding formation can be the primary predictor of potential environmental contamination [15]. Metallic deposits besides gold, silver or copper have other metals that can have the potential to affect human health [16]. Is not uncommon that gold deposits are rich in arsenic or cadmium, two elements with extensive evidence of toxicity to human beings [17]. Oil reservoirs as well can have different geochemical compositions, particularly in its formation water. Produced water or formation water is usually rich in salt and can have heavy metals such as cadmium or lithium in its composition [18]. Thus, understanding the spatial distribution and relative concentration of potentially toxic elements in precious metal and oil deposits is paramount for predicting potential risks when mining the ore or producing petroleum. Furthermore, this information should be a keystone for developing adequate environmental management plans.

Another natural factor, beyond human control, relevant to understanding the environmental risks of different mining operations is the amount of rain (precipitation) a site receives [19]. Once the rocks are extracted from the underground, the chemical composition and state of oxidation can release to the environment potentially harmful heavy metals and produce acid rock drainage [20]. Therefore, areas with high precipitation or humidity have an increased risk of acid rock drainage and mobilization of heavy metals from the reservoir rock or surrounding sterile material, in the case where there is no proper management of tailing ponds and dumps.

Heavy metals such as cadmium, lead or mercury are found in the produced water fraction of the overall petroleum production [18]. Water, as the universal solvent, in the case where it is spilled, will carry all the chemicals and metals that are dissolved and readily transport them, infiltrating the soil and contaminating nearby water bodies [21]. In this case, local precipitation and other local ecosystem characteristics such as soil cover, topography and soil composition determine the likely destiny of any heavy metals that escape the petroleum facilities. Oil spills are the most visible environmental impacts, but formation water spills or mishandling can reach further and have longer term impacts. Produced water, due to its salt content, can directly and immediately affect plants and fresh water organisms. The heavy metal content of produced water can persist in the environment and enter the food chain of the local human population and the local ecosystem [22]. In the case of the petroleum industry, we focused our analysis on a single major oilfield in the Ecuadorian Amazon basin, the Auca oilfield. This particular oilfield is responsible for 18% of Ecuadorian oil production and produces an average of 468,000 daily barrels of produced water per day.

Mining, particularly artisanal and small-scale gold mining (ASGM) also has a special heavy metal contamination risk associated with the chemical handling needed for the ore processing. This is particularly true for gold mining, where extensive Hg is used to extract gold through an amalgamation process. Mercury is a known teratogen and carcinogenic, and if effluents are not properly treated, these chemicals end up in rivers and ravines and through bio-accumulation enter the food chain affecting the health of the local community [23,24]. The use of mercury for gold amalgamation has been prohibited in Ecuador, but the reality is that its use is still widespread [25].

Ecuador has several very rich metallic deposits. Some have been exploited for many decades or even centuries by ASGM. Well-known ASGM areas in Ecuador are Nambija, Zaruma-Portovelo and Ponce-Enríquez [26–31]. Other major deposits have been recently discovered and due to their characteristics need more industrialized exploitation techniques. Deposits such as Mirador or Fruta del Norte are in the early steps of producing copper and gold, respectively, and are being managed by large multinational mining companies.

The geological and mineralogical richness of Ecuador can be explained by the diversity of geological environments and the geological processes involved, such as volcanism, which is the origin of mineral deposits [32]. Thus, most mineral deposits are located in the slopes of the Andes.

Bioaccumulation of heavy metals through food chains is a very well-documented process, where heavy metals such as Hg and Cd enter the environment [33]. Once exposed to sun, humidity and other effects, they transform into bioavailable chemical species that can be absorbed by living organisms. In each step of the food chain their absolute concentration increases, until they reach a point where they can have toxic effects, either to animals themselves or to human beings that consume contaminated products, such as fish from rivers or crops grown on contaminated soil or irrigated with contaminated water [33]. Toxicity effects of heavy metals depend on the chemical compound, element concentration, exposure route and time. Acute or chronic effects can vary from dermal lesions, neurological problems to different type of cancers [34–47]. Drinking water is one of the major recognized sources for chronic effects of heavy metals [48,49].

A key piece of information for land use planning and risk analysis from these extractive industries is to know the contamination's potential and its spatial distribution. Understanding the risks associated with the natural composition of soil, rocks and underground geological formations allows local authorities to better plan the distribution of productive activities in their territories [50]. Moreover, it also helps to establish adequate environmental guidelines and better practice requirements for local industries. This is particularly relevant for ASGM communities, that usually are in need of technical training to improve safety and environmental standards and have a suite of associated social problems that need to be addressed by the local authorities.

Thus, the objective of this study was to determine the most likely composition and potential distribution of heavy metal contamination in various metallic deposits, mainly from ASGM districts, throughout Ecuador and a major oilfield in the Ecuadorian Amazon basin. More specifically, the interest was in: (1) analyzing the specific composition of the deposits that contain precious metals in the major ASGM areas in Ecuador; (2) the spatial distributions of the different deposits and reservoirs and their variation when analyzed element by element; (3) determination of the presence and distribution of heavy metals and metalloids with potential to harm living beings in some of the major mineral deposits and reservoirs of oil fields in Ecuador; and (4) analyzing this information under the local political and administrative contexts, which are key to determining how likely it is that mismanagement of the local mines and petroleum exploitation projects will end up causing environmental degradation.

Ecuador has several large deposits of precious metals and oil reservoirs [51]. Even though there is interest from the government in developing the mining industry, there is resistance from the local communities because of fear of contamination of their water sources and agricultural lands [52]. This concern comes in part due to previous experience with Ecuadorian oil production that started in the 1960s, which has a history of environmental impacts [53]. This paper attempts to determine the potential distribution of heavy metal contamination in different habitats due to these large-impact industries. This information may be key for better environmental planning, better informed community positions and even better community relationship programs. We decided to include oil production in the analysis because this industry represents a major source of income for the Ecuadorian government and is an interesting contrast between an obligated industrialized, regulated and legal activity, and ASGM which lacks regulations, control and law enforcement. It is performed by individuals or small companies instead of major national or international companies.

2. Materials and Methods

2.1. Study Areas

We analyzed 11 mineral deposits in Ecuador; some are new prospective medium- to large-scale mining projects and others are long-term artisanal and small-scale mining (ASGM) areas. The studied areas of medium- to large-scale mining projects were: Chical, Junin-Llurimagua, Agroindustrial el Corazón, Fruta del Norte, San Carlos Panantza, Quimsacocha-Loma Larga, Río Blanco and Curipamba sur (Figure 1). On the ASGM areas, we analyzed the mineral reservoirs of Chinapintza, Bella Rica and Portovelo-Zaruma (Figure 1). These reservoirs have a long history of exploitation and a known track record of environmental problems [26–31]. In the case of the petroleum industry, we analyzed samples from 4 different points located throughout the Auca oilfield (Figure 1).

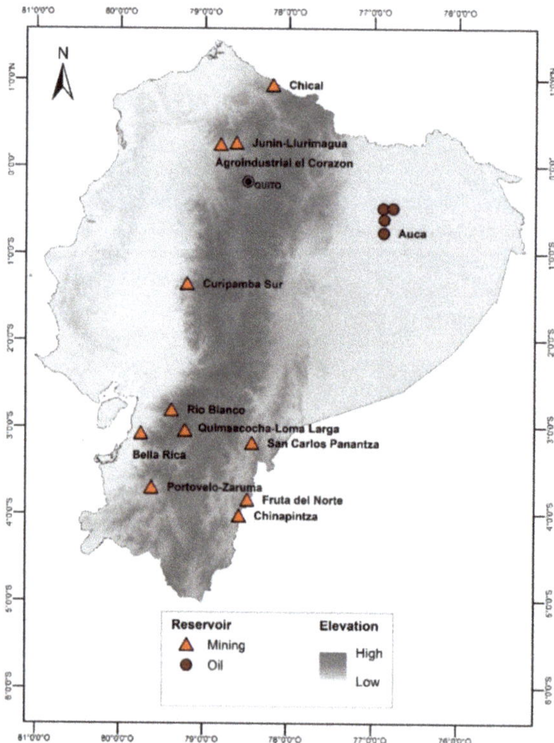

Figure 1. Map of study area. Location of the studied metallic reservoirs and studied oil field.

2.2. Data Collection

The collected data come from different sources including published research papers as primary sources, technical reports from government repositories, such as environmental impact assessments approved by the Ecuadorian Ministry of Environment, and online databases. The intent was to find specific concentrations of the different elements in the deposits, but in several cases, qualitative presence/absence information was found. Table 1 summarizes the sources where the information was obtained. An extensive and intensive search of information was conducted and compiled in a database that included information from a total of 19 elements (Table 2).

Table 1. Sources of information for the different metallic reservoirs and location coordinates (WGS84).

Reservoir Name	Latitude	Longitude	Main Metal	Current State	Reference
Agroindustrial el Corazón	0.249	−78.811	gold, copper, silver	ASGM	[54–56]
Bella Rica	3.076	−74.741	gold, silver	ASGM	[57]
Chical	0.935	−78.190	gold, copper, silver	ASGM	[58]
Chinapintza	−4.022	−78.572	gold	ASGM	[59]
Curipamba-sur	−1.356	−79.202	gold	Exploration for large scale mining	[60]
Fruta del Norte	−3.836	−78.471	gold, silver	Early large scale mining	[61]
Junin-Llurimagua	0.269	−78.623	copper, gold	Exploration for large scale mining	[62,63]
Portovelo-Zaruma	−3.698	−79.612	gold, silver	ASGM	[64]
Quimsacocha-Loma larga	−3.043	−79.220	gold	Exploration for large scale mining	[65]
San Carlos Panantza	−3.189	−78.419	copper, gold	Exploration for large scale mining	[66,67]
Nambija	0.103	−78.789	gold	ASGM	[68,69]
Rio Blanco	−2.806	−79.366	gold	Early medium scale mining	[70]

2.3. Spatial Distribution

To better represent the distribution of the different elements, the deposits and reservoirs were georeferenced for each mining area and points within the Auca oilfield, and projected using ArcGIS ver 10.7 [71], using a WGS 84 coordinate system. Maps for each element in the analysis were developed, classifying the concentration in relative ranges, and also comparing the concentration with the World Health Organization recommendations for water consumption. We chose each element to be studied by analyzing and classifying the elements and their risk according to Williams et al. [72]. We decided to compare the information of concentration of the different elements in the deposit with the WHO drinking water standards [73] only as a proxy for safety. We decided to perform this comparison with the logic that if the rocks were once mined from the deposit and the precious metals extracted, the leftovers or residues will be left in tailing dams or dumps. Therefore, these elements in contact with water might mobilize and end up in the local hydric system. The WHO standards were used as a general reference [73] for the safety of the elements themselves and the safety threshold for concentration. However, processes like oxidation, bioavailability of the element and bioaugmentation need to be further explored to assess in detail the potential effects and risks of heavy metal contamination in different areas. In the case of produced water, it was considered that there is a chance of spills and also if the water is injected it might end up in superficial aquifers and the superficial hydric system, due to cracks in the well's cement seal or connectivity between underground formations. Furthermore, the specific management of the different mining areas can be the main factor to determine if there will be major environmental impacts or not.

Table 2 summarizes the known information regarding the effects of the studied elements and their potential for bioaccumulation. This information is fundamental to understanding the potential risks the local environment may face.

Table 2. Potential health effects of the studied elements, maximum recommended concentration by the WHO [73] and potential for bioaccumulation.

Component	Maximum Recommended Concentration [mg/L] (WHO, 2017)	Health Effects (WHO, 2017)	Bioaccumulation Potential
Aluminum [Al]	0.1–0.2	Acceleration Alzheimer's disease onset	Medium [34]
Arsenic [As]	0.01	Dermal lesions (hyperpigmentation and hypopigmentation), peripheral neuropathy, skin, bladder and lung cancers and peripheral vascular disease	High [35]
Cadmium [Cd]	0.003	Kidney cancer	High [36,37]
Chromium [Cr(VI)]	0.05	Lung cancer, other types of cancer	Medium [37,38]
Cobalt [Co]	—	Lung diseases and respiratory effects (Kim et al., 2006)	Medium [37]
Copper [Cu]	2	Gastrointestinal bleeding, hematuria, intravascular hemolysis, methaemoglobinaemia, hepatocellular toxicity, acute renal failure and oliguria	High [39]
Fluoride [F⁻]	1.5	Skeletal fluorosis	Medium [40]
Iron [Fe]	0.3	Turbidity, color and bad taste to water	Low [41]
Lead	0.01	Neurodevelopmental effects, cardiovascular diseases, impaired renal function, hypertension, impaired fertility and adverse pregnancy outcomes	High [42]
Manganese [Mn]	0.1–0.5	Adverse neurological effects (WHO, 2004)	Medium [38]
Mercury [Hg]	0.006	Severe disruption of any tissue with which it comes into contact in sufficient concentration, neurological and renal disturbances	High [43]
Molybdenum	0.07	None	Low [44]
Nickel [Ni]	0.02	Carcinogenic	Medium [59]
Selenium [Se]	0.04	Gastrointestinal disturbances, discoloration of skin, decayed teeth, hair and nail loss	Medium [37]
Silver [Ag]	0.1	Argyria	Medium [37]
Vanadium [V]	—	Lung diseases and respiratory effects	Medium [37]
Zinc	3	Pulmonary distress and gastroenteritis	Medium [39]
Antimony [Sb]	0.02	Gastrointestinal mucosa irritated, abdominal cramps, diarrhea and cardiac toxicity	High [46]
Tin [Sn]	—	Acute gastric irritation	High [47]

3. Results and Discussion

3.1. General Description of Studied Metallic Reservoirs

Information from 11 metallic deposits in Ecuador was found. This will be analyzed case by case for the main mineralogical and geochemical composition.

3.1.1. Agroindustrial El Corazón

This deposit is found surrounded by andesitic lava and tuff. The most abundant precious metal is gold, which is located in siliceous hydrothermal breccia with quartz veins. There are known anomalies of iron, mercury and copper in this deposit and it is rich in sulphur. Moreover, silver is also prevalent in this deposit.

This deposit also has very high concentrations of arsenic, chrome, copper, lead, mercury, molybdenum, nickel, silver, vanadium, zinc and antimony. Moreover, the most common mineral species are: cassiterite {SnO_2}, wolframite {$CaSiO_3$-Ca_3}, molybdenite {MoS_2}, sphalerite {ZnS}, chalcopyrite {$CuFeS_2$}, magnetite {$Fe^{2+}Fe^{3+}_2O_4$}, argentite {Ag_2S}, enargite {Cu_3AsS_4}, galena {PbS}, pyrite {FeS_2}, marcasite {FeS_2} [54–56].

3.1.2. Bella Rica

The rocks in the Bella Rica are surrounded mainly by volcanic rocks, andesite, basalt, breccia and diabase. Free gold is common in the deposit, as well as multiphase veins of gold associated with quartz and carbonates. The deposit is quite rich in sulphide and arsenic.

The most common minerals found in the deposit are: pyrrhotite $(Fe_{(1-x)}S)$, arsenopyrite $\{FeAsS\}$, chalcopyrite $\{CuFeS_2\}$, epidote $\{Ca_2Fe^{3+}Al_2O(OH)\}$, wurtsite $\{ZnS\}$, galena $\{PbS\}$, hematite $\{Fe_2O_3\}$, molybdenite $\{MoS_2\}$, magnetite $\{Fe^{2+}Fe^{3+}{}_2O_4\}$, cuprite $\{Cu_2O\}$, chalcocite $\{Cu_2S\}$, covellite $\{CuS\}$, malachite $\{Cu_2CO_3(OH)_2\}$ [57].

3.1.3. Chical

Volcanic rocks surround the deposit. The deposit has quartz veins in the shape of roots embedded in the volcanic rocks. There are evident deposits of precious metals in the quartz veins, with sulphide, zinc and copper evident as well.

The most common minerals found in this reservoir are chlorite, epidote $\{Ca_2Fe^{3+}Al_2O(OH)\}$, calcite $\{CaCO_3\}$ and traces of pyrite $\{FeS_2\}$. Specifically, in the quartz veins there is up to 2% of pyrite, up to 1% chalcopyrite and less than 1% of galena and sphalerite [58].

3.1.4. Chinapintza

The mineralization occurs within an intrusive felsic volcanic complex from the late Cretaceous period. The deposit is mostly formed by epithermal deposition on quartz veins. It is very rich in sulphur associated with metals, mainly: pyrite, sphalerite, galena, arsenopyrite, pyrrhotite, chalcopyrite, bornite, tetrahedrite and malachite [59].

3.1.5. Curipamba Sur

It is located within an important hydrothermal alteration zone. It includes massive sulphur deposits. The precious metals are associated with sulphurs. The most common mineral species are: tennantite $\{Cu_{12}As_4S_{13}\}$, pyrite $\{FeS_2\}$, tetrahedrite $\{(Cu,Fe)_{12}Sb_4S_{13}\}$, galena $\{PbS\}$, chalcopyrite $\{CuFeS_2\}$, sphalerite $\{ZnS\}$, covellite $\{CuS\}$, bornite $\{Cu_5Fe S_4\}$, azurite $\{Cu_3(CO_3)_2(OH)_2\}$, arsenopyrite $\{FeAsS\}$ [60].

3.1.6. Fruta del Norte

The deposit is surrounded by andesites from the Misahualli formation and feldspar porphyry intrusions. The mineralization is characterized by quartz-sulphide veins, particularly in the central and north of the deposit, where also gold and silver are more abundant. The main minerals found are: orthoclase $\{KAlSi_3O_8\}$, rodocrosite $\{MnCO_3\}$, barite $\{BaSO_4\}$, marcasite $\{FeS_2\}$, pyrite $\{FeS_2\}$, sphalerite $\{ZnS\}$, galena $\{PbS\}$, chalcopyrite $\{CuFeS_2\}$, alabandite $\{MnS\}$, stibnite $\{SbS_3\}$, arsenopyrite $\{FeAsS\}$, acanthite $\{Ag_2S\}$, freibergite $\{Ag_6Cu_4Fe_2Sb_4S_{13}\}$, boulangerite $\{Pb_5Sb_4S_{11}\}$, jamesonite $\{Pb_4FeSb_6S_{14}\}$, valentinite $\{Sb_2O_3\}$, senarmontite $\{Sb_2O_3\}$ [61].

3.1.7. Junin-Llurimagua

The desposit is located with intrusive granodiorite bodies. The mineralization is associated with bornite $\{Cu_5FeS_4\}$, chalcopyrite $\{CuFeS_2\}$ and molibdenite $\{MoS_2\}$ [62,63].

3.1.8. Portovelo-Zaruma

The surrounding rock is formed by andesites and tuff. There are three mineralization phases, one mainly formed by quartz-pyrite-chlorite-hematite, one a quartz-pyrite-chalcopyrite and one a polymetallic quartz rich in galena and sphalerite and galena-chalcopyrite. The most common minerals are: bornite $\{Cu_5FeS_4\}$, hematite $\{Fe_2O_3\}$, tennantite $\{Cu_{12}As_4S_{13}\}$, tetrahedrite $\{(Cu,Fe)_{12}Sb_4S_{13}\}$, magnetite $\{Fe^{2+}Fe^{3+}{}_2O_4\}$, molibdenite $\{MoS_2\}$, argentite $\{Ag_2S\}$, freibergite $\{Ag_6Cu_4Fe_2Sb_4S_{13}\}$, nagyagite $\{(Pb(Pb,Sb)S_2)((Au,Te))\}$, proustite $\{Ag_3AsS_3\}$, bournonite $\{PbCuSbS_3\}$, pyrite $\{FeS_2\}$, chalcopyrite $\{CuFeS_2\}$ [64].

3.1.9. Quimsacocha-Loma Larga

The deposit is located in andesitic lava flows, with breccias and tuff. The mineralization is rich in sulphur. Thus, the most common minerals are: pyrite $\{FeS_2\}$, enstatite $\{MgSiO_3\}$, lizardite $\{Mg_3Si_2O_5(OH)_4\}$, chalcopyrite $\{CuFeS_2\}$, covelite $\{CuS\}$, luzonite $\{Cu_3AsS_4\}$, tennantite $\{Cu_{12}As_4S_{13}\}$, tetrahedrite $\{(Cu,Fe)_{12}Sb_4S_{13}\}$ [65].

3.1.10. San Carlos Panantza

The deposit is found surrounded by granite and leucogranite. The most common minerals are: chalcopyrite $\{CuFeS_2\}$, molibdenite $\{MoS_2\}$, magnetite $\{Fe^{2+}Fe^{3+}_2O_4\}$, pyrite $\{FeS_2\}$, anhidrite $\{CaSO_4\}$, gypsum $\{CaSO_4 \cdot 2H_2O\}$, chalcocite $\{Cu_2S\}$, malachite $\{Cu_2CO_3(OH)_2\}$, chrysocolla $\{(Cu,Al)_4H_4(OH)_8Si_4O_{10} \cdot nH_2O)\}$, cuprite $\{Cu_2O\}$ [66,67].

3.1.11. Río Blanco

The surrounding rock is mainly andesite feldspar, tuff and breccia with volcanic sandstone and dacitic tuff. The gold–silver mineralization is found in veins several hundred meters long. The most common minerals are: electrum $\{Au,Ag\}$, pyrite $\{FeS_2\}$, pyrrhotite $\{Fe_{1-x}S\}$, pyrargyrite $\{Ag_3SbS_3\}$, tetrahedrite $\{(Cu,Fe)_{12}Sb_4S_{13}\}$, arsenopyrite $\{(Cu,Fe)_{12}Sb_4S_{13}\}$, sphalerite $\{ZnS\}$, galena $\{PbS\}$, chalcopyrite $\{PbS\}$ [70].

3.1.12. Nambija

The host rock is a complex sequence of tuff, lapilli tuff with skarn and pyroclastic breccia. The main mineralization in the Nambija deposit is gold, with some porphyry associations of Cu-Au and Cu-Mo. The gold is usually free with a 90% purity and 7–10% silver. The most common minerals are: pyrite $\{FeS_2\}$, chalcopyrite $\{CuFeS_2\}$, pirrhotite $\{Fe_{(1-x)}S\}$, sphalerite $\{ZnS\}$ and galena $\{PbS\}$ [68,69].

3.2. Distribution of Anomalies of Element Concentration

Table 3 summarizes the concentration distribution of the studied elements. Moreover, the comparison of the element concentration with WHO standards are shown for each element.

Several anomalies were found distributed throughout Ecuador. Any value significantly higher than the other reported values is considered an anomaly. Some remarkable findings are as follows.

In the case of aluminum, it was present in Fruta del Norte in southeastern Ecuador and in the Auca oilfield in high levels. In these same places it exceeds the WHO recommendations. On the other hand, arsenic was present throughout Ecuador. The concentrations were particularly high in Quimsacocha and Agroindustrial Corazón. It exceeds the WHO limit in Quimsacocha, Bella Rica, Fruta del Norte, Agroindustrial el Corazón and Junin-Llurimagua. The fact that the values were very high in Quimsacocha is very worrisome considering that the area is a source of drinking water for Cuenca, one of the largest cities in Ecuador. This is even more concerning when analyzing the mercury concentrations (Table 3). The highest concentration of natural mercury among the analyzed sites was found at Quimsacocha. The Auca oilfield and all the northwestern sites showed presence of natural mercury in levels above the WHO threshold, and other sites had mercury presence as well. Mercury is known to bioaccumulate once it has been methylated. Therefore, these findings suggest that there might be a significant presence of mercury in the local food chains due to natural geochemical factors.

Table 3. Summary of concentrations of the studied elements in the different deposits. Present-NC indicates that qualitative information of presence was found, but no details of specific concentration. An * indicates the concentration exceeds the WHO [73] safety limits.

	Agroindustrial el Corazón	Auca	Bella Rica	Chical	Chinapintza	Curipamba-Sur
Aluminum [ppm]	Present-NC	0.86 *	-	0.1	-	-
Arsenic [ppm]	290 *	0.01	0.48 *	-	-	Present-NC
Cadmium [ppm]	Present-NC	0.13 *	-	0.02 *	-	-
Chromium [ppm]	207 *	0.05	-	10 *	-	-
Cobalt [ppm]	17 *	0.55 *	-	11 *	-	-
Cooper [ppm]	570 *	0.01	21 *	0.06	Present-NC	19,200 *
Fluoride [ppm]	-	9 *	-	-	-	-
Iron [ppm]	Present-NC	105 *	-	-	-	Present-NC
Lead [ppm]	12 *	0.006	-	17 *	16,800 *	3700 *
Manganese [ppm]	430 *	2.9 *	-	-	-	-
Mercury [ppm]	30 *	0.00003	-	0.1 *	-	-
Molybdenum [ppm]	5.44 *	0.06	-	0.1 *	-	-
Nickel [ppm]	25 *	0.64 *	-	0.05 *	-	-
Selenium [ppm]	-	0.01	-	0.005	-	-
Silver [ppm]	1.5 *	-	-	-	Present-NC	58 *
Vanadium [ppm]	24 *	0.7 *	-	8 *	-	-
Zinc [ppm]	18 *	0.85	-	0.2	30,200 *	35,200 *
Antimony [ppm]	21 *	-	-	-	-	-
Tin [ppm]	-	-	-	-	-	-

	Fruta del Norte	Junin-Llurimagua	Nambija	Portovelo-Zaruma	Quimsacocha-Loma larga	Rio Blanco	San Carlos Panantza
Aluminum [ppm]	9400 *	-	-	-	Present-NC	-	-
Arsenic [ppm]	300 *	1.1 *	-	-	2200 *	Present-NC	-
Cadmium [ppm]	-	0.1 *	-	Present-NC	Present-NC	-	-
Chromium [ppm]	-	22 *	-	-	Present-NC	-	-
Cobalt [ppm]	10 *	5.5 *	-	Present-NC	Present-NC	-	-
Cooper [ppm]	-	377 *	Present-NC	40,000 *	5900 *	Present-NC	6400 *
Fluoride [ppm]	-	-	Present-NC	-	-	-	-
Iron [ppm]	10,300 *	-	Present-NC	Present-NC	32 *	-	16,800 *
Lead [ppm]	Present-NC	3.2 *	Present-NC	40,000 *	Present-NC	-	-
Manganese [ppm]	-	-	-	-	Present-NC	-	-
Mercury [ppm]	0.03 *	0.1 *	-	Present-NC	50 *	Present-NC	-
Molybdenum [ppm]	-	2.4 *	Present-NC	-	Present-NC	Present-NC	80 *
Nickel [ppm]	10 *	9 *	-	Present-NC	Present-NC	-	-
Selenium [ppm]	-	1 *	-	Present-NC	Present-NC	-	-
Silver [ppm]	12 *	Present-NC	Present-NC	Present-NC	47 *	Present-NC	1.3 *
Vanadium [ppm]	-	94 *	-	-	-	-	-
Zinc [ppm]	0.007	63 *	Present-NC	100,000 *	182 *	Present-NC	-
Antimony [ppm]	-	-	-	Present-NC	-	Present-NC	-
Tin [ppm]	-	Present-NC	Present-NC	-	-	-	-

Another element known to bioaccumulate is cadmium. This element was present in high levels in the Auca oilfield and Junin-Llurimagua. Chical showed lower concentrations, but still exceeded the WHO limits. The Auca oilfield is located in the Ecuadorian Amazon basin, where there are numerous cocoa plantations. The contamination with cadmium, a product of produced water spills, has already started affecting cocoa bean exports because the product exceeded the maximum Cd levels allowed by the European Union [74]. To the best of our knowledge, there is no information on the effects on human beings in that area, but it is likely that they are consuming contaminated crops.

In the case of lead, it was found in very high concentrations in most sites. This is probably related to the abundance of galena in the metallic deposits. This is another element that can cause major metabolic disruption in the human body and is widely distributed in Ecuador. Chromium is present and particularly prevalent in the northern sites (Table 3). It exceeds the WHO limits in Chical, Agroinsutrial Corazón and Junin-Llurimagua. Cobalt exceeds the safety limits in two points in Auca and Chical, Agroindustrial Corazón, Junin-Llurimagua and Fruta del Norte (Table 3). Ecuador's metallic deposits tend to be very rich in copper. Therefore, in most sites the concentration is very high. The results are similar for nickel and silver (Table 3), found in most of the sites in high concentrations.

We found references of the presence of manganese only for Agroindustrial el Corazón and Auca (Table 3). However, there are reports of manganese in groundwater wells near the Bella Rica deposit [75]. In the case of fluoride, we only found information for the Auca oilfield, where it was above the WHO threshold.

Other elements such as selenium, molybdenum, vanadium and zinc were found in different sites in Ecuador with locally high values (Table 3). We found no information regarding the presence of francium, radium or actinium.

3.3. A Call to Action

Our analysis shows the widespread and prevalent presence of heavy metals throughout Ecuador. This indicates the potential for the local environment to suffer from heavy metal contamination and environmental degradation. However, this is only a risk which has not materialized in most of the studied deposits.

An important point to make is that the development of this project was hindered by the lack of published information about the geochemistry of most of the different mineral deposits and oil fields. It was particularly difficult to find published information for specific mineral deposits traditionally or currently exploited by artisanal and small-scale gold mining (ASGM), such as Chinapintza, Portovelo-Zaruma or Bella Ric, probably, because these types of mining operations do not have a formal geological exploration. A special case is the mining deposit of Nambija, located in the Ecuadorian Amazon basin and considered one of the richest gold deposits in Ecuador. Even though this deposit has been exploited for more than 50 years by artisanal gold miners, there was no information on its geochemistry available or published. It is very likely that there is significant geochemical information on these and other deposits, but that information is not widely available. Thus, studies such as this are hindered and our ability to give information for better land use planning and even improve community relations that would benefit the different mining and petroleum companies, is limited. To us, this is a call to strengthen the field of geochemistry in Ecuador and motivate the publication of the information by companies and governmental agencies.

Most of the studied areas are known deposits but there are no active mining activities to exploit them. Ecuador is interested in developing its deposits in a regulated and industrialized way. Thus, the largest deposits, such as Fruta del Norte, Mirador or Quimsacocha are either in the initial producing steps or in licensing phases to large national or international companies [76]. Thus, we still have time to avoid the environmental problems that are common in small and artisanal gold mining areas, that have been operating for more than 50 years.

Artisanal areas such as Zaruma-Portovelo, Ponce-Enríquez, Chinapintza and Nambija have a well-documented and sad record of environmental and social problems [26–31]. These very rich gold deposits have been exploited in a completely unregulated manner with no government intervention to stop or at least regulate these activities, likely due to corruption networks with interests in these activities protecting and extorting the miners [77,78].

An extreme case of illegal mining in Ecuador happened in another reservoir (not included in our analysis due to lack of information) called Buenos Aires, located near Junín-Llurimagua. In the last two years the Buenos Aires deposit became known to artisanal miners from other parts of Ecuador and in the course of one year, more than 8000 people colonized this mountainous area to extract gold illegally. Finally, in early 2020, the government evicted the illegal miners, but the environmental damage, mainly due to deforestation, was already there. Nobody knows how much gold was extracted, but it is common knowledge that the material rich in gold was transported to southern Ecuador, 500 km away, to Ponce-Enríquez and Portovelo where benefit plants operate. Thus, it is very likely that a corruption network developed to protect those shipments from police controls [79,80].

In the case of oil, the exploration and production is performed by national or international companies under a strict legal framework. In the past, the regulatory framework was very lax and there are still environmental passives (mainly old oil spills) across north eastern Ecuador. Even though nowadays the oil industry in Ecuador is improving its environmental track record, there are still oil spills and produced water spills [81]. As part of the current legal framework, the companies have the obligation to report any oil or produced water spills. In reality, it is likely this is not happening, due to the limited on-the-ground control and corruption.

A positive recent piece of news is that Ecuador joined in October 2020 the Extractive Industries Transparency Initiative (EITI) [82]. This is one of a few steps the government is taking to strengthen its institutions and prevent corruption. Valuable commodities such as gold, silver, or oil can provide the resources to underpin the economy of a developing country, such as Ecuador, but this is not possible if most of the resources are lost due to corruption [83].

Thus, the responsibility lies in the central Ecuadorian government, responsible for licensing, and in the local governments and Environmental Ministry offices, responsible for monitoring and assessing the fulfillment of the legal framework and environmental action plans. There is an urgent need to stop corruption and to strengthen the local institutions so they have the resources and, most importantly, the will to protect the local environment. Therefore, our work, besides showing the spatial distribution of these potential risks, it is a call to action to all the local stakeholders interested in exploiting these valuable resources and in protecting the local environment. It is also a call to the Ecuadorian academia, to reach out to local stakeholders, miners, farmers and environment ministry officers to collaborate on finding technical solutions to exploit these resources in a safer and sustainable manner.

To the best of our knowledge, this work is the first attempt to understand the bigger risk the natural composition of metallic deposits represents in Ecuador. Furthermore, we consider our work not as a call to stop mining, but on the contrary, this is a strong call to plan every mining and petroleum production project considering these risks. Moreover, we want to motivate the empowerment of the local municipalities to control their territories and protect their environment. It is also a call to the local municipalities to develop, update and enforce land use plans in their jurisdictions. Ecuador is a biologically rich and diverse country, and is also incredibly diverse and rich in its geological resources. We need to consider both sources of richness for future planning and development.

4. Conclusions

The major ASGM districts in Ecuador, Bella Rica, Portovelo-Zaruma, Nambija and Chinapintza, have an extensive presence of heavy metals in their deposits. Thus, the lack of formal environmental planning, control and law enforcement as well as technical mining training are major hazards for the local environment and population.

There is a need to further study the ecosystems surrounding the mineral deposits to confirm the presence of heavy metal contamination and understand if the specific local conditions are in fact allowing contaminant movement.

Ecuador is embarking on the large-scale development and exploitation of its metallic deposits. Thus, we still have time to prevent further environmental damage if the government officers and institutions fulfill their duty to oversee the operation of legal, small and large mining operations, stop illegal mining and monitor the operation of oil companies.

Author Contributions: Conceptualization, B.B. and X.D.; Data curation, M.B.O.; Formal analysis, J.L.R.-P.; Investigation, J.L.R.-P.; Visualization, M.B.O.; Writing—original draft, J.L.R.-P.; Writing—review & editing, B.B., X.D. and M.B.O. All authors have read and agreed to the published version of the manuscript.

Funding: This research was possible thanks to Escuela Politécnica Nacional grants PIMI 16-05 and PIS 17-16.

Institutional Review Board Statement: Not applicable.

Informed Consent Statement: Not applicable.

Data Availability Statement: The data presented in this study are available on request from the corresponding author.

Conflicts of Interest: The authors declare no conflict of interest.

References

1. Hagelüken, C.; Corti, C.W. Recycling of gold from electronics: Cost-effective use through 'Design for Recycling'. *Gold Bull.* **2010**, *43*, 209–220. [CrossRef]
2. Asokan, A. CRISPR genome editing in stem cells turns to gold. *Nat. Mater.* **2019**, *18*, 1038–1039. [CrossRef] [PubMed]
3. Kyaw, W.T.; Kuang, X.; Sakakibara, M. Health Impact Assessment of Artisanal and Small-Scale Gold Mining Area in Myanmar, Mandalay Region: Preliminary Research. *Int. J. Environ. Res. Public Health* **2020**, *17*, 6757. [CrossRef] [PubMed]
4. LeBlanc, M.E.; Parsons, M.B.; Chapman EE, V.; Campbell, L.M. Review of ecological mercury and arsenic bioaccumulation within historical gold mining districts of Nova Scotia. *Environ. Rev.* **2020**, *28*, 187–198. [CrossRef]
5. Okop, I.; Persaud, K. Qualitative and Quantitative Assessment of Petroleum Contaminants in Soils under Tropical Weather Conditions. *Am. J. Anal. Chem.* **2019**, *10*, 112. [CrossRef]
6. Pateda, S.M.; Sakakibara, M.; Sera, K. Lung function assessment as an early biomonitor of mercury-induced health disorders in artisanal and small-scale gold mining areas in Indonesia. *Int. J. Environ. Res. Public Health* **2018**, *15*, 2480. [CrossRef] [PubMed]
7. Zolnikov, T.R.; Ortiz, D.R. A systematic review on the management and treatment of mercury in artisanal gold mining. *Sci. Total Environ.* **2018**, *633*, 816–824. [CrossRef]
8. Perez, C.R.; Moye, J.K.; Cacela, D.; Dean, K.M.; Pritsos, C.A. Low level exposure to crude oil impacts avian flight performance: The Deepwater Horizon oil spill effect on migratory birds. *Ecotoxicol. Environ. Saf.* **2017**, *146*, 98–103. [CrossRef]
9. Lessmann, J.; Fajardo, J.; Munoz, J.; Bonaccorso, E. Large expansion of oil industry in the Ecuadorian Amazon: Biodiversity vulnerability and conservation alternatives. *Ecol. Evol.* **2016**, *6*, 4997–5012. [CrossRef]
10. Rotta LH, S.; Alcântara, E.; Park, E.; Negri, R.G.; Lin, Y.N.; Bernardo, N.; Gonçalves Mendesb, T.S.; Souza Filho, C.R. The 2019 Brumadinho tailings dam collapse: Possible cause and impacts of the worst human and environmental disaster in Brazil. *Int. J. App. Earth Obs. Geoinf.* **2020**, *90*, 102119. [CrossRef]
11. Barron, M.G.; Vivian, D.N.; Heintz, R.A.; Yim, U.H. Long-Term Ecological Impacts from Oil Spills: Comparison of Exxon Valdez, Hebei Spirit, and Deepwater Horizon. *Environ. Sci. Technol.* **2020**. [CrossRef]
12. Beyer, J.; Trannum, H.C.; Bakke, T.; Hodson, P.V.; Collier, T.K. Environmental effects of the Deepwater Horizon oil spill: A review. *Mar. Pollut. Bull.* **2016**, *110*, 28–51. [CrossRef]
13. Caballero Espejo, J.; Messinger, M.; Román-Dañobeytia, F.; Ascorra, C.; Fernandez, L.E.; Silman, M. Deforestation and forest degradation due to gold mining in the Peruvian Amazon: A 34-year perspective. *Remote Sens.* **2018**, *10*, 1903. [CrossRef]
14. Santos-Francés, F.; Martínez-Graña, A.; Alonso Rojo, P.; García Sánchez, A. Geochemical background and baseline values determination and spatial distribution of heavy metal pollution in soils of the Andes mountain range (Cajamarca-Huancavelica, Peru). *Int. J. Environ. Res. Public Health* **2017**, *14*, 859. [CrossRef] [PubMed]
15. Kacmaz, H. Assessment of heavy metal contamination in natural waters of Dereli, Giresun: An area containing mineral deposits in northeastern Turkey. *Environ. Monit. Assess.* **2020**, *192*, 1–12. [CrossRef] [PubMed]

16. Tabelin, C.B.; Igarashi, T.; Villacorte-Tabelin, M.; Park, I.; Opiso, E.M.; Ito, M.; Hiroyoshi, N. Arsenic, selenium, boron, lead, cadmium, copper, and zinc in naturally contaminated rocks: A review of their sources, modes of enrichment, mechanisms of release, and mitigation strategies. *Sci. Total Environ.* **2018**, *645*, 1522–1553. [CrossRef]

17. Rahman, Z.; Singh, V.P. The relative impact of toxic heavy metals (THMs) (arsenic (As), cadmium (Cd), chromium (Cr)(VI), mercury (Hg), and lead (Pb)) on the total environment: An overview. *Environ. Monit. Assess.* **2019**, *191*, 419. [CrossRef] [PubMed]

18. Neff, J.; Lee, K.; DeBlois, E.M. Produced water: Overview of composition, fates, and effects. In *Produced Water*; Springer: New York, NY, USA, 2011; pp. 3–54.

19. Van den Berg, G.A.; Loch, J.P.G.; Winkels, H.J. Effect of fluctuating hydrological conditions on the mobility of heavy metals in soils of a freshwater estuary in the Netherlands. *Water Air Soil Pollut.* **1998**, *102*, 377–388. [CrossRef]

20. Kusin, F.M.; Sulong, N.A.; Affandi FN, A.; Molahid, V.L.M.; Jusop, S. Prospect of abandoned metal mining sites from a hydrogeochemical perspective. *Environ. Sci. Pollut. Res.* **2020**, *28*, 2678–2695. [CrossRef]

21. Shrestha, N.; Chilkoor, G.; Wilder, J.; Gadhamshetty, V.; Stone, J.J. Potential water resource impacts of hydraulic fracturing from unconventional oil production in the Bakken shale. *Water Res.* **2017**, *108*, 1–24. [CrossRef]

22. Lee, K.; Neff, J. (Eds.) *Produced Water: Environmental Risks and Advances in Mitigation Technologies*; Springer Science & Business Media: Heidelberg, Germany, 2011.

23. Gafur, N.A.; Sakakibara, M.; Sano, S.; Sera, K. A case study of heavy metal pollution in water of Bone River by Artisanal Small-Scale Gold Mine Activities in Eastern Part of Gorontalo, Indonesia. *Water* **2018**, *10*, 1507. [CrossRef]

24. Sakakibara, M.; Sera, K. Current mercury exposure from artisanal and small-scale gold mining in Bombana, southeast Sulawesi, Indonesia—future significant health risks. *Toxics* **2017**, *5*, 7.

25. Schudel, G.; Kaplan, R.; Miserendino, R.A.; Veiga, M.M.; Velasquez-López, P.C.; Guimarães JR, D.; Bergquist, B.A. Mercury isotopic signatures of tailings from artisanal and small-scale gold mining (ASGM) in southwestern Ecuador. *Sci. Total Environ.* **2019**, *686*, 301–310. [CrossRef]

26. Marshall, B.G.; Veiga, M.M.; da Silva, H.A.; Guimarães, J.R.D. Cyanide Contamination of the Puyango-Tumbes River Caused by Artisanal Gold Mining in Portovelo-Zaruma, Ecuador. *Curr. Environ. Health Rep.* **2020**, *14*, 34. [CrossRef]

27. Schutzmeier, P.; Berger, U.; Bose-O'Reilly, S. Gold Mining in Ecuador: A cross-sectional assessment of mercury in urine and medical symptoms in miners from Portovelo/Zaruma. *Int. J. Environ. Res. Public Health* **2017**, *14*, 34. [CrossRef] [PubMed]

28. Counter, S.A.; Buchanan, L.H.; Ortega, F.; Laurell, G. Elevated blood mercury and neuro-otological observations in children of the Ecuadorian gold mines. *J. Toxicol. Environ. Health Part A* **2002**, *65*, 149–163. [CrossRef]

29. Tarras-Wahlberg, N.H. Environmental management of small-scale and artisanal mining: The Portovelo-Zaruma goldmining area, southern Ecuador. *J. Environ. Manag.* **2002**, *65*, 165–179. [CrossRef] [PubMed]

30. Appleton, J.D.; Williams, T.M.; Orbea, H.; Carrasco, M. Fluvial contamination associated with artisanal gold mining in the Ponce Enriquez, Portovelo-Zaruma and Nambija areas, Ecuador. *Water Air Soil Pollut.* **2001**, *131*, 19–39. [CrossRef]

31. Tarras-Wahlberg, N.H.; Flachier, A.; Fredriksson, G.; Lane, S.; Lundberg, B.; Sangfors, O. Environmental impact of small-scale and artisanal gold mining in southern Ecuador. *AMBIO A J. Hum. Environ.* **2000**, *29*, 484–491. [CrossRef]

32. Spikings, R.; Cochrane, R.; Villagomez, D.; Van der Lelij, R.; Vallejo, C.; Winkler, W.; Beate, B. The geological history of northwestern South America: From Pangaea to the early collision of the Caribbean Large Igneous Province (290–75 Ma). *Gondwana Res.* **2015**, *27*, 95–139. [CrossRef]

33. Ali, H.; Khan, E.; Ilahi, I. Environmental chemistry and ecotoxicology of hazardous heavy metals: Environmental persistence, toxicity, and bioaccumulation. *J. Chem.* **2019**, *2019*. [CrossRef]

34. World Health Organization. *Aluminium in Drinking-Water: Background Document for Development of WHO Guidelines for Drinking-Water Quality*; World Health Organization: Geneva, Switzerland, 2003.

35. Roy, J.S.; Chatterjee, D.; Das, N.; Giri, A.K. Substantial evidences indicate that inorganic arsenic is a genotoxic carcinogen: A review. *Toxicol. Res.* **2018**, *34*, 311–324. [CrossRef]

36. Jakimska, A.; Konieczka, P.; Skóra, K.; Namieśnik, J. Bioaccumulation of metals in tissues of marine animals, Part I: The role and impact of heavy metals on organisms. *Pol. J. Environ. Stud.* **2011**, *20*, 1117–1125.

37. Sánchez-Bayo, F.; van den Brink, P.J.; Mann, R.M. *Ecological Impacts of Toxic Chemicals*; Francisco Sanchez-Bayo, F., van den Brink, P.J., Mann, R.M., Eds.; Bentham Science Puiblishers, Ltd.: Sarja, United Arab Emirates, 2011.

38. Seenayya, G.; Prahalad, A.K. In situ compartmentation and biomagnification of chromium and manganese in industrially polluted Husainsagar Lake, Hyderabad, India. *Water Air Soil Pollut.* **1987**, *35*, 233–239. [CrossRef]

39. Delahaut, V.; Rašković, B.; Salvado, M.S.; Bervoets, L.; Blust, R.; De Boeck, G. Toxicity and bioaccumulation of Cadmium, Copper and Zinc in a direct comparison at equitoxic concentrations in common carp (Cyprinus carpio) juveniles. *PLoS ONE* **2020**, *15*, e0220485. [CrossRef] [PubMed]

40. Camargo, J.A. Fluoride toxicity to aquatic organisms: A review. *Chemosphere* **2003**, *50*, 251–264. [CrossRef]

41. World Health Organization. *Iron in Drinking-Water: Background Document for Development of WHO Guidelines for Drinking-Water Quality*; World Health Organization: Geneva, Switzerland, 2003.

42. Ara, A.; Usmani, J.A. Lead toxicity: A review. *Interdiscip. Toxicol.* **2015**, *8*, 55–64.

43. Bernhoft, R.A. Mercury toxicity and treatment: A review of the literature. *J. Environ. Public Health* **2012**, *2012*, 460508. [CrossRef]

44. Regoli, L.; Van Tilborg, W.; Heijerick, D.; Stubblefield, W.; Carey, S. The bioconcentration and bioaccumulation factors for molybdenum in the aquatic environment from natural environmental concentrations up to the toxicity boundary. *Sci. Total Environ.* **2012**, *435*, 96–106. [CrossRef] [PubMed]

45. Ogle, R.S.; Knight, A.W. Selenium bioaccumulation in aquatic ecosystems: 1. Effects of sulfate on the uptake and toxicity of selenate in Daphnia magna. *Arch. Environ. Contam. Toxicol.* **1996**, *30*, 274–279. [CrossRef]

46. Zeng, D.; Zhou, S.; Ren, B.; Chen, T. Bioaccumulation of antimony and arsenic in vegetables and health risk assessment in the superlarge antimony-mining area, China. *J. Anal. Methods Chem.* **2015**, *2015*, 909724. [CrossRef]

47. Ashraf, M.A.; Maah, M.J.; Yusoff, I. Bioaccumulation of Heavy Metals in Fish Species Collected from Former Tin Mining Catchment. *Int. J. Environ. Res.* **2012**. [CrossRef]

48. Mishra, S.; Dwivedi, S.P.; Singh, R.B. A Review on Epigenetic Effect of Heavy Metal Carcinogens on Human Health. *Open Nutraceuticals J.* **2010**, *3*, 188–193. [CrossRef]

49. Morais, S.; Garcia e Costa, F.; Pereira, M.L. Heavy Metals and Human Health. Available online: http://www.intechopen.com/books/environmental-health-emerging-issues-and-practice/heavy-metals-and-human-health (accessed on 20 August 2020).

50. Zuluaga, M.C.; Norini, G.; Ayuso, R.; Nieto, J.M.; Lima, A.; Albanese, S.; De Vivo, B. Geochemical mapping, environmental assessment and Pb isotopic signatures of geogenic and anthropogenic sources in three localities in SW Spain with different land use and geology. *J. Geochem. Explor.* **2017**, *181*, 172–190. [CrossRef]

51. Wacaster, S. The Mineral Industry of Ecuador. In *Minerals Yearbook Area Reports: International Review 2013 Latin America and Canada*; USGS: Reston, VA, USA.

52. Avcı, D.; Fernández-Salvador, C. Territorial dynamics and local resistance: Two mining conflicts in Ecuador compared. *Extr. Ind. Soc.* **2016**, *3*, 912–921.

53. Valladares, C.; Boelens, R. Extractivism and the rights of nature: Governmentality, 'convenient communities' and epistemic pacts in Ecuador. *Environ. Politics* **2017**, *26*, 1015–1034. [CrossRef]

54. Carrasco Lara, S.P. Paragénesis de la Mineralización Metálica del Sistema de Vetas y Diseminado del Área Minera "El Corazón", Cantón Cotacachi, Provincia de Imbabura. Bachelor's Thesis, Escuela Politécnica Nacional, Quito, Ecuador, 2019.

55. Dávila Ortiz, L.E. Diseño de Explotación del Mineral Existente en el Bloque "Cascada", del Área Minera "El Corazón", Ubicada en la Parroquia Garcia Moreno, Cantón Cotacachi, Provincia de Imbabura. Bachelor's Thesis, Universidad Central del Ecuador, Quito, Ecuador, 2017.

56. Robalino Cando, A.P. Caracterización litológica, mineralógica y estructural del sistema de vetas del área minera el corazón, cantón Cotacachi, provincia de Imbabura. Bachelor's Thesis, Universidad Central del Ecuador, Quito, Ecuador, 2018.

57. Vega Oyola, A.G. Calculo de Reservas de la Veta "Paraiso" Mina Paraiso-Distrito Ponce Enriquez. Bachelor's Thesis, Escuela Superior Politécnica del Litoral, Guayaquil, Ecuador, 2013.

58. Bioterra CIa. Ltda. Estudio de Imapcto Ambiental Ex-Ante Para las Fases de Exploración, Explotación y Beneficio Simultáneo de Minerales Métalicos Bajo el Régimen de Pequeña Minería del Área Minera Magadalena (CÓD. 400229). Available online: https://maecalidadambiental.files.wordpress.com/2019/06/eia-y-pma-el-chical-version-1.5.1_compressed.pdf (accessed on 15 September 2020).

59. Cisneros, S.; Hermel, L. Caracterización Petrogenética y Alteración Hidrotermal de Rocas Subvolcánicas del Sistema Epitermal oro–Plata de la Zona Condor Camp, Distrito Minero de Chinapintza, Provincia de Zamora Chinchipe. Bachelor's Thesis, Universidad Central del Ecuador, Quito, Ecuador, 2020.

60. Vallejo, C.; Soria, F.; Tornos, F.; Naranjo, G.; Rosero, B.; Salazar, F.; Cochrane, R. Geology of El Domo deposit in central Ecuador: A VMS formed on top of an accreted margin. *Miner. Depos.* **2016**, *51*, 389–409. [CrossRef]

61. Leary, S.; Sillitoe, R.H.; Stewart, P.W.; Roa, K.J.; Nicolson, B.E. Discovery, geology, and origin of the Fruta del Norte epithermal gold-silver deposit, southeastern Ecuador. *Econ. Geol.* **2016**, *111*, 1043–1072. [CrossRef]

62. ENAMI, EP. Portafolio de PROYECTOS 2016. Empresa Nacional Minera del Ecuador. Available online: https://www.enamiep.gob.ec/transparencia/2016/portafolioweb.pdf (accessed on 20 August 2020).

63. ENTRIX AMÉRICAS, S.A. Estudio Complementario al Estudio de Impacto Ambiental ExAnte y Plan de Manejo Ambiental para la Fase de Exploración Avanzada de Minerales Metálicos del Área Minera LLURIMAGUA. Empresa Nacional Minera. Available online: https://www.enamiep.gob.ec/wp-content/uploads/downloads/2019/01/EIA-LLurimagua-Complementario.pdf (accessed on 20 August 2020).

64. Vikentyev, I.; Banda, R.; Tsepin, A.; Prokofiev, V.; Vikentyeva, O. Mineralogy and formation conditions of Portovelo-Zaruma gold-sulphide vein deposit, Ecuador. *Geochem. Mineral. Petrol.* **2005**, *43*, 148–154.

65. Scholtz, E.; Gagnon, D.; Frost, D.; Del Carpio, S.; Masun, K.; Kalenchuk, K.; Liu, H.; Kaplan, P.; Chubb, D.; Correia, L. Technical Report Inv. Metals, Inc. Loma Larga Project Azual Province, Ecuador. Available online: https://www.invmetals.com/wp-content/uploads/2019/03/J01834-PM-REP-002-NI-43-101_Loma-Larga-20190114-FINAL.pdf (accessed on 20 August 2020).

66. Drobe, J. Panantza Copper Project Southeast Ecuador Update on Inferred Resource Estimate 43-101 Technical Report Corriente Resource Inc. Available online: http://www.corriente.com/media/PDFs/news/technical_reports/43_101_Tech_report_Panantza.pdf (accessed on 15 September 2020).

67. Drobe, J.; Hoffert, J.; Fong, R.; Haile, J.; Rokosh, J.; Panantza and San Carlos Copper Project. Preliminary Assesment Report Corriente Resources Inc. Available online: http://www.corriente.com/media/PDFs/news/technical_reports/20071030_PSC_Prelim_Assessment_Report.pdf (accessed on 15 September 2020).
68. Yautibug Guagcha, G.W. Geología y Metalogénia del Área Minera Sultana, Provincia de Zamora Chinchipe-Ecuador. Bachelor's Thesis, Escuela Politécnica Nacional, Quito, Ecuador, 2009.
69. León Peñafiel, P.J. Estabilización de Taludes en el Sector "El Tierrero" de la Mina Nambija. Bachelor's Thesis, Universidad Central del Ecuador, Quito, Ecuador, 2015.
70. Bineli-Betsi, T.; Chiaradia, M. *The Low-Sulphidation Au-Ag Deposit of Rio Blanco (Ecuador): Geology, Mineralogy, Geochronology and Isotope (S, Pb) Geochemistry*; Swiss Academy of Sciences: Bern, Switzerland, 2006.
71. ESRI. *ArcGIS Desktop: Release 10.7*; Environmental Systems Research Institute: Redlands, CA, USA, 2019.
72. Williams, T.M.; Dunkley, P.N.; Cruz, E.; Acitimbay, V.; Gaibor, A.; Lopez, E.; Baez, N.; Aspden, J.A. Regional geochemical reconnaissance of the Cordillera Occidental of Ecuador: Economic and environmental applications. *Appl. Geochem.* **2000**, *15*, 531–550. [CrossRef]
73. World Health Organization. *Guidelines for Drinking-Water Quality: Fourth Edition Incorporating the First Addendum*; WHO: Geneva, Switzerland, 2017.
74. Argüello, D.; Chavez, E.; Lauryssen, F.; Vanderschueren, R.; Smolders, E.; Montalvo, D. Soil properties and agronomic factors affecting cadmium concentrations in cacao beans: A nationwide survey in Ecuador. *Sci. Total Environ.* **2019**, *649*, 120–127. [CrossRef] [PubMed]
75. Johnson, W.P.; (Departamento de Metalurgia Extractiva, Escuela Politécnica Nacional, Quito, Ecuador); Diaz, X.; (Departamento de Metalurgia Extractiva, Escuela Politécnica Nacional, Quito, Ecuador). Personal communication, 2020.
76. ARCOM. Catastro Minero del Ecuador. Available online: http://geo.controlminero.gob.ec:1026/geo_visor/ (accessed on 15 September 2020).
77. Vangsnes, G.F. The meanings of mining: A perspective on the regulation of artisanal and small-scale gold mining in southern Ecuador. *Extr. Ind. Soc.* **2018**, *5*, 317–326.
78. Warnaars, X.; Bebbington, A. Negotiable Differences? Conflicts over Mining and Development in South East Ecuador. In *Natural Resource Extraction and Indigenous Livelihoods: Development Challenges in an Era of Globalization*; Dorset Press: Dorchester, UK, 2014; pp. 109–128.
79. Plan, V. La Zona Minera de Imbabura es (de Nuevo) una Bomba de Tiempo. Available online: https://www.planv.com.ec/confidenciales/confidencial-sociedad/la-zona-minera-imbabura-nuevo-una-bomba-tiempo (accessed on 10 December 2020).
80. Plan, V. Los Mineros de Buenos Aires: "La Policía Nos Extorsionaba". Available online: https://www.planv.com.ec/investigacion/investigacion/mineros-buenos-aires-la-policia-nos-extorsionaba (accessed on 10 December 2020).
81. Rivera-Parra, J.L.; Vizcarra, C.; Mora, K.; Mayorga, H.; Dueñas, J.C. Spatial distribution of oil spills in the north eastern Ecuadorian Amazon: A comprehensive review of possible threats. *Biol. Conserv.* **2020**, *252*, 108820. [CrossRef]
82. EITI. Ecuador se une al EITI. Available online: https://eiti.org/es/news/ecuador-se-al-eiti (accessed on 10 December 2020).
83. Lyall, A. A moral economy of oil: Corruption narratives and oil elites in Ecuador. Culture. *Theory Crit.* **2018**, *59*, 380–399. [CrossRef]

International Journal of
***Environmental Research
and Public Health***

MDPI

Article

Sustainability of Cocoa (*Theobroma cacao*) Cultivation in the Mining District of Ponce Enríquez: A Trace Metal Approach

Carolina Ramos [1,2,*], **Jeny Ruales** [2], **José Luis Rivera-Parra** [3], **Masayuki Sakakibara** [4,5,6] **and Ximena Díaz** [7]

[1]   Área de Ambiente y Sustentabilidad, Universidad Andina Simón Bolivar, Toledo N22-80 (Plaza Brasilia),
     Quito 170525, Ecuador
[2]   Departamento de Ciencia de Alimentos y Biotecnología DECAB, Escuela Politécnica Nacional,
     Ladrón de Guevara E11-253, Quito 170525, Ecuador
[3]   Departamento de Petróleos, Escuela Politécnica Nacional, Ladrón de Guevara E11-253, Quito 170525, Ecuador
[4]   Department of Earth Science, Graduate School of Science and Engineering, Ehime University,
     Matsuyama 790-8577, Japan
[5]   Faculty of Collaborative Regional Innovation, Ehime University, Matsuyama 790-8577, Japan
[6]   Research Institute for Humanity and Nature, Kyoto 603-8047, Japan
[7]   Departamento de Metalurgia Extractiva, Escuela Politécnica Nacional, Ladrón de Guevara E11-253,
     Quito 170525, Ecuador
*   Correspondence: carolina.ramos@epn.edu.ec or caritocr_19@hotmail.com

Citation: Ramos, C.; Ruales, J.; Rivera-Parra, J.L.; Sakakibara, M.; Díaz, X. Sustainability of Cocoa (*Theobroma cacao*) Cultivation in the Mining District of Ponce Enríquez: A Trace Metal Approach. *Int. J. Environ. Res. Public Health* **2022**, *19*, 14369. https://doi.org/10.3390/ijerph192114369

Academic Editor: Richard A. Lord

Received: 21 September 2022
Accepted: 27 October 2022
Published: 3 November 2022

Publisher's Note: MDPI stays neutral with regard to jurisdictional claims in published maps and institutional affiliations.

Abstract: Historically, cocoa (*Theobroma cacao*) has been one of Ecuador's most important export crops. In the Ponce Enriquez district, artisanal and small gold mining (ASGM), and quarrying account for 42% of economic activities, while agriculture and livestock farming account for 30%, making the analysis of their synergy and interaction key to understanding the long term viability of the different activities. In this study, we evaluated the concentration of potentially toxic metals in different parts of the cocoa plant and fruit, in relation to mining activities within the area. Gold extraction generates pollution, including potentially toxic metals such as mercury (Hg), cadmium (Cd), arsenic (As), copper (Cu), lead (Pb) and zinc (Zn). In order to understand the mobility of these metals within the cocoa plant and fruit, the analysis was conducted separately for leaves, pod, husk and cocoa bean. Concentrations of the target metals in the different plant parts and soil were measured using ICP-MS, and the mobility and risk factors were calculated using the transfer factor (TF) and the risk ratio (HQ). The results suggest that Zn, Cd and Cu are indeed moving from the soil to cocoa leaves and beans. Furthermore, the results show that the concentrations of toxic metals in the different parts of the cocoa fruit and plant, particularly in the cocoa bean, which is used for chocolate manufacture, are not higher than those regulated by FAO food standards, as is the case of Cd, which is limited to 0.2 mg Cd/kg and in the samples analyzed does not exceed this limit. Even though the concentration of these metals does not exceed the safety standard, the presence of these potentially hazardous metals, and the fact they are absorbed by this important local crop, are worrying for the long-term sustainability of cocoa cultivation in the area. Therefore, it is fundamental to monitor the local environment, understanding the distribution of heavy metal pollution, and work with the local authorities in landscape management to minimize the exposure of crops to ASGM pollution.

Keywords: cocoa; contaminant mobility; contamination; health risk; toxic metals

1. Introduction

Cocoa (*Theobroma cacao*), an ancestral crop widespread around the American tropical countries, presents a historical and cultural richness. Initial references date from the first years after the arrival of the Spaniards, hence its importance for farmers and for trade [1].

Ecuador is a cocoa producing country with more than 240 years in the international market. Cocoa beans export significantly and have contributed to the country's economy, especially in its first and second "Cocoa Boom" [2]. It is currently the fourth cocoa producing country in the world, with 300,000 tons per year [3].

The environmental impact of gold extraction generated by artisanal and small gold mining (ASGM) is a global concern. Thus, several countries are trying to mitigate the environmental and health problems that arise, particularly if mercury amalgamation is used for gold purification [4]. In Ecuador, gold mining has existed since pre-Columbian times [5] and is currently carried out in a small scale, which varies between artisanal mining related to illegal practices to small scale formal mining [6]. The migration process of artisanal miners to the Ponce Enrique mining district was a boom in the 1980s, which triggered a concern for its potential detrimental effects and motivated a search for responsible and environmentally sustainable activities [5].

Since mining is seen as an activity that generates significant economic income for a community, preventing its impact on water and soil quality, biodiversity, and the basis of human life, should be a priority [7]. The main resource affected is water, because mining modifies the composition and quality of local rivers, causing an increase of dissolved solids, metalloids and cations concentrations rendering it unsafe for human use, especially since these same water sources are used for irrigation and human consumption [8,9].

This study focuses on one of the main crops within the study area, cocoa. In order to evaluate the sustainability of cocoa cultivation in the area, we started with the identification of environmental and anthropic factors that affect production, such as land use, the availability of irrigation water, and other economic activities that are carried out in the area [10]. One of those activities is artisanal and small-scale metallic mining, which generates significant environmental impacts [9,11]. Moreover, there is evidence that gold mining in the Ponce Enriquez county has led to the contamination of the surface water systems in the area [9,12], generating problems in the quality of the water. Use of this water becomes potentially dangerous for human and environmental health, generating concerns for the sustainability of cocoa and other crops production.

The permanent demand for food, water and ecosystem services induces impacts on the structure, distribution and functioning of resources, generating constant pressure due to human activities, causing a deterioration of the ecosystem, and jeopardizing the availability of food [13]. Therefore, considering a major local crop and the potential presence of contaminants is an interesting approach for understanding its long term sustainability, and to start understanding the sustainability of the local economic fabric.

There are international standards, such as CODEX STAN 193-1995, that regulate the content of metals in cocoa, such as lead and cadmium. Moreover, the European Union standard regulation N° 488/2014 came into force on 1 January 2019, which includes limits for metal concentrations in chocolate, one of cocoa's main products. [14]. Therefore, metal concentration analysis became important for cocoa exports. Furthermore, there is evidence that other metals such as mercury or magnesium, which are known to cause detrimental health effects, can be absorbed by the cocoa plant and end up in its products, consequently with potential health risks [15].

In general, for an analysis of major and trace elements of natural or anthropogenic origin, those classified as toxic, which are related to potential health risks, such as Hg related to Minamata syndrome, and As related to skin keratosis, etc., will be used [16].

In studies in Indonesia, the concentration of Hg in soil in the mining area is classified as critical (Hg > 0.3 ppm), as well as in plant tissue (Hg > 0.03 ppm). Strategic regulation is therefore recommended to protect people from mercury exposure, to maintain agricultural activity and to minimize environmental damage and threats from mining production [17].

This study aims to assess the sustainability of cocoa (*Theobroma cacao*) cultivation in the mining district of Ponce Enriquez. Understanding the sustainability of cocoa cultivation is essential when analyzing economic activities in Ponce Enriquez because two activities, mining and agricultural production (cocoa and banana), are the basis of the local economy and contribute to the development of the community.

2. Materials and Methods

2.1. Study Area

The present research was carried out in the Azuay province, in the mining district of Camilo Ponce Enriquez. The simple random model was used for sampling distribution. Eighteen productive cocoa farms were identified and sampled (Figure 1). The identified main water basin that would influence cocoa crops in the sampled areas is the Gala River. The Chico River and Siete River, which are the main tributaries to the main basin, are directly affected by ASGM activities.

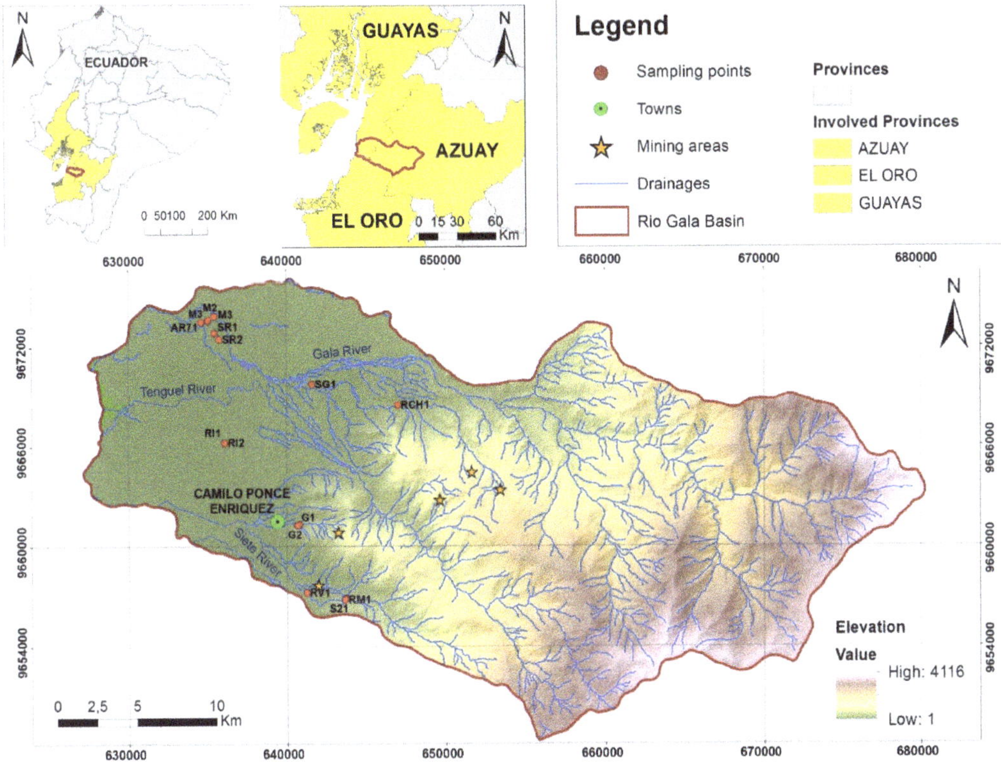

Figure 1. Macro location of sampling points. Mines near the sampling site. Sampling sites.

The study area has a tropical climate. The average annual temperature is 27° and the average annual rainfall is 364 mm, the average humidity is 80% and the UV Index is 6. There are two main seasons, rainy (October to April) and dry (May to September) [12].

2.2. Field Sampling

The cocoa farms are scattered throughout the Camilo Ponce Enriquez mining district. Sampling was conducted in the months of March and July 2019. Each sampling site was identified and georeferenced in situ. Samples were taken from 18 farms identified in a random sampling, as indicated in Figure 2. Soil samples were taken near the sampled plant at 20 cm and 50 cm depth with a drill. These samples were stored in resealable bags and kept refrigerated (4 °C). The fruit samples were further separated into three subsamples: almonds (defined as the cocoa bean without the husk), shell and husk. These samples were frozen at −80 °C, together with leaves from each sampled plant.

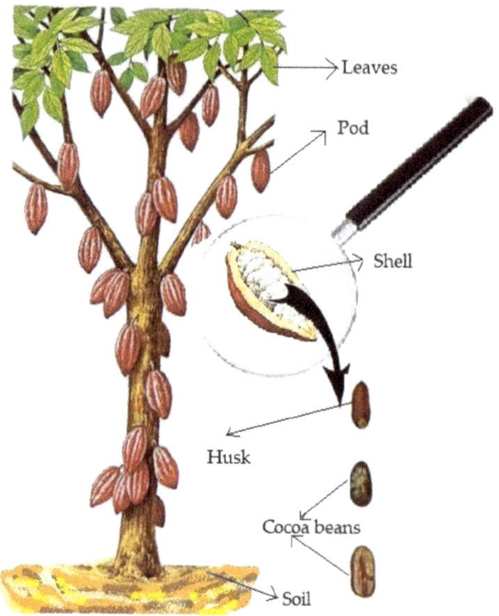

Figure 2. Parts of cocoa plant sampled.

2.3. Laboratpry Analysis

All the samples were freeze-dried (LEYBOLD freeze drier, model ELEKTROVER GT2) in a process lasting 72 h. The dried samples were crushed in an agate mortar, which was previously and thoroughly cleansed to avoid cross contamination. The crushed and homogenized material was used for subsequent acid digestion in a microwave oven (MILESTONE ETHOS UP) using 500 mg of sample with 9 mL of nitric acid (65% w/w, Fisher Chemical, Waltham, MA, USA), and 1 mL of hydrogen peroxide. The samples were gauges to 50 mL with type I water and stored in Falcon tubes at 4 °C.

Soil samples were dried at room temperature (19 °C on average) for 4 weeks. The dried samples were pulverized and then acid digested in a microwave oven (MILESTONE ETHOS UP). Acid digestion was performed by mixing 150 mg of homogenized soil sample with 6 mL of HNO_3 (65% w/w, Fisher Chemical) and 2 mL of hydrochloric acid HCl (37% w/w, Merck, Rahway, NJ, USA). The extract obtained was gauged to 50 mL with type I water. The extract was transferred to a falcon tube and stored under refrigerated storage at 4 °C.

In the case of mercury analysis, it was performed by means of cold vapor atomic fluorescence spectroscopy. Major and trace metals analysis were performed in an Agilent 7500 ce ICP-MS (Agilent Technologies, Inc., Santa Clara, CA, USA). Interferences were minimized by collision or reaction with gas in a collision cell. Indium (10 ng/L equivalent concentration) was used as internal standard.

Quality control was carried out using the EPA Contract Laboratory Program Statement of Work for Superfund Analytical Methods (Multi-Media, Multi-Concentration), SFAM01.0 Exhibit D Inorganic Methods, Inductively Coupled Plasma—Mass Spectrometry Metals Analysis, released in May 2019.

The samples used for QA/QC (quality assurance/quality control) included an initial calibration blank (ICB), initial calibration verification (ICV), CRQL check standard (CRI), continuing calibration verification (CCV), continuing calibration blank (CCB) and interference check sample (ICS). For each 10 samples, a duplicate, spike, spike duplicated, serial dilution, CCV and CCB were run.

2.4. Data Analysis

All statistical tests were conducted using IBM SPSS Statistics 25.0 Core System, EEUU. The SPSS (Statistical Package for the Social Sciences) software was used for statistical analysis on the concentrations for each sample (soil at 20 cm, soil at 50 cm, leaves, husks, shells and cocoa beans). In order to evaluate if there is a significant difference between the soil samples, an analysis of the means was performed. Metal mobility through the plant and fruit was determined by linear correlation through the different plant parts along a metabolic gradient, going from the leaves to the most internal parts of the fruit.

For the mobility analysis, the transfer factor (TF) was used. Equation (1) describes TF as the ratio between metal concentration in the cocoa beans or leaves and the concentration obtained in the soil sample. If TF is greater than 1, it means that the metal has the capacity to move from the soil to the aerial part of the plant being analyzed [18]:

$$TF = \frac{C_{plant}\left(\frac{mg}{kg}\right)}{C_{soil}\left(\frac{mg}{kg}\right)} \tag{1}$$

where

C_{plant} = Metal concentration in different parts of the plant.

C_{soil} = Concentration of metal in the soil.

In order to estimate the risk to human health the chronic daily intake index (CDI) was calculated using Equation (2) [17]:

$$CDI = C\left(\frac{\mu g}{L}\right) * \frac{DI\left(\frac{L}{day}\right)}{BW\ (kg)} \tag{2}$$

where

C = Concentration of the metal of interest,

DI = Average daily intake

BW = Average body weight

To calculate the chronic daily intake (CDI), the average body weight reported by [19] are used (70 kg for adults and 15 kg for children). Similarly, the average daily intake used is 2 L/day for adults and 1 L/day for children. The units of measurement are expressed in L considering that the density of the liquid resulting from acid digestion is 1 kg/L.

The risk ratio HQ (Equation (3)) is calculated, where RfD represents the reference dose of toxicity of an element. When calculating the HQ value, if this value is less than 1, the population is healthy and safe, regarding metals such as Cd, Pb, Cr and Fe [17]. This ratio is relevant as it helps us to identify metals that potentially become health hazards:

$$HQ = \frac{CDI\left(\frac{\mu g}{kg*day}\right)}{RfD(mg/(kg*day))} \tag{3}$$

where

CDI = Chronic daily intake

RfD = Toxicity reference dose of an element

3. Results

3.1. Metal Content

Our analysis focused on Hg, Cd, As, Cu, Pb and Zn. These elements were chosen considering they have been determined as contaminants generated from mining activity and classified as potentially toxic [16]. Figure 3 plots these metals in a range of 0 to 1.2 ppm concentration.

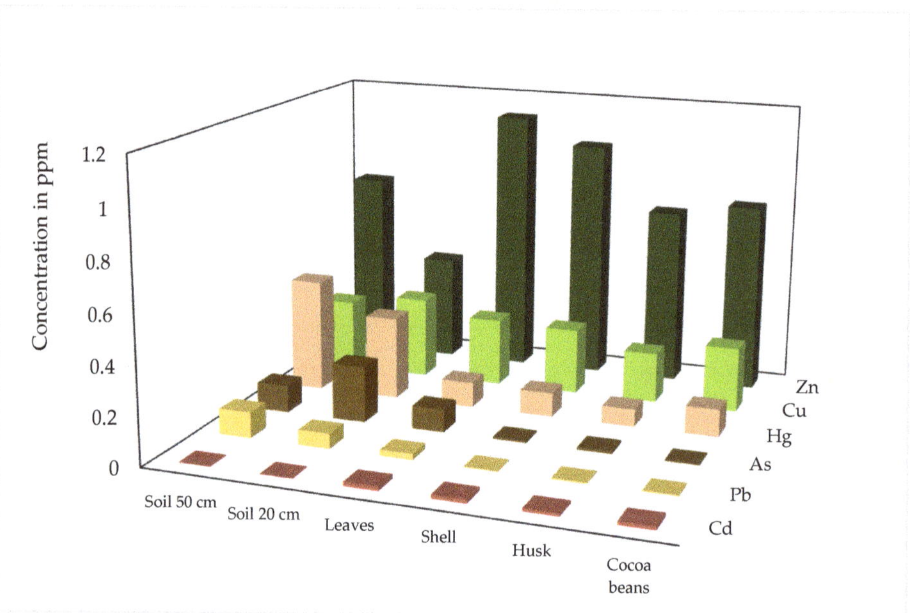

Figure 3. Metal content in soil and different parts of the cocoa plant.

Metals such as Hg and Cd present greater content in cocoa beans than in the husk, leaves and shell, generating concerns for their presence in processed products such as chocolate (Figure 3).

In the case of Cu, concentration values are maintained throughout the crop system (Figure 3). Although its concentration is not higher than the safety standards, it should be monitored regularly because at high concentrations in cocoa beans it can cause different alterations in plants and human health, such as a decrease in growth rate, hypochromic anemia and diarrhea, among others [20].

Zn is a micronutrient needed for plant growth that directly affects the quality of cocoa beans [21]. Its distribution in the plant components is not uniform, with higher concentrations in leaves and lower concentrations in surface soils (Figure 3).

The concentration of these metals (Zn, Cu, Hg, As, Pb and Cd) in the leaves of the crop highlights the importance of management practices. Regular pruning is practiced during agricultural work, and this plant material is deposited around the plant itself. Thus, the absorbed metals may be incorporated back into the soil and the plant itself.

3.2. Mobility Analysis

To analyze the mobility of metals within the plant, the transfer factor (TF) is calculated. The transfer factor varies from 0 to approximately 3.5 between cocoa beans and the soil (Figure 4). Metals presenting mobility capacity are mainly Cd, Zn and Cu. Hg showed high mobility in only one sample.

It is important to analyze mobility between the soil and the cocoa beans because the soil is consumed and becomes the medium for the metals to enter the organism, and during the production cycle the cocoa beans remain on the plant for a short time, which indicates how quickly the metals are mobilized in the plant.

Regarding the transfer factor calculated from the concentration of the leaves and the soil, Figure 5 shows that mainly Zn, Cd and Cu could move from the soil to the leaves. These values of the transfer of metals to the leaves are used as an alternative for a bioremediation proposal. Considering that metals accumulate in leaves, this plant material can be pruned and removed to prevent the transferred metals from returning to the soil, as Pb and Hg seem to be mobilized in few of the samples from the soil to the leaves.

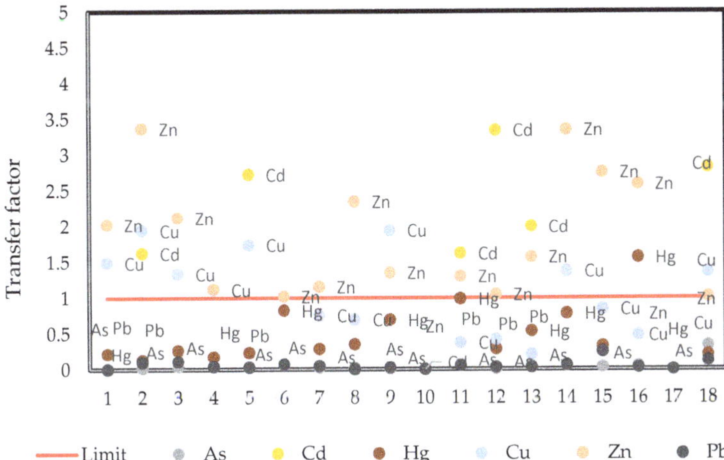

Figure 4. Transfer factor for cocoa beans-soil. The red line represents the limit above which metals are observed to be able to be transferred in the plant.

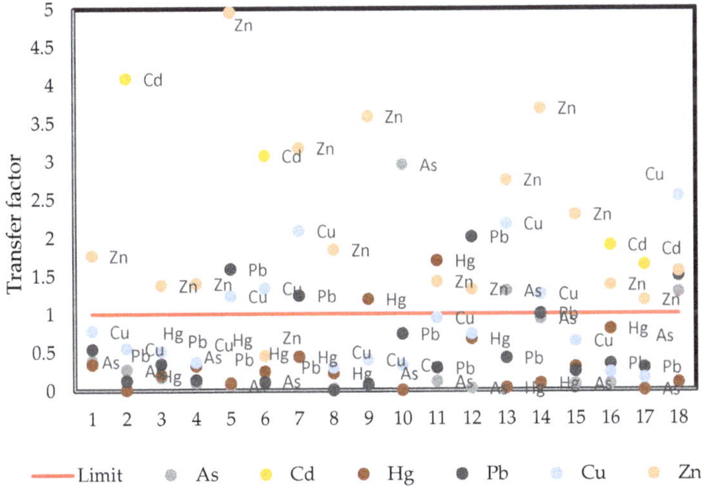

Figure 5. Transfer factor for leaves-soil. The red line represents the limit above which metals are observed to be able to be transferred in the plant.

The maximum permitted concentration limits of metals for plant material, according to the FAO, establish standards for Cd (0.2 mg/kg) and Zn (99.4 mg/kg) [22].

With regards to Cd (Figure 6), the concentration in the culture system is below the permitted level (0.2 mg Cd/kg) [22] for the samples analyzed. Although the concentrations are below the permitted level, they should be monitored to prevent anthropogenic conditions from altering these levels and making the concentration toxic to humans. The observed higher Cd concentration in cocoa beans than in the soil, as well as in the case of the leaves, allow us to conclude that Cd is being bioaccumulated during its productive cycle in these parts of the plant.

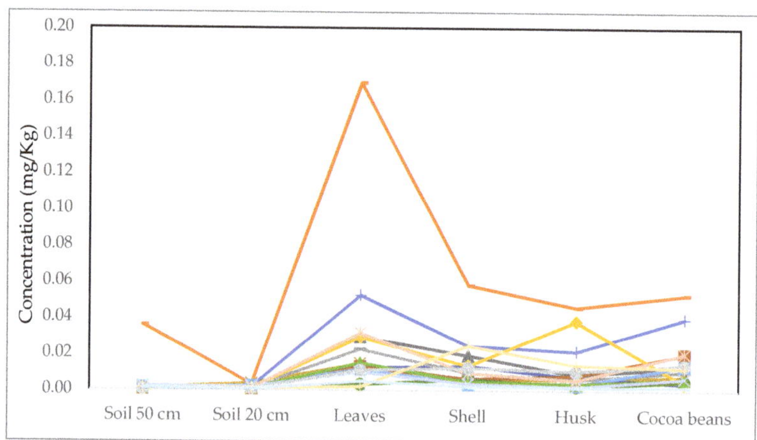

Figure 6. Cd content in soil and different parts of cocoa plant, the concentration is observed for 18 samples. The limit is 0.2 mg/kg, and no sample exceeds this level.

The concentration of Zn in the crop system (Figure 7) is below the permitted level for this metal (99.4 mg Zn/kg) [22]. It is observed that for some cases the concentration in the cocoa beans is higher than in the soil, as in the case of the leaves, concluding that it is being bioaccumulated during its productive cycle in these parts of the plant.

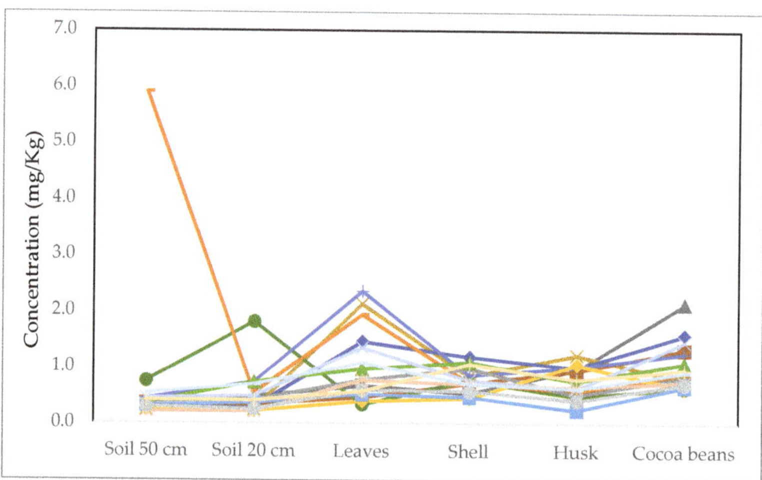

Figure 7. Zn content in soil and different parts of cocoa plant, the concentration is observed for 18 samples. The limit is 99.4 mg/kg, and no sample exceeds this level.

However, previous studies show that Zn is an element that moves little in the plant but has many important functions. The structure and functionality of many enzymes depends on the presence of Zn in plants, so the concentration present in the plant does not affect it [23].

3.3. Risk Analysis

Table 1 presents the data obtained from the risk analysis of the calculated metals of interest. Based on the risk ratio for the non-cancer risk, HQ (hazard quotient), the oral toxicity reference dose, RfD, according to the USEPA were considered [17] for the different metals. Results show HQ values less than one in all cases, which indicate that the exposed population can be considered safe from a non-carcinogenic risk, when ingesting the metal concentration found in the cocoa beans. For the calculation of chronic daily intake (CDI); the average body weight reported by [19] for adults is 70 kg and for children 15 kg is used. For the case of average daily intake was used for adults 2 L/day and for children 1 L/day.

Table 1. Hazard quotients and total hazard index for heavy metals.

	Cd	Pb
$\overline{C}$ (µg/L)	1.31×10^{-5}	3.56×10^{-6}
IDU Adults	1.07×10^{-5}	2.91×10^{-6}
CDI Children	5.84×10^{-5}	1.58×10^{-5}
RfD (mg/(kg-day))	5×10^{-4}	3.5×10^{-3}
HQ Adults	2.14×10^{-2}	8×10^{-4}
HQ kids	11.67×10^{-2}	4.5×10^{-3}

4. Discussion

The metal concentrations analyzed were not higher than those established in food safety standards. However, frequent monitoring should be carried out to ensure that those concentrations do not increase and become a health problem. Specifically, the concentration of toxic metals in cocoa beans, although they do not exceed the permitted toxic levels, should be monitored, mainly because during the transformation process to chocolate, they might increase. This is similar to what happens in the case of Pb in rice, where levels increased during processing [24]. Additionally, the concentration in all waste material generated in cocoa processing should be monitored, specifically in cocoa husks, because they are used as raw material for packaging, animal feed and in some cases as an alternative source of fiber for food [25]. So, they may become a potential hazard for consumers if the toxic metal concentrations exceed the permitted levels after transformation.

Considering that Cd is regulated for chocolate, the main processed product of cocoa, it is necessary to generate alternatives to reduce the absorption of this metal by the plant. A common agricultural task is pruning; this activity consists of removing plant material and placing it as soil cover near the plant, helping the crop due to the flow of mineral nutrients produced in the biogeochemical cycles present in natural or agroforestry systems and reducing evaporation [26]. As a natural phytoremediation mechanism, it is recommended to remove the leaves when pruning and not to leave them on the soil, to avoid Cd and other metals being released back to the substrate where the plants are growing. This removed plant material should be treated, considering its concentration of heavy metal concentration. An alternative within our study area is to deposit the material in the tailing ponds managed by ASGM projects, where it could go through composting processes as a pretreatment, then be incinerated, which will significantly reduce the contaminated biomass [27]. However, a balance must be obtained between removing toxic metals present in the leaf litter and taking advantage of the nutrients. Furthermore, phytoremediation processes must be continually evaluated because they may suffer interference in their efficacy due to the appearance of new contaminants from emerging technologies containing trace metals [28].

The presence of toxic metals in soil, water and air has different origins, the natural ones need to be monitored to assess their influence on the environment; but those of anthropogenic origin such as those generated in industries, mining activities, agriculture, etc., generate an urgent need to reduce the amount, before it is too late. Therefore, sustainability measures should focus on reducing the use of fertilizers and pesticides [29].

The analysis of the composition and concentration of toxic metals of natural origin depends on the type of rock, as well as the environmental conditions that cause weathering processes. Metals from the geological environment present in the plant matter can be Cd, Pb, Co, Mn, Cr, Zn, Cu, Ni, Hg and Sn [30]. Therefore, it is suggested to evaluate the origin of Cd, Pb and other metals, that are observed to be mobilized in the plant. Metal content in water in highly mineralized areas, such as the Camilo Ponce Enriquez mining district, are expected to be naturally high, because of natural geochemistry and the weathering process [31]. Therefore, the high metal content in the soil and corps may also be affected by water irrigation.

The major concern regarding the presence of heavy metal pollution in the different environmental compartments in the area is worrisome, from a public health perspective, due to a potential bioaccumulation and long-term toxicity. Orchard plants, such as spinach, tomatoes and peppers are prone to incorporate toxic metals into their tissues, becoming a risk for local farmers [31].

Even though toxic metals and metalloids are silently present in all environmental compartments, their detrimental effects can be felt through generations and even be exported as part of a crop like cocoa beans. Monitoring and controlling the environmental problems, in areas such as Camilo Ponce Enriquez, could make it possible to speak of a coexistence between agriculture and mining [32]. High impact activities such as mining are keystones of our modern way of life. Therefore, understanding its effects on complex land use matrixes, such as our study area, opens the possibility to test the management and technological alternatives to achieve long term sustainability, which would support the local economy and the community well-being.

5. Conclusions

The analysis of toxic metals is important because of their bioaccumulation. Bioaccumulation implies an increase in the concentration of an organism over time, relative to the concentration in the environment.

The concentration of the metals analyzed were not higher than those mentioned in food standards, however, frequent monitoring should be carried out to prevent the metal presence becoming a health problem.

Of the metals analyzed, Cd was found to be mobile within the plant. As part of the agricultural work, it is recommended to remove the leaves when pruning and not leave them on the ground, as a natural phytoremediation mechanism.

Caring for the environment is important for the future of mankind, but it is crucial that remediation measures are continually being evaluated to ensure they work effectively and help to care for natural resources.

The occurrence of toxic metals in soil, water and air has different origins; the natural ones need to be evaluated on the influence on the environment and those of anthropogenic origin create an urgent need to reduce them before it is too late to achieve crop sustainability.

From the mobility and risk factor values, it is observed that Zn, Cd and Cu move from the soil to leaves and cocoa beans. In addition, the results show that there are concentrations of toxic metals in different parts of the cocoa fruit and plant, particularly in the cocoa bean. This generates concern for the cocoa processing industry, because their products (chocolate) could have concentrations higher than those regulated by FAO food standards, due to bioaccumulation.

Soil concentrations and external conditions (water source, suspended solids in the air) should be monitored to try to determine the origin of increased concentrations of these metals in cocoa beans, compared to the soil.

Author Contributions: Conceptualization, C.R. and J.R.; Data curation, C.R.; Formal analysis, C.R., J.R. and J.L.R.-P.; Investigation, C.R., J.R., J.L.R.-P. and X.D.; Visualization, C.R.; Writing—original draft, C.R.; Writing—review & editing, C.R., J.R., X.D. and X.D.; visualization, M.S. All authors have read and agreed to the published version of the manuscript.

Funding: This research was funded by the PIMI 16-05, and PIE-DPET-PNUD-2020 projects carried out at Escuela Politécnica Nacional. It was partially financed by VLIR-UOS TEAM project EC2018TEA461A105.

Institutional Review Board Statement: Not applicable.

Informed Consent Statement: Not applicable.

Data Availability Statement: Even though the data is not publicly available it can be requested to the corresponding author.

Acknowledgments: To the DECAB and DEMEX laboratories in which this research was carried out and who facilitated the use of equipment, reagents and materials. We want to express our gratitude to Bella Rica Cooperative and Somilor S.A. al the Camilo Ponce Enríquez mining district, as well as the local partners that made this work possible. We would like to thank of the University of Utah, USA, Department of Geology and Geophysics, for technical support. This research was funded by the Research Institute for Humanity and Nature (RIHN; a constituent member of NIHU), Project No. RIHN 14200102.

Conflicts of Interest: The authors declare no conflict of interest.

References

1. Zúñiga, A. Riqueza, prestigio y ofrenda divina: Los usos del cacao en el mundo nahua. *Hosp. ESDAI* **2020**, *37*, 35–54.
2. Vassallo, M. Diferenciación y agregado de valor en la cadena ecuatoriana del cacao. *Inf. Investig.* **2016**, *1*, 23–43.
3. Abad, A.; Acuña, C.; Naranjo, E. El cacao en la Costa ecuatoriana: Estudio de su dimensión cultural y económica. *Estud. Gestión. Rev. Int. Adm.* **2020**, *7*, 59–83.
4. Gunson, A.J.; Marcello, M. Mercury and artisanal mining in China. *Environ. Pract.* **2004**, *6*, 109–120. [CrossRef]
5. Lozada, D. La Conminución en la Minería Pequeña y Artesanal en Ecuador. Available online: https://www.researchgate.net/publication/292983759 (accessed on 3 May 2022).
6. Ecuador, E.L.; Tributario, R.A.; De, R.R.; Interno, R.T.; Los, R.R. *Ley Reformatoria Para la Equidad Tributaria del Ecuador*; Asamblea Constituyente, 2008. Available online: https://www.gob.ec/sites/default/files/regulations/2018-11/ley%20equidad.pdf (accessed on 3 May 2022).
7. Betancourt, Ó.; Barriga, R.; Davée, J.; Cueva, E.; Betancourt, S. Impacts on Environmental Health of Small-Scale Gold Mining in Ecuador. *Ecohealth Res. Pract.* **2012**, *1*, 119–130.
8. Evaluación del Marco de Políticas Mineras del IGF. Available online: https://www.iisd.org/system/files/publications/ecuador-mining-policy-framework-assessment-es.pdf (accessed on 12 April 2022).
9. Carling, G.T.; Diaz, X.; Ponce, M.; Perez, L.; Nasimba, L.; Pazmino, E.; Rudd, A.; Merugu, S.; Fernandez, D.P.; Gale, B.K.; et al. Particulate and dissolved trace element concentrations in three southern Ecuador rivers impacted by artisanal gold mining. *Water Air Soil Pollut.* **2013**, *224*, 1415. [CrossRef]
10. Duguma, B.; Gockowski, J.; Bakala, J. Desafíos biofísicos y oportunidades para el cultivo sostenible de cacao (Theobroma cacao Linn.) en sistemas agroforestales de Africa Occidental y Central. *Rev. Agroforestería Américas* **1999**, *6*, 7–10.
11. Pantoja, F.H.; Pantoja, S.D. Problemas y desafíos de la minería de oro artesanal y en pequeña escala en Colombia. *Rev. Fac. Cienc. Económicas* **2016**, *24*, 147–160.
12. Velásquez-López, P.C.; López, I.Y.; Rivera, M.F. Estimation of the ecological and human health risk of mercury in a mangrove area of the La Puntilla estuary, El Oro province, southern Ecuador. *Bol. Investig. Mar. Costeras* **2020**, *49*, 81–100. [CrossRef]
13. Challenger, A.; Dirzo, R. Tendencias de cambio y estado de la biodiversidad, los ecosistemas y sus servicios. *Factores Cambio Estado Biodivers.* **2009**, *11*, 205–210.
14. Barraza, F.; Schreck, E.; Lévêque, T.; Uzu, G.; López, F.; Ruales, J.; Prunier, J.; Marquet, A.; Maurice, L. Cadmium bioaccumulation and gastric bioaccessibility in cacao: A field study in areas impacted by oil activities in Ecuador. *Environ. Pollut.* **2017**, *229*, 950–963. [CrossRef]
15. Casteblanco, J.A. Heavy metals remediation with potential application in cocoa cultivation. *Granja* **2018**, *27*, 21–35. [CrossRef]
16. Williams, M. Algunas aplicaciones de la información geoquímica de la cordillera occidental del Ecuador. In *Geoquímica y Ambiente*, 1st ed.; UPC PRODEMINCA: Guayas, Ecuador, 2001; pp. 82–86.
17. Muhammad, M.; Shah, T.; Khan, S. Health risk assessment of heavy metals and their source apportionment in drinking water of Kohistan region, northern Pakistan. *Microchem. J.* **2011**, *98*, 334–343. [CrossRef]

18. Mesa-Pérez, M.; Díaz-Rizo, O.; Tavella, M.; Baqué, D.; Sanchez-Pérez, J. Soil-to-Plant Transfer Factors of Rare Earth Elements in Rice (*Oryza sativa* L.). *Rev. Cienc. Técnicas Agropecu.* **2018**, *27*, 2071-0054.

19. Onyele, O.; Anyanwu, E. Human Health Risk Assessment of Some Heavy Metals in a Rural Spring, Southeastern Nigeria. *Afr. J. Environ. Nat. Sci. Res.* **2018**, *1*, 15–23.

20. Londoño, L.; Londoño, P.; Muñoz, F. Los Riesgos de los Metales Pesados en la Salud Humana y Animal. *Biotecnoloía Sect. Agropecu. Agroind.* **2016**, *14*, 145–153. [CrossRef]

21. De Araujo, Q.R.; Baligar, V.C.; Loureiro, G.D.A.; de Souza Júnior, J.O.; Comerford, N.B. Impact of soils and cropping systems on mineral composition of dry cacao beans. *J. Soil Sci. Plant Nutr.* **2017**, *17*, 410–428. [CrossRef]

22. Mensah, E.; Kyei-Baffour, N.; Ofori, E.; Obeng, G. Influence of human activities and land use on heavy metal concentrations in irrigated vegetables in Ghana and their health implications. *Appropr. Technol. Environ. Prot. Dev. World* **2009**, *1*, 9–14.

23. Cuenca, G.; Cáceres, A.; Oirdobro, G.; Hasmy, Z.; Urdaneta, C. Las micorrizas arbusculares como alternativa para una agricultura sustentable en áreas tropicales. *Interciencia* **2007**, *32*, 23–29.

24. Román-Ochoa, Y.; Choque, G.; Tejada, T.; Yucra, H.; Durand, A.; Hamaker, B. Heavy metal contamination and health risk assessment in grains and grain-based processed food in Arequipa region of Peru. *Chemosphere* **2021**, *274*, 129792. [CrossRef]

25. Handojo, L.; Triharyogi, H.; Indarto, A.; Indarto, A. Residuos de cáscara de grano de cacao como materia prima potencial para el polvo de fibra dietética. *Int. J. Recycl. Org. Waste Agric.* **2019**, *8*, 485–491. [CrossRef]

26. Hosseini, S.; Gallart, M.; Singh, K.; Hannet, G.; Komolong, B.; Yinil, D.; Field, D.; Muqaddas, B.; Wallace, H. Leaf litter species affects decomposition rate and nutrient release in a cocoa plantation. Agriculture. *Ecosyst. Environ.* **2022**, *324*, 107705.

27. Ghosh, M.; Singh, S. A review on phytoremediation of heavy metals and utilization of it's by products. *Asian J. Energy Environ.* **2005**, *6*, 214–231.

28. Guerra Sierra, B.E.; Muñoz Guerrero, J.; Sokolski, S. Phytoremediation of Heavy Metals in Tropical Soils an Overview. *Sostenibilidad* **2021**, *13*, 2574. [CrossRef]

29. Ntiamoah, A.; Afrane, G. Environmental impacts of cocoa production and processing in Ghana: Life cycle assessment approach. *J. Clean. Prod.* **2008**, *16*, 1735–1740. [CrossRef]

30. Sandeep, G.; Vijayalatha, K.R.; Anitha, T. Metales pesados y su impacto en cultivos hortícolas. *Rev. Int. Estud. Químicos* **2019**, *7*, 1612–1621.

31. Kacmaz, H. Assessment of heavy metal contamination in natural waters of Dereli, Giresun: An area containing mineral deposits in northeastern Turkey. *Environ. Monit. Assess.* **2020**, *192*, 91. [CrossRef]

32. Malone, A.; Smith, N.; Zeballos, E. Coexistence and conflict between artisanal mining, fishing, and farming in a Peruvian boomtown. *Geoforum* **2021**, *120*, 142–154. [CrossRef]

International Journal of
Environmental Research and Public Health

Article

Economic Impacts on Human Health Resulting from the Use of Mercury in the Illegal Gold Mining in the Brazilian Amazon: A Methodological Assessment

Leonardo Barcellos de Bakker [1,*], Pedro Gasparinetti [2], Júlia Mello de Queiroz [3] and Ana Claudia Santiago de Vasconcellos [4]

1   Leonardo B. Bakker Assessoria, São Clemente Street, Rio de Janeiro 254, Rio de Janeiro 22260-004, Brazil
2   Conservation Strategy Fund, Av. Churchill 129, Rio de Janeiro 20020-050, Brazil; pedro@conservation-strategy.org
3   Julia Queiroz Consultoria Desenvolvimento Verde, Maria Angelica Street, Rio de Janeiro 382, Rio de Janeiro 22461-152, Brazil; julia.melloqueiroz@gmail.com
4   Laboratory of Professional Education in Health Surveillance, Joaquim Venâncio Polytechnic School of Health, Oswaldo Cruz Foundation, Rio de Janeiro 21040-900, Brazil; ana.vasconcellos@fiocruz.br
*   Correspondence: leonardo.bakker@gmail.com

Citation: de Bakker, L.B.; Gasparinetti, P.; de Queiroz, J.M.; de Vasconcellos, A.C.S. Economic Impacts on Human Health Resulting from the Use of Mercury in the Illegal Gold Mining in the Brazilian Amazon: A Methodological Assessment. *Int. J. Environ. Res. Public Health* **2021**, *18*, 11869. https://doi.org/10.3390/ijerph182211869

Academic Editors: Masayuki Sakakibara, Win Thiri Kyaw and José Luis Rivera Parra

Received: 23 September 2021
Accepted: 9 November 2021
Published: 12 November 2021

Publisher's Note: MDPI stays neutral with regard to jurisdictional claims in published maps and institutional affiliations.

Abstract: Artisanal small-scale gold mining (ASGM) in the Amazon results in the dumping of tons of mercury into the environment annually. Despite consensus on the impacts of mercury on human health, there are still unknowns regarding: (i) the extent to which mercury from ASGM can be dispersed in the environment until it becomes toxic to humans; and (ii) the economic value of losses caused by contamination becomes evident. The main objective of this study is to propose a methodology to evaluate the impacts of ASGM on human health in different contexts in the Brazilian Amazon. We connect several points in the literature based on hypotheses regarding mercury dispersion in water, its transformation into methylmercury, and absorption by fish and humans. This methodology can be used as a tool to estimate the extent of environmental damage caused by artisanal gold mining, the severity of damage to the health of individuals contaminated by mercury and, consequently, can contribute to the application of fines to environmental violators. The consequences of contamination are evaluated by dose-response functions relating to mercury concentrations in hair and the development of the following health outcomes: (i) mild mental retardation, (ii) acute myocardial infarction, and (iii) hypertension. From disability-adjusted life years and statistical life value, we found that the economic losses range from 100,000 to 400,000 USD per kilogram of gold extracted. A case study of the Yanomami indigenous land shows that the impacts of mercury from illegal gold mining in 2020 totaled 69 million USD, which could be used by local authorities to compensate the Yanomami people.

Keywords: mercury; methylmercury; artisanal small-scale gold mining; Amazon; human health; economic valuation

1. Introduction

The Minamata disaster in Japan, in which thousands of people were seriously impacted by mercury dumped by various industries, culminated in the Minamata Convention in 2013. As a result, mercury's use has been restricted [1] and it is now considered by the World Health Organization (WHO) as one of the six most dangerous substances to health due to its high toxicity and the risks it poses to human health and the environment [2].

Mercury is a heavy metal widely distributed across the planet and is therefore classified as a global pollutant [3,4]. This metal can be found in nature in three main chemical forms: ionic mercurial forms (e.g., Hg[II]), in its elemental form (e.g., Hg0), and in organomercurial forms (e.g., methylmercury) [5]. Although all mercurial forms have the potential to cause toxic health effects on people, methylmercury is the most dangerous [6]. This organomercurial species affects the central nervous system, causing neurobehavioral effects, motor coordination disorders, and cardiovascular diseases [7–9]. Since it chronically affects the population, its effects can arise over many years and cause severe damage to an entire generation. This mercurial form is especially harmful to pregnant women due to the fact that the fetal brain is more sensitive to the action of methylmercury, causing many neurodevelopment problems to occur including mental retardation, learning delays, visual and auditory alterations, and other harmful effects [10–12].

Despite its damage to human health, mercury is still widely used in legal and illegal gold mining in Brazil, an activity that has been growing every year due to the high gold price and lack of inspections [13]. The situation becomes even more serious since this increase is largely concentrated in indigenous areas, mainly affecting the Yanomami and Munduruku traditional territories [14,15]. Gold mining uses mercury during the amalgamation process, which unites mercury with gold. Although much of the mercury is reused in the process, some is lost and is dispersed in rivers, soils, and the atmosphere [16–20]. The Brazilian Ministry of the Environment estimated that, in 2016, between 18.5 and 221 tons of mercury were lost during gold mining in Brazil (in the form of both emissions to the atmosphere and release in rivers and soils) [21].

While the repression of illegal activities can imply the imposition of fines related to damages, there are still important methodological bottlenecks to establish responsibilities, and no standardized approach to evaluate damages. Methodological bottlenecks depend on several factors. Firstly, it is necessary to differentiate between natural mercury (i.e., particles mineralized present in soil and sediment) and additional mercury from human activities (i.e., mercury that is intentionally released in the environment) [22–27]. Secondly, there are other potential sources of mercury, such as deforestation from agriculture and cattle ranching [28–34]. Although some studies argue that these activities can have a greater aggregate impact on mercury release into the environment than small-scale gold mining [35–43], global statistics indicate that 30% of mercury in the environment results from anthropogenic activities, of which small-scale gold mining accounts for 37% of all releases and is a major source of contamination [44]. The third factor is related to the difficulty in attributing responsibility to specific mines. Although there is evidence of increased mercury concentration in the Amazonian population, this increase is generated by the combined effect of several illegal gold mines. Moreover, besides the mines existing today, there are mines that ceased to exist decades ago, leaving a cumulative impact [45,46]. Therefore, the accountability of specific gold mines has become a great challenge given the complexity of the mercury cycle.

The mercurial form used in the gold ore extraction process is metallic mercury, also known as elemental mercury (Hg^0). The fraction of metallic mercury that is not recovered during the extraction process contaminates the atmosphere and rivers in the Amazon region. Once released into aquatic systems, a part of metallic mercury is oxidized and can be methylated by the action of microorganisms or abiotic factors. This process gives rise to the most dangerous mercury species, methylmercury (MeHg). Methylmercury is biomagnified along the aquatic food chain, contaminating fish and other organisms used for food such as turtles, crabs, shrimp and alligators. Furthermore, much of the toxicity of methylmercury is due to its high neurotoxic potential and its ability to overcome the blood-brain and placentary barriers.

A vast body of literature has analyzed the increase in contamination levels in the Amazon population [47–58]. Vasconcellos et al. [49] found an average methylmercury hair level of 7.0 µg/g in the Munduruku indigenous community in Tapajós and Vega et al. [58] observed that the Yanomami indigenous community, also in Brazil, had hair methylmercury levels higher than 6.0 µg/g, which is far above the maximum recommended level of 1.0 µg/g by the United States Environmental Protection Agency (U.S.EPA) [59] and 2.3 µg/g recommended by The Food and Agriculture Organization of the United Nations (FAO/WHO) [60]. Despite evidence of the contamination of the population, to the best of our knowledge, no study in the world has attributed the relative share of responsibility of artisanal and small-scale gold mining (ASGM) extraction to the health impacts of the population.

On the other hand, the available literature has extensively documented the impact of mercury on health outcomes. For example, the cohort studies conducted in the Faroe Islands and New Zealand indicate that even in low doses, the consumption of mercury-contaminated fish during pregnancy can cause important cognitive alterations in children [61,62]. In this sense, mercury's potential neurotoxic effects in children and adults of the Amazon have been analyzed in some studies [63–68]. The most common effects in children are cognitive problems, neurodevelopmental impairment, and psychomotor disorders. Depending on the mercury exposure level in the prenatal period, the child may be born with mild mental retardation. Axelrad et al. [69] showed that for each additional 1.0 µg/g of methylmercury in maternal hair a reduction of 0.18 IQ points is expected in the child. According to Vasconcellos et al. [10], the methylmercury hair concentrations detected in women of reproductive age in the Amazon region are high enough to cause the emergence of cases of mild mental retardation. In adults, decreased visual field, neurobehavioral, and motor coordination disorders are most frequently reported [68]. Meanwhile, literature on the impact of mercury on increases in cardiovascular disease is still not unanimous [70,71], but there is evidence of this relationship in non-Amazonian countries [8,9,47,48]. Salonen et al. [8] showed an increased risk of myocardial infarction of 69% in men over 40 years when hair mercury levels were above or equal to 2.0 µg/g, compared to men with levels below to 2.0 µg/g.

Several studies have addressed the relationship between extracted gold kilograms and effects on human health [72–76]. For evaluating public policies associated with mercury impacts on human health, we used a combination of non-monetary indicator–disability adjusted life years (DALY) [77–79] and monetary indicator–value of statistical life (VSL) [80], based on willingness to pay for risk reduction [81,82]. Many studies use this relationship between VSL and DALY. Neumann et al. [83] review studies that depart from the analysis in DALY to assess the impact through a monetary indicator. Fan et al. [84] estimate COVID-19's impact from DALY and Statistical Value of Life. Grandjean and Bellanger [85] calculate disease burden associated with environmental chemical exposures, including methylmercury worth US$15 billion for Europe and the United States.

Many authors have studied methylmercury ingestion by humans via contaminated fish consumption, analyzed the impact of this contaminant on human health, and established relationships between average mercury intake and negative effects on human health [8,10,47,48]. However, these studies do not distinguish the potential origin of mercury (i.e., whether it is natural or additional mercury). For this reason, we sought to develop a methodology that would allow us to analyze mercury dispersal by gold mines, making it possible to link the human health impact per gold mine from the input unit, such as impacted hectares or gram of gold extracted.

The main literature gaps addressed here are: (i) the potential area of mercury dispersal in the water, as well as the estimation of the maximum affected population, given the average level of contamination and fish consumption; (ii) the definition of dose-response functions for different health outcomes and population groups; and (iii) the valuation of health impacts of mercury in monetary terms. Therefore, we seek to link a chain of events, concepts, and value estimates from the literature through a series of hypotheses, which allow us to relate the use of mercury in small-scale gold mines to health outcomes and their negative economic values.

Considering all of the gaps mentioned, the main objective of this study was to develop a precise methodology to estimate the health damage caused exclusively by the mercury used in ASGM (that is, avoiding counting the natural mercury present in Amazonian soils). For estimation calculations, the amount of gold that is extracted in a given location and the amount of mercury used and discarded during the process were considered. The study provides value estimates that can support social impact assessment, define fines for illegal gold miners, and provide parameters for the evaluation of public policies related to inspection and prevention of this activity. The methodology construction contributes to the incorporation of impacts on human health by decision makers and expands the discussion on the effects of illegal mining activities in a strategic area such as the Amazon for economic development. This study is the first to estimate the average impacts that a gold mine can generate on human health due to the use of mercury and its ingestion from fish consumption.

2. Materials and Methods

This section, based on an extensive literature review, establishes the hypotheses to make the connections between average mercury use, its dispersion in the environment, potential absorption by humans, and its effects on human health. The study area is the Brazilian Amazon, which concentrates 93% of all small-scale mining in Brazil [86]. Gold mining is more concentrated in the states of Pará and Mato Grosso, in addition to significant impacts in the state of Roraima, and more specifically in Yanomami Indigenous Land.

We establish links between several factors, using average values from the literature, in a chain which can be divided into two major goals: how mercury used in gold mining is dispersed in water until it reaches humans, and the quantification of the impact on human health from mercury ingestion.

To complete the logical line presented in Figure 1 below, an extensive bibliographic review was carried out on various topics such as biophysics, biochemistry, epidemiology, and public health.

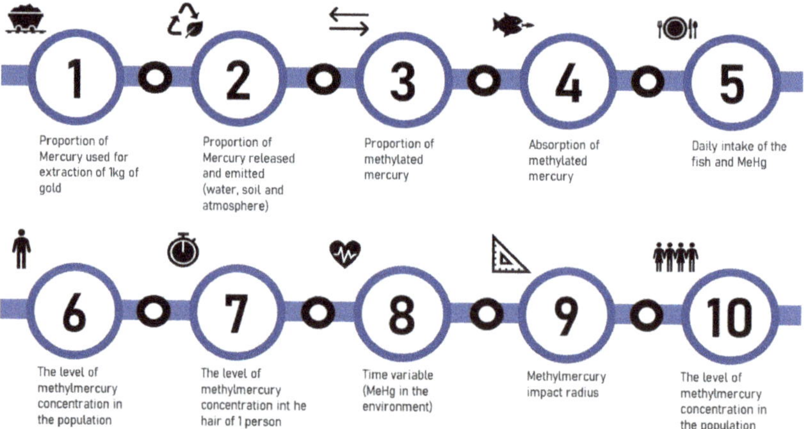

Figure 1. Logical line for relating the existence of gold mining to human health outcomes.

The study begins with the presentation of the relationship between the use of mercury from gold mines and the effects of increasing methylmercury concentration in the hair of the affected population. Then, the study assesses which adverse effects on human health are considered when there is an increase in the level of methylmercury in the hair.

2.1. Dispersion of Mercury Used by Gold Mining and the Extent of Exposure and the Health Risk the Affected Population

We aim to relate the use of mercury by gold mining with the increased mercury concentration in the hair of the affected population. We divided into two main specific objectives. The first is to explain how the mercury used in gold mining is dispersed in water until it reaches humans. The second is to describe how mercury from gold mining and human mercury intake affects the population.

2.1.1. Variables That Define How the Mercury Used in Gold Mining Is Dispersed in the Water until It Reaches Humans

The following variables have the objective of explaining which variables are important in defining how the mercury used in gold mining is dispersed in the water until it reaches humans. For this, the next three described variables are the proportion of mercury used for each kilogram of mined gold, the proportion of mercury released in water and soil and emitted into the atmosphere, and the proportion of mercury that transforms into methylmercury (methylation). The last section presents a summary of these variables.

Proportion of Mercury Used for Each Kilogram of Mined Gold

Mercury is used in the amalgamation process, which, when combined with gold, forms a metallic alloy called amalgam. The literature shows that the proportion of mercury used in the extraction process can vary in both Brazil and worldwide due to different factors (such as different yields for gold extraction). This proportion can vary considerably from 1.3–8.0 g of mercury for each g of gold extracted [17,72,74,87–89]. An average ratio (Hg:Au) of 2.6:1 demonstrated by Castilho and Domingos [17] is assumed here, since it was obtained as an average from different gold mines in Brazil.

Proportion of Mercury Released in Water and Soil and Emitted into the Atmosphere

The mercury used in gold mining is dispersed in both soils and rivers, as well as in the atmosphere. This study analyzes only the release of mercury in aquatic environments, since the objective is to understand the effects of methylmercury on human health from ingesting contaminated fish. Therefore, we sought to review the literature that indicates the proportion of mercury released in soil and water. This proportion can vary from 12–35%, with the lowest proportion (12%) being a conservative scenario [17–20,88]. It is important to mention that such studies already consider environmental controls such as filtering in the amalgamation process (which recovers 50% of the mercury).

Proportion of Mercury That Transforms into Methylmercury (Methylation)

When mercury is dumped into an aquatic environment, part of it is transformed into an organic mercury compound called methylmercury, which is about 30 times more toxic than inorganic mercury [90] and is ingested via the consumption of contaminated fish. Once ingested by humans, methylmercury is rapidly absorbed by the gastrointestinal tract and widely distributed throughout the body, including reaching the central nervous system, which can cause IQ loss in children and cardiovascular diseases [8–10,47,48,72–76].

Given the knowledge of the amount of mercury released into water, another process widely discussed in the literature is the proportion of mercury that transforms into methylmercury (MeHg).

The literature shows that the methylation process occurs in different contexts in the Amazon, with conditions such as low pH and high levels of dissolved organic carbon favoring mercury oxidation and methylation [90]. Several studies have shown that the methylation rate can vary between 3% and 22% [91–97]. Conservatively, the choice for the lowest methylation rate indicates that 3% of mercury released into the water will change to methylmercury over time.

Variable's Interaction to Explain How Mercury Used in Gold Mining Is Dispersed in the Water until it Reaches Humans

The formula below demonstrates the results found from Section 2.1.1:

$$X = A * B * C * D \tag{1}$$

where:

X = mercury used by mining, is released into aquatic environments and undergoes methylation

A = gold amount extracted by mining (kilos)
B = proportion of mercury used for extraction of each kilogram of gold (%)
C = proportion of mercury released in the water (%)
D = methylation rate (%)

2.1.2. Variables That Describe How Mercury from Gold Mining and Human Mercury Intake Define the Affected Population

After estimating the amount of mercury that is dispersed in the environment and potentially consumed by humans, we estimated how this amount will be consumed: (i) by individuals who eat contaminated fish daily; and (ii) by the number of individuals that may absorb this amount of mercury across their lives. Therefore, we need to understand how many people may be exposed (up to an average contamination level) by a consuming a given additional amount of mercury throughout their lifetime.

The average long-term contamination level of one individual is estimated based on the daily contaminated fish consumption, which gives us the total amount of mercury that will lead one individual to the negative health outcomes considered. Based on this individual total consumption, we can estimate how many individuals could be impacted at the same contamination level within a given impact area.

Methylmercury Absorption by Fish

Methylmercury is a chemical substance that is absorbed by the trophic chain in aquatic environments [98]. Through biomagnification, substances or elements in living organisms travels through the food webs and accumulates at the highest trophic level, differing between predatory and non-predatory species [99–101].

As an organic form of Hg, MeHg has extremely neurotoxic effects and is readily accumulated in biota due to its lipophilic and protein-binding properties [102,103]. A series of studies have shown the average levels of contamination of aquatic species in the Brazilian Amazon [49,52–58,104–109]. For example, contamination can reach 0.13 to 2.85 µg Hg/g for certain fish species [104] and Dórea et al. [55] detected mean mercury levels of 0.578 g Hg/g piscivorous fish and 0.052 g Hg/g non-piscivorous fish in the upper Tapajós basin.

To use recommended values at a global level, this study chose the Codex Alimentarius [110], jointly developed by the Food and Agriculture Organization of the United Nations (FAO) and the World Health Organization (WHO). According to Codex [110], the maximum permitted mercury levels are 1.0 µg/g and 0.5 µg/g for predatory and non-predatory fish trade, respectively.

Average Daily Intake of Fish and Methylmercury for Different Profiles in Brazil

The Amazon region is historically known for eating fish, whether in riverside and indigenous populations or large urban centers. Riverside dwellers, for example, eat more fish, with an average consumption of 189 to 243 g of fish per day [18,49,50,109,111–113]. Meanwhile, the indigenous population have an average daily fish consumption of 100 g per day [49] from fishing. Finally, the urban population, such as in Belém city (Pará state), have a more diversified diet with other proteins and, therefore, consume an average of 57 g per day [114].

To calculate the average daily mercury intake per person, we described the average weight of an individual in each population profile as 70 kg for urban people [115], 65 kg for riverside dwellers [49], and 53 kg for indigenous people [50].

Chronic ingestion of mercury-contaminated fish by the Amazon population increases health risks.

The formula below represents how average daily intake is calculated:

$$I = P * \left[\frac{Cm * Cont)}{W} \right] \tag{2}$$

where:

I = average mercury daily intake (μg/kg bw (Bw = body weight)/day)

P = proportion of urban, indigenous, and riverside populations in the municipality affected by gold mining. Therefore, a balance between these factors is necessary.

Cm = average fish consumption per day (g/day) for population type

Cont = average contamination in fish (μg/g fish) (depends on fish absorption of methylmercury presented in Section 2.1.2)

W = individual weight (kg) for type of population

The average daily intake is related to contextual factors of the location that is being affected by gold mining, such as the proportion of indigenous, riverside, and urban populations.

Variable Time for Methylmercury Ingestion by Fish and Humans

Several studies have demonstrated the instability and complexity of methylation and demethylation [91–97]. Bisinoti and Jardim [116] demonstrated that all mercury-containing river and lake sediments are dangerous since confined mercury can remain active as a substrate for methylation for approximately 100 years. Thus, the conservative hypothesis is that mercury is bioavailable in the environment for 50 years and can cause harm to human health for this duration.

Based on this hypothesis, it is possible to quantify the mercury consumption for an average individual over 50 years using the following formula:

$$Z = Y * T \tag{3}$$

where:

Z = average mercury intake per person over 50 years (g/50 years per person)

T = time (50 years = 18,250 days)

Y = average mercury daily intake per person (g/day per person) (conversion based on average individual weight (kg) and average mercury daily intake (μg/kg bw/day)

Therefore, it is assumed that the daily mercury intake by each social group in the region (riverside, indigenous, and urban population) will chronically occur over 50 years. In other words, individuals with an average daily intake of 0.76 g/kg/day will have a total intake of 0.9 g over 50 years, resulting in an average increase in mercury contamination of 5.0 μg/g μg/g (hair). Such information is essential to understand how methylmercury released by gold mining will be distributed among fish consumers in the region.

Knowledge regarding mercury release by gold mining and subsequent impacts on fish and daily mercury intake by humans, when associated with the time variable, requires a hypotheses on the proportion of methylmercury that will be absorbed during this period. We hypothesize that all bioavailable methylmercury from gold mining will be consumed by humans, that is, will be consumed between 0.22% and 4.5% of the total mercury used by ASGM. This hypothesis does not imply that all methylmercury will be instantly absorbed by humans, but rather chronically absorbed over 50 years by humans at the top of the trophic chain. This type of hypothesis is needed due to the literature gap on tracing mercury molecules from ASGM until human absorption.

Changes in the Methylmercury Hair Concentration Level

Methylercury will accumulate in the hairs of people who consume of fish contaminated by mercury. This clear relationship between contaminated fish consumption and methylmercury accumulation in hair was demonstrated by the Poulin and Gibbs [11]. According to the World Health Organization [117], an average daily intake of 0.1 μg/kg of methylmercury leads to a concentration of 1.0 μg/g of methylmercury in hair. Thus, it is possible to estimate how gold mining influences methylmercury concentration in hair.

$$F = \frac{I}{0.1} \tag{4}$$

where:

F = hair methylmercury concentration (in ug/g)
I = average mercury daily intake (μg/kg bw/day)

Population Impacted by Increased Methylmercury Concentration in Hair

Knowledge about the amount of mercury released by gold mining, methylated in the aquatic environment (Section 2.1.1), and distributed over time (50 years) with the average intake per person (grams of mercury in 50 years) (Section 2.1.2), contributes to defining the population affected by mercury used in gold mining as the formula can be presented as follows:

$$\text{Pop} = \frac{X}{Z} \tag{5}$$

where:

Pop = population affected by mercury contamination from gold mining
Z = average mercury intake per person
X = estimated amount of methylmercury from mining that reaches the top of the trophic chain.

Socioeconomic characterization of the affected population is considered, where groups with higher fish consumption are expected to have higher contamination values. The average daily intake of mercury through fish consumption is differentiated by the riverside population, which has a higher fish consumption, and urban population, having a lower fish consumption and a lower chance of being exposed.

Additionally, we considered the population density as a limiting factor of the total number of people that may be exposed within a given radius. Mercury distribution in river and tributaries is influenced by the distance from gold mines [118]. Therefore, we stipulate a limit for mercury impact that depends on the population size in the neighboring mining area. The amount of mercury close to the source of contamination is high and decreases as the analysis distance increases, indicating low concentrations far from the analysis point [119,120]. Studies have also shown that, when assessing the amount of mercury in cities close to gold mining areas, the mercury concentration in the hair of the population living near the mines was greater than in people living far from the gold mines [121,122]. However, river confluence events, where one river flows into another (whose mercury concentration is higher), may indicate a pattern of increasing mercury concentration after a certain distance [123,124].

Several studies present the average distance that metallic mercury can travel in rivers, ranging from 4 to 100 km [36,124–130], as a function of river characteristics (flood events, rain, and increased water flow). In the Amazonian context, studies such as Roulet et al. [36] have shown that the significant impact radius is approximately 50 km downstream. However, it should be noted that organic mercury (methylmercury) can travel longer distances as it is absorbed by fish, which can migrate up to 2000 km, such as Brachyplatystoma (*Piratinga*) [131–136]). Therefore, we conservatively assumed that mercury will be dispersed to a radius of 100 km (that is, we did not consider the long distances traveled by some species of fish since there is heterogeneity of migration depending on the species and it would be complex to standardize for the Amazon region).

The effects of this release are limited to the number of people within a 100 km radius. A highly contaminated region with a low population density means that few humans will be affected, although it continues to have a significant impact on the region's fauna. Likewise, urban areas close to the center of contamination have a high population density, potentially causing damage to human health of more individuals.

Variable's Interaction to Explain How Mercury Used in Gold Mining Is Dispersed in the Water until It Reaches Humans

The formula below demonstrates these results found from Section 2.1.2:

$$F = \left[\frac{(Cm * Cont)}{W} \right] * \frac{P}{0.1} \tag{6}$$

where:

F = hair methylmercury concentration (in ug/g)

I = average mercury daily intake (μg/kg bw/day)

P = proportion of urban, indigenous, and riverside populations in the municipality affected by gold mining. Therefore, a balance between these factors is necessary.

Cm = average fish consumption per day (g/day) for population type

Cont = average contamination in fish (μg/g fish) (depends on fish absorption of methylmercury presented in Section 2.1.2)

W = individual weight (kg) for type of population

2.1.3. Mercury Dispersion Summary Formula

All of the logical links described above can be summarized as follows. It is possible to relate mercury use by gold mining and its respective loss in the environment until it reaches the human body and affects the population:

$$Pop = \frac{X\,(A,\,B,\,C,\,D)}{Z\,(Cm,\,Cont,\,P,\,W,\,I,\,T)} \tag{7}$$

where:

Pop = population affected by mercury contamination from gold mining

X = mercury used by mining, is released into aquatic environments and undergoes methylation (depends on the proportion of mercury used for extraction of each kg of gold, and the proportion of mercury released in the water; the methylation rate

Z = average mercury intake per person in 50 years (g/50 years per person) (depends on the absorption of methylated mercury (100%).

A = gold amount extracted by mining (kilos)

B = proportion of mercury used for extraction of each kilogram of gold (%)

C = proportion of mercury released in the water (%)

D = methylation rate (%)

I = average mercury daily intake (μg/kg bw/day)

P = proportion of urban, indigenous, and riverside populations in the municipality affected by gold mining. Therefore, a balance between these factors is necessary.

Cm = average fish consumption per day (g/day) for population type

Cont = average contamination in fish (μg/g fish) (depends on fish absorption of methylmercury presented in Section 2.1.2)

W = individual weight (kg) for type of population

T = time variable (MeHg in the environment) = 50 years (multiplied by mercury intake per person).

2.2. Quantifying Impacts on Human Health from Mercury Ingestion

The health economics literature quantifies the impact on health using the disability adjusted life years (DALY) index to compare the impact of different health problems. The DALY index weighs health measures of mortality and morbidity in one equivalent measurement unit: time (years), considering the severity, magnitude, and duration of the problem [137]. Different knowledge areas are considered to quantify the gold mining mercury impact in terms of increases in the probability of developing: (i) mild mental retardation in children, (ii) myocardial infarction, and (iii) hypertension.

To calculate the DALY, it is necessary to know the following variables: discount rate, age, weight, disability weight, disease duration, and incidence rate which can be seen in the next sections.

2.2.1. Mild Mental Retardation Impact Caused by IQ Loss in Children

The next section relates mercury release in gold mining to IQ loss in children and mild mental retardation due to maternal ingestion of contaminated fish, a health outcome that leads to loss of productivity and income from the birth of the infected child to death [11]. Axelrad et al. [69] demonstrated a linear relationship between loss of points on the IQ scale and increases in mercury concentration in maternal hair, in which 1.0 µg/g of mercury (MeHg) in the mother's hair corresponds to a loss of 0.18 IQ points in the child. Considering that IQ values in the general population have a normal distribution (Gaussian curve) and 95% of individuals have IQ values between 70 and 130, the IQ loss caused by mercury exposure during the prenatal period may cause mild mental retardation in individuals who would be born with IQ values close to 70.

Given the total affected population, as described in Section 2.1.3, we use the number of live births of 19 live births per thousand inhabitants in the North of Brazil. That is, out of a population of 1000 affected people within a radius of 100 km, around 19 babies will be born alive. It is possible to estimate the number of live births impacted by mercury release in the mine.

To calculate the DALY related to IQ loss in children, we highlighted the variables based on a literature review on the subject. One of these parameters is a discount rate which can be defined with the objective of assigning less importance relative to years lost in the future than to years of life lost in the present, given that a human being, in general, has short-term rather than long-term preferences [138,139].

We chose to use the 3% discount rate as it is applied in health economics studies [140], in environmental projects [59], in the calculation of the social cost of carbon [141] and in for social projects in Latin American countries [142]. Another variable is the age weight that is the age weight that corresponds to society's preferences, since less value is given to healthy years of life lost during childhood and old age, due to the low productivity common to these stages of life. The weight of age varies in a range from zero (without weight) to one (100% of weight), being relevant as a weighting factor so that greater weights are not attributed to cases of death in young individuals. The third parameter is the incidence rate (number of cases per thousand people) which is calculated by the Mercury Spreadsheet [11] from the knowledge of the mean concentration of mercury in the hair and the standard deviation associated with the knowledge of the number of affected people). The fourth variable for calculating the DALY is the disability weight which is the result of some studies that create scenarios for individuals to declare their preferences and, therefore, the different outcomes are compared by patients or specialists, creating a ranking [143]. The disability weight can range from 0 to 1, where 0 is a healthy situation and 1 corresponds to death. In the specific case of mild mental retardation in children due to mercury ingestion, according to the WHO [143], the weight was 0.361. The fifth parameter is the year onset of disability and duration that are fundamental in weighing the impacts, since years lived with disability or premature death are counted. In the specific case of IQ loss, the outcome starts in the first year of the child's life and remains throughout life. As in the North region of Brazil, there is a life expectancy of 72 years, thus meaning a disease duration of 72 years.

The monetary measurement of DALY has been widely discussed in several studies, such as Kenkel [82] and Hammit and Robinson [144], who proposed that 1 DALY corresponds to the annualized value of statistical life. This means that the monetary measurement of DALY can reach values above 200,000 USD per DALY [145]. We use the recommendation of the World Health Organization [146], which suggests that one year of healthy life lost (DALY unit) corresponds to 3 GDP per capita, that is, 20,600 USD in Brazil in 2020 [147]. Thus, the mild mental retardation in children due to the extraction of 10 kg of gold corresponds to 10,000 USD in Brazil if all of the average values described throughout the paper are observed.

2.2.2. Cardiovascular Diseases

The association between contaminated fish consumption and cardiovascular diseases considers that mercury in fish muscle, when absorbed by the human gastrointestinal tract, interferes with lipid peroxidation and can cause atherosclerosis. This condition can lead to increased blood pressure [47] and acute myocardial infarction [9]. On the other hand, some studies have not found a relationship between mercury and cardiovascular disease, although they suggest the need for studies on such a relationship [148–152].

This section presents the parameters used to describe the relationship between mercury concentration in hair due to the use of mercury in gold mining and two cardiovascular diseases: acute myocardial infarction and arterial hypertension.

Acute Myocardial Infarction Attributable to Mercury Exposure

For acute myocardial infarction, Salonen et al. [8] found that an individual with a hair mercury concentration of ≥ 2.0 µg/g has a 69% higher risk of acute myocardial infarction than individuals with a concentration of less than 2.0 µg/g. This relative risk presented by Salonen et al. [8] was adjusted for confounding factors, such as alcohol consumption, smoking, and lifestyle factors, and refers to the probability of incidence of acute myocardial infarction, fatal or non-fatal, in Finnish men over 40 years.

1. Calculating Acute Myocardial Infarction Burden Disease Attributable to Mercury Ingestion from Gold Mining

The disease burden methodology in the gold mining context in the Amazon was adapted from a study by Salonen et al. [8]. We first calculated the attributable fraction based on the relative risk estimated by the study of Salonen et al. [8] in Finland. Studies such as Rockhill et al. [153], Fewtrell et al. [149], and Porta [154] presented formulas to estimate the attributable fraction from the relative risk, calculated at 1.69, in Salonen et al. [8] for an exposure to mercury above 2 µg/g. Based on this, we estimated that the risk of myocardial infarction occurrence [9] is 0.4, that is, 40% of myocardial infarction cases can be attributed exclusively to mercury exposure ≥ 2.0 µg/g. This paper assumes as a hypothesis, based on evidence from field measurements [10,49,50], that, due to the high mercury intake, the entire affected population will be at risk of an average mercury concentration above 2.0 µg/g.

To estimate the "Number of Infarction Cases Attributable to Mercury Exposure (≥ 2.0 µg/g Hg)" it is necessary to multiply the total number of infarcts in the sample and the attributable fraction.

$$Ip = \frac{Afi}{Ti} \tag{8}$$

where:

Ip = people infarcted due to mercury levels > 2.0 µg/g

Afi = attributable fraction (infarction

Ti = total number of infraction cases

The same gender and age cut out made by Salonen et al. [8] was used: men over 40 years, which represents 12% of the population of the Brazilian Amazon [155]. Therefore, if data from Datasus [156] are observed, in 2015–2020, approximately 0.16% of this population would be at risk of hospitalization due to infarction in the North region.

As the present study uses the study by Salonen et al. [8] as a basis, we opted for the conservative premise that the year onset of disability is equal to the youngest age of the sample in Salonen et al. [8] (that is, 40 years old). It should be noted that data from Datasus [156] are probably underestimated since the Brazilian health system cannot compute information from isolated areas in the northern region of Brazil.

To adapt to the Amazon context the regional life expectancy was set as 67 years for men. As we consider that the average age of the infarction is 40 years, this means that individuals will live with a disability from 40 years of age to 67 years of age (that is, they will have lived with such disability for 27 years).

Figure 2 demonstrates the logical chain built above:

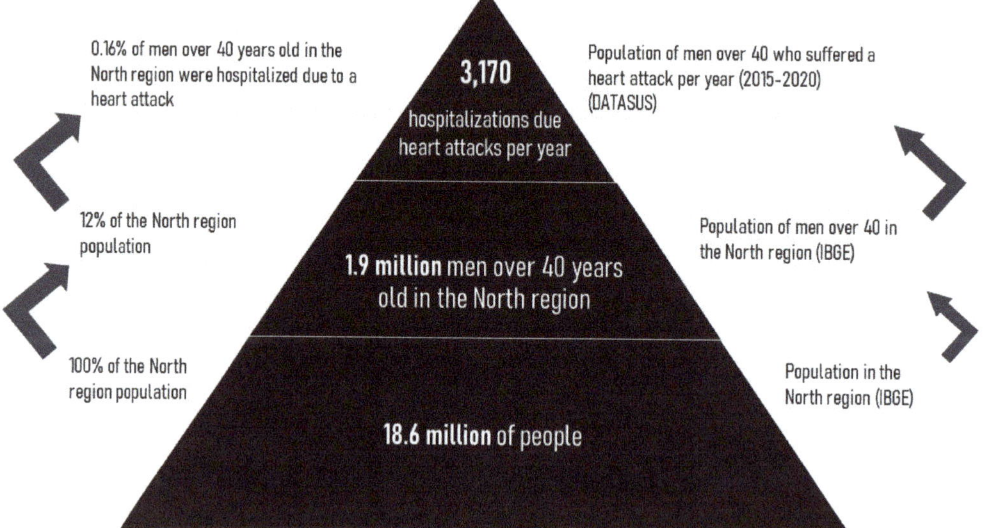

Figure 2. Outcome of myocardial infarction associated with mercury ingestion.

Values for the probability of risk accumulated over the years are based on Zaletel-Kragelj and Bozikov [157], who estimated the cumulative risk of mercury-associated myocardial infarction at 1.61%.

The estimation of the number of infarction cases associated with mercury over the years for the male population over 40 years old and the accumulated infarction risk associated with mercury over time is multiplied.

$$Hm = Cri * Mp \tag{9}$$

where:

Cri = cumulative risk of mercury-associated myocardial infarction = 1.61%

Hm = male population over 40 in the region who will be hospitalized in 27 years for mercury ingestion

Mp = male population over 40 years in region

2. Variables for Calculating the DALY and Monetary Impact of Mercury-Associated Acute Myocardial Infarction from Gold Mines

The DALY for mercury-associated acute myocardial infarction was based on the following parameters: 3% discount rate, 100% weight for age, disability weight for acute myocardial infarction of 0.439 [143], disability onset at age 40, disability duration of 27 years (assuming life expectancy of 67 years), and an incidence rate of 1.9 cases of infarction for every 1000 people.

The resulting value is given in years lived with disability for the extraction of gold per kg. For example, 10 kg of gold can generate, on average, the impact of 8.5 years lived with disability or 174,000 USD at 20,600 USD for each DALY [147].

Arterial Hypertension Disease Attributable to Mercury Exposure

High blood pressure has long been recognized as a major risk factor for cardiovascular diseases. A recent analysis suggests that the burden of high blood pressure has increased over the past three decades [158,159]. In addition to traditional risk factors for hypertension, such as high salt intake and overweight/obesity, environmental exposure to heavy metals can also play an important role [160–162]. Although the mechanisms by which mercury induces hypertension are not fully understood, plausible explanations include oxidative stress and inflammation, which promote endothelial and renal dysfunction and binding of selenium-related enzymes. Hu et al. [47] included a systematic review, building a meta-analysis both with general studies and with the occupational population exposed.

1. Methodology for Calculating the Hypertension Burden Disease Attributable to Mercury Ingestion from Gold Mining

The hypertension disease burden methodology in the context of gold mining in the Amazon fundamentally involves adaptation to the study by Hu et al. [47], with the definition of all applied premises being relevant. The first adaptation to the study by Hu et al. [47] consists of the estimate of the attributable fraction from the odds ratio (OR) of 1.35, given by the meta-analysis for mercury exposure. Since the OR is analogous to the relative risk, it is assumed that they are similar, as shown in studies such as Bonita et al. [163].

Although Hu et al. [17] presented studies for the Amazon context, such as Fillion et al. [9], with an OR of 3.8, indicating a high concentration of mercury in the Brazilian Amazon population), we adopted, conservatively, the OR of the meta-analysis, that is, 1.35, since this is a comprehensive study review on the relationship between hypertension and mercury intake. Therefore, it is possible to quantify the attributable fraction using the following equation:

$$FAP = \frac{(OR - 1)}{OR} \tag{10}$$

Based on this understanding, we estimated the risk of arterial hypertension occurrence [47] to be 0.26, that is, 26% of cases of arterial hypertension would be due exclusively to mercury exposure ≥ 2.0 µg/g.

To estimate the "Number of Hypertension Cases Attributable to Mercury Exposure (≥ 2.0 µg/g Hg)" we multiplied the total number of hypertension cases in the sample by the attributable fraction.

$$Hp = Afh * Th \tag{11}$$

where:

Hp = number of hypertension cases attributable to mercury exposure (≥ 2.0 µg/g Hg)
Afh = attributable fraction (hypertension)
Th = total number of hypertension cases

Unlike the myocardial infarction outcome, the literature does not indicate a greater or lesser hypertension risk depending on gender (male or female); that is, the population over 20 years should only be evaluated as that is the year in which hypertension begins to be observed [47]. Figure 3 summarizes the logical lines built above.

To attribute the fraction of this outcome to fish intake, it is necessary to use the attributable fraction calculated as 26% of the risk associated with mercury. Using the methodology of Zaletel-Kragelj and Bozikov [157], the cumulative hypertension risk associated with mercury was estimated to be 1.21%.

Based on the knowledge of the temporality of the outcome, it is feasible to estimate the number of cases of hypertension associated with mercury over the years. For this, the population over 20 years in the region that will be hospitalized over 52 years of exposure and the accumulated risk of hypertension associated with mercury over time is multiplied.

$$Hp = Crh * Pp \tag{12}$$

where:

Crh = cumulative risk of mercury-associated hypertension

Hp = population over 20 years in the region who will be hospitalized in 52 years for mercury ingestion

Pp = population over 20 years in region

2. Variables for Calculating DALY and Monetary Impact of High Blood Pressure Associated with Mercury from Gold Mining

To calculate the DALY related to arterial hypertension, we considered the following parameters: discount rate of 3%; 100% weight for age; disability weight of 0.246 [143]; year onset of disability at the age of 20 years, with a duration of 52 years to meet the 72 years of life expectancy in the northern region of Brazil.

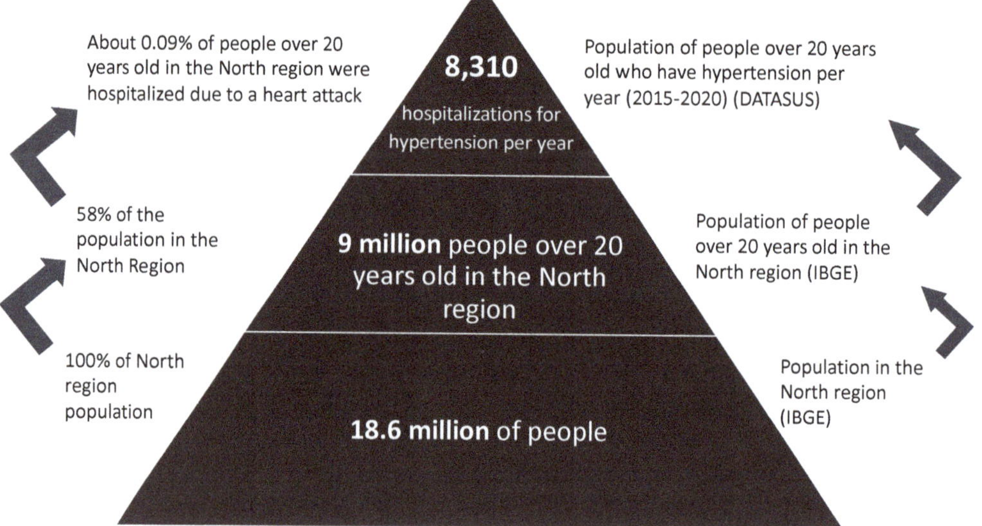

Figure 3. Hypertension scenario in Northern Brazil.

3. Results

3.1. Results Presented from Methodology

The methodology presented above consists of the first estimate of the relationship between the use of mercury by gold mining and the negative effects on human health. Defining the amount of mercury used per kilogram of gold mined (Section 2.1.1), the proportion of mercury loss to the environment (Section 2.1.1), and the methylation rate (Section 2.1.1) means that, on average, between 0.22% and 4.5% of the total mercury used by gold mining turns into methylmercury, entering the trophic chain and therefore affecting human health. Such an amount may seem small, but the effects on human health are varied and extremely harmful. The result is the first estimation that quantifies the release of mercury into the water, which contributes to combat arguments such as that deforestation, is the main cause of the release of natural mercury into the environment.

Given the knowledge of the amount of mercury released by gold mining, the present study demonstrated that the population context is essential to define the impacts on human health. Therefore, it is assumed that the daily mercury intake depends on social group affected in the region (e.g., riverside, indigenous, and urban population) As an example of a population with indigenous and riverine population, an average daily intake of 1.2 µg/kg bw/day corresponds to an average concentration in hair of 12 µg/g, which corroborates other studies. Bastos et al. [18], identified an average mercury concentration of 9.81 µg/g in 45 riverine communities. Vasconcellos et al. [49] detected mercury levels above 6.0 µg/g in hair samples in Munduruku indigenous communities in the Pará state. However, in urban populations, a daily mercury intake of 0.4 µg/kg bw/day corresponds to an average concentration of mercury in hair of 4.0 µg/g; greater than the 1.0 µg/g recommended level by the North American Environmental Protection Agency [59] and 2.3 µg/g by the United Nations Food and Agriculture Organization [60].

Likewise, the study demonstrated that the context of the affected population is not restricted to these population characteristics, it is also important to consider population density and the distance of the population to the gold mining. This is explained due to the fact that areas with low population density and far from gold mining will have a limited effect on the health of the population, while regions with high population density and close to a radius of less than 100 km has a higher probability of consumption the methylmercury released by the ASGM.

The Table 1 below seeks to summarize the ranges between the variables used in the model:

Table 1. Summary of variables, unit of measurement, range values, and source used in this article.

Variables	Unit of Measurement	Range	Source
Distance of mercury dispersion in rivers	kilometers	4–100 km (Dispersal of mercury can be carried out by fish that travel long distances up to 5000 km [131–136]. Conservatively, the dispersion of mercury by fish is not considered).	[30,124–130]
Proportion of mercury used for each kilo of gold extracted	Mercury: Gold	1.6:1–5:1	[17,72,74,87]
Proportion of mercury released into water	Percentage	7%–21%	[17–20]
Methylation rate in water	Percentage	2%–22%	[91–97]
Average mercury contamination of fish	Microgram of mercury per gram of fish	0.13–3 (µg Hg/g of fish)	[49,52–58,104–110]

By defining the characteristics that influence the level of contamination and the number of people affected, the study made use of existing studies that addressed the relationship between exposure to mercury and impacts on human health [9,10,47]. In the outcome of mild mental retardation in children, there is already wide acceptance about the relationship between level of contamination and lost IQ points [69]. Therefore, the study applied the methodology already developed by Poulin and Gibb [11] that uses the DALY indicator to quantify the loss of well-being.

Meanwhile, for cardiovascular outcomes (arterial hypertension [47] and myocardial infarction [9]), adaptations were necessary since there is no direct relationship between the level of contamination and the impact on human health. Therefore, deepening the theme of epidemiology was necessary, being one of the contributions of the article.

As a result, in a population of 100,000 impacted people, around 193 men over 40 years of age will have a myocardial infarction associated with mercury ingestion. The incidence rate of this outcome is given by the number of cases in every 1000 affected people. That is, 1.9 hospitalization cases for infarction associated with mercury in every 1000 affected people. Similarly, the results in an affected population of 100,000 people will average about 700 people with high blood pressure in the population over 20 years of age.

3.2. Human Health Impact in the Indigenous Territory Due to Illegal Gold Mining

The methodology was applied to evaluate the negative impacts of illegal miners occupying the Yanomami Indigenous Land (YIL), an area located mainly in the Brazilian Amazon in the states of Roraima and Amazonas. YIL is the largest indigenous land in

Brazil, with an area of 96,000 km^2 and a total population of 26,780 indigenous people. Although the Brazilian constitution prohibits economic activity on indigenous lands, the main potential threat faced by YIL is invasion by illegal miners. It is estimated that more than 25,000 miners live and work illegally in the territory.

The variables used to present the results are described in the following Table 2.

Table 2. Summary of variables, unit of measurement, range values, and source used in Yanomami case study.

Variables	Unit of Measurement	Value	Source
Area impacted by gold mining in 2020	km^2	5 km^2	[164]
Average individual weight of the indigenous population	Kilogram	53.2 kg	[49]
Average daily consumption of fish per rural individual	Grams of fish per person per day	100 g/person/day (Average between indigenous (100 g/person/day) [49] and Riverside (189 g/person/day) [50])	[49]
Average population density	Inhabitants by km^2	2 inhab/km^2	[156]
Urban population (state of Roraima as a whole)	Percentage	76%	[156]
Rural population (state of Roraima as a whole)	Percentage	24%	[156]
Distance of mercury dispersion in rivers	kilometers	100 km	[116]
Proportion of mercury used for each kilo of gold extracted	Mercury: Gold	2.6:1	[17]
Proportion of mercury released into water	Percentage	13%	[20]
Methylation rate in water	Percentage	2%	[91]
Average mercury contamination of fish	Microgram of mercury per gram of fish	0.5 (µg Hg/g of fish)	[110]

Despite the variability of each parameter shown in Table 1, we considered conservative parameters in the literature in Table 2, showing that the risk to human health can potentially be greater than what is being presented with this estimate. A more pessimistic scenario, following the precautionary principle, with higher parameters, demonstrates a potential for greater harm to the population's health.

In 2020, 5 km^2 were degraded by illegal ASGM [165], which, considering average productivity of 1.7 kg of gold per hectare in Brazilian Amazon [166], would use around 2.2 tons of mercury for 863 kg of gold production. As a result, we estimated that approximately 32 kg of mercury was released into local rivers, which could affect 44,000 people. Using our methodology, we estimated that 307 people would develop hypertension problems, 85 acute myocardial infarction, and 4 mild mental retardation. The economic value of these human health damages would total 69 million USD, divided into: (a) 1 million USD due to IQ loss; (b) 15 million USD due to acute myocardial infarction; and (c) 52 million USD due to increases in hypertension problems. This estimated value may be used by local authorities to set compensation for Yanomami people.

4. Discussion

The methodology developed in this paper is the first of its kind to assess the impact on human health caused by mercury used in gold mining. To achieve this methodology, it was necessary to understand the complexity of the mercury cycle and the ASGM processes. First, using this methodology, we presented evidence against the argument that the source of the impacting mercury is deforestation, which releases natural mercury present in the forest [23–26]. The mobilization of natural mercury by deforestation and forest fires represents a relevant impact [30–43], but the use of mercury by ASGM represents the greatest participation in the release of mercury [44]. In addition, it is important to remember that the artisanal mining activity also causes the mobilization of natural mercury due to the process of excavating the soil and sediment from the rivers [91]. Thus, we can conclude this deforestation contributes to the increase in mercury circulation in the Amazon. However, our focus on this paper is related to impact of mercury released directly by gold mining.

Given that we consider only the additional mercury released from ASGM, another complexity arises from the difficulty in attributing responsibility to specific mines. Such complexity can be explained by the ability of mercury to remain bioavailable for long periods [116], bringing cumulative impacts from ASGM exploration [45,46,72–78]. Thus, the understanding that there is an increase in methylmercury concentration in the hair of a population [49,50,69–73,123] is not enough information to make a specific gold mine responsible, since this increase in concentration can be explained by the history of exploration of other mines that released mercury into the environment. Therefore, the judicialization process of a specific illegal gold mine becomes more challenging as there are several illegal ASGMs in the Amazon [167].

Given the whole context of mercury and the ASGM, this article proposes an innovative methodology that proposes a linear relationship based on the amount of mercury used by ASGM and its adverse effects on human health. For this, an extensive literature review was performed that tracks the average mercury use of ASGM [17,72,74,87,88], mercury disposal in the environment [17–20], fish consumption [18,49,50,108,110–113], and the level of potential contamination harmful to human health [52–58]. The study is also innovative compared to other studies [10,47–49] observed due to the fact that the impacts on human health vary according to the context analysed, such as fish intake in the population and demographic density. This means that a replicable methodology was developed which was adaptable to the different contexts observed within the Amazon

After the study related the use of mercury in gold mining and the average increase in mercury concentration in the population, we explored the literature relating gold extracted and negative effects on human health [72–74]. Kahhat et al. [74] used Usetox software, which characterizes chemical impacts on human health and freshwater ecotoxicity and scales an impact of 2 non-cancerous cases and 0.0192 cancerous cases for each kilogram of gold extracted. Similarly, Gulley [72] and Spadaro and Rabl [73] assessed the IQ loss impact for each kilogram of gold from the calibrated benefit transfer approach employed in environmental valuation literature estimates of the impact of mercury on global earnings to twelve gold mining sites around the world. Gulley [72] used studies [37,39,40] that map quantities of mercury emissions into lost earnings due to fetal IQ loss to produce monetary estimates of the impact of mercury emissions. According to Gulley [72], the weighted average estimates an impact of US$ 7300 per kilogram of mercury released into the environment, and this value may increase to the upper limit of US$ 22,300 depending on the assumptions adopted. However, such methodologies are different from the proposal presented in this article, which links mercury use at the beginning of the chain, from the mercury loss in the aquatic environment, to assess the impact on human health. The methodology presented measures the loss of well-being caused using mercury by gold mining. However, other studies [168–171] value the impact of mercury by the cost of remediation of mercury in vegetation.

We must reinforce that before the development of this methodology, the Brazilian institutions responsible for setting fines for illegal ASGM did not measure the impact on human health from the use of mercury in the mines. This methodological gap is filled with the tool developed in this article, and based on this, institutions such as the Federal Police and Federal Public Prosecution in Brazil have instruments to prosecute illegal gold mining damages in order to stop the advance of this activity in the Amazon.

Nevertheless, this article recognizes the limitations of the relationship between mercury use in gold mining and its impacts on human health. Given the lack of studies that assess the factors that influence the response time between changes in deposition and changes in methylmercury concentrations in fish, in this study we needed to assume the temporal effect of mercury release and fish bioaccumulation over 50 years. In addition, the

Amazon region is complex and diverse. The model presented is simplified and does not consider local differences, such as river color and water flow, that could impact the local mercury cycle. Such characteristics need to be further studied.

We did not find any study in Brazil that relates mercury concentration to myocardial infarction. The study by Salonen et al. [8], carried out in Finland, was the only related study found in the literature. Even though both regions have a high intake of fish contaminated by mercury, the physical differences between the populations in the Amazon and in Finland should be further studied for further adaptations.

The limitations presented throughout the article can be overcome with new studies of biophysical mercury dispersion until it reaches humans and possible impacts to human health from increased mercury intake by humans. Long-term monitoring studies that collect mercury concentrations in water, sediments, and fish are needed, particularly in the Amazon. These efforts can lead to long-term data records that can be compared to predictions. Likewise, additional studies should be carried out to assess fish consumption rates in pregnant women, women of childbearing age, and men.

Finally, the methodology presented in this article does not address all impacts on human health since only the outcomes related to the release of mercury in the aquatic environment are measured, not considering the impacts of air exposure due to the inhalation of mercury in the atmosphere (especially in miners) [78,172,173]. The mercury cycle is very complex and, for this reason, the article does not measure the effects caused by the mercury emission into the atmosphere, which can travel long distances. In addition, the article focused on three negative health effects observed in the literature. However, it is possible that there are other health problems associated with exposure to mercury via ingestion of contaminated fish. In this sense, it can be said that the impacts presented are underestimated.

5. Conclusions

This article is the first scientific work whose objective was to propose a methodology to quantify the average economic health impact of the extraction of gold by ASGM in Brazil (using the DALY indicator) and convert them into monetary losses (using the VSL indicator).

The application of this methodology to estimate the impacts of mining in Yanomami territory revealed that the 5 km^2 illegally deforested for ASGM in 2020 would result in severe damages to human health. Based on this, we estimate that around 32 kg of mercury were spilled into rivers, affecting up to 44,000 people. We estimated that 307 people will develop hypertension problems from this activity, 85 people will develop acute myocardial infarction, and 4 children would have born with mild mental retardation. These effects are related only to the presence of small-scale gold mining in 2020, which demonstrates the significant and growing impact with the expansion of this activity in Brazil in recent years. The economic health impact of these health outcomes could reach up to 69 million USD for the 2020 spills alone.

The standardization of mercury impacts assessment is essential for supporting the containment of illegal gold mining activities in the Brazilian Amazon. The developed methodology contributes to the work of control agencies such as the Brazilian Public Prosecutors Office and the Federal Police, who are already using this methodology to estimate fines in the estimation of compensation, showing reliability on the part of these institutions in the scientific results presented. In addition, it contributes to the speed of judicialization in the fight against this illegal activity, and it may support policymakers to plan investments in command-and-control that can prevent the expansion of illegal ASGM.

Author Contributions: Conceptualization, L.B.d.B., P.G.; methodology, L.B.d.B. and P.G.; software, L.B.d.B.; validation, A.C.S.d.V. and J.M.d.Q.; formal analysis, L.B.d.B.; investigation, L.B.d.B.; data curation, L.B.d.B.; writing—original draft preparation, L.B.d.B.; writing—review and editing, A.C.S.d.V.

and P.G.; visualization, A.C.S.d.V.; supervision, P.G.; project administration, P.G.; funding acquisition, Porticus Foundation. All authors have read and agreed to the published version of the manuscript.

Funding: This research was funded by PORTICUS FOUNDATION, grant number "GR-070717".

Institutional Review Board Statement: Ethical review and approval were waived for this study because we did not do any fieldwork with a sample of people. In other words, although we are evaluating the results in humans, we didn't contact any specific population to reach the results. We have mapped several studies in the literature and the results are based solely on what these studies present.

Informed Consent Statement: Patient consent was waived because we did not do any fieldwork with a sample of people. In other words, although we are evaluating the results in humans, we didn't contact any specific population to reach the results. We have mapped several studies in the literature and the results are based solely on what these studies present.

Data Availability Statement: Data Availability in http://calculadora.conservation-strategy.org/#/ (accessed on 4 November 2021).

Acknowledgments: Conservation Strategy Fund (CSF-Brasil) thanks the Federal Public Prosecution and the Department of Expertise, Research, and Analysis of the Attorney General's Office for their partnership in carrying out this work, especially the Federal Attorney and Coordinator of the Amazon Task Force, Ana Carolina Haliuc Bragança, and the Supervisor of Expertise in Economics at the Federal Public Prosecution, José Jorge Júnior. We would like to thank Gustavo Geisel, Federal Police expert, for the technical support. To Suzane Girondi Culau Merlo for the legal support on environmental pain and suffering. To WWF-Brasil, for providing a database on mercury contamination studies in the Amazon. To the more than 30 experts consulted during the research and to the speakers at the methodological discussion seminars. To the Porticus Foundation, for the financial support that allowed this work to be carried out.

Conflicts of Interest: The authors declare no conflict of interest.

References

1. UN Environment. Minamata Convention on Mercury: Text and Annexes. 2019. Available online: https://www.mercuryconvention.org/Portals/11/documents/Booklets/COP3-version/Minamata-Convention-booklet-Sep2019-EN.pdf (accessed on 30 July 2020).
2. Sebastián-Rubiano, G. *The Amazon Biome in the Face of Mercury Contamination: An Overview of Mercury Trade, Science, and Policy in the Amazonian Countries*; Surkin, J., Carrizosa, J., Guío, C., Pon, J., Eds.; WWF: Gland, Switzerland, 2018; p. 165.
3. Driscoll, C.T.; Mason, R.P.; Chan, H.M.; Jacob, D.J.; Pirrone, N. Mercury as a global pollutant: Sources, pathways, and effects. *Environ. Sci. Technol.* **2013**, *47*, 4967–4983. [CrossRef]
4. Sakamoto, M.; Nakamura, M.; Murata, K. Mercury as a global pollutant and mercury exposure assessment and health effects. *Nihon Eiseigaku Zasshi Jpn. J. Hyg.* **2018**, *73*, 258–264. [CrossRef] [PubMed]
5. Selin, N.E. Global biogeochemical cycling of mercury: A review. *Annu. Rev. Environ. Resour.* **2009**, *34*, 43–63. [CrossRef]
6. Dos Santos, A.A.; Hort, M.A.; Culbreth, M.; López-Granero, C.; Farina, M.; Rocha, J.B.; Aschner, M. Methylmercury and brain development: A review of recent literature. *J. Trace Elem. Med. Biol.* **2016**, *38*, 99–107. [CrossRef] [PubMed]
7. Lacerda, E.M.D.C.B.; Souza, G.D.S.; Cortes, M.I.T.; Rodrigues, A.R.; Pinheiro, M.C.N.; Silveira, L.C.D.L.; Ventura, D.F. Comparison of visual functions of two Amazonian populations: Possible consequences of different mercury exposure. *Front. Neurosci.* **2020**, *13*, 1428. [CrossRef]
8. Salonen, J.T.; Seppänen, K.; Nyyssönen, K.; Korpela, H.; Kauhanen, J.; Kantola, M.; Tuomilehto, J.; Esterbauer, H.; Tatzber, F.; Salonen, R. Intake of mercury from fish, lipid peroxidation, and the risk of myocardial infarction and coronary, cardiovascular, and any death in eastern Finnish men. *Circulation* **1995**, *91*, 645–655. [CrossRef] [PubMed]
9. Fillion, M.; Mergler, D.; Passos, C.J.S.; Larribe, F.; Lemire, M.; Guimarães, J.R.D. A preliminary study of mercury exposure and blood pressure in the Brazilian Amazon. *Environ. Health* **2006**, *5*, 1–9. [CrossRef]
10. Vasconcellos, A.C.S.; Barrocas, P.R.G.; Ruiz, C.M.V.; Mourão, D.D.S.; Hacon, S.D.S. Burden of Mild Mental Retardation attributed to prenatal methylmercury exposure in Amazon: Local and regional estimates. *Cienc. Saude Coletiva* **2018**, *23*, 3535–3545. [CrossRef]
11. Poulin, J.; Gibb, H.; Prüss-Üstün, A.; World Health Organization. Mercury: Assessing the Environmental Burden of Disease at National and Local Levels; World Health Organization: Geneva, Switzerland, 2008. *Int. J. Environ. Res. Public Health* **2021**, *18*, 7940.
12. Bose-O'Reilly, S.; McCarty, K.M.; Steckling, N.; Lettmeier, B. Mercury exposure and children's health. *Curr. Probl. Pediatric Adolesc. Health Care* **2010**, *40*, 186–215. [CrossRef]

13. Siqueira-Gay, J.; Soares-Filho, B.; Sanchez, L.E.; Oviedo, A.; Sonter, L.J. Proposed legislation to mine Brazil's indigenous lands will threaten Amazon forests and their valuable ecosystem services. *One Earth* **2020**, *3*, 356–362. [CrossRef] [PubMed]
14. Ferrante, L.; Andrade, M.B.T.; Leite, L.; Silva-Junior, C.A.; Lima, M.; Coelho-Junior, M.G.; Da Silva Neto, E.C.; Campolina, D.; Carolino, K.; Diele-Viegas, L.M.; et al. Brazils Highway BR-319: The road to the collapse of the Amazon and the violation of indigenous rights. *DIE ERDE J. Geogr. Soc. Berl.* **2021**, *152*, 65–70.
15. Ferrante, L.; Fearnside, P.M. Brazil's new president and "ruralists" threaten Amazonia's environment, traditional peoples and the global climate. *Environ. Conserv.* **2019**, *46*, 261–263. [CrossRef]
16. Lacerda, L.D. Contaminação por mercúrio no Brasil: Fontes industriais vs garimpo de ouro. *Química Nova* **1997**, *20*, 196–199. [CrossRef]
17. Castilhos, Z.C.; Domingos, L.M. Inventário Nacional de Emissões e Liberações de Mercúrio no âmbito da Mineração Artesanal e de Pequena Escala no Brasil. *Ministério Do Meio Ambiente* **2018**, *1*, 56.
18. Bastos, W.R.; Gomes, J.P.O.; Oliveira, R.C.; Almeida, R.; Nascimento, E.L.; Bernardi, J.V.E.; de Lacerda, L.D.; da Silveira, E.G.; Pfeiffer, W.C. Mercury in the environment and riverside population in the Madeira River Basin, Amazon, Brazil. *Sci. Total Environ.* **2006**, *368*, 344–351. [CrossRef]
19. Goix, S.; Maurice, L.; Laffont, L.; Rinaldo, R.; Lagane, C.; Chmeleff, J.; Menges, J.; Heimbürger, L.-E.; Maury-Brachet, R.; Sonke, J.E. Quantifying the impacts of artisanal gold mining on a tropical river system using mercury isotopes. *Chemosphere* **2019**, *219*, 684–694. [CrossRef]
20. López, V.; Colón, P. Mercury in Artisanal and Small Scale Gold Mining: Identifying Strategies to Reduce Environmental Contamination in Southern Ecuador. Ph.D. Thesis, University of British Columbia, Vancouver, BC, Canadá, 2010.
21. Ministério do Meio Ambiente. Inventário Mapeia Mercúrio nos Garimpos do Brasil. Available online: https://www.gov.br/mma/pt-br/noticias/noticia-acom-2018-06-3041 (accessed on 4 November 2021).
22. Wasserman, J.; Hacon, S.; Wasserman, M. O ciclo do mercúrio no ambiente amazônico. *Mundo Vida* **2001**, *2*, 46–53.
23. Yang, J.; Zhu, W.; Qu, W.; Yang, J.; Wang, J.; Zhang, M.; Li, H. Selenium Functionalized Metal–Organic Framework MIL-101 for Efficient and Permanent Sequestration of Mercury. *Environ. Sci. Technol.* **2019**, *53*, 2260–2268. [CrossRef]
24. Fostier, A.H.; Melendez-Perez, J.J.; Richter, L. Litter mercury deposition in the Amazonian rainforest. *Environ. Pollut.* **2015**, *206*, 605–610. [CrossRef]
25. Silva, G.; Bisinoti, C.; Fadini, P.; Magarelli, G.; Jardim, W.; Fostier, H. Major Aspects of the Mercury Cycle in the Negro River Basin, Amazon. *J. Braz. Chem. Soc.* **2009**, *20*, 1127–1134. [CrossRef]
26. Wang, X.; Yuan, W.; Lin, C.J.; Zhang, L.; Zhang, H.; Feng, X. Climate and vegetation as primary drivers for global mercury storage in surface soil. *Environ. Sci. Technol.* **2019**, *53*, 10665–10675. [CrossRef] [PubMed]
27. Melendez-Perez, J.; Fostier, A.; Carvalho, J.; Windmoller, C.; Santos, J.; Carpi, A. Soil and biomass mercury emissions during a prescribed fire in the Amazonian rain forest. *Atmos. Environ.* **2014**, *96*, 415–422. [CrossRef]
28. Veiga, M.M.; Meech, J.A.; Oñates, N. Mercury Pollution from Deforestation. *Nature* **1994**, *368*, 816–817. [CrossRef]
29. Lacerda, L. Amazon mercury emissions. *Nature* **1995**, *374*, 20–21. [CrossRef]
30. Roulet, M.; Lucotte, M.; Farella, N.; Serique, G.; Coelho, H.; Passos, C.J.S.; da Silva, E.D.; De Andrade, P.S.; Mergler, D.; Guimaraes, J.R.D.; et al. Effects of recente human colonization on the presence of Mercury in Amazoioan ecosystem. *Wait. Air Soil Pollut.* **1999**, *112*, 297–313. [CrossRef]
31. Friedli, H.R.; Arellano, A.F.; Cinnirella, S.; Pirrone, N. Initial Estimates of Mercury Emissions to the Atmosphere from Global Biomass Burning. *Environ. Sci. Technol.* **2009**, *43*, 3507–3513. [CrossRef]
32. Michelazzo, P. Emissões de mercúrio originárias das queimadas da floresta amazônica e de canaviais. 2007. 121p: il. Tese (doutorado)-Universidade Estadual de Campinas, Instituto de Quimica, Campinas, SP. Available online: http://www.repositorio.unicamp.br/handle/REPOSIP/249440 (accessed on 12 July 2021).
33. Fostier, A.H.; Forti, M.C.; Jardim, W.F.; Junior, C.; de Andrade, J. Impacts of Deforestation on Mercury Cycle in the Brazilian Amazonian Region. In Proceedings of the 10th International Conference on Mercury as a Global, Halifax, NS, Canada, 24–29 July 2021; Volume 1, pp. 1–3.
34. Fostier, A.; Forti, M.; Guimaraes, J.R.; Melfi, A.; Boulet, R.; Espírito Santo, C.; Krug, F. Mercury fluxes in a natural forested Amazonian catchment Serra do Navio, Amapa State, Brazil. *Sci. Total Environ.* **2000**, *26*, 201–211. [CrossRef]
35. Guimaraes, J.R.D. Mercury in the Amazon: Problem or opportunity? A commentary on 30 years of research on the subject. *Elem. Sci. Anth* **2020**, *8*, 032. [CrossRef]
36. Roulet, M.; Lucotte, M.; Canuel, R.; Rheault, I.; Tran, S.; Gog, Y.D.F.; Farella, N.; Vale, R.S.D.; Passos, C.J.S.; Silva, E.D.J.D.; et al. Distribution and partition of total mercury in waters of the Tapajós River Basin, Brazilian Amazon. *Sci. Total Environ.* **1998**, *213*, 203–211. [CrossRef]
37. Roulet, M.; Lucotte, M.; Canuel, R.; Farella, N.; Courcelles, M.; Guimaraes, J.R.D.; Mergler, D.; Amorim, M. Increase in mercury contamination recorded in lacustrine sediments following deforestation in the central Amazon. *Chem. Geol.* **2000**, *165*, 243–266. [CrossRef]
38. Farella, N.; Lucotte, M.; Louchouarn, P.; Roulet, M. Deforestation modifying terrestrial organic transport in the Rio Tapajos, Brazilian Amazon. *Org. Geochem.* **2001**, *32*, 1443–1458. [CrossRef]

39. Farella, N.; Davidson, R.; Lucotte, M.; Daigle, S. Nutrient and mercury variations in soils from family farms of the Tapajos region (Brazilian Amazon): Recommendations for better farming. *Ecosyst. Environ.* **2007**, *120*, 449–462. [CrossRef]
40. Patry, C.; Davidson, R.; Lucotte, M.; Béliveau, A. Impact of forested fallows on fertility and mercury content in soils of the Tapajós River region, Brazilian Amazon. *Sci. Total Environ.* **2013**, *458*, 228–237. [CrossRef] [PubMed]
41. Béliveau, A.; Lucotte, M.; Davidson, R.; Paquet, S.; Mertens, F.; Passos, C.J.; Romana, C.A. Reduction of soil erosion and mercury losses in agroforestry systems compared to forests and cultivated fields in the Brazilian Amazon. *J. Environ. Manag.* **2017**, *203*, 522–532. [CrossRef] [PubMed]
42. Lacerda, L.D.; de Souza, M.; Ribeiro, M.G. The effects of land use change on mercury distribution in soils of Alta Floresta, Southern Amazon. *Environ. Pollut.* **2004**, *129*, 247–255. [CrossRef] [PubMed]
43. Almeida, M.D.; Lacerda, L.D.; Bastos, W.R.; Herrmann, J.C. Mercury loss from soils following conversion from forest to pasture in Rondônia, Western Amazon, Brazil. *Environ. Pollut.* **2005**, *137*, 179–186. [CrossRef]
44. United Nations Environment Programme-UNEP (2013). *Global Mercury Assessment 2013: Sources, Emissions, Releases and Environmental Transport*; UNEP Chemicals Branch: Geneva, Switzerland, 2013.
45. Bishop, K.; Shanley, J.B.; Riscassi, A.; de Wit, H.A.; Eklöf, K.; Meng, B.; Mitchell, C.; Osterwalder, S.; Schuster, P.F.; Webster, J.; et al. Recent advances in understanding and measurement of mercury in the environment: Terrestrial Hg cycling. *Sci. Total Environ.* **2020**, *721*, 137647. [CrossRef]
46. Streets, D.; Horowitz, H.; Lu, Z.; Levin, L.; Thackray, C.; Sunderland, E. Five hundred years of anthropogenic mercury: Spatial and temporal release profiles. *Environ. Res. Lett.* **2019**, *14*, 84004. [CrossRef]
47. Hu, X.F.; Singh, K.; Chan, H.M. Mercury Exposure, Blood Pressure, and Hypertension: A Systematic Review and Dose–response Meta-analysis. *Environ. Health Perspect.* **2018**, *126*, 076002. [CrossRef]
48. Rice, G.E.; Hammitt, J.K.; Evans, J.S. A Probabilistic Characterization of the Health Benefits of Reducing Methyl Mercury Intake in the United States. *Environ. Sci. Technol.* **2010**, *44*, 5216–5224. [CrossRef] [PubMed]
49. Vasconcellos, A.C.S.; Hallwass, G.; Bezerra, J.G.; Aciole, A.N.S.; Meneses, H.N.M.; Lima, M.O.; Jesus, I.M.; Hacon, S.S.; Basta, P.C. Health Risk Assessment of Mercury Exposure from Fish Consumption in Munduruku Indigenous Communities in the Brazilian Amazon. *Int. J. Environ. Res. Public Health* **2021**, *18*, 7940. [CrossRef] [PubMed]
50. Hacon, S.D.S.; Oliveira-Da-Costa, M.; Gama, C.D.S.; Ferreira, R.; Basta, P.C.; Schramm, A.; Yokota, D. Mercury Exposure through Fish Consumption in Traditional Communities in the Brazilian Northern Amazon. *Int. J. Environ. Res. Public Health* **2020**, *17*, 5269. [CrossRef] [PubMed]
51. Oliveira Santos, E.C.; de Jesus, I.M.; da Silva Brabo, E.; Loureiro, E.C.B.; da Silva Mascarenhas, A.F.; Weirich, J.; Cleary, D. Mercury exposures in riverside Amazon communities in Para, Brazil. *Environ. Res.* **2000**, *84*, 100–107. [CrossRef]
52. Hacon, S.; Barrocas, P.R.; Vasconcellos, A.C.S.D.; Barcellos, C.; Wasserman, J.C.; Campos, R.C.; Azevedo-Carloni, F.B. An overview of mercury contamination research in the Amazon basin with an emphasis on Brazil. *Cad. Saúde Pública* **2008**, *24*, 1479–1492. [CrossRef]
53. Faial, K.; Deus, R.; Deus, S.; Neves, R.; Jesus, I.; Santos, E.; Brasil, D. Mercury levels assessment in hair of riverside inhabitants of the Tapajós River, Pará State, Amazon, Brazil: Fish consumption as a possible route of exposure. *J. Trace Elem. Med. Biol.* **2015**, *30*, 66–76. [CrossRef]
54. Freitas, J.S.; Lacerda, E.; Maria, C.; Rodrigues, D., Jr.; Corvelo, T.C.O.; Silveira, L.C.L.; Souza, G.S. Mercury exposure of children living in Amazonian villages: Influence of geographical location where they lived during prenatal and postnatal development. *An. Acad. Bras. Ciências* **2019**, *91* (Suppl. 1). Available online: https://www.scielo.br/j/aabc/a/LR7bb3VDGJVvxxjr9pNqYxM/?format=pdf&lang=en (accessed on 21 March 2021). [CrossRef]
55. Dórea, J.G.; Barbosa, A.C.; Ferrari, Í.; De Souza, J.R. Fish consumption (Hair Mercury) and nutritional status of Amazonian Amer-Indian Children. *Am. J. Hum. Biol. Off. J. Hum. Biol. Assoc.* **2005**, *17*, 507–514. [CrossRef]
56. Dórea, J.G.; Marques, R.C. Mercury levels and human health in the Amazon Basin. *Ann. Hum. Biol.* **2016**, *43*, 349–359. [CrossRef]
57. Dorea, J.G.; Barbosa, A.C.; Ferrari, I.; Souza, J.R. Mercury in hair and in fish consumed by riparian women of the Rio Negro, Amazon, Brazil. *Int. J. Env. Health Res.* **2003**, *13*, 239–248. [CrossRef]
58. Vega, C.; Orellana, J.; Oliveira, M.; Hacon, S.; Basta, P. Human Mercury Exposure in Yanomami Indigenous Villages from the Brazilian Amazon. *Int. J. Environ. Res. Public Health* **2018**, *15*, 1051. [CrossRef] [PubMed]
59. USEPA. *Mercury Update: Impact on Fish Advisories*; EPA-823-F01-011; Office of Water: Washington, DC, USA, 2001.
60. The Food and Agriculture Organization of the United Nations (FAO). *Summary and Conclusions of the Sixty-First Meeting of the Joint FAO/WHO Expert Committee on Food Additives (JECFA)*; World Health Organization: Rome, Italy, 2003; pp. 18–22. Available online: https://apps.who.int/iris/bitstream/handle/10665/42849/WHO_TRS_922.pdf?sequence=1&isAllowed=y (accessed on 2 May 2021).
61. Weihe, P.; Grandjean, P.; Jørgensen, P.J. Application of hair-mercury analysis to determine the impact of a seafood advisory. *Environ. Res.* **2005**, *97*, 201–208. [CrossRef] [PubMed]
62. Crump, K.S.; Kjellström, T.; Shipp, A.M.; Silvers, A.; Stewart, A. Influence of prenatal mercury exposure upon scholastic and psychological test performance: Benchmark analysis of a New Zealand cohort. *Risk Anal.* **1998**, *18*, 701–713. [CrossRef]
63. Santos Serrão de Castro, N.; de Oliveira Lima, M. Hair as a biomarker of long-term mercury exposure in Brazilian Amazon: A systematic review. *Int. J. Environ. Res. Public Health* **2018**, *15*, 500. [CrossRef]

64. Reuben, A.; Frischtak, H.; Berky, A.; Ortiz, E.J.; Morales, A.M.; Hsu-Kim, H.; Pendergast, L.L.; Pan, W.K. Elevated hair mercury levels are associated with neurodevelopmental deficits in children living near artisanal and small-scale gold mining in Peru. *Geo. Health* **2020**, *4*, e2019GH000222. [CrossRef] [PubMed]
65. Costa Junior, J.M.F.; Lima, A.A.D.S.; Rodrigues Junior, D.; Khoury, E.D.T.; Souza, G.D.S.; Silveira, L.C.D.L.; Pinheiro, M.D.C.N. Emotional and motor symptoms in riverside dwellers exposed to mercury in the Amazon. *Rev. Bras. De Epidemiol.* **2017**, *20*, 212–224. [CrossRef] [PubMed]
66. Marques, R.C.; Bernardi, J.V.; Abreu, L.; Dórea, J.G. Neurodevelopment outcomes in children exposed to organic mercury from multiple sources in a tin-ore mine environment in Brazil. *Arch. Environ. Contam. Toxicol.* **2015**, *68*, 432–441. [CrossRef]
67. Marques, R.C.; Bernardi, J.V.; Cunha, M.P.; Dórea, J.G. Impact of organic mercury exposure and home delivery on neurodevelopment of Amazonian children. *Int. J. Hygen Environ. Health* **2016**, *219*, 498–502. [CrossRef] [PubMed]
68. Crespo-Lopez, M.E.; Augusto-Oliveira, M.; Lopes-Araújo, A.; Santos-Sacramento, L.; Yuki Takeda, P.; Macchi, B.M.; Do Nascimento, J.L.M.; Maia, C.S.F.; Lima, R.R.; Arrifano, G.P. Mercury: What can we learn from the Amazon? *Environ. Int.* **2021**, *146*, 106223. [CrossRef]
69. Axelrad, D.A.; Bellinger, D.C.; Ryan, L.M.; Woodruff, T. Dose–Response Relationship of Prenatal Mercury Exposure and IQ: An Integrative Analysis of Epidemiologic Data. *Environ. Health Perspect.* **2007**, *115*, 609–615. [CrossRef]
70. Dórea, J.G.; de Souza, J.R.; Rodrigues, P.; Ferrari, Í.; Barbosa, A.C. Hair mercury (signature of fish consumption) and cardiovascular risk in Munduruku and Kayabi Indians of Amazonia. *Environ. Res.* **2005**, *97*, 209–219. [CrossRef]
71. Bélangera, M.; Miraul, M.; Dewailly, E.; Plante, M.; Berthiaume, L.; Noël, M.; Julien, P. Seasonal mercury exposure and oxidant-antioxidant status of James Bay sport fishermen. *Metab. J.* **2008**, *57*, 630–636. [CrossRef] [PubMed]
72. Gulley, A.L. Valuing environmental impacts of mercury emissions from gold mining: Dollar per troy ounce estimates for twelve open-pit, small-scale, and artisanal mining sites. *Resour. Policy* **2017**, *52*, 266–272. [CrossRef]
73. Spadaro, J.V.; Rabl, A. Global Health Impacts and Costs Due to Mercury Emissions. *Risk Anal.* **2008**, *28*, 603–613. [CrossRef] [PubMed]
74. Kahhat, R.; Parodi, E.; Larrea-Gallegos, G.; Mesta, C.; Vázquez-Rowe, I. Environmental impacts of the life cycle of alluvial gold mining in the Peruvian Amazon rainforest. *Sci. Total Environ.* **2019**, *662*, 940–951. [CrossRef]
75. Rice, G.; Hammit, J.K. Economic Valuation of Human Health Benefits of Controlling Mercury Emissions from U.S. Coal-Fired Power Plants. NESCAUM—Northeast States for Coordinated Air Use Management. 2005. Available online: https://www.nescaum.org/documents/rpt050315mercuryhealth.pdf (accessed on 10 August 2020).
76. Transande, L.; DiGangi, J.; Evers, D.; Petrlik, J.; Buck, D.; Samánek, J.; Beeler, B.; Turnquist, M.; Regan, K. Economic implications of mercury exposure in the context of th global mercury treaty: Hair mercury levels and estimated lost economic productivity in selected developing countries. *J. Environ. Manag.* **2016**, *183*, 229–235. [CrossRef] [PubMed]
77. Fox-Rushby, J. Disability adjusted life years (DALYs) for decision-making? An overview of the literature. In *Health Economics Research Group (HERG) Dept of Life Sciences Research Papers*; Office of Health Economics: London, UK, 2002.
78. Steckling, N.; Bose-O'Reilly, S.; Pinheiro, P.; Plass, D.; Shoko, D.; Drasch, G.; Bernaudat, L.; Siebert, U.; Hornberg, C. The burden of chronic mercury intoxication in artisanal small-scale gold mining in Zimbabwe: Data availability and preliminary estimates. *Environ. Health* **2014**, *13*, 111. [CrossRef] [PubMed]
79. Swain, E.; Jakus, P.; Rice, G.; Lupi, F.; Maxson, P.; Pacyna, J.; Penn, A.; Spiegel, S.; Veiga, M. Socioeconomic Consequences of Mercury Use and Pollution. Royal Swedish Academy of Sciences. *Ambio* **2007**, *36*, 45–61. [CrossRef]
80. Viscusi, W.K.; Aldy, J.E. The value of a statistical life: A critical review of market estimates throughout the world. *J. Risk Uncertainty* **2003**, *27*, 5–76. [CrossRef]
81. Cameron, T.; Deshazo, J.R.; Johnson, E. *Willingness to Pay for Health Risk Reductions: Differences by Type of Illness*; Working Paper; Department of Economics, University of Oregon: Eugene, OR, USA, 2011.
82. Kenkel, D. WTP- and QALY-Based Approaches to Valuing Health for Policy: Common Ground and Disputed Territory. *Environ. Resour. Econ.* **2006**, *34*, 419–437. [CrossRef]
83. Neumann, P.J.; Sanders, G.D.; Russell, L.B.; Siegel, J.E.; Ganiats, T.G. *Cost-Effectiveness in Health and Medicine*, 2nd ed.; Oxford University Press: New York, NY, USA, 2016.
84. Fan, C.-Y.; Fann, J.C.-Y.; Yang, M.-C.; Lin, T.-Y.; Chen, H.-H.; Liu, J.-T.; Yang, K.-C. Estimating global burden of COVID-19 with disability-adjusted life years and value of statistical life metrics. *J. Formos. Med. Assoc.* **2021**, *120*, S106–S117. [CrossRef] [PubMed]
85. Grandjean, P.; Bellanger, M. Calculation of the disease burden associated with environmental chemical exposures: Application of toxicological information in health economic estimation. *Environ. Health* **2017**, *16*, 123. [CrossRef] [PubMed]
86. Mapbiomas. Small-Scale Gold Mining in Brazil. 2021. Available online: https://plataforma.brasil.mapbiomas.org/ (accessed on 15 October 2021).
87. Swenson, J.J.; Carter, C.E.; Domec, J.C.; Delgado, C.I. Gold mining in the Peruvian Amazon: Global prices, deforestation, and mercury imports. *PLoS ONE* **2011**, *6*, e18875. [CrossRef]
88. Kutter, Vinicius Tavares; Castilhos, Zuleica Carmen. Inventário do uso e Emissões de Mercúrio em Mineração Artesanal de pequena escala de ouro no Brasil [resultados preliminares]. VI Jornada do Programa de Capacitação Institucional–PCI/CETEM–21 de setembro de. 2017. Available online: https://www.wwf.org.br/?65922/estudo-estima-emissoes-mercurio-garimpos-ouro-brasil (accessed on 4 November 2021).

89. Beckvar, N.; Field, J.; Salazar, S.; Hoff, R. Contaminants in aquatic habitats at hazardous waste sites: Mercury. *NOAA Tech. Memo. NOS ORCA* **1996**, *100*, 74.

90. Ikingura, J.; Akagi, H.; Mujumba, J.; Messo, C. Environmental assessment of mercury dispersion, transformation and bioavailability in the Lake Victoria Goldfields, Tanzania. *J. Environ. Manag.* **2006**, *81*, 167–173. [CrossRef] [PubMed]

91. Lino, A.S.; Kasper, D.; Guida, Y.S.; Thomaz, J.R.; Malm, O. Total and methyl mercury distribution in water, sediment, plankton and fish along the Tapajós River basin in the Brazilian Amazon. *Chemosphere* **2019**, *235*, 690–700. [CrossRef] [PubMed]

92. Vieira, M.; Bernardi, J.V.; Dórea, J.G.; Rocha, B.C.; Ribeiro, R.; Zara, L.F. Distribution and availability of mercury and methylmercury in different waters from the Rio Madeira Basin, Amazon. *Environ. Pollut.* **2018**, *235*, 771–779. [CrossRef] [PubMed]

93. Jardim, W.F.; Bisinoti, M.C.; Fadini, P.S.; da Silva, G.S. Mercury Redox Chemistry in the Negro River Basin, Amazon: The Role of Organic Matter and Solar Light. *Aquat. Geochem.* **2010**, *16*, 267–278. [CrossRef]

94. Richter, L. Especiação Química e Composição Lsotópica de Elementos Traço No Ambiente: Arsênio No Pantanal e Mercúrio na Floresta Amazônica. Ph.D. Thesis, Universidade Estadual de Campinas, Campinas, Brazil, 2020.

95. Lázaro, W.; Díez, S.; Bravo, A.; Silva, C.; Ignácio, A.; Guimarães, J. Cyanobacteria as regulators of methylmercury production in periphyton. *Sci. Total Environ.* **2019**, *668*, 723–729. [CrossRef]

96. Guimaraes, J.; Malm, O.; Pfeiffer, W. A simplified radiochemical technique for measurements of net mercury methylation rates in aquatic systems near gold mining areas, Amazon, Brazil. *Sci. Total Environ.* **1995**, *175*, 151–162. [CrossRef]

97. Brito, B.C.; Forsberg, B.R.; Kasper, D.; Amaral, J.H.; de Vasconcelos, M.R.; de Sousa, O.P.; Cunha, F.A.; Bastos, W.R. The influence of inundation and lake morphometry on the dynamics of mercury in the water and plankton in an Amazon floodplain lake. *Hydrobiologia* **2017**, *790*, 35–48. [CrossRef]

98. Wu, P.; Kainz, M.; Bravo, A.; Akerblom, F.; Sonesten, L.; Bishop, K. The importance of bioconcentration into the pelagic food web base for methylmercury biomagnification: A meta-analysis. *Sci. Total Environ.* **2019**, *646*, 357–367. [CrossRef] [PubMed]

99. Kelly, D.J.; Budd, K.; Lefebvre, D.D. Biotransformation of mercury in pH-stat cultures of eukaryotic freshwater algae. *Arch. Microbiol.* **2007**, *187*, 45–53. [CrossRef] [PubMed]

100. Peterson, L.R.; Trivett, V.; Baker, A.J.; Aguiar, C.; Pollard, A.J. Spread of metals through an invertebrate food chain as influenced by a plant that hyperaccumulates nickel. *Chemoecology* **2003**, *13*, 103–108. [CrossRef]

101. Passos, C.J.S.; Mergler, D.; Fillion, M.; Lemire, M.; Mertens, F.; Guimarães, J.R.D.; Philibert, A. Epidemiologic confirmation that fruit consumption influences mercury exposure in riparian communities in the Brazilian Amazon. *Environ. Res.* **2007**, *105*, 183–193. [CrossRef]

102. International Programme on Chemical Safety (IPCS): Environmental Health Criteria 101, Methylmercury. World Health Organization, Geneva. 1990. Available online: http://www.inchem.org/documents/ehc/ehc/ehc101 (accessed on 10 January 2020).

103. Ulrich, S.; Tanton, T.; Abdrashitova, S. Mercury in the Aquatic Environment: A Review of Factors Affecting Methylation. Critical Reviews in Environmental Science and Technology. *Crit. Rev. Environ. Sci. Technol.* **2001**, *31*, 241–293. [CrossRef]

104. Nevado, J.B.; Martín-Doimeadios, R.R.; Bernardo, F.G.; Moreno, M.J.; Herculano, A.M.; Do Nascimento, J.L.M.; Crespo-López, M.E. Mercury in the Tapajós River basin, Brazilian Amazon: A review. *Environ. Int.* **2010**, *36*, 593–608. [CrossRef]

105. Bastos, W.; Dórea, J.; Bernardi, J.; Lauthartte, L.; Mussy, M.; Lacerda, L.; Malm, O. Mercury in fish of the Madeira river (temporal and spatial assessment); Brazilian Amazon. *Environ. Res.* **2015**, *140*, 191–197. [CrossRef]

106. Barbosa, A.C.; De Souza, J.; Dórea, J.G.; Jardim, W.F.; Fadini, P.S. Mercury Biomagnification in a Tropical Black Water, Rio Negro, Brazil. *Arch. Environ. Contam. Toxicol.* **2003**, *45*, 235–246. [CrossRef]

107. Malm, O.; Castro, M.B.; Bastos, W.R.; Branches, F.J.; Guimarães, J.R.; Zuffo, C.E.; Pfeiffer, W.C. An assessment of Hg pollution in different goldmining areas, Amazon Brazil. *Sci. Total Environ.* **1995**, *175*, 127–140. [CrossRef]

108. Brabo, E.D.S.; Santos, E.D.O.; Jesus, I.M.D.; Mascarenhas, A.F.; Faial, K.F. Mercury levels in fish consumed by the Sai Cinza indigenous community, Munduruku Reservation, Jacareacanga County, State of Para, Brazil. *Cad. Saúde Pública* **1999**, *15*, 325–332. [CrossRef] [PubMed]

109. Passos, C.J.S.; Da Silva, D.S.; Lemire, M.; Fillion, M.; Guimaraes, J.R.D.; Lucotte, M.; Mergler, D. Daily mercury intake in fish-eating populations in the Brazilian Amazon. *J. Exp. Sci. Environ. Epidemiol.* **2008**, *18*, 76–87. [CrossRef] [PubMed]

110. Organização das Nações Unidas para Agricultura e Alimentação (FAO). *Codex Genereal Standard for Contaminants and Toxins in Food and Feed*; FAO: Rome, Italy, 1995; Available online: https://www.fao.org/fao-who-codexalimentarius/sh-proxy/en/?lnk=1&url=https%253A%252F%252Fworkspace.fao.org%252Fsites%252Fcodex%252FStandards%252FCXS%2B193-1995%252FCXS_193e.pdf (accessed on 25 April 2021).

111. Isaac, V.J.; Almeida, M.C. *El Consumo de Pescado en la Amazonía Brasileña*; FAO: Rome, Italy, 2011; ISBN 978-92-5-307029-9.

112. Cerdeira, R.G.P.; Ruffino, M.I.; Isaac, V.J. Consumo de pescado e outros alimentos pela população ribeirinha do lago grande de Monte Alegre, PA-Brasil. *Acta Amaz.* **1997**, *27*, 213–227. [CrossRef]

113. Batista, V.D.; Isaac, V.J.; Viana, J.P. *Exploração e Manejo Dos Recursos Pesqueiros da Amazônia. A Pesca e os Recursos Pesqueiros na Amazônia Brasileira*; Ibama/ProVárzea: Manaus, Brazil, 2004; pp. 63–151.

114. Mangas, F.P.; Rebello, F.K.; dos Santos, M.A.S.; Martins, C.M. Caracterização do perfil dos consumidores de peixe no município de Belém, estado do Pará, Brasil. *Rev. Em Agronegócio E Meio Ambiente* **2016**, *9*, 839–857. [CrossRef]

115. Instituto Brasileiro de Geografia e Estatística-IBGE. Pesquisa de Orçamentos Familiares. Estimativas Populacionais das Medianas de Altura e Peso de Crianças, Adolescentes e Adultos, Por Sexo, Situação do Domicílio e Idade—Brasil e Grandes Regiões—Região Norte. 2008. Available online: https://www.ibge.gov.br/estatisticas/sociais/saude/24786-pesquisa-de-orcamentos-familiares-2.html?=&t=microdados (accessed on 13 June 2021).

116. Bisinoti, M.C.; Jardim, W.F. O comportamento do metilmercúrio (metilHg) no ambiente. *Química Nova* **2004**, *27*, 593–600. [CrossRef]

117. World Health Organization—WHO. Guidance for identifying populations at risk from mercury exposure. In *UNEP DTIE Chemicals Branch and WHO Department of Food Safety, Zoonoses and Foodborne Diseases*; WHO: Geneva, Switzerland, 2008; Available online: https://wedocs.unep.org/bitstream/handle/20.500.11822/11786/IdentifyingPopnatRiskExposuretoMercury_2008Web.pdf?sequence=1&isAllowed=y (accessed on 10 October 2020).

118. Kocman, D.; Horvat, M. 2011.Non-point source mercury emission from the Idrija Hg-mine region: GIS mercury emission model. *J. Environ. Manag.* **2011**, *92*, 2038–2046. [CrossRef]

119. Olivero-Verbel, J.; Caballero-Gallardo, K.; Turizo-Tapia, A. Mercury in the gold mining district of San Martin de Loba, South of Bolivar (Colombia). *Environ. Sci. Pollut. Res.* **2015**, *22*, 5895–5907. [CrossRef]

120. Chen, X.; Ji, H.; Yang, W.; Zhu, B.; Ding, H. Speciation and distribution of mercury in soils around gold mines located upstream of Miyun Reservoir, Beijing, China. *J. Geochem. Explor.* **2016**, *163*, 1–9. [CrossRef]

121. Langeland, A.L.; Hardin, R.D.; Neitzel, R.L. Mercury levels in human hair and farmed fish near artisanal and small-scale gold mining communities in the Madre de Dios River Basin, Peru. *Int. J. Environ. Res. Public Health* **2017**, *14*, 302. [CrossRef]

122. Olivero-Verbel, J.; Caballero-Gallardo, K.; Negrete-Marrugo, J. Relationship Between Localization of Gold Mining Areas and Hair Mercury Levels in People from Bolivar, North of Colombia. *Biol. Trace Element Res.* **2011**, *144*, 118–132. [CrossRef]

123. Barbieri, F.L.; Gardon, J. Hair mercury levels in Amazonian populations: Spatial distribution and trends. *Int. J. Health Geogr.* **2009**, *8*, 71. [CrossRef] [PubMed]

124. Appleton, J.; Williams, T.; Breward, N.; Apostol, A.; Miguel, J.; Miranda, C. Mercury contamination associated with artisanal gold mining on the island of Mindanao, the Philippines. *Sci. Total Environ.* **1999**, *228*, 95–109. [CrossRef]

125. van Straaten, P. Human exposure to mercury due to small scale gold mining in northern Tanzania. *Sci. Total Environ.* **2000**, *259*, 45–53. [CrossRef]

126. Telmer, K.H.; Daneshfar, B.; Sanborn, M.S.; Kliza-Petelle, D.; Rancourt, D.G. The role of smelter emissions and element remobilization in the sediment chemistry of 99 lakes around the Horne smelter, Quebec. *Geochem. Explor. Environ. Anal.* **2006**, *6*, 187–202. [CrossRef]

127. Green, C.S.; Lewis, P.J.; Wozniak, J.R.; Drevnick, P.E.; Thies, M.L. A comparison of factors affecting the small-scale distribution of mercury from artisanal small-scale gold mining in a Zimbabwean stream system. *Sci. Total Environ.* **2019**, *647*, 400–410. [CrossRef] [PubMed]

128. Tomiyasu, T.; Kono, Y.; Kodamatani, H.; Hidayati, N.; Rahajoe, J.S. The distribution of mercury around the small-scale gold mining area along the Cikaniki river, Bogor, Indonesia. *Environ. Res.* **2013**, *125*, 12–19. [CrossRef]

129. Scarlat, A. Mercury Contamination in the Amazon Basin. Unpublished. 2014.

130. Diringer, S.E.; Feingold, B.J.; Ortiz, E.J.; Gallis, J.A.; Araújo-Flores, J.M.; Berky, A.; Pan, W.K.Y.; Hsu-Kim, H. River transport of mercury from artisanal and small-scale gold mining and risks for dietary mercury exposure in Madre de Dios, Peru. *Environ. Sci. Process. Impacts* **2015**, *17*, 478–487. [CrossRef] [PubMed]

131. Sousa, R.G.C.; Humston, R.; Freitas, C.E.C. Movement patterns of adult peacock bass Cichla temensis between tributaries of the middle Negro River basin (Amazonas–Brazil): An otolith geochemical analysis. *Fish. Manag. Ecol.* **2016**, *23*, 76–87. [CrossRef]

132. Pfeiffer, W.; Malm, O.; Souza, C.; Lacerda, L.; Silveira, E.; Bastos, W. Mercury in the Madeira River ecosystem, Rondônia, Brazil. *For. Ecol. Manag.* **1991**, *38*, 239–245. [CrossRef]

133. Resende, E.K.; Catella, A.C.; Nascimento, F.L.; Palmeira, S.D.S.; Pereira, R.A.C.; Lima, M.D.S.; de Almeida, V.L.L. *Biologia do Curimbatá (Prochilodus Lineatus), pintado (Pseudoplatystoma Corruscans) e Cachara (Pseudoplatystoma fasciatum) na Bacia Hidrográfica do rio Miranda, Pantanal do Mato Grosso do Sul, Brasil*; EMBRAPA-CPAP: Corumbá, MS, USA, 1996; 75p.

134. Barthem, R.B.; de Brito Ribeiro, M.C.L.; Petrere, M., Jr. Life strategies of some long-distance migratory catfish in relation to hydroelectric dams in the Amazon Basin. *Biol. Conserv.* **1991**, *55*, 339–345. [CrossRef]

135. Nunes, M.U.S.; Hallwass, G.; Silvano, R.A.M. Fishers' local ecological knowledge indicate migration patterns of tropical freshwater fish in an Amazonian river. *Hydrobiologia* **2019**, *833*, 197–215. [CrossRef]

136. Oliveira, R.C.; Dórea, J.G.; Bernardi, J.V.E.; Bastos, W.R.; Almeida, R.; Manzatto, G. Fish consumption by traditional subsistence villagers of the Rio Madeira (Amazon): Impact on hair mercury. *Ann. Hum. Biol.* **2010**, *37*, 629–642. [CrossRef] [PubMed]

137. Brazilian Ministry of Health. Glossário Temático de Economia da Saúde. 3ª Edição. 2013. Available online: http://bvsms.saude.gov.br/bvs/publicacoes/glossario_tematico_economia_saude.pdf (accessed on 19 February 2021).

138. Murray, C.J. Quantifying the burden of disease: The technical basis for disability-adjusted life years. *Bull. World Health Organ.* **1994**, *72*, 429–445. [PubMed]

139. Murray, C.J.; Lopez, A.D. Measuring the Global Burden of Disease. *N. Engl. J. Med.* **2013**, *369*, 448–457. [CrossRef]

140. Haacker, M.; Hallett, T.; Atun, R. On discount rates for economic evaluations in global health. *Health Policy Plan.* **2019**, *35*, 107–114. [CrossRef]

141. Nordhaus, W. Estimates of the social cost of carbon: Background and results from the RICE-2013 model and alternative approaches. *J. Assoc. Environ. Resour. Econ.* **2014**, *1*, 273–312.
142. Moore, M.A.; Boardman, A.E.; Vining, A.R. Social Discount Rates for Seventeen Latin American Countries: Theory and Parameter Estimation. *Public Finance Rev.* **2019**, *48*, 43–71. [CrossRef]
143. The World Health Organization—WHO. Global Burden of Disease 2004 Update: Disability Weights for Diseases and Conditions. Available online: https://www.who.int/healthinfo/global_burden_disease/GBD2004_DisabilityWeights.pdf?ua=1 (accessed on 29 June 2020).
144. Hammitt, J.K.; Robinson, L.A. The Income Elasticity of the Value per Statistical Life: Transferring Estimates between High and Low Income Populations. *J. Benefit-Cost Anal.* **2011**, *2*, 1–29. [CrossRef]
145. Bosworth, R.; Hunter, A.; Kibria, A. *The Value of a Statistical Life: Economics and Politics*; STRATA: Logan, UT, USA, 2017.
146. The World Health Organization—WHO. Macroeconomics and health: Investing in health for economic development. In *Report of the Comission on Macroeconomics and Health*; WHO: Geneva, Switzerland, 2001.
147. Instituto Brasileiro de Geografia e Estatística—IBGE. GDP per Capital Brazil. 2020. Available online: https://www.ibge.gov.br/explica/pib.php (accessed on 10 January 2021).
148. Roman, H.A.; Walsh, T.L.; Coull, B.A.; Dewailly, E.; Guallar, E.; Hattis, D.; Mariën, K.; Schwartz, J.; Stern, A.H.; Virtanen, J.K.; et al. Evaluation of the Cardiovascular Effects of Methylmercury Exposures: Current Evidence Supports Development of a Dose–Response Function for Regulatory Benefits Analysis. *Environ. Health Perspect* **2011**, *119*, 607–614. [CrossRef]
149. Fewtrell, L.J.; Pruss-Ustun, A.; Landrigan, P.; Ayuso-Mateos, J.L. Estimating the global burden of disease of mild mental retardation and cardiovascular diseases from environmental lead exposure. *Environ. Res.* **2004**, *94*, 120–133. [CrossRef]
150. Oliveira, G.M.; Brant, L.C.; Polanczyk, C.A.; Biolo, A.; Nascimento, B.R.; Malta, D.C.; Souza, M.D.; Soares, G.P.; Xavier, G.F.; Machline-Carrion, M.J. Estatística Cardiovascular—Brasil 2020. *Arq. Bras. Cardiol.* **2020**, *115*, 308–439. [CrossRef] [PubMed]
151. Karagas, M.R.; Choi, A.L.; Oken, E.; Horvat, M.; Schoeny, R.; Kamai, E.M.; Cowell, W.; Grandjean, P.; Korrick, S. Evidence on the Human Health Effects of Low-Level Methylmercury Exposure. *Environ. Health Perspect.* **2012**, *120*, 799–806. [CrossRef] [PubMed]
152. Mozaffarian, D.; Shi, P.; Morris, J.S.; Spiegelman, D.; Grandjean, P.; Siscovick, D.S.; Willett, W.C.; Rimm, E.B. Mercury exposure and risk of cardiovascular disease in two US cohorts. *N. Engl. J. Med.* **2011**, *364*, 1116–1125. [CrossRef] [PubMed]
153. Rockhill, B.; Newman, B.; Weinberg, C. Use and misuse of population attributable fractions. *Am. J. Public Health* **1998**, *88*, 15–19. [CrossRef]
154. Porta, M. International Epidemiological Association. In *A Dictionary of Epidemiology, 5 th ed.*; Oxford University Press: New York, NY, USA, 2008.
155. Instituto Brasileiro de Geografia e Estatística—IBGE. Population Density and Proportion of Urban and Rural Population. 2010. Available online: https://censo2010.ibge.gov.br/sinopse/index.php?dados=8 (accessed on 26 June 2021).
156. Datasus. Morbidade de Hipertensão Arterial E Infarto Do Miocárdio Agudo NA Região Norte Do Brasil. 2021. Available online: http://tabnet.datasus.gov.br/cgi/tabcgi.exe?sih/cnv/niuf.def (accessed on 11 March 2021).
157. Zaletel-Kragelj, L.; Bozikov, J. Methods and tools in public health: A Handbook for Teachers, Researchers and Health Professionals. In Proceedings of the Forum for Public Health in South Eastern Europe, North Macedonia. 2010. Available online: https://www.researchgate.net/publication/256011831_METHODS_AND_TOOLS_IN_PUBLIC_HEALTH_A_Handbook_for_Teachers_Researchers_and_Health_Professionals_Title_Address_for_correspondence (accessed on 4 November 2021).
158. Forouzanfar, M.H.; Liu, P.; Roth, G.A.; Ng, M.; Biryukov, S.; Marczak, L. Global burden of hypertension and systolic blood pressure of at least 110 to 115 mm Hg, 1990–2015. *JAMA* **2017**, *317*, 165–182. [CrossRef]
159. Abhyankar, L.N.; Jones, M.R.; Guallar, E.; Navas-Acien, A. Arsenic Exposure and Hypertension: A Systematic Review. *Environ. Health Perspect.* **2012**, *120*, 494–500. [CrossRef] [PubMed]
160. Eum, K.-D.; Lee, M.-S.; Paek, D. Cadmium in blood and hypertension. *Sci. Total Environ.* **2008**, *407*, 147–153. [CrossRef]
161. Houston, M.C. Role of Mercury Toxicity in Hypertension, Cardiovascular Disease, and Stroke. *J. Clin. Hypertens.* **2011**, *13*, 621–627. [CrossRef]
162. Navas-Acien, A.; Guallar, E.; Silbergeld, E.K.; Rothenberg, S.J. Lead Exposure and Cardiovascular Disease—A Systematic Review. *Environ. Health Perspect.* **2007**, *115*, 472–482. [CrossRef]
163. Bonita, R.; Beaglehole, R.; Kjestrom, T. Epidemiologia Básica. 2nd Edition. 2010. Available online: https://apps.who.int/iris/bitstream/handle/10665/43541/9788572888394_por.pdf?sequence=5 (accessed on 28 May 2021).
164. Instituto Socioambiental. Cicatrizes na Floresta: Evolução do Garimpo Illegal na TI Yanomami em 2020. Hutukara Associação Yanomami Associação Wanasseduume Ye'kwana. Roraima, Brazil. 2021. Available online: https://acervo.socioambiental.org/acervo/documentos/cicatrizes-na-floresta-evolucao-do-garimpo-ilegal-na-ti-yanomami-em-2020 (accessed on 1 April 2021).
165. RAISG—Amazon Network of Georeferenced Socioenvironmental Information. Available online: https://www.amazoniasocioambiental.org/pt-br/mapas/ (accessed on 10 February 2021).
166. Gasparinetti, P.; De Bakker, L.; Queiroz, J.; Vilela, T. Economic valuation of artisanal small-scale gold mining impacts: A framework for value transfer application. *Resour. Policy J.* **2021**, in press.
167. World Wide Fund for Nature—WWF. Mercury Observatory. Available online: https://panda.maps.arcgis.com/apps/Cascade/index.html?appid=e74f4c219b3428b8e4bce4d7295f210 (accessed on 21 July 2021).

168. CID PUCESE-PRAS. *Informe de Valoración de Pasivos Socios Ambientales Vinculados a la Actividad Minera Aurífera Ilegal en El Norte de Esmeraldas*; Technical report; Pontifical Catholic University of Ecuador Sede Esmeraldas: Esmmeraldas City, Ecuador, 2011.
169. Miranda, J. Estimativa da Quantidade de Mercúrio Usada em Uma Mina de Ouro Abandonada e Avaliação das Técnicas de Remediação da Área Contaminada, em Descoberto—MG. Ph.D. Thesis, Federal University of Ouro Preto, Ouro Preto, Brazil, 2019.
170. Wan, X.; Lei, M.; Chen, T. Cost–benefit calculation of phytoremediation technology for heavy-metal-contaminated soil. *Sci. Total Environ.* **2016**, *563–564*, 796–802. [CrossRef] [PubMed]
171. Román-Danobeytia, F.; Huayllani, M.; Michi, A.; Ibarra, F.; Loayza-Muro, R.; Vásquez, T.; Rodríguez, L.; Carcía, M. Reforestation with four native tree species after abandoned goldmining in the Peruvian Amazon. *Ecol. Eng.* **2015**, *85*, 39–46. [CrossRef]
172. Steckling, N.; Devleesschauwer, B.; Winkelnkemper, J.; Fischer, F.; Ericson, B.; Krämer, A.; Hornberg, C.; Fuller, R.; Plass, D.; Bose-O'Reilly, S. Disability Weights for Chronic Mercury Intoxication Resulting from Gold Mining Activities: Results from an Online Pairwise Comparisons Survey. *Int. J. Environ. Res. Public Health* **2017**, *14*, 57. [CrossRef] [PubMed]
173. Steckling, N.; Tobollik, M.; Plass, D.; Hornberg, C.; Ericson, B.; Fuller, R.; Bose-O'Reilly, S. Global Burden of Disease of Mercury Used in Artisanal Small-Scale Gold Mining. *Ann. Glob. Health* **2017**, *83*, 234–247. [CrossRef] [PubMed]

International Journal of
Environmental Research and Public Health

Article

Impairment in Working Memory and Executive Function Associated with Mercury Exposure in Indigenous Populations in Upper Amazonian Peru

Alycia K. Silman [1,2,*], Raveena Chhabria [3], George W. Hafzalla [3], Leahanne Giffin [3], Kimberly Kucharski [3], Katherine Myers [3], Carlos Culquichicón [4], Stephanie Montero [4], Andres G. Lescano [4], Claudia M. Vega [5], Luis E. Fernandez [2,5,6,7], Miles R. Silman [2,6], Michael J. Kane [8] and John W. Sanders [3]

[1] Department of Psychology, Wake Forest University, Winston-Salem, NC 27109, USA
[2] Center for Energy, Environment, and Sustainability, Wake Forest University, Winston-Salem, NC 27109, USA
[3] Wake Forest School of Medicine, Winston-Salem, NC 27101, USA
[4] Emerge, Emerging Diseases and Climate Change Research Unit, School of Public Health and Administration, Universidad Peruana Cayetano Heredia (UPCH), San Martin de Porres 15102, Peru
[5] Centro de Innovación Científica Amazónica, Puerto Maldonado 17001, Peru
[6] Department of Biology, Wake Forest University, Winston Salem, NC 27109, USA
[7] Carnegie Amazon Mercury Project, Department of Global Ecology, Carnegie Institution for Science, 260 Panama Street, Stanford, CA 94305, USA
[8] Department of Psychology, University of North Carolina at Greensboro, Greensboro, NC 27412, USA
* Correspondence: silmanak@wfu.edu

Citation: Silman, A.K.; Chhabria, R.; Hafzalla, G.W.; Giffin, L.; Kucharski, K.; Myers, K.; Culquichicón, C.; Montero, S.; Lescano, A.G.; Vega, C.M.; et al. Impairment in Working Memory and Executive Function Associated with Mercury Exposure in Indigenous Populations in Upper Amazonian Peru. *Int. J. Environ. Res. Public Health* **2022**, *19*, 10989. https://doi.org/10.3390/ijerph191710989

Academic Editors: Masayuki Sakakibara, Win Thiri Kyaw, José Luis Rivera Parra and Paul B. Tchounwou

Received: 1 July 2022
Accepted: 30 August 2022
Published: 2 September 2022

Publisher's Note: MDPI stays neutral with regard to jurisdictional claims in published maps and institutional affiliations.

Abstract: The Matsigenka people living traditional lifestyles in remote areas of the Amazon rely on a fish-based diet that exposes them to methylmercury (MeHg) at levels that have been associated with decreased IQ scores. In this study, the association between Hg levels and working memory was explored using the framework of the Multicomponent Model. Working memory tasks were modified to fit the culture and language of the Matsigenka when needed and included measures for verbal storage (Word Span) visuospatial storage (Corsi Block Task) and a measure of executive functions, the Self-Ordered Pointing Task (SOPT). An innovation of the Trail Making Tests A & B (TMT A & B) was pilot tested as another potential measure of executive functions. The mean hair Hg levels of 30 participants, ages 12 to 55 years, from three different communities (Maizal, Cacaotal and Yomibato) was 7.0 ppm (sd = 2.40), well above the World Health Organization (WHO) limit for hair of 2.0 ppm and ranged from 1.8 to 14.2 ppm, with 98% of a broader sample of 152 individuals exceeding the WHO limit. Hair Hg levels showed significant associations with cognitive performance, but the degree varied in magnitude according to the type of task. Hg levels were negatively associated with executive functioning performance (SOPT errors), while Hg levels and years of education predicted visuospatial performance (Corsi Block accuracy). Education was the only predictor of Word Span accuracy. The results show that Hg exposure is negatively associated with working memory performance when there is an increased reliance on executive functioning. Based on our findings and the review of the experimental research, we suggest that the SOPT and the Corsi Block have the potential to be alternatives to general intelligence tests when studying remote groups with extensive cultural differences.

Keywords: methylmercury; working memory; executive functions; indigenous population; environmental exposure; Matsigenka; Manu National Park; Amazon Basin

1. Introduction

Mercury (Hg) is a heavy metal and neurotoxin that is damaging to humans and wildlife and is a persistent global environmental pollutant that can be emitted from both natural and anthropogenic sources [1]. As an element, it is a contaminant that can be exceptionally long-lived, taking tens of thousands of years to be cleared from landscapes by moving into fauna and flora, being exported to other ecosystems, or becoming permanently buried [2]. The largest contributor of mercury to the environment is artisanal and small-scale gold mining (ASGM), which accounts for over 37% of the global total of mercury emitted from all sources [3].

All chemical forms of Hg are toxic to humans. Exposure to several forms can cause deleterious effects on the central and peripheral nervous systems, cardiovascular system, urinary system, immune system, skin, and lungs [4,5]. Of most concern is the organic and highly bioavailable form, methylmercury (MeHg) [6,7]. Primarily produced in aquatic environments through the methylation of elemental mercury by microorganisms, MeHg is a potent neurotoxin that readily accumulates in living organisms and biomagnifies within food webs, becoming enriched in high trophic levels of freshwater and marine ecosystems [8–10]. Humans with diets that consume such higher trophic level organisms, such as Amazonian indigenous populations [11] are at elevated risk of methylmercury exposure through diet [12].

The Matsigenka in the upper Madre de Dios watershed in Southeastern Peru is one group of people susceptible to Hg poisoning because of their reliance on fish as a main source of protein [11,13,14]. Estimated to number around 1000, the Matsigenka families living inside the protected areas of Manu National Park (MNP) are geographically restricted and culturally traditional relative to the lives and practices of Matsigenka in nearby regions [15]. They maintain a lifestyle that includes the practice of swidden agriculture, hunting with bow and arrow, fishing, and gathering [13,14] with the bulk of their animal protein coming from migratory fish known to have high concentrations of mercury [16]. The Peruvian government maintains primary schools in all villages and health posts in the most inhabited communities but living conditions inside the park are restricted to protect the ecosystems of MNP and the Matsigenka culture and include prohibitions on the types of everyday objects and practices that one would readily find in the nearest towns outside of the park [17]. The improvement of education, health, nutrition, and basic services (drinkable water and solar electricity) has led to migration and integration of groups of Matsigenka living in isolation in the headwaters of the river. There are other groups in voluntary isolation that are neither contacted nor registered, but their number is unknown.

Studies with Amazonian peoples show that high levels of Hg are associated with decreased cognitive functioning. Impairments are typically assessed with Intelligence or IQ tests that are a battery of standardized measures summarized into a composite score and normalized against a known population. Recent work has showed that for Peruvian children each one unit increase in log hair mercury levels was associated with a 2.59-point decrease in the IQ index for General Cognitive Ability [18]. Similarly, studies of children in the Amazon region of Brazil reported that for every 10 ppm Hg increase there was a decrease of half a standard deviation in estimated IQ scores [19]. These dose-related effects have also been reported outside of Amazonia regions, such as the Faroe Islands where longitudinal work has found that every 10-ppm increase in Hg is associated with a 1.8 to 2.2-point mean decrease in IQ score [20]. The findings are consistent with other studies conducted across the globe [21–23] and provide strong evidence (with exceptions reported from studies in the Seychelle Islands [23,24] and the United States [25] that exposure to MeHg negatively impacts the general intelligence of humans.

Despite their breadth and validity, standardized intelligence tests have long been suspected of educational, cultural, and language biases [26] and present unique challenges when being used with indigenous people. Discerning how impairment is related to elemental toxins becomes difficult or impossible when used to assess people whose environment and lifestyle does not match the origin of the test, and when the regionally validated norms used for evaluation force participants to use a second language and counting system [27].

These challenges are particularly striking when considering the culturally isolated and in some cases recently contacted Amazonian indigenous populations. For example, the forward and backwards digit span task [28], included in many intelligence batteries or as a measure of short-term memory [19], requires the recall of digits from a base-10 number system. Some Amazonian indigenous groups use a "one, two, many" counting system that does not include discrete indicators for items of three or above [29,30]. Remote communities vary greatly in how much formal schooling is available, leading to differences in instruction in Spanish and the use of Arabic numerals. When IQ tests are administered in a second language and with a less familiar set of knowledge, the true scores of participants may be obscured and could exaggerate deficits due to environmental toxins. The concern has not gone unnoticed [31,32] and are supported by studies that report differences between urban Amazonian groups and rural indigenous people on intelligence tests [18,33].

One way to minimize cultural bias in cognitive testing is to shift away from using broad standardized test batteries and instead focus on select cognitive processes that are well-understood in contemporary theory and can be measured with tools tailored for a specific population. To this end, the construct of working memory is promising for a closer examination of how mercury may affect intelligence and cognitive functioning, and previous Hg investigations with Amazonian children that have measured working memory using sub-scales of IQ tests have found them to be sensitive to levels of hair Hg levels [18,19]. Working memory has been conceptualized in many ways [34–36] from a domain-free model based on time [37] to one based on attention resources [38]. (For a comparison among models, see [34–36]. One of the earliest and most productive models is the Baddeley Multicomponent Model that presents working memory as a network of interacting components including short-term memory storage for different types of information (verbal and visual-spatial) to be held temporarily or rehearsed. Operating concurrently with the storage components is a "central executive" that is used when tasks demand attention and cognitive control beyond passive storage and rehearsal [39–41].

The experimental work from the Multicomponent Model and others has led to results suggesting that working memory is key to understanding the fluid aspects of intelligence that are assessed in standardized IQ tests [42]. Kane et al. ([43], p. 170) describe working memory span tasks as measures that "tap a very general—and very important—cognitive primitive" that contributes to individual scores in general intelligence factors. More recently, Shipstead et al. ([44], p. 773) argued that "…working memory capacity and fluid intelligence arise from similar cognitive mechanisms but are reliant on these mechanisms to different degrees". In support of this view, correlations between measures of complex working memory tasks and measures of general fluid intelligence or "Gf" are less than perfect, but robust (r = 0.59, [45]; r = 0.65, [45]). Although not equivalent to IQ scores [46], select working memory measures that involve the use of executive functions [47] may provide coarse estimates of associations between Hg and the cognitive processes underlying the IQ score. Considering these relationships among assessment tools could provide an effective workaround to measure cognitive impairments when there are extensive cultural differences among people living in areas of with Hg exposures.

In this study, we first sought to measure hair Hg levels of Matsigenka residents living in three villages within the restricted area of Manu National Park in Madre de Dios, Peru. Testing inside MNP has been limited given the isolation and restrictions imposed in the area, but assessment is important for a full understanding of how MeHg is reaching communities living in the headwaters far from, but still connected by river and migratory fish to, ASGM activity. Second, we sought to understand how exposure to MeHg may impact human cognitive functions by focusing on the construct of working memory in lieu of traditional IQ tests. Three tasks were used to measure working memory in this study, one for each component of storage and executive functions: verbal short-term memory (Word Span), visuospatial short-term memory (Corsi Block) and central executive processing (Self-Ordered Pointing Task or SOPT). These tasks were chosen because they have a strong tradition within working memory research and the stimuli and instructions can be easily adapted for the Matsigenka as needed. Additionally, the tasks allow for a comparison of performance across components that employ central executive processes to varying degrees. 'Simple' storage measures like the verbal word span should not rely on executive functions [47] and therefore, performance should not be associated with Hg levels. In contrast, a 'complex' working memory measure, like the SOPT, requires executive functions [47] and performance is expected to show associations with Hg levels in a similar way as an IQ test. Visuospatial storage tasks, such as the Corsi Block, are designed to capture capacity of the visual and spatial domain, but there is ample evidence that they elicit executive functions to similar extents as the more 'complex' working memory tasks [48–50]; thus, performance is also expected to be associated with Hg levels.

A fourth measure in this study is a pilot test for a new version of the Trail Making Test A & B (TMT-A, TMT-B). The TMT A & B is a clinical assessment of executive functions that is sensitive to a variety of brain injuries [51]. The version created for the Matsigenka communities replaces alphabet and numerical stimuli with non-verbal stimuli and alters the instructions to match cultural norms, but still demands the use of executive functions in the form of set shifting [52]. As a pilot test, the results from the TMT are considered separately from the other measures, but the same predictions regarding executive processes still apply and decreased performance is expected to be associated with higher levels of Hg levels.

The objective for this study is to learn about the levels of MeHg exposure for people who are living in the headwaters of the Manu River located inside the restricted zone of MNP and to investigate how that exposure is associated with impairment of cognitive functions. The cognitive measures were chosen or designed specifically for the Matsigenka residents. This study is novel because our approach forgoes the traditional IQ test and focuses singly on the construct of working memory using findings from the cognitive experimental literature for prediction. We highlight the relationship between central executive functions of working memory and those underlying intelligence testing. This cross-discipline work is important because it offers a strategy for studying how exposure to Hg is linked to cognitive impairment even when investigating people who live in isolated areas with cultures and languages that are distinct enough to render an IQ score uninterpretable [53].

2. Materials and Methods

Study site and population. Manu National Park (MNP) is Earth's highest biodiversity park, consisting of 1.7 M ha of forested landscapes in the tropical Andes and adjacent Amazonian lowlands in SE Peru. Participants in this study come from three Matsigenka communities living along the Manu River inside the MNP: Maizal, Cacaotal and Yomibato, (Figure 1). All three communities are located within the upper Madre de Dios watershed, and collectively are situated about 180 Km away from any urban areas. Two cohorts of data collection are presented in this study. The first data collection was in June 2017 and was only from the community of Maizal. This collection was intended to test levels of Hg in hair samples from residents and to test the feasibility of the modified SOPT task with a Matsigenka sample. A total of 38 individuals, ages ranging from age 1 to 65 years, were given physicals and samples of blood and hair were collected. Twelve adults, ages between 22 and 65 years, were also administered the SOPT task in an abbreviated form to learn if the task had potential as a tool for assessing executive function. Two participants in the Maizal community in 2017 did not know their age but were determined during the physical exam and interview to be over 12 years old.

The second data collection was in June 2018 and included the Maizal community as well as the Cacaotal and Yomibato communities. A total of 114 individuals across the three communities were given a physical and neurological examination (Mini Mental Status Exam and Cranial Nerve Exam), and samples of blood and hair were collected. Participants who were ages 12 years or older and who provided a full set of responses to the Word Span task, Corsi Block task, and SOPT (n = 30). Three individuals from the Cacaotal community did not know their age or birthday but were judged to be at least 12 years old based on measures of size, maturation, and appearance.

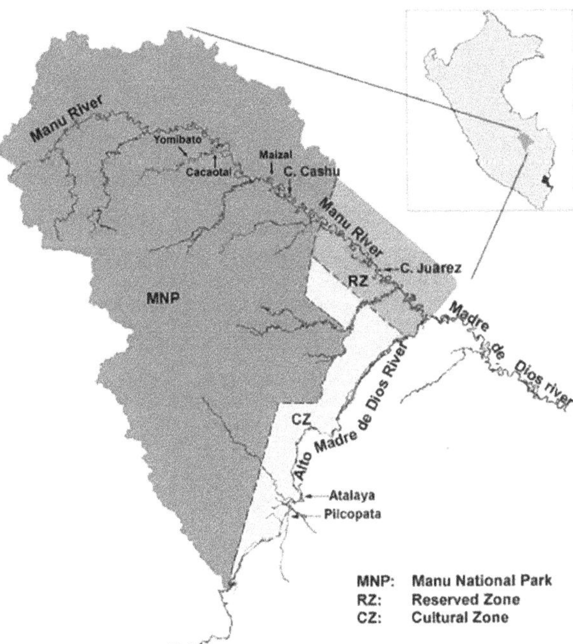

Figure 1. Map of Manu National Park, Peru. Location of Cacaotal, Yomibato, and Maizal are indicated by red stars on map.

For the pilot test of the TMT, the sample size is smaller. The TMT was given to a subset of 19 participants who were able to successfully complete the practice trials. The eligible individuals were from the villages of Cacaotal and Yomibato; none of the residents from the village of Maizal were able to complete the practice trials. A comparison of mean Hg hair levels conducted between the group included in the TMT sample (n = 19, mean hair Hg level = 6.43 ppm (sd = 3.80)) and those excluded from the TMT sample (n = 11, mean hair Hg level = 4.63 ppm (sd = 2.92)) did not show that there was a significant difference between groups, t = -1.45, p = 0.07.

For the remainder of this paper, the descriptions of protocol, data presented in the Results section, and interpretations come from the second cohort of data collection in 2018 from the residents of Maizal, Caocatal and Yomibato unless explicitly stated to be from the Maizal 2017 cohort or to include participants from both data collections.

Consent Process. Written informed consent was obtained from participants ages 18 or older with signature and/or fingerprint (in case of illiteracy) and for those under the age of 18 years, written consent was obtained from the parent or accompanying adult family member, as well as an informed assent for participants ages 12 or older. Ethics approval to conduct research on human subjects was granted through Wake Forest University Medical School Institutional Review Board (Human Subjects: IRB000044673) and the Institutional Ethics Committee of the Universidad Peruana Cayetano Heredia, Lima, Peru (N° 100806). The authorization to conduct the study in 2017 and 2018 within the Manu National Park was provided by the Peruvian National Protected Areas Service (Servicio Nacional de Áreas Naturales Protegidas por el Estado, or SERNANP by its Spanish acronym).

Mercury Assessment. Hair samples of approximately 0.5 g were collected from each participant. Hair was cut close to the occipital area of the scalp with stainless steel scissors, placed in paper envelopes inside of zip lock plastic bags with silica, and stored at room temperature. Hair samples were analyzed for total mercury (THg) using EPA Method 7473 (Mercury in Solids and Solutions by Thermal Decomposition, Amalgamation, and

Atomic Absorption Spectrophotometry; EPA 1998) on a Milestone DMA-80 dual-cell Direct Mercury Analyzer at the Mercury and Environmental Chemistry Laboratory (LAMQA) laboratory in Puerto Maldonado, Madre de Dios, Peru. Exposure to THg was used as a proxy of exposure to MeHg as more than 90% of Hg in hair is MeHg [54].

Physical Exam and Neurological Assessment. Central nervous system assessment included measures of head circumference and a comprehensive cranial nerve exam. Peripheral nervous system assessment involved tests for abnormalities in reflexes, strength, sensation, tremor, and gingivitis. Motor systems dysfunction was screened for by testing balance, bilateral coordination, upper extremity coordination, visual motor control, visuospatial organization, and upper extremity speed. During the examination, participants' height was measured using a stadiometer and their weight was assessed using an electronic digital scale to calculate the body mass index. They were also interviewed about their education levels, family status, and diet and nutrition. From these assessments, two participants were determined ineligible for participation in the cognitive tasks due to potential mental disability and visual abnormalities. Another two participants were determined ineligible because they were not residents MNP but were only visiting temporarily.

Fish Consumption. During the interview, participants were asked, (1) "How frequently do you eat fish?" with responses ranging from "every day" to "every 5 days", "never" or "other" as well as (2) "How many times per day do you eat fish generally?" with responses ranging from "1 time per day" to 5 times per day", "never" or "other". These two responses were used to create an index of total weekly fish consumption by dividing days in a week by the reported frequency (question 1), multiplied by the reported daily intake (question 2).

Blood tests. About 10 uL of blood samples from finger-prick were collected. Then, hemoglobin was analyzed in a portable photometric device; HemoCue® Hb 201 DM system (HemoCue® AB, Ängelholm, Sweden), which allowed it to measure 0–25.6 g/dL. Assessment of anemia was considered as <13.5 g/dL for males and <12 g/dL for females.

Cognitive assessments. Measures for each component of working memory are described below along with any adaptations made from the original, or most commonly used, measure. The cognitive tasks were administered in fixed order (as presented). All instructions and verbal stimuli were provided in the Matsigenka language and were administered through a native Matsigenka translator who traveled with the research team to translate Matsigenka into Spanish for test administration.

Word Span Task. Verbal short-term memory is commonly measured in intelligence batteries by presenting a short list of stimuli, such as numbers (as in the Digit Span task) or words (as in the Word Span task) and asking participants to recall the stimuli immediately after presentation. The Word Span is used in some intelligence batteries instead of Digit Span, particularly those designed for children such as the Woodcock-Johnson test [55]. For Matsigenka participants, the stimuli were lists of high-frequency, unrelated words as stimuli presented in the Matsigenka language. Words were chosen from three noun categories that would have multiple exemplars found within the environment and interactions of the daily lives of the Matsigenka living in MNP: Nature (14 items), Human (10 items), Functional Object (12 items). The list of thirty-six words were translated into both Spanish and Matsigenka and were determined by bilingual Matsigenka translators to be of regular frequency use in the everyday lexicon of the communities sampled (list presented in Supplemental Section S1). Word Span tests were administered in four set sizes (3, 4, 5, and 6 words), with two trials of each presented in ascending order. The test administrator read the words aloud at the rate of approximately one word per second. The participant was asked to orally recall the items in order. An accuracy score was calculated with full credit for recall of words in the correct order and half credit for words recalled out of order per trial. Scores per trial were summed and divided by the set size and then averaged across trials per participant.

Corsi Block Task. The original Corsi Block Task was developed to assess hemispheric specialization between verbal and spatial processing [56,57] but variations have since been created for clinical diagnosis, experimental research, and intelligence testing. A traditional paper form of the Corsi Block Task was chosen here for its longstanding use across testing situations [58]. A single paper for presentation of locations was used across all trials. Stimuli were nine two-dimensional squares printed on paper in an asymmetrical pattern (see Figure 2a). The test administrator tapped a sequence of squares and participants immediately responded by tapping the same squares in order. Each participant's responses were recorded by the test administrator on a separate response form. One practice trial of three squares was given before test trials began. If a participant responded with error, the same practice trial was repeated until the participant responded correctly. Test trials were of four different set sizes presented in ascending order (3, 4, 5, and 6), with two trials of each for a total of eight trials. An accuracy score was calculated with full credit for recall of locations in the correct order and half credit for locations recalled out of order per trial. Scores per trial were summed and divided by the set size and then averaged across trials per participant.

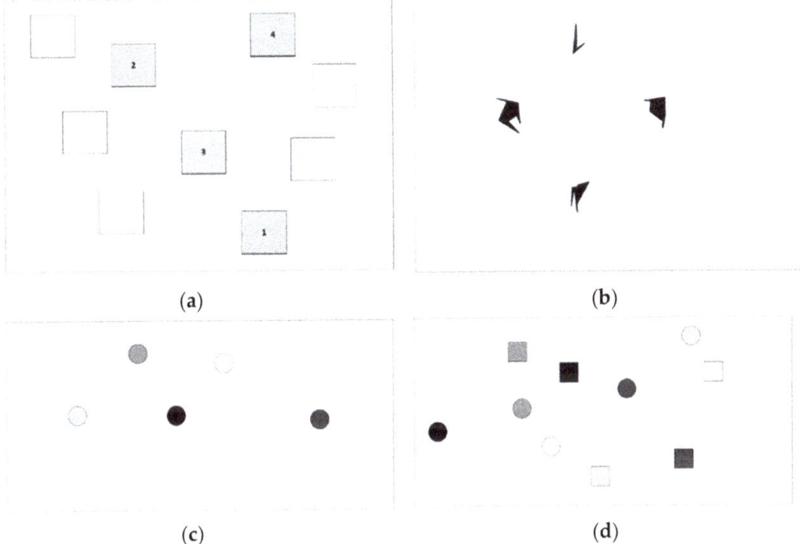

Figure 2. (**a–d**) Example Stimuli for Cognitive Tasks: (**a**) Example of Corsi Block set size 4. Numbers in square are used to demonstrate a sequential pattern but were not displayed to the participant. (**b**) Example of SOPT set size 4 with 4 different "Attneave" shapes, (**c**) Example of TMT A "Shades" presented in random locations, (**d**) Example of TMT B "Shades & Shapes" presented in random locations.

Self-Ordered Pointing Task (SOPT). The SOPT was originally developed to look for associations between frontal lobe functioning and working memory [51] and is now considered a well validated measure of central executive processes [59]. For Matsigenka living in remote areas of the Amazon with highly variable schooling, the advantages of the SOPT include that the test is free of language and number stimuli, can be administered without the use of a computer, and speeded response times are not required. The last two issues, administration without need for a computer and speeded response, are preferred testing formats given the context of the Matsigenka lifestyle and environment which does not involve computers or timed behaviors (for discussion on cultural influences in cognitive tasks including speed, see [60]).

During the SOPT, participants were presented with a set size of either 4, 6, or 8 shapes on a single sheet of paper (see Figure 2b for example). Stimuli in this version of the task was abstract Attneave shapes [61]. The use of shapes that are abstract is a departure from other versions that have been used in cultural adaptations of the task [62] but were chosen for two reasons: (1) to avoid participants' use of a verbal strategy that may reduce the use of executive functions, and (2) to choose stimuli that could potentially be used with other groups of people being studied because of Hg exposure. During the practice and test trials, the same Attneave shapes were presented on each page during a trial but with the locations of each shape shifted between pages. Locations of shapes were random within an area on the page. As the test administrator turned pages, participants were to point to a new item on each page (i.e., an item not yet pointed to on that trial), but traditional instructions telling participants to point to a different location each time was not given to participants. (After data collection, response forms were checked for this strategy but there were no participants that pointed to the same location repeatedly.) Two trials of the set size of three were presented as practice prior to beginning test trials. Set sizes were given in ascending order with two trials per each set-size of 4, 6, and 8 items with errors totaled across all trials. The total number of errors was summed across set sizes and trials for the dependent variable, rather than accuracy in keeping with the scoring method used in the cognitive literature.

This version of the SOPT was piloted in 2017 with a small sample of adults in the Maizal community (N = 12). For this data collection, only the highest set size successfully completed set size, or "span" score was recorded. All 12 Maizal participants were able to correctly respond to at least one trial at the lowest level (set 4). Furthermore, included in that version was an additional set size of 10 shapes per page, but this level was determined to be too difficult and was removed from testing for the 2018 data collection.

Trail Making Test. Another well-regarded executive task, the TMT [63] is commonly used in clinical assessments of executive functions (also referred to as "cognitive flexibility" [64,65]. The TMT was first published as part of the Army Individual Test Battery in 1944 and was later adopted for research and diagnosis of hemispheric and frontal lobe dysfunction [66]. In the original TMT, participants are asked to connect numbers in ascending order (Version A or TMT-A) and then to alternate between connecting numbers and letters in alphabetical/numerical order (Version B or TMT-B) and the time to complete each trial is recorded. Time to complete (in seconds) can then be calculated as a difference score 64or ratio [65] between the B and A forms. For the Matsigenka, a novel version of the TMT was created to avoid the use of an alphabet or numbers. For Trail Making Test A "Shades" (Figure 2c), shaded circles replaced letters so that participants were asked to connect circles using a pen on a response sheet, ordering from lightest to darkest. For Trail Making Test B "Shades and Shapes" (Figure 2d), numbers were replaced with shapes. Participants were asked to alternate between connecting shades and shapes. For example, in Figure 2d, the participant would start at the white circle, move to the white square, then the lightest gray circle, then the lightest gray square until complete with the black circle and black square. The alternating between shapes does not match the original B version with regard to complexity because the number stimuli require participants to advance to different numerals in ascending order and the shapes remain constant between connections. It is possible that this could have decreased the difficulty of the "Shades and Shapes" version in comparison to the original TMT B with letters and numbers.

After two practice trials with three shades/shapes each, the test trials began with five trials of TMT A "Shades" followed by five trials of TMT B "Shades and Shapes". Trials were scored as being successfully completed if all connections between shades and/or shapes were accurate. Typically, the TMT is a speeded task and cut-offs for deficiencies and impairments are determined by the number of seconds an individual requires to perform the task perfectly. For our versions, the subjects were instructed to emphasize accuracy and there was no mention of completing the task as quickly as possible.

Other versions of the TMT have been created that are language-free and use accuracy for the outcome measure but are still not appropriate for the Matsigenka because the modifications involve the use of numbers and/or involve memorizing stimuli [67]. Although using accuracy as a dependent variable eliminates the possibility of comparing outcomes with other studies using a traditional TMT task, it does not seem to change the underlying cognitive processes being captured, as demonstrated by a culturally adapted version of the TMT developed to assess Northern Aboriginal people in Northern Australia [62].

Data Analysis

Descriptive statistics of all demographics, health, diet, and hair Hg levels and cognitive task variables were summarized across and between each community. Community comparisons with post hoc analysis and effect size calculations were calculated for most variables including cognitive tasks and Hg levels. For the TMT task, a *t*-test was done on the Hg levels between participants qualifying for the test trials and those ineligible for the test trials. Correlations and confidence intervals were calculated between all variables within the sample size of 30 participants and again separately for the sample of 19 participants who took the TMT tasks as part of the pilot test for the two versions. Because the three main predictor variables of Hg levels, Age, and Education were correlated, Variance Inflation Factors were calculated for each before running regression analyses. The missing ages for participants were replaced with the average age for the relevant community. The association between hair mercury levels and cognitive processing for three working memory tasks: Word Span, Corsi Block, and SOPT were tested using Ordinary Least Squares multiple linear regression. For each outcome, the same model was run using the covariates of Hg levels, Years of Education, and Age. All analyses were performed in JMP 16.0.0 (SAS Institute, Inc., Cary, NC, USA) for Windows software.

3. Results

3.1. Demographics and Health Indicators

The mean and frequency outcomes for each community are presented in Table 1 along with the summary statistics for the total sample across communities.

Table 1. Summary Statistics for Demographics, Health and Diet Indicators, and Cognitive Tasks.

Community		All Participants	Cacaotal	Maizal	Yomibato
n		30	11	3	16
Gender	F (M)	18 (12)	6 (05)	3 (0)	9 (7)
Age	average years (sd)	29 (13.8)	28 * (9.6)	45 (16.2)	26 (14.7)
Education					
	none	8	3	2	4
	primary	12	6	1	5
	secondary	10	2	0	7
	mean years	4.77 (3.8)	4.82 (3.68)	1.33 (2.31)	5.38 (3.99)
Fish Consumption					
	Total per Week	4.10 (2.4)	4.58 (1.8)	3.73 (2.9)	3.98 (2.5)
Body Mass Index					
	mean (sd)	22.93 (2.72)	22.8 (2.42)	24.38 (3.25)	22.75 (2.92)
	<20	4	1	0	3
Anemia					
	Hemoglobin	11.77 (1.42)	11.8 (1.56)	11.5 (1.65)	12.15 (1.25)
	Males < 13.5, Females < 12	22	6	1	12
Hg (ppm)					
	mean (sd)	7.05 (2.40)	6.04 (2.43)	11.49 (2.40)	3.61 (2.38)
	min	1.81	2.84	9.67	1.81
	max	14.21	11.42	14.21	11.43
Word Span					
	mean accuracy (sd)	0.52 (0.19)	0.45 (0.19)	0.30 (0.11)	0.60 (0.16)
Corsi Block Span					
	mean accuracy (sd)	0.59 (0.27)	0.47 (0.25)	0.25 (0.14)	0.73 (0.25)
SOPT Errors					
	mean errors (sd)	4.23 (1.65)	4.72 (1.38)	4.67 (0.55)	3.81 (1.90)

* Three participants from Cacaotal did not know their age.

There were no physical or neurological impairments detected in any of the 30 participants included in the sample of cognitive measures, but some health indicators did suggest malnutrition and anemia. The mean BMI was 22.23 (sd = 3.46), with eight participants being underweight (BMI < 20) and the mean hemoglobin level was 11.81 (sd = 1.42), with 22 participants presenting anemia (females < 12 g/dL; males < 13.5 g/dL). Anemia is a variable of interest because it can be used as a proxy for chronic malnutrition [68–70], which is associated with poorer performance on intelligence measures [71]. Tests between communities did not reveal differences for BMI, F (2, 27) = 0.45, p = 0.64, ω^2 = 0, or Hemoglobin, F (2, 27) = 1.5, p = 0.234, ω^2 = 0.03. The calculated index of Fish Consumption 'Total per Week' did not show significant differences between communities, F (2, 27) = 0.29, p = 0.75, ω^2 = 0. Education levels varied, with some participants from each community reporting no schooling (27%) and most participants reported either having attended some primary school education (40%) or some secondary school (33%). When counted in years of schooling from 0 to 12, the mean was 4.77 years (sd = 3.8). The community of Maizal reported the lowest mean years in education (1.33 years) followed by Cacaotal (4.82) and the highest mean years of education were in Yomibato (5.38). One participant in Yomibato reported having finished secondary school and was now serving as a teacher for the community. The comparisons between communities for years of education were not significant, F (2, 27) = 1.45, p = 0.25, ω^2 = 0.03.

3.2. Hair Mercury Levels

Each participant's Hg level was the average of two samples of hair strand analyses except for eight participants whose Hg level was based on the analysis from a single hair sample. Hg levels were high, with the mean Hg level across communities of 7.05 ppm, exceeding the WHO [72,73] recommended limits by 3.5×, with 98% of a broader sample of 152 individuals (across 2017 and 2018 data collections) exceeding the WHO limit, having mean Hg levels above the World Health Organization threshold limit of 2.0 µg/g. The exceptions were three individuals from Yomibato, who had levels of 1.81, 1.92, and 1.99 µg/g. An ANOVA test showed that Hg levels between groups were significant, F (2, 27) = 14.44, p < 0.01, ω^2 = 0.47. Maizal residents had the highest mean Hg levels. Both collection years (2017 and 2018) found that Hg levels approached six times that of the threshold level [1,2]. The 2018 collection from Maizal had mean hair Hg levels of 11.49 ppm and the 2017 collection from Maizal (n = 38) was Hg level of 11.9 ppm (sd = 3.19, min = 2.4 ppm, max = 16.5 ppm). Figure 3. shows the observed hair Hg levels per village and includes the 2017 and 2018 data points for Maizal.

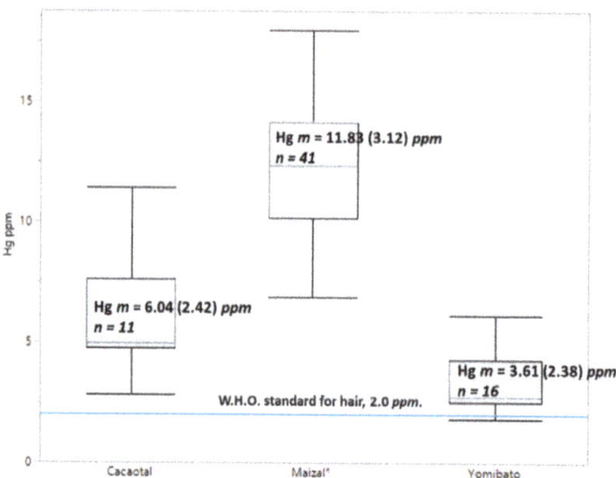

Figure 3. Hg levels per village with both the 2017 and 2018 cohort for the village of Maizal. Boxes show mean, interquartile range, and 2.5–97.5 percentiles.

3.3. Cognitive Tasks

3.3.1. Word Span

Across the three villages, the mean accuracy of words recalled was 0.52 (sd = 0.19). At the lowest set size of three items the mean accuracy was 0.87 (sd = 0.25) and 80% of participants were able to recall the words perfectly on at least one trial beyond the practice exercises. (Comparisons between set sizes for Word Span and subsequent tasks are presented in Supplemental Section S2). Differences in Word Span accuracy between communities were significant, F (2, 27) = 5.60, $p < 0.01$, $\omega^2 = 0.23$, as observed between Yombiato and Maizal ($p = 0.018$, Tukey's HSD) but were not significance between Yomibato and Cacaotal ($p = 0.07$; Tukey's HSD).

3.3.2. Corsi Block

Across communities, the mean accuracy for recall of block locations was 0.59 (sd = 0.27). Comparisons of mean Corsi Block accuracy between communities was significant, F (2, 27) = 8.67, $p < 0.01$, $\omega^2 = 0.34$, and post hoc tests confirmed differences between Yombiato and Maizal ($p < 0.01$, Tukey's HSD) and between Yomibato and Cacaotal ($p = 0.01$, Tukey's HSD).

3.3.3. Self-Ordered Pointing Test

The SOPT was scored by summing the number of errors across trials and set sizes rather than the accuracy (as was done with the Word Span and Corsi Block). The mean error score was 4.23 (sd = 1.65). At the lowest level, the mean error score was 0.47 (sd = 0.63) and 60% participants were able to complete the trial without any errors. Comparisons of mean SOPT errors between communities were not significant, F (2, 27) = 1.43, $p = 0.25$, $\omega^2 = 0.03$.

3.3.4. Trail Making Task A and B

The sample for the TMT analyses was the participants who were age 12 or older, were able to complete the Word Span task, Corsi Block Task, and SOPT, and who also completed the practice trials for both the TMT A and TMT B task. Using these criteria for inclusion, the sample size decreased from 30 to 19 participants all from the communities of Cacaotal and Yomibato (see Table 2 for summary statistics and TMT scores for this sample). In the traditional version of the TMT task, the primary dependent variable is time (in seconds) to complete the task and a difference score is calculated between the time for Part A and time for Part B with the resulting score capturing the time to complete the "task switching" required in Part B [74]. This version of the TMT record accuracy instead of time. Difference scores for accuracy were not calculated because the accuracy on Part A "Shades" and Part B "Shades and Shapes" were correlated, $r = 0.48$, $p = 0.03$, and would be unreliable as a measure of performance. Instead, accuracy was considered separately for each part of the TMT and for the relationship to other variables such as Hg levels.

TMT A "Shades". Mean proportion of correct responses across all trials and villages was 0.68 (sd = 0.35). Two participants were not able to answer any of the test trials correctly despite being able to complete the prior practice trials. Across all five trials, 42% of participants responded perfectly. Differences between the two communities, Yomibato and Cacaotal, were not significant, $t(17) = 0.03$, $p = 0.48$, d = 0.05.

TMT B "Shades & Shapes". Mean proportion of correct responses across all trials and villages was 0.31 (sd = 0.32). Six participants were not able to answer any of the test trials correctly after successful completion of the practice trials. Only one participant completed all five trials of Part B with perfect accuracy. Performance between the two communities, Yomibato and Cacaotal, were significantly different, $t(17) = 3.05$, $p < 0.01$, d = 0.72 and paired t-test between TMT A and TMT B responses confirmed the difference between the two versions, $t(18) = 8.56$, $p < 0.01$, d = 1.17.

Table 2. Summary Statistics for Trail Making Test Sample Pilot Test.

Community		All Participants *	Yomibato	Cacaotal
n		19	14	5
Gender	F (M)	10 (9)	7 (7)	3 (2)
Age	mean years	23 (12.9)	22 (12.61)	21 (5.58)
Education				
	mean years	6.63 (3.50)	6.29 (3.85)	7.6 (2.30)
Hg (ppm)				
	mean (sd)	4.63 (2.92)	3.56 (2.54)	7.12 (2.64)
	min	1.92	1.92	4.75
	max	11.43	11.43	11.42
Trail Making Test A "Shades"				
	mean percent accuracy (sd)	0.68 (0.35)	0.66 (0.34)	0.68 (0.42)
Trail Making Test B "Shapes & Shades"				
	mean percent accuracy (sd)	0.31 (0.32)	0.34 0(.29)	0.08 (0.11)

* Maizal participants were not administered the TMT A & B.

3.4. Correlations among Demographic Variables, Health, and Diet Indicators

The correlation matrix for all variables is presented in Table 3 and significant results are listed below along with 95% confidence intervals (CI). Hair Hg levels were positively correlated with Age, $r = 0.50$, [CI 0.17, 0.73], $p < 0.01$, but did not show a correlation with measures of Hemoglobin, $r = -0.13$, [CI -0.47, 0.24], $p = 0.48$, Body Mass Index, $r = 0.11$, [CI -0.26, 0.45], $p = 0.57$, or Fish Consumption as measured by the index Total per Week, $r = -0.13$, [CI -0.47, 0.24], $p = 0.48$. Education was negatively associated with Age, $r = -0.47$, [CI -0.71, -0.13], $p < 0.01$, indicating that the younger members of each community have more years of schooling than the older members. The measures of Fish Consumption, BMI, or Hemoglobin were not associated with any variables.

Table 3. Correlation Matrix for All Variables across Communities.

	Age	Education	BMI	Hemoglobin	Fish Consumption	Hg	Word Span	Corsi Block	SOPT Errors
Age	1.00								
Education	−0.47 *	1.00							
BMI	0.26	0.29	1.00						
Hemoglobin	−0.01	0.04	0.16	1.00					
Fish Consumption	0.04	0.12	−0.14	−0.02	1.00				
Hg	0.50 *	−0.21	0.11	−0.13	−0.13	1.00			
Word Span	−0.56 *	0.63 *	−0.08	0.31	−0.13	−0.38 *	1.00		
Corsi Block	−0.44 *	0.59 *	−0.01	0.28	−0.04	−0.56 *	0.62 *	1.00	
SOPT Errors	0.10	−0.33	−0.20	−0.28	0.21	0.41 *	−0.34	−0.31	1.00

* $p < 0.05$.

3.5. Correlations among Demographic Variables and Cognitive Tasks

Word Span accuracy was negatively related to Age, $r = -0.56$, [CI -0.76, -0.25] $p = 0.01$, and positively related to Education, $r = 0.63$, [CI 0.35, 0.81] $p < 0.01$. Corsi Block accuracy was also negatively related to Age, $r = -0.44$, [-0.69, -0.09] $p = 0.01$, and positively to Education, $r = 0.59$, [0.30, 0.79] $p < 0.01$. The correlation between errors on the SOPT and Education approached significance, $r = -0.33$, [CI $= -0.62$, 0.32] $p = 0.07$. Finally, Word Span accuracy and the Corsi Block accuracy were related, $r = 0.62$, [CI 0.33, 0.80] $p < 0.01$. The associations with Age and Education are not surprising given that declines in working memory with increased age are well-established [75] and the influence of education on intelligence tests with mercury exposed samples is also documented [18].

3.6. Correlations between Cognitive Tasks and Hair Hg Levels

Pearson's correlations between each of the cognitive tasks and Hg levels can be found in Table 3. Hair Hg levels showed a significant negative relationship with Words Span accuracy, $r = -0.38$ [CI -0.65, -0.03], $p = 0.04$, as well as Corsi Block accuracy, $r = -0.56$ [CI -0.76, -0.25], $p < 0.01$ indicating that increased Hg levels are associated with poorer performance. The correlation between errors on the SOPT and Hair Hg levels was a significant positive, $r = 0.41$ [0.06, 0.67], $p = 0.03$, indicating that Hg levels increased so too did errors (i.e., poorer performance on the SOPT).

A separate correlation analysis was conducted for the sample of participants who participated in the TMT A and B versions and found that accuracy in both versions had a negative relationship to Hair Hg levels, but neither were significant, (TMT B "Shades and Shapes", $r = -0.35$, [CI -0.69, 0.12] $p = 0.14$, and TMT A "Shades", $r = -0.36$, [CI -0.70, 0.11] $p = 0.13$. Accuracy on the TMT A and B did correlate with each other, $r = 0.48$, [CI 0.03, 0.77], $p = 0.04$, but there were no other significant correlations between either TMT versions and the other demographic variables (age, education) or the health indicators (BMI, Hemoglobin). The Version B of the TMT showed significant correlations with the Corsi Block accuracy, $r = 0.60$, [CI 0.20, 0.83] $p < 0.01$ and a negative relationship with SOPT errors, $r = -0.55$, [CI -0.80, -0.13] $p = 0.01$ but, not with Word Span Accuracy, $r = 0.31$, [CI -0.16, 0.67], $p = 0.20$.

To understand how specific health and demographic factors may contribute to the observed relationships between Hg levels and the working memory components, a multiple regression analysis was conducted for each of the three cognitive outcome tasks (Word Span, Corsi Block, and SOPT). The models were limited to three predictor variables because of the small sample size across communities; the same three predictors were used for each of the cognitive tasks to allow for comparisons of fit and contribution. The first predictor, hair Hg levels, was chosen to test the hypothesis that Hg levels are associated with the components of working memory to different degrees, depending on the involvement of executive functions. The next predictors, Education and Age, were prioritized based on the strength and significance of their correlations with Hg levels. Although the three predictors were correlated, Variance Inflation Factors (VIF) were low, showing that multicollinearity was not an issue: Age (VIF = 1.63), Education (VIF = 1.29), Hg level (VIF = 1.33). Variables not included in the model were the health indicators of Fish Consumption, BMI, and Hemoglobin because they showed no significant correlation to other variables. Before running the model, the three participants in the Cacaotal community whose ages were unknown, were interpolated using the group mean age of 28 years. A regression analysis was not conducted on the TMT A and B data because of the smaller sample size and because the data collection was considered a pilot test of the measure.

The multiple linear regression results for each of the three cognitive tasks are presented in Table 4. For the Word Span Task, the model using Age, Education, and Hg levels as predictors was able to explain 50% of the variance in accuracy, but only one variable, Education, was a significant predictor.

Table 4. Multiple Linear Regression Models for Working Memory Components and Predictors.

y	Model Fit	x	b	95% CI	p
Word Span Accuracy	$R^2 = 0.50$, Adj $R^2 = 0.44$	Age	-0.00	$(-0.01, 0.00)$	0.16
	$F(3,26) = 8.65, p < 0.01$	Education	0.02	$(0.01, 0.04)$	0.01
		Hg	-0.01	$(-0.03, 0.01)$	0.35
Corsi Block Accuracy	$R^2 = 0.55$, Adj $R^2 = 0.49$	Age	0.00	$(-0.01, 0.01)$	0.83
	$F(3,26) = 10.44, p < 0.01$	Education	0.04	$(0.01, 0.06)$	0.00
		Hg	-0.04	$(-0.06, -0.01)$	0.01
SOPT Errors	$R^2 = 0.29$, Adj $R^2 = 0.21$	Age	-0.04	$(-0.09, 0.01)$	0.14
	$F(3,26) = 3.57, p = 0.03$	Education	-0.16	$(-0.32, 0.00)$	0.05
		Hg	0.23	$(0.05, 0.42)$	0.02

The same model predicting Corsi Block accuracy was able to explain 55% of the observed variance. Education and Hg levels, but not age, were significant predictors of

Corsi Block accuracy. The model for SOPT errors were less predictive than for the other two cognitive tasks with 29% of variance explained. Hg levels significantly predicted errors on the SOPT when controlling for Age and Education. The predictor of Education approached significance when controlling for Age and Hg levels. The predictor of Age was not significant when controlling for Education and Hg levels. Importantly, responses on the Corsi Block and SOPT showed a continuous and quantitative response to Hg level across the range measured but was not significant for Word Span (Figure 4a–c).

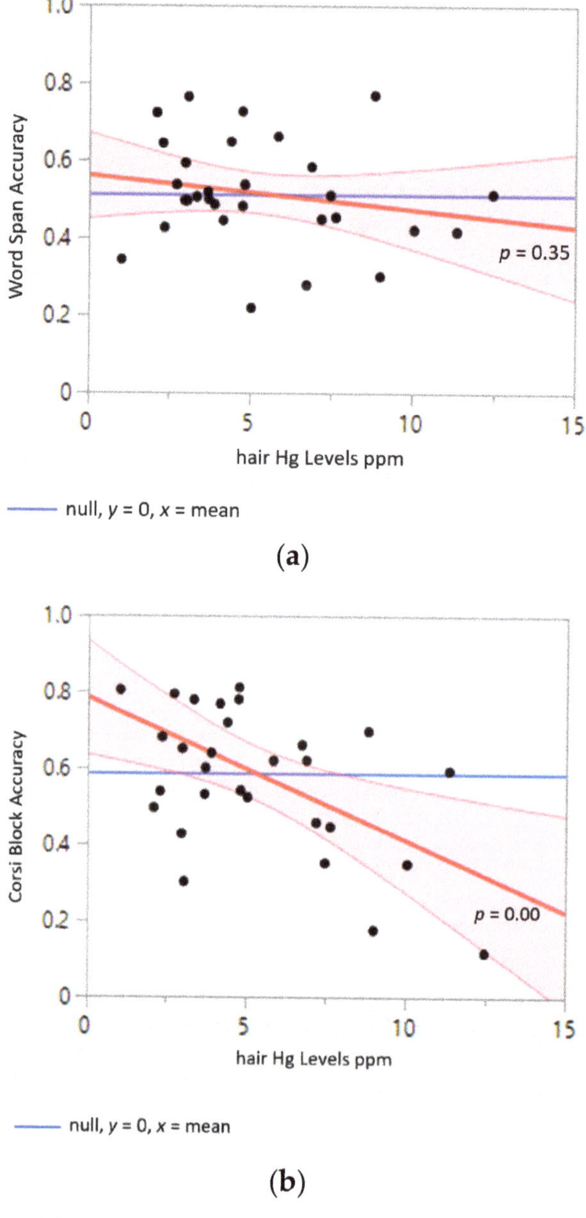

(a)

(b)

Figure 4. *Cont.*

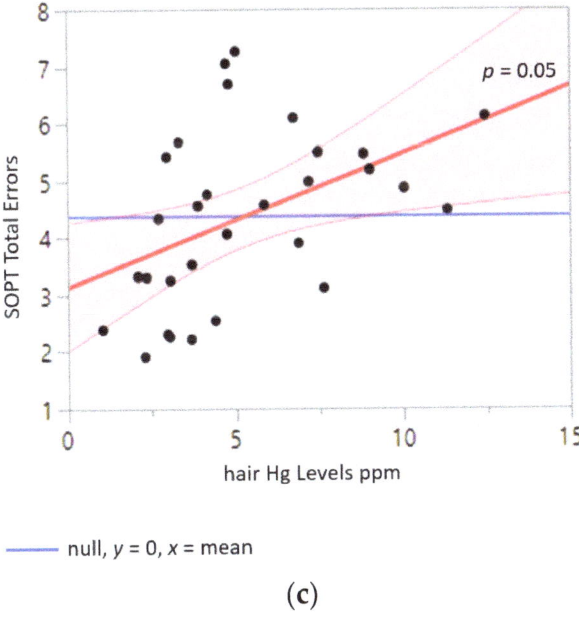

(c)

Figure 4. Added variable plots [76] showing the unique effect of a cognitive task (red line and associated 95% CI) as compared to a model assuming the model contains all the other terms (blue line) for each cognitive task but no relationship between the dependent and independent variable. (**a**) Word Span Accuracy vs. Hg, (**b**) Corsi Block Accuracy vs. Hg, (**c**) SOPT Total Errors vs Hg.

4. Discussion

The levels of methylmercury in hair of thirty participants from three different indigenous communities along the Manu River were found to be above the maximum threshold of 2.0 ppm in all but three individuals, with levels exceeding the WHO limit [72,73] by an average of 3.5×. The results point to widespread elevated Hg levels in even the most remote indigenous populations living in watersheds with illegal artisanal and small-scale gold mining—in this case between 300 and 400 km upstream from the nearest mine. This study also explored the association between methylmercury exposure and working memory using cognitive tasks chosen to measure the short-term storage and executive functions. Previous studies exploring the impact of Hg exposure have mostly relied on IQ tests or working memory sub-scales that are likely to be influenced by culture, education, and language and would not be appropriate for the Matsigenka living inside the restricted area of MNP [18,19]. Here, some tasks were modified by changing stimuli, instructions, and outcome variables to make the tasks more consistent with traditional Matsigenka culture. Results suggest that Hg exposure may impair cognitive processes that rely on executive functioning, and that these effects are seen at even relatively lower levels of Hg exposure and increase monotonically with increasing Hg concentrations in hair. The findings highlight the risk to Amazonian indigenous populations, especially those living in areas impacted by ASGM where mining activity increases levels of Hg in fish that are consumed as a main part of the indigenous diet [77]. Below, we discuss the results within the framework of the Multicomponent Model, cultural considerations, their relevance to work using IQ tests, and how these tasks may be used in future studies and contribute to the Hg assessment field more broadly.

4.1. Hg Levels

Outcomes from the Hg assays establish that the mercury occurring in the environment exists at detectable and high concentrations in the Matsigenka residents of MNP. Across two collection years and the broad sample of hair Hg levels, 98% percent of the participants in this study showed levels of Hg hair above the 2.0 ppm threshold [ref]. For the second collection year, hair Hg levels were different between the three villages: Yomibato had the lowest mean levels (3.61 ppm), Cacaotal was almost double (6.04 ppm), and in the village of Maizal measures were almost doubled again (11.49 ppm). The relatively lower Hg levels in the residents of Yomibato and Cacaotal are consistent with levels reported by other studies of Amazonian people living outside of the park [9] while the elevated levels in the Maizal community are consistent with the range found in the 2017 data collection, which averaged just under 12 ppm with a larger sample (n = 41). The communities did not differ on fish consumption nor was the diet indicator (Total per Week) correlated with hair Hg level. This finding is not surprising because our measure of fish consumption did not account for the type or size of fish eaten. The trophic status of fish and where it is caught is known to be related to the level of Hg passed on to humans via fish consumption [8,9]. Each of the three communities fish in the part of the river that is nearest and most accessible to them, which may also influence outcomes. For instance, the community of Maizal catch fish in the main portion of the river where it may be easier to catch larger predatory fish such as catfish. Yomibato and Cacaotal are located on a small tributary that gives them access to different and smaller fish. A closer analysis of the diet of each of the communities is needed before the role of fish consumption in Hg levels can be determined.

4.2. Hg Levels and Working Memory Tasks

The association between Hg exposure and cognition was examined for separate theoretical components of working memory and the measures for each were based on well-established assessment tools that have been validated in experimental and clinical research but were modified specifically for the Matsigenka. Importantly, for prediction and hypothesis testing, the tasks also involved differing degrees of central executive functions.

4.2.1. Verbal Short-Term Memory and Cultural Considerations

The word span task was used to measure verbal short-term memory and all participants demonstrated understanding and competency on practice trials and in the smallest set size, demonstrating participants' ability to hold information temporarily. Recall declined as set size increased, demonstrating that the task was challenging enough to avoid ceiling effects in performance. Zero-order correlation between hair Hg levels and Word Span accuracy was moderate and significant, yet when fit to the regression model for Word Span that controlled for Education and Age, Hg level was no longer a significant predictor variable (see Figure 4a). At first glance, this may appear to be evidence for the null, that Hg has no effect on working memory; however, the finding is consistent with the Multicomponent Model and the hypothesis that Hg levels are associated with compromised central executive functions rather than the storage and recall of words [41].

Despite the simplicity of the word span task, consideration of the measure reveals the nuances of measuring cognitive abilities across diverse cultures. For instance, changes in how the results are scored and evaluated might determine if a score is interpreted as below "normal" or impaired. Many clinical assessments of short-term memory capacity, use an outcome that is reported as a "span" score or the average number of items that a person can temporarily hold and then recall (for discussion on span vs. accuracy calculations, see [78]. The average span score is reliably reported to be around seven for short-term memory [79] and four for working memory [80].

While it may seem straightforward to calculate the average span size among the sampled communities and compare it to these well-known metrics, that approach could be misleading when applied to indigenous groups like the Matsigenka. There are multiple known influences from language and culture that can introduce variation into a span score. One well documented is the Word Length Effect (WLE) that is the fluctuation in span size across languages depending upon the amount of time it takes to pronounce or rehearse words or numbers in a sequence [81–83]. For the Matsigenka, the WLE might occur if a participant's span size in the Matsigenka language is different than their span size when speaking in a second language, Spanish. Span differences due to the WLE would not reflect an impaired ability to store and recall, but rather articulation time for the words in the sequence due to the structure of the language and the participant's fluency.

Another cultural factor that could contribute to misleading differences when using a clinical norm to assess impairment is the counting systems of Amazonian groups. The traditional Matsigenka counting system categorizes items above three or four into a group of "many". It is possible that the strategy of categorizing multiple items with the same descriptor of "many" indicates that a different rehearsal strategy is used when temporarily storing information [79,80]. The explanation has not been tested, but seems reasonable since anthropological writings about the Matsigenka have provided descriptions of individuals "losing count" of discreet number of objects when in a large amount, such as the account of a Matsigenka child over the age of six listing their possessions:

> *If he says he has three needles, he has three. He begins to lose count only as numbers mount above five; like all his neighbors, he tends to remember in increments of five or ten and he can indicate these increments by opening his fists and flashing his fingers the right number of times.* ([84]. Families of the Forest, 2003, p. 153)

In a similar way, a culture's system for categorizing information for memory could influence the internal processes of storing and retaining verbal-numerical items in memory. That is, the Matsigenka may or may not rely on articulatory rehearsal strategy, which is the repeating of words/digits between the presentation of the next stimuli and recall. If so, they could have decreased span size due to strategy, but not necessarily due to capacity.

The word span task used in this study was intended to measure verbal short-term storage of the Matsigenka in their own language and without comparison to a control-group norm, but the results cannot inform about potential word length effects, the influence of a different counting system on storage capacity, or a use of a strategy that is not articulatory rehearsal. None of these possibilities can be addressed with the current data but could influence the overall storage capacity estimates.

Although the word span task allows for illustrations of cultural diversity in psychological measurement, it can also be noted that none of these factors should have influenced the between-community differences that the current study found in Word Span accuracy scores and would only apply to those studies using normative data from outside the test group's culture. The finding of differences between the communities as reported here are due to the factors measured (i.e., Hg levels or education) or others not yet explored. Further the pattern is consistent with other studies testing verbal short-term memory in a simple and complex format [19].

4.2.2. Visuospatial Short-Term Memory

A negative relationship between Hg and the Corsi Block Task accuracy showed that the measure was sensitive to varied levels of Hg exposure. Higher hair Hg levels were associated with poorer visuospatial retention and recall, and this association remained significant after controlling for Age and Education (Figure 4c). However, this result does not necessarily mean that the storage capacity of visuospatial items is impaired exclusively. Although the task is often considered a measure of short-term storage in clinical assessment and as conceptualized in the Multicomponent Model, cognitive research provides ample evidence that the executive-processing contributions is higher than verbal span tasks, and on par with some domain-general tasks that are marked as measures of executive functions [48]. The association with Hg levels may be due to the impairment to the executive processes that are recruited during a visuospatial task and are the same that underlie the executive functions captured by IQ scores.

The evidence of Hg's association with poorer visuospatial performance is partly consistent with other studies of Amazonian peoples. Santos-Lima et al.'s work (2020) with Amazonian children in Brazil found relationships between Hg hair levels and performance on the Corsi Block when using the backward recall version but not with the forward version. The order of recall (forward or backward) is thought to increase a task's complexity by requiring the participant to reverse the sequence during recall and thus, causing the central executive to be more engaged [85]. While it is subjectively true that a backward task feels more difficult to participants, it may not be due to central executive contributions, as demonstrated in factor analysis studies on types of working memory tasks [42]. One possible reason for the difference in findings is that the children in previous Amazonian studies were living in more populated areas and recruited through their local schools. The fact that schools could organize recruitment for samples of children may indicate that there is greater stability, resources, and quality of formal education that can influence a child's development of test scores or even teach memory strategy that is advantageous for such tests. With the current data, conclusion about the effects of school on task performance, but Education does seem to explain variance in cognitive scores in a way that is unique from Hg levels, as shown by the linear regression outcomes.

In the Faroe Islands, a cohort of residents was tested with a similar measure of visuospatial short-term memory, the Spatial Span, but did not find evidence that the Hg levels measured at birth negatively impacted recall in early adolescents [86] or young adulthood [20]. Instead, Spatial Span scores at age 14 years showed an unexpected positive correlation with Hg cord blood samples. It is not clear why there are disparate outcomes for correlations between visuospatial memory and Hg levels across studies, especially those found in opposite directions. Some reasons to consider are the scoring system used for the task, types of mercury assays during development (cord blood at birth vs. current hair levels), range of Hg detected within the sample, and larger sample sizes. Following up with more studies to understand how Corsi Block performance relates to hair Hg levels is worth exploring and could help to bridge our understanding of how visuospatial processing should be considered during assessment. For the Matsigenka studied in the current sample, poorer performance on the Corsi Block was associated with hair Hg levels and could be an important tool for assessing cognitive functions, providing an alternative path to capturing estimates of general cognitive impairment for Hg studies in ASGM active areas.

4.2.3. Executive Functions

The major hypothesis offered in the cognitive literature and tested here, was supported by the results from the central executive task, the SOPT. The SOPT error score was positively correlated to Hg levels, so that participants with higher Hg levels were also more likely to make errors when remembering the previously selected shape. The association remained significant when Age and Education were controlled in a regression model (see Figure 4c). The SOPT task, though simple in form and instruction, has not been previously used in Hg studies but is often used in lab studies exploring models of working memory and intelligence. Similar to the Corsi Block, these findings in combination with the experimental cognitive literature showing that the SOPT correlates with intelligence tests [59], suggest that the SOPT could be a suitable substitute when intelligence tests are not appropriate for use in remote areas with indigenous people.

Previous studies with general working memory indices and executive tasks are consistent with the SOPT results. Studies with children in Amazonian Peru [18] and Brazil [19] found that working memory sub-scales included in IQ tests were correlated with Hg levels after controlling for covariates like socioeconomic measures and education. The finding that education is a common predictor across studies and even for each cognitive task in this study is interesting and highlights the influence of formal schooling on cognitive assessments (although it is also possible that cognitive abilities measured by our assessments might instead influence educational attainment). Just like the Word Span and Corsi Block tasks in this study, Education was correlated with SOPT error scores ($p = 0.05$) when Hg and Age were controlled. The measure of Education was years in school (up to 12th grade), which included a full range of scores (0 to 12 years), but it is unlikely that this metric captured the type or quality of education available in each of the communities. Based on our observations, there appeared to be disparate levels of educational resources among communities. While one village, Yomibato, had a building designated as classroom space for lessons; Cacaotal did not have classrooms, nor did they seem to have the basic supplies that would be expected in a classroom environment. Yomibato was also the only community with a drinkable water source close to the school. These resource differences are not reflected in participants' reports of years in school. In other words, school attendance up to 4th grade in Cacaotal should not be assumed to be equal to attendance up to 4th grade in Yomibato. To account for differences in quality of education, it would be helpful to have additional indicators of community resources and organization.

Finally, this study presents data from a pilot test for a version of The Trail Making Test specifically created for the Matsigenka inside MNP as a measure of executive functions. There was evidence in the pilot test sample that the TMT "Shapes and Shades" was measuring executive functions: TMT B "Shapes and Shades" was significantly related to Corsi Block accuracy and the SOPT errors, but not to Word Span accuracy. Although the TMT B "Shades and Shapes" seems promising as a measure of executive functions, it is not certain that it could be used for detecting Hg associated impairment because the negative relationship between hair Hg levels and TMT B accuracy was not significant, $r = -0.35$, $p = 0.12$. Other issues for the TMT A & B pilot were that the tasks proved difficult for participants across communities; some could not successfully complete the practice trial and only one participant successfully finished both Parts A & B without error.

It could be that the shaded stimuli chosen for the TMT tasks were difficult to distinguish between, making the task visually challenging and also caused the two versions (A and B) to become conflated. Forms A & B are intended to capture different cognitive processes [87] with the simple version (A) replacing letters for shades, and the complex version (B) replacing numbers with shapes (see Figure 2c,d). The TMT B "Shades and Shapes" did require participants to alternate between decision sets, executive function. However, the results of both Part A and Part B in this study were correlated, which suggests that the underlying processes were similar or overlapping. This may be due to fact that the second dimension in Part B, the shapes, did not change incrementally as the numbers do in the original version. Overall, the TMT "Shapes and Shades" should be further investigated as an assessment tool, ideally in a larger sample and with more varied stimuli.

4.3. Limitations and Future Directions

A limitation of this work was the modest number of participants with fully completed cognitive tasks and health assessments. The numbers reflect the challenges of collecting data from remote regions. Even so, the three villages considered for this project were chosen for their location within the most remote areas of the protected zone of the park and at different points along the Manu River. Each community visited could be described as small communities with some degree of illiteracy present in each. These factors make assessment of cognitive impairment difficult and associations with Hg could possibly be obscured by other factors. Despite the sample size we had reasonable power (70%) to detect correlations as small as $r = 0.40$. The sample size may have also led to the sample not being representative of the Matsigenka population living inside the park, although this is difficult to know for certain. There are records of communities, especially the younger residents who have attended school, but the ages and other demographics of adults are sometimes unknown, and it is difficult to determine how many people are living in total within the boundaries of the park.

Another limitation of this study, albeit anticipated and considered, was the lack of existing norms of data and/or a proper control group for the communities visited. With indigenous people in Amazonia, especially those in the restricted area of MNP, there are reasons to closely consider comparisons to matched groups from the surrounding regions. Using such comparison groups, especially those too far outside areas with Hg exposure may have introduced error associated with differences of nutrition, general health, education, and familiarity with formal testing. This is a familiar problem to researchers in the field [18,88] and the problem may demand multiple strategies from different research programs to achieve a clear understanding of the more subtle impacts associated with Hg. For the Matsigenka of MNP, we decided against using an outside control group and, instead, focused on the dose-related relationship of Hg and cognitive performance. This design may be the soundest approach and is not unlike the types of follow-up and secondary analyses that are found in benchmark studies of the field. For example, in the Faroe Island cohort studies, Grandjean et al. [88], created a control group from the children whose mothers had low mercury exposure. Similarly, Dos Santos-Lima et al. [19] used the top and bottom quartile of Hg levels in their participant sample for inferential analyses.

Outside of limitations, this study is bolstered by the theoretical framework provided by experimental cognitive research and offers a different approach to learning how Hg exposure may be impacting Amazonian communities. The tasks used for working memory were tailored to our sample and showed the relationships predicted by the Multicomponent Model and other literature on working memory. The findings encourage a closer look at how intelligence tests versus cognitive construct measures can be used in future assessments in isolated populations. More specifically, measuring working memory components may be a preferred index over IQ scores when there is sufficient cultural divergence between IQ batteries and participants. The results in this study suggest two options for detecting Hg impairment when working with isolated indigenous groups: the Corsi Block Test and the SOPT. The Corsi Block Test because it is more related to executive process than a simple storage task [48] and the SOPT because it is known to capture the unique executive abilities of strategic responding, internal organization and updating of information, and behavior regulation [51]. Moreover, the SOPT error scores correlate with tasks that are cornerstones of intelligence tests, such as the WAIS-III Block Design and Spatial Span subtests [89]. Both the Corsi Block Test and SOPT represent components of working memory that may be impaired by exposure to Hg and are also appropriate for use with indigenous people.

Future directions for this research can include more exploration of culturally modified cognitive tasks for Amazonian groups. Greater reference to the experimental literature on cognition can increase testing options for Hg studies when general intelligence tests and standardized comparison groups are not feasible in remote areas. Other areas of cognitive research should be explored for use in Hg studies including a deeper investigation into the types of executive functions sensitive to Hg exposure, such as updating, sequencing, and shifting of attention [34]. The research on executive functions beyond assessment of indigenous groups, a future direction for research can be to use current findings on intelligence for the development of intervention programs aimed at ameliorating or strengthening cognitive processes in ways that would be beneficial to Amazonia children and adults. Intervention programs can continue to be investigative while also addressing the immediate need for groups living among ASGM activity and other areas of Hg exposure.

5. Conclusions

The levels of methylmercury in hair of thirty participants from three different indigenous communities along the Manu River were found to be above the maximum threshold of 2.0 ppm [72,73] in all but three individuals, with levels exceeding the WHO limit by an average of 3.5×. The results point to widespread elevated Hg levels in even the most remote indigenous populations living in watersheds with illegal artisanal and small-scale gold mining—in this case between 300 and 400 km upstream from the nearest mine. This study also explored the association between methylmercury exposure and working memory using tasks chosen to measure the short-term storage and executive functions. Previous studies exploring the impact of Hg exposure have mostly relied on IQ tests or working memory sub-scales that are likely to be influenced by culture, education, and language and would not be appropriate for the Matsigenka living inside the restricted area of MNP [18,19]. Here, some tasks were modified by changing stimuli, instructions, and outcome variables to make the tasks more consistent with traditional Matsigenka culture. Results suggest that Hg exposure may impair cognitive processes that rely on executive functioning, and that these effects are seen at even relatively lower levels of Hg exposure and increase monotonically with increasing Hg concentrations in hair. The findings highlight the risk to Amazonian populations, especially those living in areas impacted by ASGM where mining activity increases levels of Hg in fish that are consumed as a main part of the diet [77,90].

Supplementary Materials: The following supporting information can be downloaded at: https://www.mdpi.com/article/10.3390/ijerph191710989/s1, Section S1: Word Span Stimuli in English, Spanish, and Matsigenka., Section S2: Analysis for Set Size Differences across Communities for Word Span, Corsi Block, and SOPT Errors.

Author Contributions: Conceptualization, L.E.F., J.W.S., M.R.S., C.M.V., A.K.S.; methodology, A.K.S., M.J.K.; formal analysis, A.K.S.; resources, A.K.S., M.R.S., L.E.F., C.M.V., M.J.K.; data curation, R.C., G.W.H., L.G., K.K., K.M., S.M., C.M.V.; writing—original draft preparation, A.K.S., M.R.S., L.E.F.; writing—review and editing, M.R.S., M.J.K., G.W.H., S.M., C.M.V., L.E.F.; supervision, J.W.S.; project administration, C.M.V., C.C., S.M., A.G.L.; funding acquisition, J.W.S., A.G.L. All authors have read and agreed to the published version of the manuscript.

Funding: This research was funded by the Wake Forest Medical School "Whitaker Fund" and the Wake Forest Medical Student Research Program. Funding was also provided by the United States Agency for International Development (USAID) under the terms of USAID/Wake Forest University Cooperative Agreement No. AID-527-A-16-00001 for L.E.F. and C.M.V. The contents do not necessarily reflect the views of USAID or the United States Government. A.G.L., S.M. and C.C. were also supported by training grant D43 TW007393 awarded by the Fogarty International Center of the US National Institutes of Health.

Institutional Review Board Statement: The study was conducted in accordance with the Declaration of Helsinki and approved by the Institutional Review Board of Wake Forest University Medical School Institutional Review Board (Human Subjects: IRB000044673) and the Institutional Ethics Committee of the Universidad Peruana Cayetano Heredia, Lima, Peru (SIDISI N° 100806).

Informed Consent Statement: Informed consent was obtained from all subjects involved in the study. Written informed consent was obtained from participants ages 18 or older with signature and/or fingerprint (in case of illiteracy) and for those under the age of 18 years, written consent was obtained from the parent or accompanying adult family member, as well as an informed assent for participants ages 12 or older.

Data Availability Statement: The data presented in this study are available on request from the corresponding author. The data are not publicly available due to protection of participant privacy.

Acknowledgments: The authors gratefully thank all the members of the Matsigenka communities of Maizal, Cacaotal and Yomibato for their patience, support and participation in this study. We also thank John Terborgh and Lisa Davenport for the initial discussions about examining mercury in Manu National Park, and Cesar Flores, Roxana Arauco and the team at Cocha Cashu Biological Station/San Diego Zoo Global for their kind support for our field crews. We also thank David JX Gonzalez's valuable support during the 2017 field expedition. We thank Juvenal Silva, Oscar Mujica, Jorge Curo, and Eustaquio Cahuaniri at the Frankfurt Zoological Society, and Ana Maria Morales and Elias Malaga at the Centro de Estudios, Investigaciones y Servicios en Salud Publica for logistical support for the field expeditions, and for the support of the Madre de Dios Direccion Regional de Salud Madre de Dios, the Peruvian Ministry of Culture, and the SERNANP Manu National Park management team for the years of support for this study. We thank all the members of the river and terrestrial transport crews, without which we could not do our work. Finally, we thank the entire CINCIA team for their support in all forms throughout the study, and in particular, Cesar Ascorra and Marta Torres for their guidance, and Jesus Alferes and Julio Araújo-Flores field coordination.

Conflicts of Interest: The authors declare no conflict of interest. The funders had no role in the design of the study; in the collection, analyses, or interpretation of data; in the writing of the manuscript, or in the decision to publish the results.

References

1. Pirrone, N.; Cinnirella, S.; Feng, X.; Finkelman, R.B.; Friedli, H.R.; Leaner, J.; Mason, R.; Mukherjee, A.B.; Stracher, G.B.; Streets, D.G.; et al. Global mercury emissions to the atmosphere from anthropogenic and natural sources. *Atmos. Chem. Phys.* **2010**, *10*, 5951–5964. [CrossRef]
2. Singer, M.B.; Aalto, R.; James, L.A.; Kilham, N.E.; Higson, J.L.; Ghoshal, S. Enduring legacy of a toxic fan via episodic redistribution of California gold mining debris. *Proc. Natl. Acad. Sci. USA* **2013**, *110*, 18436–18441. [CrossRef] [PubMed]
3. Driscoll, D.; Sorensen, A.; Deerhake, M. A multidisciplinary approach to promoting healthy subsistence fish consumption in culturally distinct communities. *Health Promot. Pract.* **2012**, *13*, 245–251. [CrossRef] [PubMed]
4. Chen, C.Y.; Driscoll, C.T. Integrating mercury research and policy in a changing world. *Ambio* **2018**, *47*, 111–115. [CrossRef] [PubMed]
5. Driscoll, C.T.; Mason, R.P.; Chan, H.M.; Jacob, D.J.; Pirrone, N. Mercury as a Global Pollutant: Sources, Pathways, and E ffects. *Environ. Sci. Technol.* **2013**, *47*, 4967–4983. [CrossRef] [PubMed]
6. Hsu-Kim, H.; Kucharzyk, K.H.; Zhang, T.; Deshusses, M.A. Mechanisms regulating mercury bioavailability for methylating microorganisms in the aquatic environment: A critical review. *Environ. Sci. Technol.* **2013**, *47*, 2441–2456. [CrossRef]
7. Ullrich, S.M.; Tanton, T.W.; Abdrashitova, S.A. Mercury in the aquatic environment: A review of factors affecting methylation. *Crit. Rev. Environ. Sci. Technol.* **2001**, *31*, 241–293. [CrossRef]
8. Azevedo-Silva, C.E.; Almeida, R.; Carvalho, D.P.; Ometto, J.P.; de Camargo, P.B.; Dorneles, P.R.; Azeredo, A.; Bastos, W.R.; Malm, O.; Torres, J.P. Mercury biomagnification and the trophic structure of the ichthyofauna from a remote lake in the Brazilian Amazon. *Environ. Res.* **2016**, *151*, 286–296. [CrossRef]
9. Bastos, W.R.; Dórea, J.G.; Bernardi, J.V.E.; Lauthartte, L.C.; Mussy, M.H.; Lacerda, L.D.; Malm, O. Mercury in fish of the Madeira river (temporal and spatial assessment), Brazilian Amazon. *Environ. Res.* **2015**, *140*, 191–197. [CrossRef]
10. Callister, S.M.; Winfrey, M.R. Microbial methylation of mercury in upper Wisconsin river sediments. *Water Air Soil Pollut.* **1986**, *29*, 453–465. [CrossRef]
11. Hacon, S.D.S.; Oliveira-Da-Costa, M.; Gama, C.D.S.; Ferreira, R.; Basta, P.C.; Schramm, A.; Yokota, D. Mercury exposure through fish consumption in traditional communities in the Brazilian northern Amazon. *Int. J. Environ. Res. Public Health* **2020**, *17*, 5269. [CrossRef] [PubMed]
12. Basu, N.; Horvat, M.; Evers, D.C.; Zastenskaya, I.; Weihe, P.; Tempowski, J. A state-of-the-science review of mercury biomarkers in human populations worldwide between 2000 and 2018. *Environ. Health Perspect.* **2018**, *126*, 106001. [CrossRef] [PubMed]
13. Dufour, D.L.; Piperata, B.A.; Murrieta, R.S.S.; Wilson, W.M.; Williams, D.D. Amazonian foods and implications for human biology. *Ann. Hum. Biol.* **2016**, *43*, 330–348. [CrossRef] [PubMed]
14. Ohl-Schacherer, J.; Shepard, G.H.; Kaplan, H.; Peres, C.A.; Levi, T.; Yu, D.W. The sustainability of subsistence hunting by matsigenka native communities in Manu National Park, Peru. *Conserv. Biol.* **2007**, *21*, 1174–1185. [CrossRef]
15. Bunce, J.; Minaya, C. Matsigenka and Colonos-Lowland, Peru. 2022. Available online: https://www.eva.mpg.de/ecology/fieldwork/matsigenka-and-colonos/ (accessed on 1 June 2022).
16. Diringer, S.E.; Feingold, B.J.; Ortiz, E.J.; Gallis, J.A.; Araújo-Flores, J.M.; Berky, A.; Pan, W.K.Y.; Hsu-Kim, H. River transport of mercury from artisanal and small-scale gold mining and risks for dietary mercury exposure in Madre de Dios, Peru. *Environ. Sci. Process. Impacts* **2014**, *17*, 478–487. [CrossRef]
17. Shepard, G.H.; Rummenhoeller, K.; Ohl-Schacherer, J.; Yu, D.W. Trouble in paradise: Indigenous populations, anthropological policies, and biodiversity conservation in Manu National Park, Peru. *J. Sustain. For.* **2010**, *29*, 252–301. [CrossRef]
18. Reuben, A.; Frischtak, H.; Berky, A.; Ortiz, E.J.; Morales, A.M.; Hsu-Kim, H.; Pendergast, L.L.; Pan, W.K. Elevated hair mercury levels are associated with neurodevelopmental deficits in children living near artisanal and small-scale gold mining in Peru. *Geohealth* **2020**, *4*, e2019GH000222. [CrossRef]
19. Dos Santos-Lima, C.; Mourão, D.D.S.; de Carvalho, C.F.; Souza-Marques, B.; Vega, C.M.; Gonçalves, R.A.; Argollo, N.; Menezes-Filho, J.A.; Abreu, N.; Hacon, S.D.S. Neuropsychological effects of mercury exposure in children and adolescents of the Amazon region, Brazil. *NeuroToxicology* **2020**, *79*, 48–57. [CrossRef]
20. Debes, F.; Weihe, P.; Grandjean, P. Cognitive deficits at age 22 years associated with prenatal exposure to methylmercury. *Cortex* **2016**, *74*, 358–369. [CrossRef]
21. Crump, K.S.; Kjellström, T.; Shipp, A.M.; Silvers, A.; Stewart, A. Influence of prenatal mercury exposure upon scholastic and psychological test performance: Benchmark analysis of a New Zealand cohort. *Risk Anal.* **1998**, *18*, 701–713. [CrossRef]
22. Grandjean, P.; Weihe, P.; White, R.F.; Debes, F.; Araki, S.; Yokoyama, K.; Murata, K.; Sørensen, N.; Dahl, R.; Jørgensen, P.J. Cognitive deficit in 7-year-old children with prenatal exposure to methylmercury. *Neurotoxicol. Teratol.* **1997**, *19*, 417–428. [CrossRef]
23. Davidson, P.W.; Myers, G.J.; Cox, C.; Axtell, C.; Shamlaye, C.; Sloane-Reeves, J.; Cernichiari, E.; Needham, L.; Choi, A.; Clarkson, T.W.; et al. Effects of prenatal and postnatal methylmercury exposure from fish consumption on neurodevelopment: Outcomes at 66 months of age in the Seychelles Child Development Study. *JAMA* **1998**, *280*, 701–707. [CrossRef] [PubMed]

24. Myers, G.J.; Davidson, P.W.; Cox, C.; Shamlaye, C.F.; Palumbo, D.; Cernichiari, E.; Sloane-Reeves, J.; Wilding, G.E.; Kost, J.; Huang, L.-S.; et al. Prenatal methylmercury exposure from ocean fish consumption in the Seychelles child development study. *Lancet* **2003**, *361*, 1686–1692. [CrossRef]

25. Wang, Y.; Chen, A.; Dietrich, K.N.; Radcliffe, J.; Caldwell, K.L.; Rogan, W.J. Postnatal exposure to methyl mercury and neuropsychological development in 7-year-old urban inner-city children exposed to lead in the United States. *Child Neuropsychol.* **2014**, *20*, 527–538. [CrossRef] [PubMed]

26. Walker, A.J.; Batchelor, J.; Shores, E.A. Effects of education and cultural background on performance on WAIS-III, WMS-III, WAIS-R and WMS-R measures: Systematic review. *Aust. Psychol.* **2009**, *44*, 216–223. [CrossRef]

27. Ortiz, S.; Ochoa, S.H.; Dynda, A.M. Testing with culturally and linguistically diverse populations: Moving beyond the verbal-performance dichotomy into evidence-based practice. In *Contemporary Intellectual Assessment: Theories, Tests, and Issues*; Guilford Press: New York, NY, USA, 2012; pp. 526–552.

28. Aben, B.; Stapert, S.; Blokland, A. About the distinction between working memory and short-term memory. *Front. Psychol.* **2012**, *3*, 301. [CrossRef]

29. Shepard, G. Ethnozoological classification in Machiguenga, an Arawakan language. In *The Journal Of Amazonian Languages*; Ladefoged, P., Kaufman, T., Payne, D., Pullum, G., Rodrigues, A.D.I., Faco, M., Wetzels, L., Eds.; University of Pittsburgh: Pittsburgh, PA, USA, 1997.

30. Everett, C.; Madora, K. Quantity recognition among speakers of an anumeric language. *Cogn. Sci.* **2012**, *36*, 130–141. [CrossRef] [PubMed]

31. Counter, S.A.; Buchanan, L.H.; Ortega, F. Neurocognitive screening of mercury-exposed children of andean gold miners. *Int. J. Occup. Environ. Health* **2006**, *12*, 209–214. [CrossRef]

32. Chevrier, C.; Sullivan, K.; White, R.F.; Comtois, C.; Cordier, S.; Grandjean, P. Qualitative assessment of visuospatial errors in mercury-exposed Amazonian children. *Neurotoxicology* **2009**, *30*, 37–46. [CrossRef]

33. Tavares, L.M.B.; Câmara, V.M.; Malm, O.; Santos, E.C.D.O. Performance on neurological development tests by riverine children with moderate mercury exposure in Amazonia, Brazil. *Cad. Saúde Pública* **2005**, *21*, 1160–1167. [CrossRef]

34. Miyake, A.; Shah, P. *Models of Working Memory: Mechanisms of Active Maintenance and Executive Control*; Cambridge University Press: Cambridge, UK, 1999.

35. Logie, R.; Camos, V.; Cowan, N. *Working Memory: State of the Science*; Oxford University Press: Oxford, UK, 2021.

36. Cowan, N. The many faces of working memory and short-term storage. *Psychon. Bull. Rev.* **2017**, *24*, 1158–1170. [CrossRef] [PubMed]

37. Barrouillet, P.; Camos, V. The time-based resource-sharing model of working memory. In *Working Memory*; Oxford University Press: Oxford, UK, 2020; pp. 85–115. [CrossRef]

38. Engle, R.W. Working memory capacity as executive attention. *Curr. Dir. Psychol. Sci.* **2002**, *11*, 19–23. [CrossRef]

39. Baddeley, A.D.; Hitch, G. Working memory. In *Psychology of Learning and Motivation—Advances in Research and Theory*; Elsevier: Amsterdam, The Netherlands, 1974; Volume 8, pp. 47–89. [CrossRef]

40. Baddeley, A.D. *Essentials of Human Memory. Taylor & Francis*; Psychology Press: London, UK, 1999.

41. Baddeley, A.D.; Hitch, G.J.; Allen, R.J. A multicomponent model of working memory. In *Working Memory: State of the Science*; Logie, R., Camos, V., Cowan, N., Eds.; Oxford University Press: Oxford, UK, 2021.

42. Engle, R.W.; Laughlin, J.E.; Tuholski, S.W.; Conway, A.R.A. Working memory, short-term memory, and general fluid intelligence: A latent-variable approach. *J. Exp. Psychol. Gen.* **1999**, *128*, 3091999. [CrossRef] [PubMed]

43. Kane, M.J.; Bleckley, M.K.; Conway, A.R.A.; Engle, R.W. A controlled-attention view of working-memory capacity. *J. Exp. Psychol. Gen.* **2001**, *130*, 169–183. [CrossRef]

44. Shipstead, Z.; Yonehiro, J. The domain-specific and domain-general relationships of visuospatial working memory to reasoning ability. *Psychon. Bull. Rev.* **2016**, *23*, 1504–1512. [CrossRef]

45. Süß, H.-M.; Oberauer, K.; Wittmann, W.W.; Wilhelm, O.; Schulze, R. Working-memory capacity explains reasoning ability and a little bit more. *Intelligence* **2002**, *30*, 261–288. [CrossRef]

46. Conway, A.R.; Kane, M.J.; Engle, R.W. Working memory capacity and its relation to general intelligence. *Trends Cogn. Sci.* **2003**, *7*, 547–552. [CrossRef] [PubMed]

47. Shipstead, Z.; Harrison, T.L.; Engle, R.W. Working memory capacity and fluid intelligence. *Perspect. Psychol. Sci.* **2016**, *11*, 771–799. [CrossRef]

48. Kane, M.J.; Hambrick, D.Z.; Tuholski, S.W.; Wilhelm, O.; Payne, T.W.; Engle, R.W. The generality of working memory capacity: A latent-variable approach to verbal and visuospatial memory span and reasoning. *J. Exp. Psychol. Gen.* **2004**, *133*, 189–217. [CrossRef]

49. Rudkin, S.J.; Pearson, D.G.; Logie, R.H. Executive processes in visual and spatial working memory tasks. *Q. J. Exp. Psychol.* **2007**, *60*, 79–100. [CrossRef]

50. Hamilton, C.; Coates, R.; Heffernan, T. What develops in visuo-spatial working memory development? *Eur. J. Cogn. Psychol.* **2003**, *15*, 43–69. [CrossRef]

51. Petrides, M.; Milner, B. Deficits on subject-ordered tasks after frontal-and temporal-lobe lesions in man. *Neuropsychologia* **1982**, *20*, 249–262. [CrossRef]

52. Ardila, A.; Moreno, S. Neuropsychological test performance in Aruaco Indians: An exploratory study. *J. Int. Neuropsychol. Soc.* **2001**, *7*, 510–515. [CrossRef] [PubMed]
53. Fonseca, M.D.F.; Dórea, J.G.; Bastos, W.R.; Marques, R.C.; Torres, J.P.; Malm, O. Poor psychometric scores of children living in isolated riverine and agrarian communities and fish-methylmercury exposure. *NeuroToxicology* **2008**, *29*, 1008–1015. [CrossRef] [PubMed]
54. Berglund, M.; Lind, B.; Björnberg, K.A.; Palm, B.; Einarsson, Ö.; Vahter, M. Inter-individual variations of human mercury exposure biomarkers: A cross-sectional assessment. *Environ. Health* **2005**, *4*, 20. [CrossRef] [PubMed]
55. Schrank, F.; Decker, S.; Garruto, J. *Essentials of WJ IV Cognitive Abilities Assessment*; John Wiley & Sons: New York, NY, USA, 2016.
56. Corsi, P.M. Short Title Memory and the Medial Temporal Region of the Brain. Ph.D. Thesis, McGill University, Montreal, QC, Canada, 1972.
57. Milner, B. Interhemispheric differences in the localization of psychological processes in man. *Br. Med. Bull.* **1971**, *27*, 272–277. [CrossRef] [PubMed]
58. Kessels, P.R.C.; van Zandvoort, M.J.E.; Postma, A.; Kappelle, L.J.; de Haan, E.H.F. The Corsi Block-Tapping Task: Standardization and Normative Data the corsi block-tapping task. *Appl. Neuropsychol.* **2000**, *7*, 252–258. [CrossRef]
59. Ross, T.P.; Hanouskova, E.; Giarla, K.; Calhoun, E.; Tucker, M. The reliability and validity of the self-ordered pointing task. *Arch. Clin. Neuropsychol.* **2007**, *22*, 449–458. [CrossRef]
60. Ardila, A. Cultural values underlying psychometric cognitive testing. *Neuropsychol. Rev.* **2005**, *15*, 185–195. [CrossRef]
61. Attneave, F.; Arnoult, M.D. The quantitative study of shape and pattern perception. *Psychol. Bull.* **1956**, *53*, 452–471. [CrossRef]
62. Rock, D.; Price, I.R. Identifying culturally acceptable cognitive tests for use in remote northern Australia. *BMC Psychol.* **2019**, *7*, 62. [CrossRef] [PubMed]
63. Bowie, C.R.; Harvey, P.D. Administration and interpretation of the trail making test. *Nat. Protoc.* **2006**, *1*, 2277–2281. [CrossRef] [PubMed]
64. Kortte, K.B.; Horner, M.D.; Windham, W.K. The trail making test, part B: Cognitive flexibility or ability to maintain set? *Appl. Neuropsychol.* **2002**, *9*, 106–109. [CrossRef]
65. Arbuthnott, K.; Frank, J. Trail making test, part B as a measure of executive control: Validation using a set-switching paradigm. *J. Clin. Exp. Neuropsychol.* **2000**, *22*, 518–528. [CrossRef]
66. Stuss, D.T.; Bisschop, S.M.; Alexander, M.P.; Levine, B.; Katz, D.; Izukawa, D. The trail making test: A study in focal lesion patients. *Psychol. Assess.* **2001**, *13*, 230–239. [CrossRef] [PubMed]
67. Barncord, S.W.; Wanlass, R.L. The symbol trail making test: Test development and utility as a measure of cognitive impairment. *Appl. Neuropsychol.* **2001**, *8*, 99–103. [CrossRef]
68. Weinhouse, C.; Ortiz, E.J.; Berky, A.J.; Bullins, P.; Hare-Grogg, J.; Rogers, L.; Morales, A.-M.; Hsu-Kim, H.; Pan, W.K. Hair mercury level is associated with anemia and micronutrient status in children living near artisanal and small-scale gold mining in the Peruvian Amazon. *Am. J. Trop. Med. Hyg.* **2017**, *97*, 1886–1897. [CrossRef]
69. Anticona, C.; Sebastian, M.S. Anemia and malnutrition in indigenous children and adolescents of the Peruvian Amazon in a context of lead exposure: A cross-sectional study. *Glob. Health Action* **2014**, *7*, 22888. [CrossRef]
70. Rice, K.M.; Walker, E.M., Jr.; Wu, M.; Gillette, C.; Blough, E.R. Environmental mercury and its toxic effects. *J. Prev. Med. Public Health* **2014**, *47*, 74–83. [CrossRef]
71. Pizzol, D.; Tudor, F.; Racalbuto, V.; Bertoldo, A.; Veronese, N.; Smith, L. Systematic review and meta-analysis found that malnutrition was associated with poor cognitive development. *Acta Paediatr.* **2021**, *110*, 2704–2710. [CrossRef]
72. WHO. Evaluations of the Joint FAO/WHO Expert Committee on Food Additives (JECFA). Available online: https://apps.who.int/food-additives-contaminants-jecfa-database/Home/Chemical/3083 (accessed on 1 August 2022).
73. Joint FAO/WHO Expert Committee on Food Additives (JECFA). Methylmercury. Summary and conclusions. In Proceedings of the 67th Joint FAO/WHO Expert Committee on Food Additives, Rome, Italy, 20–29 June 2006.
74. Salthouse, T.A. What cognitive abilities are involved in trail-making performance? *Intelligence* **2011**, *39*, 222–232. [CrossRef] [PubMed]
75. Salthouse, T.A. The aging of working memory. *Neuropsychology* **1994**, *8*, 535–543. [CrossRef]
76. Cook, R.D.; Weisberg, S. *Residuals and Influence in Regression*; Chapman and Hall: New York, NY, USA, 1982.
77. Langeland, A.L.; Hardin, R.D.; Neitzel, R.L. Mercury levels in human hair and farmed fish near artisanal and small-scale gold mining communities in the Madre de Dios River Basin, Peru. *Int. J. Environ. Res. Public Health* **2017**, *14*, 302. [CrossRef]
78. Cowan, N.; Elliott, E.M.; Saults, J.S.; Morey, C.C.; Mattox, S.; Hismjatullina, A.; Conway, A.R. On the capacity of attention: Its estimation and its role in working memory and cognitive aptitudes. *Cogn. Psychol.* **2005**, *51*, 42–100. [CrossRef] [PubMed]
79. Miller, G.A. The magical number seven, plus-or-minus two or some limits on our capacity for processing information. *Psychol. Rev.* **1956**, *2*, 175–202. [CrossRef]
80. Cowan, N. The magical number 4 in short-term memory: A reconsideration of mental storage capacity. *Behav. Brain Sci.* **2001**, *24*, 87–114. [CrossRef] [PubMed]
81. Elliott, J.M. Forward digit span and articulation speed for Malay, English, and two Chinese dialects. *Percept. Mot. Ski.* **1992**, *74*, 291–295. [CrossRef]

82. Naveh-Benjamin, M.; Ayres, T.J. Digit span, reading rate, and linguistic relativity. *Q. J. Exp. Psychol. Sect. A* **1986**, *38*, 739–751. [CrossRef]
83. Ellis, N.C.; Hennelly, R.A. A bilingual word-length effect. *Br. J. Psychol.* **1980**, *71*, 43–51. [CrossRef]
84. Johnson, A. *Families of the Forest: The Matsigenka Indians of the Peruvian Amazon*; University of California Press: Berkeley, CA, USA, 2003. [CrossRef]
85. Donolato, E.; Giofrè, D.; Mammarella, I.C. Differences in verbal and visuospatial forward and backward order recall: A review of the literature. *Front. Psychol.* **2017**, *8*, 663. [CrossRef]
86. Debes, F.; Budtz-Jørgensen, E.; Weihe, P.; White, R.F.; Grandjean, P. Impact of prenatal methylmercury exposure on neurobehavioral function at age 14 years. *Neurotoxicol. Teratol.* **2006**, *28*, 363–375. [CrossRef] [PubMed]
87. Martin, T.A.; Hoffman, N.M.; Donders, J. Clinical utility of the trail making test ratio score. *Appl. Neuropsychol.* **2003**, *10*, 163–169. [CrossRef] [PubMed]
88. Grandjean, P.; Weihe, P.; White, R.F.; Debes, F.P. *Cognitive Performance of Children Prenatally Exposed to "Safe" Levels of Methylmercury*; Elsevier: Amsterdam, The Netherlands, 1998.
89. Stebbins, G.T. Neuropsychological testing. In *Textbook of Clinical Neurology*, 3rd ed.; Elsevier: Amsterdam, The Netherlands, 2007; pp. 539–557. [CrossRef]
90. Gonzalez, D.J.X.; Arain, A.; Fernandez, L.E. Mercury exposure, risk factors, and perceptions among women of childbearing age in an artisanal gold mining region of the Peruvian Amazon. *Environ. Res.* **2019**, *179 Pt A*, 108786. [CrossRef] [PubMed]

International Journal of
Environmental Research
and Public Health

Case Report

Lead Poisoning among Male Juveniles Due to Illegal Mining: A Case Series from South Africa

Thokozani Patrick Mbonane [1,*], Angela Mathee [1,2,3], André Swart [1] and Nisha Naicker [1]

1. Department of Environmental Health, Faculty of Health Sciences, University of Johannesburg, Johannesburg 2000, South Africa; Angie.Mathee@mrc.ac.za (A.M.); andres@uj.ac.za (A.S.); nnaicker@uj.ac.za (N.N.)
2. Environment and Health Research Unit, South African Medical Research Council, Johannesburg 2000, South Africa
3. School of Public Health, Faculty of Health Sciences, University of Witwatersrand, Johannesburg 2000, South Africa
* Correspondence: tmbonane@uj.ac.za; Tel.: +27-(011)-559-6240

Abstract: Illegal mining is a major public health and societal concern. Recent scientific evidence indicates elevated blood–lead levels in illegal gold miners and associated communities. Yet, there is little research in this regard from low- to middle-income countries (LMICs), where illegal mining is growing. This case series is extracted from a cross-sectional study of lead exposure in incarcerated juveniles in greater Johannesburg. From survey records (blood–lead levels and questionnaires), three males had elevated blood–lead levels and presented with health conditions and behavioural problems putatively linked with lead poisoning. Based on the record review, all three juveniles were in a secure facility due to illegal mining-related activities. All three cases had high blood–lead levels and demonstrated a tendency toward aggressive or violent behaviour. They also presented with conditions associated with lead poisoning, such as anaemia, respiratory illness, abdominal disorders, and musculoskeletal conditions. Juveniles involved in illegal mining are at risk of exposure to heavy metals such as lead, and there is a need for relevant preventative action and health care programmes in this group.

Keywords: illegal mining; elevated blood levels; artisanal and small-scale gold mining; para-occupational; environmental health

Citation: Mbonane, T.P.; Mathee, A.; Swart, A.; Naicker, N. Lead Poisoning among Male Juveniles Due to Illegal Mining: A Case Series from South Africa. *Int. J. Environ. Res. Public Health* **2021**, *18*, 6838. https://doi.org/10.3390/ijerph18136838

Academic Editors: Masayuki Sakakibara, Win Thiri Kyaw and José Luis Rivera Parra

Received: 8 June 2021
Accepted: 23 June 2021
Published: 25 June 2021

Publisher's Note: MDPI stays neutral with regard to jurisdictional claims in published maps and institutional affiliations.

1. Introduction

Environmental and para-occupational lead exposure is an under-recognized societal issue in low- to middle-income countries [1]. Children are particularly affected, including adolescent males [2]. Blood–lead levels (BLLs) have been linked to societal and behavioural issues such as aggression, anti-social behaviour, delinquency, and violent or criminal behaviour [2–7]. Research has highlighted elevated BLLs in adolescent males who are prone to involvement in acts of violence or criminal [4,5]. Elevated BLLs have also been linked to the number of arrests and the type of crime committed [8]. Studies from high-income countries have shown that children in juvenile systems have elevated blood–lead levels compared to children who have never been in conflict with the law [5].

Involvement in informal or artisanal mining has been associated with lead exposure [9,10]. Artisanal gold mining is regarded as an illegal economic activity in many low- to middle-income countries, including South Africa [11,12]. Elevated blood–lead levels have been found in as many as 92.5% of children aged under six years within an artisanal mining community [10]. Boys residing in communities near artisanal and small-scale gold mining activities are at greater risk of having high blood–lead levels than girls [13]. Similar findings were found near the lead–zinc mine, where eight children had BLLs exceeded

427.8 µgdL^{-1} [14]. However, reductions in BLLs have been seen in a community where safer artisanal mining practices have been adopted [15].

Artisanal mining has been growing in several African countries, with growing concerns over the health of those involved [16]. South Africa has a long history of gold mining, increasing artisanal gold mining in recent decades [17]. Despite this, there is a paucity of information on exposure to toxic substances and the health of this vulnerable group that often includes adolescents and children. In South Africa, children and young adults found guilty of criminal acts are confined to secure facilities [18]. They may receive access to services such as education, recreation, and limited public-health programmes (curative health only) as part of their rehabilitation programmes before being released to a parent or guardian [19]. These services are designed to address mental, social, and community factors that contribute to criminal behaviour, but seldom address preventable environmental exposures as contributors to violent or criminal behaviour.

This study presents a series of three cases of elevated blood–lead levels related to artisanal mining in male juveniles held in secure facilities in South Africa.

2. Materials and Methods

From October 2020 to February 2021, venous blood samples were collected by a qualified and registered nurse from three children/juveniles at the correctional centres. Approximately 50–100 µL of blood were collected into an anticoagulant ethylene diamine tetra–acetic acid sterile test tube and transported on the same day to the laboratory. The blood samples were divided in the laboratory to determine blood–lead levels and haemoglobin. Information on the criminal behaviour of participants was retrieved from their records at the incarceration facilities. A semi-structured questionnaire was administered to participants to obtain information on socio-demographic factors, self-reported health issues, and environmental exposures. The Youth Self-Report scale was used to screen for tendencies to aggression or violent behaviour [20,21].

The study was approved by the Faculty of Health Sciences, Research Ethics Committee from the University of Johannesburg. The ethics approval number is (REC-241112-035I). The researchers sought consent from young males of 18 years or older. The parents and legal guardians of the young males were also approached for consent. The three cases reported here are from an ongoing study [22].

3. Findings

3.1. Case Report 1

An 18-year-old male had a blood–lead level of 48.11 µg/dL, with a haemoglobin level of 7.9 g/dL (severe anaemia). Before coming to the incarceration centre, he had lived in an informal settlement with a communal water supply. He was of Mozambican birth and raised by a single parent (his mother). He started working as an artisanal gold miner at the age of 15 years in South Africa. He was a digger (colloquially referred to as "moles"), who went underground to dig for gold and handled explosives for underground blasting. From time to time, he served as a "drainer" as well. A drainer is responsible for refining gold from the dirt retrieved from underground. He was arrested and incarcerated for illegal gold mining in Randfontein.

The participant indicated that he was required to work for periods of up to two weeks underground without coming to the surface. While underground, he was not provided with masks or any other personal protective gear to provide protection against dust or contaminated air. He usually used a cloth or discarded personal respiratory safety equipment that he had found. While underground, he ate only canned foods and bread. Occasionally, he received food from a "runner" (also called messengers) responsible for removing the waste material generated underground. He smoked marijuana and used other illicit substances.

There were no health outcomes recorded in his file; however, he complained of severe headaches, abdominal pain, constipation, painful muscles, weak joints, and a persistent cough. He has shown tendencies of violent behaviour in the last six months. He indicated that he had been shoved or pushed, hit, emotionally/verbally abused, and robbed, mainly underground. Lastly, he had regularly threatened someone with a sharp object to protect himself and the gold that he had discovered while underground.

3.2. Case Report 2

An 18-year-old male with blood–lead levels of 45.87 µg/dL and Hb of 7.1 g/dL (severe anaemia). Prior to his incarceration, he had lived in an informal settlement with a communal water supply. He had been born in Mozambique and lived with his unemployed mother until relocating to South Africa. He was recruited by an older male family friend to smuggle explosives for use in the informal gold-mining sector. In time he became a messenger and a drainer.

He had never been given personal protective gear. He used discarded safety goggles, boots and cloth as a mask; most of these materials he had found the refinery sites where discovered gold was smelted prior to sale.

There was no medical history captured in the participant's file. He indicated having experienced chest pains, constipation, persistent cough, joint weakness, and muscle pains in the last 12 months and feeling tired most of the time since he had been in the facility. He admitted to being violent towards other gangs in the area where they mined gold. In the last six months, he has verbally abused, threatened to hurt, robbed, and used a sharp object to assault someone from a rival group.

3.3. Case Report 3

A 15-year-old immigrant boy had a blood–lead level of 35.76 µg/dL and haemoglobin of 8.6 g/dL, classified as severe anaemia. He had been orphaned at a young age and left his home in Mozambique due to physical abuse from a relative. Prior to being incarcerated, he had lived in a South African informal settlement with his brothers, who had introduced him to illegal gold mining. He had been working as an explosives handler and as a digger when needed.

The most prolonged period during which he remained underground was one week. He was not provided with personal protective clothing and ate only canned foods and bread while underground.

He did not report any medical conditions at the time of entry into the detention centre, but during the recent interview complained of itchy skin, pain in his fingers, and watery eyes after working with chemicals. He did not report any significant violent behavioural tendencies, apart from quickly becoming irritated. He reported having emotionally abused others during the six months before the interview.

4. Discussion

This case series gives insight into various environmental, social, and health risks faced by illegal adolescent gold miners in South Africa. With blood–lead levels of 35.76, 45.87, and 48.11 µg/dL, which exceed the current reference level of the Centers for Disease Control in the USA—of 5 µg/dL—by factors between 7 and 10, all three cases had been highly exposed to environmental lead. Their work in the field of artisanal mining is most likely the source of their lead exposure. The three participants involved informal gold-mining processes such as digging, transporting food supplies into the underground mining areas, and processing the soil retrieved from underground to extract gold residues (as a drainer). The drainer is the most exposed to heavy metals and other chemicals used in the gold-extraction process. This process involves the use of sodium cyanide, sulphuric acid, and nitric acid. All three cases reported a lack of personal protective equipment when underground or involved in other mining processes.

Previous studies have related high blood–lead levels to illegal mining or other arti-sanal activities [10,23,24], including a fatal lead-poisoning case [25]. According to CDC guidelines, the three individuals should be on a treatment programme for lead poisoning.

The cases variably reported a range of non-specific ill-health symptoms that have been associated with lead exposure, including pain and weakness in their muscles and joints, constipation, severe headache, and abdominal pain.

Lead poisoning is a known neurotoxicant that can influence behaviour in particular ways, due to its ability to impair the nervous system [26]. All three cases showed signs and tendencies toward aggression or violent conduct, which have been linked to elevated lead levels [2,4,27]. Two participants have threatened to hurt, robbed, and assaulted someone with a sharp weapon, while all three have verbally/abused someone recently. This behaviour was similar to evidence from a non-exposed population with elevated blood–lead levels in South Africa [4]. Nkomo and colleagues found that young males with elevated blood–lead levels were prone to violent behaviour such as assault, robbery, and assault using a weapon with the intent to cause harm [4].

The working conditions of the cases were also highly insanitary, including a lack of water, sanitation, ventilation, and waste disposal systems. Such conditions have the potential to be highly detrimental to the subjects' health [28]. In addition, the accounts given to the interviewer illustrate the marginalization and vulnerability of the cases, for example, in terms of being in a foreign country (they were all of Mozambican origin and had come to South Africa as adolescents or children), their young age and work as child labourers in the artisanal mining sector, and their subjection to threats and violence underground over the gold they had found.

Young artisanal miners are a highly marginalized and vulnerable group regarding their abysmal and perilous working conditions and their hazardous environmental exposures, social dangers, and compromised health. They find themselves beyond the safety net of public health and social systems. There is a greater need for collaboration between different sectors such as health, social development, educational, mineral resources, economic development, and criminal justice. The action might include introducing safer mining practices at a small-scale level through regulating artisanal mining, developing a lead-poisoning screening tool and implementation of cases to the public health officials. Meanwhile, the juveniles who are in the secure facilities should be educated on the dangers of illegal mining. Incarcerated juveniles receive numerous services, such as counselling, the opportunity to continue with their studies (from primary to secondary levels), and to take skills training programmes. Once the juveniles have left the facilities they may be aided go back to school and access the social grant system. In addition, health education on lead hazards amongst public officials dealing with male juveniles and pro-environmental intervention is required to address environmental-related activities linked to societal and health issues in LMICs.

5. Conclusions

The study has highlighted the impact of illegal gold mining on young males' livelihoods, especially their education, health, and violent behaviour. The young males had blood–lead levels that require public health action, either curative or preventive. The juveniles were involved in illegal activities for survival that contribute to high blood–lead levels. There is a need for a holistic approach to address societal issues such as crime. This report identifies the need for lead poisoning monitoring and a surveillance programme on the exposed population or communities.

Author Contributions: Conceptualisation, T.P.M.; methodology, T.P.M.; validation, A.M., A.S. and N.N.; investigation, T.P.M.; formal analysis, T.P.M.; writing—original draft preparation, T.P.M.; writing—review and editing, A.M., A.S. and N.N.; visualisation, T.P.M.; supervision, A.M., A.S. and N.N.; project administration, T.P.M.; funding acquisition, T.P.M. All authors have read and agreed to the published version of the manuscript.

Funding: The National Research Foundation funded this research.

Institutional Review Board Statement: The study was conducted according to the guidelines of the Declaration of Helsinki and approved by the Research Ethics Committee of the University of Johannesburg, Faculty of Health Science (REC-241112-035 and 27 June 2018).

Informed Consent Statement: Written, informed consent has been obtained from the patient(s) to publish this paper before commencing data collection.

Data Availability Statement: The data presented in this study can be made available on request from the corresponding author.

Acknowledgments: The authors acknowledge the participants for agreeing to be part of the study. Research assistants that assisted with data collecting. The officials from the department of Gauteng Social Development for permitting us to conduct the study. The NRF for the funding to ensure that the study is conducted and completed.

Conflicts of Interest: The authors declare no conflict of interest.

References

1. Tong, S.; von Schirnding, Y.E.; Prapamontol, T. Environmental lead exposure: A public health problem of global dimensions. *Bull. World Health Organ.* **2000**, *78*, 1068–1077. [CrossRef]
2. Nkomo, P.; Naicker, N.; Mathee, A.; Galpin, J.; Richter, L.M.; Norris, S.A. The association between environmental lead exposure with aggressive behavior, and dimensionality of direct and indirect aggression during mid-adolescence: Birth to Twenty Plus cohort. *Sci. Total Environ.* **2018**, *612*, 472–479. [CrossRef]
3. Needleman, H.L.; Riess, J.A.; Tobin, M.J.; Biesecker, G.E.; Greenhouse, J.B. Bone lead levels and delinquent behavior. *J. Am. Med. Assoc.* **1996**, *275*, 363–369. [CrossRef]
4. Nkomo, P.; Galpin, J.; Mathee, A.; Richter, L.M.; Norris, S.A.; Naicker, N. The association between elevated blood lead levels and violent behavior during late adolescence: The South African Birth to Twenty Plus cohort. *Environ. Int.* **2017**, *109*, 136–145. [CrossRef] [PubMed]
5. Beckley, A.L.; Caspi, A.; Broadbent, J.; Harrington, H.; Houts, R.M.; Poulton, R.; Ramrakha, S.; Reuben, A.; Moffitt, T.E. Association of Childhood Blood Lead Levels With Criminal Offending. *JAMA Pediatr.* **2018**, *172*, 166–173. [CrossRef] [PubMed]
6. Olympio, K.P.K.; Oliveira, P.V.; Naozuka, J.; Cardoso, M.R.A.; Marques, A.F.; Günther, W.M.R.; Bechara, E.J.H. Surface dental enamel lead levels and antisocial behavior in Brazilian adolescents. *Neurotoxicol. Teratol.* **2010**, *32*, 273–279. [CrossRef] [PubMed]
7. Fergusson, D.M.; Boden, J.M.; Horwood, L.J. Dentine lead levels in childhood and criminal behaviour in late adolescence and early adulthood. *J. Epidemiol. Community Health* **2008**, *62*, 1045–1050. [CrossRef]
8. Wright, J.P.; Lanphear, B.P.; Dietrich, K.N.; Bolger, M.; Tully, L.; Cecil, K.M.; Sacarellos, C. Developmental lead exposure and adult criminal behavior: A 30-year prospective birth cohort study. *Neurotoxicol. Teratol.* **2021**, 106960. [CrossRef]
9. Street, R.A.; Mathee, A.; Tanda, S.; Hauzenberger, C.; Naidoo, S.; Goessler, W. Recycling of scrap metal into artisanal cookware in the informal sector: A public health threat from multi metal exposure in South Africa. *Sci. Total Environ.* **2020**, *699*, 134324. [CrossRef] [PubMed]
10. Getso, K.I.; Hadejia, I.S.; Sabitu, K.; Nguku, P.M.; Poggensee, G.; Aliyu, H.M.; Yalwa, H.; Sani-Gwarzo, N.; Oyemakinde, A. Prevalence and Determinants of Childhood Lead Poisoning in Zamfara State, Nigeria. *J. Health Pollut.* **2014**, *4*, 1–9. [CrossRef]
11. Rabiu, S.; Abubakar, M.G.; Sahabi, D.M.; Makusidi, M.A.; Dandare, A. Co-Exposure to Lead and Mercury among Artisanal Gold Miners. *Asian J. Environ. Ecol.* **2020**, *11*, 1–8. [CrossRef]
12. Mhlongo, S.E.; Amponsah-Dacosta, F. A review of problems and solutions of abandoned mines in South Africa. *Int. J. Mining Reclam. Environ.* **2016**, *30*, 279–294. [CrossRef]
13. Kaufman, J.A.; Brown, M.J.; Umar-Tsafe, N.T.; Adbullahi, M.B.; Getso, K.I.; Kaita, I.M.; Sule, B.B.; Ba'aba, A.; Davis, L.; Nguku, P.M.; et al. Prevalence and Risk Factors of Elevated Blood Lead in Children in Gold Ore Processing Communities, Zamfara, Nigeria, 2012. *J. Health Pollut.* **2016**, *6*, 2–8. [CrossRef]
14. Yabe, J.; Nakayama, S.M.M.; Ikenaka, Y.; Yohannes, Y.B.; Bortey-Sam, N.; Oroszlany, B.; Muzandu, K.; Choongo, K.; Kabalo, A.N.; Ntapisha, J.; et al. Lead poisoning in children from townships in the vicinity of a lead-zinc mine in Kabwe, Zambia. *Chemosphere* **2015**, *119*, 941–947. [CrossRef] [PubMed]
15. Gottesfeld, P.; Meltzer, G.; Costello, S.; Greig, J.; Thurtle, N.; Bil, K.; Mwangombe, B.J.; Nota, M.M. Declining blood lead levels among small-scale miners participating in a safer mining pilot programme in Nigeria. *Occup. Environ. Med.* **2019**, *76*, 849–853. [CrossRef]
16. Hilson, G.; Zolnikov, T.R.; Ortiz, D.R.; Kumah, C. Formalizing artisanal gold mining under the Minamata convention: Previewing the challenge in Sub-Saharan Africa. *Environ. Sci. Policy* **2018**, *85*, 123–131. [CrossRef]
17. Mhlongo, S.E.; Amponsah-Dacosta, F.; Muzerengi, C.; Gitari, W.M.; Momoh, A. The impact of artisanal mining on rehabilitation efforts of abandoned mine shafts in Sutherland goldfield, South Africa. *Jàmbá J. Disaster Risk Stud.* **2019**, *11*, 688. [CrossRef] [PubMed]

18. Ratshidi, T.M. *Juvenile Justice and Prisons in South Africa*; University of Cape Town: Cape Town, South Africa, 2012. Available online: http://hdl.handle.net/11427/4129 (accessed on 3 May 2021).
19. Van Hout, M.-C.; Mhlanga-Gunda, R. Prison health situation and health rights of young people incarcerated in sub-Saharan African prisons and detention centres: A scoping review of extant literature. *BMC Int. Health Hum. Rights* **2019**, *19*, 17. [CrossRef]
20. Naicker, N.; Richter, L.; Mathee, A.; Becker, P.; Norris, S.A. Environmental lead exposure and socio-behavioural adjustment in the early teens: The birth to twenty cohort. *Sci. Total Environ.* **2012**, *414*, 120–125. [CrossRef]
21. Achenbach, T.M.; Rescorla, L. *Manual for the ASEBA School-Age Forms & Profiles: An Integrated System of Multi-Informant Assessment*; Aseba: Burlington, VT, USA, 2001; ISBN 0938565737.
22. Mbonane, T.P.; Mathee, A.; Swart, A.; Naicker, N. A study protocol to determine the association between lifetime lead exposure and violent criminal behaviour in young males in conflict with the law. *BMC Public Health* **2019**, *19*, 932. [CrossRef]
23. Street, R.A.; Goessler, W.; Naidoo, S.; Shezi, B.; Cele, N.; Rieger, J.; Ettinger, K.; Reddy, T.; Mathee, A. Exposure to lead and other toxic metals from informal foundries producing cookware from scrap metal. *Environ. Res.* **2020**, *191*, 109860. [CrossRef]
24. Bello, O.; Naidu, R.; Rahman, M.M.; Liu, Y.; Dong, Z. Lead concentration in the blood of the general population living near a lead-zinc mine site, Nigeria: Exposure pathways. *Sci. Total Environ.* **2016**, *542*, 908–914. [CrossRef] [PubMed]
25. Dooyema, C.A.; Neri, A.; Lo, Y.; Durant, J.; Dargan, P.I.; Swarthout, T.; Biya, O.; Gidado, S.O.; Haladu, S.; Sani-Gwarzo, N.; et al. Outbreak of Fatal Childhood Lead Poisoning Related to Artisanal Gold Mining in Northwestern Nigeria, 2010. *Environ. Health Perspect.* **2012**, *120*, 601–607. [CrossRef]
26. Sanders, T.; Liu, Y.; Buchner, V.; Tchounwou, P.B. Neurotoxic effects and biomarkers of lead exposure: A review. *Rev. Environ. Health* **2009**, *24*, 15–45. [CrossRef] [PubMed]
27. Naicker, N.; de Jager, P.; Naidoo, S.; Mathee, A. Is There a Relationship between Lead Exposure and Aggressive Behavior in Shooters? *Int. J. Environ. Res. Public Health* **2018**, *15*, 1427. [CrossRef]
28. Tirima, S.; Bartrem, C.; von Lindern, I.; von Braun, M.; Lind, D.; Anka, S.M.; Abdullahi, A. Environmental Remediation to Address Childhood Lead Poisoning Epidemic due to Artisanal Gold Mining in Zamfara, Nigeria. *Environ. Health Perspect.* **2016**, *124*, 1471–1478. [CrossRef] [PubMed]

International Journal of
*Environmental Research
and Public Health*

Review

A Bibliometric Analysis of the Scientific Research on Artisanal and Small-Scale Mining

Fernando Morante-Carballo [1,2,*], Néstor Montalván-Burbano [3,4], Maribel Aguilar-Aguilar [4,*]
and Paúl Carrión-Mero [4,5]

[1] Facultad de Ciencias Naturales y Matemáticas (FCNM), ESPOL Polytechnic University,
Guayaquil 09015863, Ecuador

[2] Geo-Recursos y Aplicaciones (GIGA), ESPOL Polytechnic University, Guayaquil 09015863, Ecuador

[3] Department of Economy and Business, University of Almería, Carr. Sacramento s/n, La Cañada de San
Urbano, 04120 Almeria, Spain; nmb218@inlumine.ual.es

[4] Centro de Investigaciones y Proyectos Aplicados a las Ciencias de la Tierra (CIPAT),
ESPOL Polytechnic University, Guayaquil 09015863, Ecuador; pcarrion@espol.edu.ec

[5] Facultad de Ingeniería en Ciencias de la Tierra, Campus Gustavo Galindo, ESPOL Polytechnic University,
Guayaquil 09015863, Ecuador

* Correspondence: fmorante@espol.edu.ec (F.M.-C.); maesagui@espol.edu.ec (M.A.-A.)

Abstract: Mineral resource exploitation is one of the activities that contribute to economic growth and the development of society. Artisanal and small-scale mining (ASM) is one of these activities. Unfortunately, there is no clear consensus to define ASM. However, its importance is relevant in that it represents, in some cases, the only employment alternative for millions of people, although it also significantly impacts the environment. This work aims to investigate the scientific information related to ASM through a bibliometric analysis and, in addition, to define the new lines that are tending to this field. The study comprises three phases of work: (i) data collection, (ii) data processing and software selection, and (iii) data interpretation. The results reflect that the study on ASM developed intensively from 2010 to the present. In general terms, the research addressed focuses on four interrelated lines: (i) social conditioning factors of ASM, (ii) environmental impacts generated by ASM, (iii) mercury contamination and its implication on health and the environment, and (iv) ASM as a livelihood. The work also defines that geotourism in artisanal mining areas is a significant trend of the last decade, explicitly focusing on the conservation and use of the geological and mining heritage and, in addition, the promotion of sustainable development of ASM.

Keywords: mining; artisanal mining; small-scale mining; environment; bibliometric analysis

Citation: Morante-Carballo, F.;
Montalván-Burbano, N.;
Aguilar-Aguilar, M.; Carrión-Mero, P.
A Bibliometric Analysis of the
Scientific Research on Artisanal and
Small-Scale Mining. *Int. J. Environ.
Res. Public Health* **2022**, *19*, 8156.
https://doi.org/10.3390/
ijerph19138156

Academic Editors: Masayuki
Sakakibara, Win Thiri Kyaw and José
Luis Rivera Parra

Received: 6 May 2022
Accepted: 30 June 2022
Published: 3 July 2022

Publisher's Note: MDPI stays neutral
with regard to jurisdictional claims in
published maps and institutional affiliations.

1. Introduction

Mining is a type of extractive activity considered to be one of the most important sources of metals and non-metals [1,2]. This activity is not always carried out by large-scale companies or industrial machinery; being called small-scale or artisanal mining. Small-scale mining (SSM) was first defined by the United Nations (UN) as: "Any single mining operation that has an annual raw material production of 50,000 metric tonnes or less, measured at the mine entrance" [3]. However, despite referring to the production magnitude or exploitation size, this concept differs at the level of countries and institutions. For example, in Brazil, the National Department of Mineral Research (DNPM) defines SSM as an operation that produces between 10,000 t/a (tonnes per year) and 100,000 t/a of ore [4]. On the other hand, in Ecuador, according to the mining law, the SSM exploits and processes up to 300 tons of ore per day (tpd) [5].

The SSM can be developed technically (conventional) or in a rudimentary way. When the operation of the SSM is conventional, it is characterised by being developed under a legal situation and the technical application of mechanised exploitation, as well as being processed with engineering criteria and feasibility studies that guarantee the results of mineral production [6]. On the other hand, when the operation is carried out through simple and rudimentary techniques to extract ore without conventional ecological and engineering principles, it is called artisanal mining (AM) [7]. Currently, no country has clear regulations defining activities classified as AM, and almost all policies only refer to the size of the operation [8]. Hilson [9] describes that artisanal mining exploitation involves "intense labour activity located in remote and isolated sites using rudimentary techniques, low technological knowledge, low degree of mechanization and low levels of environmental, health and safety awareness". This term refers to the rudimentary type of exploitation, regardless of whether the mine is small or large [10].

Artisanal mining and small-scale mining are used synonymously to refer to mining activity carried out by individuals or small groups with low technology or machinery [11]. Considering their close relationship, the legislations of developing countries refer to the term "artisanal and small-scale mining (ASM)" as "individuals, groups, families or mining cooperatives with minimal or no mechanization, often in the informal (illegal) sector of the market" [12] (Figure 1).

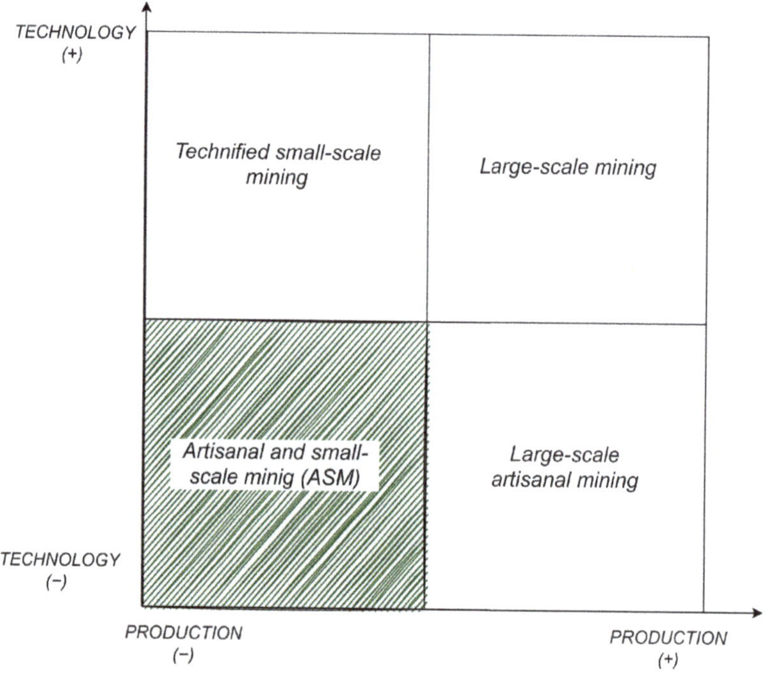

Figure 1. Schematic representation of ASM as a term that unites SSM and AM.

However, the definition of ASM is not uniform across many jurisdictions. Although there is still no internationally agreed upon definition of ASM, country-specific definitions reflect relevant situations and developments at the local level [13]. According to Seccatore et al. [7], "the term ASM is widely used to refer to those small or large operations that use rudimentary techniques to extract gold that operate legally or illegally and that are not on the radar of many companies mining companies, governments and international environmental agencies". Various authors have studied and characterised this type of activity [12,14–23].

In general, ASM is an activity that exploits small deposits, has poor capital, lacks standards to ensure health and safety, is labour intensive, and has a significant environmental impact [14]. According to [24], millions of people worldwide are dedicated to primitive mineral extraction (ASM). Most ASM operators mine precious metals and stones [25]. Other mineable materials, such as minerals, include diamonds, columbite-tantalite, and bauxite [26–28].

Recent studies have focused on large and medium-scale mining effects, updating sustainable and environmentally responsible production techniques [29–31]. However, the effects produced by ASM are still a reality due to economic, legislative, and technological limitations [22,32,33]. Furthermore, ASM has witnessed a massive expansion worldwide, employing millions of people [14,34] and producing 15–20% of the world's mineral production [7]. In addition, the areas where activities related to small-scale mining are located are studied, among other topics, from a geological point of view. In particular, in works oriented to the definition of the type of existing deposit (e.g., [35–37]), the characterization of the existing minerals of interest (e.g., [38–40]), and to the proposal of efficient exploitation alternatives (e.g., [41–43]).

Artisanal mining is driven by poverty, growing as an economic activity and adopted as a promising, and in many cases unique, alternative income [44]. However, ASM continues to develop without regulatory control in most developing countries, generating social and environmental problems in which crime, child labour, soil erosion, mercury contamination, and mining conflicts stand out [45]. The leading solution proposed by academics and professionals consists of improving ASM's environmental, technical, and socioeconomic performance by implementing regulations that organize and formalize the sector, respecting miners' rights [12,19,34].

Several literature review studies related to ASM mainly focus on systematic reviews of specific topics. Some examples are review studies about its relationship with poverty [24,46], agriculture [47], operator health [48], ecological problems [49,50], health risks [51,52], mercury contamination [6], mercury management and treatment [53], and water contamination [54], among others.

To date, no holistic analysis of ASM is recorded. This is possible with a bibliometric study that allows for knowing the structure and evolution of this field of research. Bibliometric analysis is a method that assesses the structure and trends of research in a specific body of literature [55–60], commonly used to categorize aspects of science as journals, institutions, universities, authors, and most contributing countries [61]. According to [62], this type of study is important for (i) obtaining a comprehensive overview of the subject under investigation, (ii) identifying knowledge gaps, (iii) defining novel lines in research, and (iv) positioning their contributions in the researched field. Bibliometric analysis can use two procedures: (i) analysis of scientific production, which leads to an evaluation of the impact of the field being investigated in the study and its scientific actors (authors, institutions, countries) [63,64]; and (ii) bibliometric mapping combined with clustering techniques that allow for evaluating of the cognitive structure and behaviour of the scientific field through the analysis of research fields, disciplines, and themes [65,66].

Based on the above, and considering the conflict (similarity and variation of definitions between SSM and AM), the following research question arises: How should we organize information to carry out a comprehensive analysis of the evolution and trends of the scientific production of the SSM and AM?

In this study, the term ASM is considered as a holistic concept that integrates SSM and AM as synonyms of low-production mining activity, characterised by the low-quality technology used and intensive labour. For this reason, the objective of this study is to analyse the existing literature base related to ASM through bibliometric methods that allow for the definition of the main areas being investigated, patterns, trends, and the proposal of new lines of research.

The article consists of six main sections: the introduction (Section 1), which includes a review of scientific literature related to ASM in the world; materials and methods (Section 2), which describes the procedure used in this study; results (Section 3), in which the results obtained from the analysis and processing of the database are presented; discussion (Section 4), which lies in exposing the importance of the study and the determination of future lines of research; the conclusions (Section 5), which include the limitations of the study; and finally, the references used which support this research.

2. Materials and Methods

Bibliometric research, a meta-analytic literature research tool, was conducted in this study [57,67]. This type of study is about analyzing (mapping) the structure, evolution, and research trends of a specific database [55,56,58–60,68] through parameters such as authorship, citations, keywords, journal, and affiliations [61,69].

For the bibliometric analysis of a specific field of research, it is necessary to use bibliometric maps [70,71], which can be viewed in different software (e.g., Bibexcel, CitNetExplorer, CiteSpace, CoPalRed, HistCite, Net-work Workbench Tool, SciMAT, Sci2Tool, VantagePoint, and VOSviewer). This study used the VOSviewer software [65] to build bibliometric networks in order to facilitate the analysis of the intellectual structure using various parameters obtained from scientific publications [72]. The research contemplates a systematic process distributed in three phases (Figure 2): (i) data collection, (ii) data processing and software selection, and (iii) data interpretation.

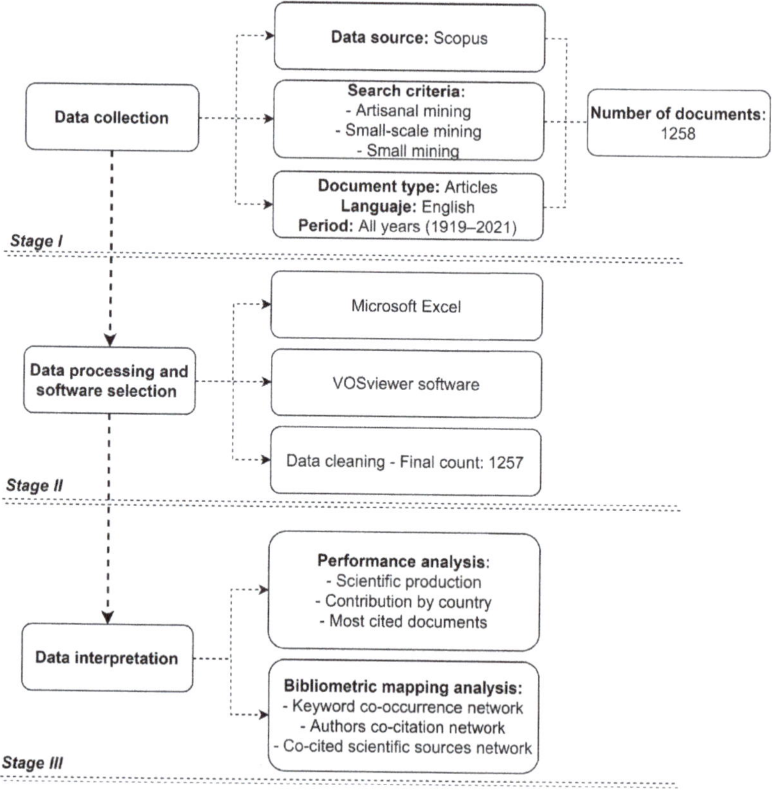

Figure 2. General methodological scheme of the study.

2.1. Data Collection

Most of the research literature on small-scale mining is closely related to artisanal mining [34,73,74]. Furthermore, scientific contributions on artisanal and small-scale mining (ASM) generally expose case studies, mainly in developing countries, in which small-scale mining is a term frequently used to refer to artisanal mining activity [25,75]. Therefore, considering this relationship, the following search terms are considered in this study: (i) artisanal mining, (ii) small-scale mining, and (iii) small mining. The selected terms will allow for the obtaining of a complete literary body on the subject for its later bibliometric analysis.

Quality databases with accurate and consistent information are essential [76,77]. Therefore, the Scopus database was selected for the search, as it is considered one of the central databases with great coverage, facilitating the study and comparison of different scientific fields [78–83]. In addition to its comprehensive coverage and ease in the tools provided for bibliometric analysis, in this specific study (artisanal and small-scale mining), we considered the main reason for the extensive coverage of Scopus in terms of scientific production related to geosciences [84–86].

Scopus constitutes an indexed and well-organised database of scientific production, with tools that allow the export of metadata [63,80,87]. In addition, it provides a series of data on scientific publications such as authors, institutions, countries, number of citations, and research areas [78,80,88,89]. An important aspect to consider in selecting the database is that the growth in the coverage of journals from Latin America and the Caribbean indexed within the Scopus database [90,91] strengthens the analysis carried out in different areas.

The search was carried out on 8 November 2021, using the terms previously defined in the titles, abstracts, and keywords of the different existing publications in Scopus. The initial search equation used was: ((TITLE-ABS-KEY ("artisanal mining") OR TITLE-ABS-KEY ("small scale mining") OR TI-TLE-ABS-KEY ("small mining"))), with a result of 1665 documents. Subsequently, the database was delimited through inclusion and exclusion criteria according to the analysis to be carried out. As a first criterion, it was considered appropriate to exclude the year 2022 and carry out the study with documents published up to the present (search date). Subsequently, the number of documents was limited to articles, since the results obtained from the initial search equation yielded more than 75% of documents as articles. Finally, considering that the English language is the most frequent in scientific publications [92], the initial search of the investigated area indicated that more than 90% of documents are written in English; the study was limited to documents in that language. The final database represents 1258 documents, which will be the basis for processing phase two of the study.

2.2. Data Processing and Software Selection

The data processing and software selection phase begins with extracting data from the Scopus database through a Microsoft Excel spreadsheet. The software uses data analysis and error elimination [93–95] and evaluated the investigated area's scientific production [96]. Specifically, the downloaded database contains authors, titles, keywords, years, number of citations, and abstracts. Then, a cleaning and error elimination process is carried out [97,98], eliminating repeated and incomplete data for this research, obtaining 1257 documents to analyse.

With the adjusted database, we construct two-dimensional bibliometric networks, which define the research structure of the field being studied using the VOSviewer software (Version 1.6.17) [65]. The software is freely available and is used as the primary tool for constructing detailed bibliometric maps through simple graphs [70,99,100]. This software is used in different scientific areas such as medicine [101–104], management [105–109], natural and cultural resources [110–112], and geosciences [68,113–116], among others.

2.3. Data Interpretation

The investigated field analysed the results through (i) performance analysis and (ii) scientific mapping [117]. The first analysis makes it possible to determine the evolution of scientific production and its impact by evaluating parameters such as authors, year, affiliations, journals, and countries [118–120]. The subsequent analysis (scientific mapping) allows for the definition of different relationships between the analysed variables, obtaining information at the micro-level (co-occurrence of author keywords), meso-level (co-citation of authors) and macro-level (journal co-citation) [94,121]. Specifically, the objective of the analysed approaches was to identify the main areas of research on ASM for the definition of new lines based mainly on innovative, sustainable, and affordable research.

3. Results

3.1. Performance Analysis

3.1.1. Scientific Production Analysis

Research studies related to artisanal and small-scale mining (ASM) began in 1919, with the study of Wormleighton [122], which marked the interest in this type of research on sewage and drainage works in a mining district. However, the first five decades (until 1979) of research in this field are scarce, with eight articles representing 0.63% of the total scientific production of ASM. Due to these reasons, excluding these years from the production analysis is considered pertinent.

This analysis is divided into three periods distributed by decades: period I (from 1981 to 2000), period II (from 2001 to 2010), and period III (from 2011 to 2021) (Figure 3). For period I, two decades are grouped (1981–2000) due to the low number of published articles.

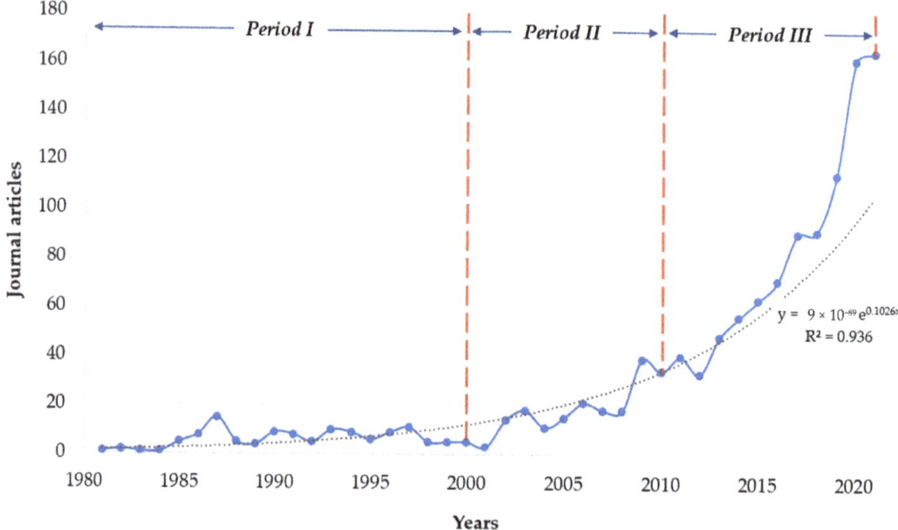

Figure 3. The behavior of ASM scientific research over time (1981–2021).

Period I (1981–2000): This research period of ASM in the world begins with 124 scientific articles, in which a similar production trend can be observed every five years (Figure 3). It is essential to highlight that in 1987 the highest production was obtained within the analysed period (Figure 3), with 15 published articles. In general, this first period marks the beginning of ASM research. The primary study topics focus on the contribution of small-scale mining to world mineral production [123], as well as its contribution to the socioeconomic development of developing countries [124]. Likewise, case studies of small-scale mining [125–129], the role of the government in promoting small-scale mining [130], and the need for government policies [131,132] are presented.

Within this research period, the authors also expose the importance and characteristics of small-scale mining [133], as well as the primary technical considerations that reduce the human and environmental risk [134], despite its limitations [135]. Likewise, it is possible to observe studies focused on the pollution problems of small-scale mining [136,137] (e.g., in water [138], soil [139], and environment [140,141]), seismicity inductions [142,143], and mining-associated diseases [144].

Period II (2001–2010): This decade is characterised by significant growth in research, with a total of 191 articles representing 15.15% of the ASM research field. In 2009 there was a peak in research with 39 publications (Figure 3). Ranging from 2001 to 2010, ASM research is related to mining environmental management [145–148] and the need for mining legislation [34,149–152] that will solve environmental pollution problems [42,153–158]. During this period, studies on illegal mining are also visible [159–162], which generate land-use conflicts due to small and large-scale mining [163,164]. On the other hand, it is essential to highlight the increase in the scientific production of gold ASM, in which the scarce legislation [147,152,154,165,166], problems of health in people [167–169], and the inclusion of women in this type of activity [170] are emphasised.

Period III (2011–2021): Finally, the third period analysed is characterised by an exponential growth in scientific production related to ASM, with a total of 911 articles representing 74.21% of the total documents analysed (Figure 3). The average annual production exceeds 80 articles, with a peak in 2020 (161 articles) and 2021 (164 articles) investigated, defining ASM as a booming research field. As mentioned in previous periods, this field of research is generally related to lines such as pollution [49,171–177], agriculture problems caused by ASM [178–181], the association of ASM with poverty [182–184], mining conflicts [185,186], informal/illegal ASM [187–189], and the influence of ASM on water quality [190,191], among others. However, this period is characterised by an intensive growth in the scientific contribution to solving mining conflict problems through ASM formalization strategies [45,192–198], in addition to contributing to research focused on strategies for reducing environmental pollution [199,200] and health risk mitigation [201].

3.1.2. Regional and Country Contribution

According to the authors' different affiliations, the contribution by country indicates that, worldwide, 46 countries contribute to research related to ASM (Figure 4). In general, four countries stand out due to their high scientific production: the United States (210 articles), United Kingdom (209 articles), Canada (133 articles), and Ghana (109 articles) (Figure 5). In addition, these countries are characterised by a high number of citations compared to the other contributing nations, with the United Kingdom standing out as the most cited country worldwide on topics related to ASM (94,929 citations) (Table 1).

Table 1. Top 10 countries by the number of documents.

Ranking	Country	Region	Documents	Citations
1	United States	América	210	3989
2	United Kingdom	Europa	209	6440
3	Canada	América	133	2891
4	Ghana	África	109	1792
5	Australia	Oceanía	83	1076
6	China	Asia	71	1508
7	Germany	Europa	67	1209
8	Brazil	América	63	931
9	South Africa	África	57	394
10	Belgium	Europa	56	1307

According to the affiliation obtained, it is essential to note that the top 10 countries that contributed the most in the field can be differentiated (Table 1), highlighting the participation of developed countries such as the United States, United Kingdom, and Canada, leaders in ASM research throughout the world.

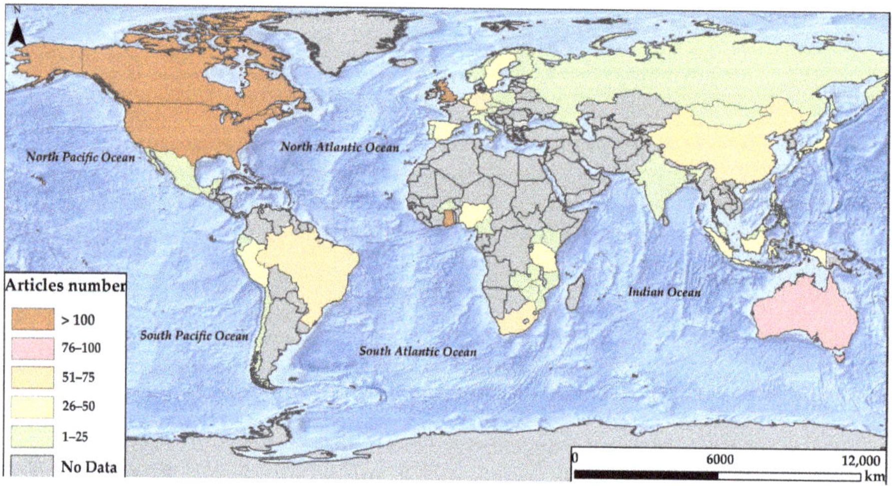

Figure 4. Contribution of studies related to ASM by country.

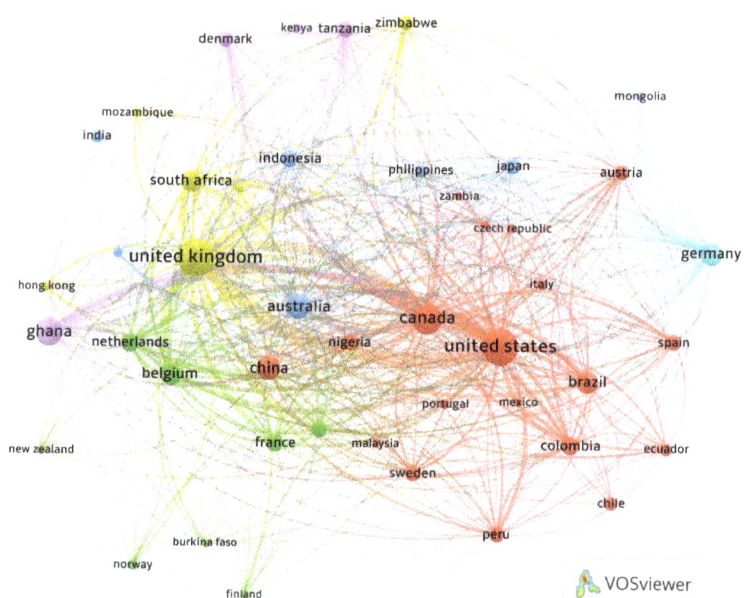

Figure 5. Country contribution bibliometric map in ASM. The nodes' size varies depending on the number of documents per country, related through links in which their collaboration is reflected.

The behaviour of collaboration between countries, based on affiliation data, indicates that the United States, Canada, Australia, Germany, Austria, and Spain are the countries with the most significant collaboration (each one collaborates with 45 different countries). The United States, the country with the highest production, contributes to 45 countries, of which Canada, Ghana, and Germany stand out. When analyzing the United States production, the studies focus on issues related to the impact that ASM generates on the environment [49,156,157,202,203], health implications [168,171,173,204–206], the effects of AMS on socioeconomic factors [24,184,207], and the inclusion of women in jobs related to this type of activity [23,170,208]. Strengthening its studies of problems associated with ASM, the United States also generated contributions in the areas focused on the need for ASM regulations [195,209–211], as well as ASM risk and contamination mitigation alternatives [201,212–214].

Although the United States is the country with the highest scientific production, the United Kingdom, with only one less article, far exceeds the number of citations in its works. These studies include the socioeconomic impacts of ASM in developing countries and strategies focused on the sector's sustainability [9], environmental problems of small-scale gold mining [42], poverty-driven informal artisanal gold mining [73], and ASM reforms [215]. This analysis also includes the study of the dependence on mercury as an agent of poverty in artisanal gold mining [216] and the pollution generated in these communities [217]. Studies on strategies to eradicate illegal artisanal mining are also included [162].

Canada, occupying third place in the contribution of ASM articles, makes contributions focused on African or South American countries. The investigations are related to the current use of mercury in ASM [7] and the proposal of actions focused on the reduction of these types of emissions [218], as well as the responsibility of miners, governments, and organizations in the search for solutions to pollution problems [41,219,220]. There are also studies related to the role of ASM formalization in Africa [34].

3.1.3. Journal Performance

The analysis included 468 journals in which 1257 scientific articles were published (database analysed) related to ASM. Table 2 shows the top 10 of the most outstanding journals, with 401 articles representing 31.9% of the total.

Table 2. Top 10 journals with the highest number of publications.

Ranking	Journal	Country	Documents Number	Representation	Citations	SJR *	Cite Score
1	*Resources Policy*	United Kingdom	116	9.2	2912	1.276	6.3
2	*Extractive Industries and Society*	The Netherlands	82	6.5	951	0.999	4.2
3	*Journal of Cleaner Production*	United Kingdom	40	3.2	1384	1.937	13.1
4	*Natural Resources Forum*	United Kingdom	37	2.9	843	0.646	2.9
5	*Science of the Total Environment*	The Netherlands	31	2.5	1492	1.795	10.5
6	*International Journal of Environmental Research and Public Health*	Switzerland	27	2.1	450	0.747	3.4
7	*Minerals and Energy—Raw Materials Report*	United Kingdom	18	1.4	70	0.143	-
8	*Geoforum*	United Kingdom	17	1.4	472	1.584	5.5
9	*World Development*	United Kingdom	17	1.4	820	2.386	8.4
10	*Environmental Research*	United States	16	1.3	677	1.460	7.9

* SJR data was obtained from Scimago Journal & Country Rank.

Resources Policy is the leading journal in scientific publications in the analysed field with 116 articles representing 9.2% of the total. This journal is the most cited worldwide, with 2912 citations. The top five studies with the highest citations (Banchirigah [162],

Hilson [221], Siegel & Veiga [34], (Mohammed Banchirigah [215], and y Geenen [193]) focus on formalization and poverty related to ASM in Africa. Based on its citations (163), the most relevant study was developed by Banchirigah [162] in Ghana. The study argues for the need to eradicate illegal mining through formalization, work alternatives, and government and military intervention. On the other hand, the journal *Science of the Total Environment*, occupying fifth place in the production of ASM, represents the second most cited journal (1492 citations). The two most cited articles correspond to the one carried out by Hylander and Goodsite [157] (191 citations) and de Cordy et al. [41] (162 citations), which discuss mercury contamination from ASM and the costs involved in remediating the environment.

3.1.4. Frequently Cited Documents

Citation analysis exposes a given article's influence by the citation it receives in another articles [222]. The scientific production for ASM globally (1257 articles) presents 20,579 citations. Table 3 presents the top 10 of the most cited documents with 1776 citations, representing 8.63% of the total. The established ranking is characterised by documents published in 2005.

Table 3. Top 10 most cited documents.

Ranking	Authors	Year	Title	Citations	Journal
1	Bebbington et al. [223]	2008	Mining and Social Movements: Struggles Over Livelihood and Rural Territorial Development in the Andes	292	*World Development*
2	Xiao et al. [173]	2017	Soil heavy metal contamination and health risks associated with artisanal gold mining in Tongguan, Shaanxi, China	196	*Ecotoxicology and Environmental Safety*
3	Hilson & Potter [73]	2005	Structural adjustment and subsistence industry: Artisanal gold mining in Ghana	194	*Development and Change*
4	Hylander & Goodsite [157]	2006	Environmental costs of mercury pollution	191	*Science of the Total Environment*
5	Banchirigah [162]	2008	Challenges with eradicating illegal mining in Ghana: A perspective from the grassroots	163	*Resources Policy*
6	Cordy et al. [41]	2011	Mercury contamination from artisanal gold mining in Antioquia, Colombia: The world's highest per capita mercury pollution	162	*Science of the Total Environment*
7	Fisher [224]	2007	Occupying the margins: Labour integration and social exclusion in artisanal mining in Tanzania	149	*Development and Change*
8	Veiga et al. [218]	2006	Origin and consumption of mercury in small-scale gold mining	149	*Journal of Cleaner Production*
9	Hilson [221]	2009	Small-scale mining, poverty and economic development in sub-Saharan Africa: An overview	141	*Resources Policy*
10	Bose-O'Reilly [167]	2008	Mercury as a serious health hazard for children in gold mining areas	139	*Environmental Research*

The study by Bebbington et al. [223] is the most cited article (292 citations), with intervention of authors from the United Kingdom, the United States, Ecuador, and Peru. In his study, reference is made to the influence of social movements against mining investment in Latin America. Mainly two case studies are exposed (Ecuador and Peru), in which it is evident how social activities can significantly modify the form and effects of the extractive industry

Second place is occupied by Xiao et al. [173], with the presence of authors from China and the United States. The research analyses soil contamination from artisanal gold mining in China and its implications for human health and environmental wellbeing by assessing heavy metal levels in soil and plants. Likewise, within its objectives, the identification of plants that promote the phytoremediation of the area is addressed.

Finally, the third most cited article related to ASM is the work developed by Hilson and Potter [73], authors from the United Kingdom. Their scientific contribution focuses on analysing Ghana's National Structural Adjustment Program (SAP) as a driver in the growth of informal artisanal gold mining driven by poverty.

3.2. Intellectual Structure Analysis

3.2.1. Co-Occurrence Author Keyword Network

The co-occurrence analysis of author words allows for the formation of connections and the building of a domain structure based on keywords [225]. The analysis included a process of cleaning and filtering the information, obtaining 90 keywords. Table 4 shows the top 15 words with the highest occurrence in the area studied, highlighting "artisanal and small-scale mining", "mercury", and "mining" as the top three most frequent keywords in ASM studies.

Table 4. The 15 main words with the highest occurrence in ASM studies.

Ranking	Keywords	Occurrences	Links	Total Link Strength
1	artisanal and small-scale mining	597	88	764
2	mercury	109	41	198
3	mining	80	49	98
4	gold	60	39	129
5	formalization	48	35	101
6	livelihood	38	24	71
7	poverty	36	23	73
8	heavy metals	34	21	53
9	sustainability	25	14	32
10	conflict	23	20	51
11	environment	21	20	52
12	mercury pollution	21	14	30
13	gender	20	19	44
14	sustainable development	20	20	34
15	galamsey	18	19	34

The bibliometric map obtained grouped the 90 keywords into nodes of different colours grouped into four clusters that represent the main research areas of ASM (Figure 6). The nodes' size varies depending on the number of occurrences of each keyword, and they are related through links in which the thickness represents a better relationship.

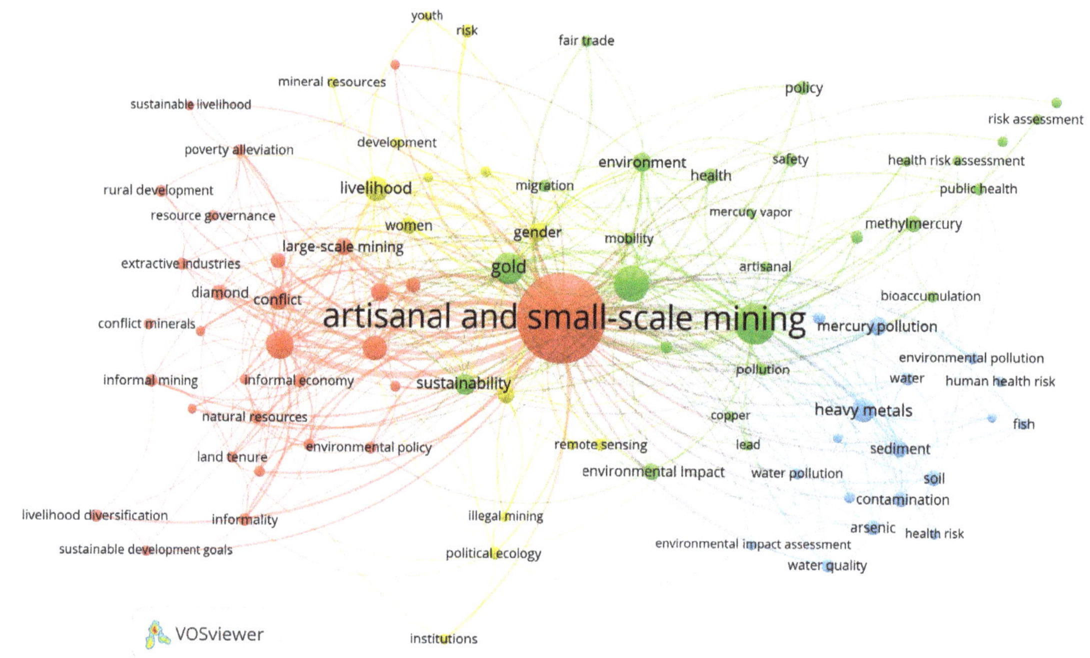

Figure 6. Author keyword co-occurrence bibliometric map in ASM.

Cluster 1 (Red Colour): Social Conditioning Factors of the ASM

The social conditioning factors of ASM represent one of the research areas aimed at understanding how poverty drives the development of this type of activity as a subsistence alternative, which entails informality [189], conflict [185,186], child labour [226], and women's labour [227]. Likewise, the link between mining and agricultural activity in rural areas with low economic resources is exposed as the primary source of subsistence for people [27,180,228]. Considering this type of problem, it is evident how formalization represents a considerable challenge [186] and is regarded as a tool that allows for regulating, controlling, and effectively supporting ASM operators [34,45,197,209,229]. However, several case studies show that formalization in various countries aggravates mining conflicts, informality, poverty, illegality, and state control [193,230–232]; entrenching poverty without achieving sustainable development [233].

Given this situation, research developed to establish strategies in ASM that allow for achieving sustainable development [234] through an analysis of social, political, economic, environmental, and health aspects [235–237]. Some examples of this type of action are: (i) the implementation of design thinking and multi-criteria decision analysis of ASM [238], (ii) national minerals policies and stakeholder participation in planning decisions [239], (iii) collaboration between the LSM and ASM, for the benefit of the communities [240], and (iv) integration of scientific and local knowledge in the planning of the remediation of contamination by ASM [214,241].

Cluster 2 (Blue Colour): ASM Environmental Impacts

Artisanal and small-scale gold mining (ASSGM) is the most developed activity in ASM. In this area of research, significant production of environmental and health impacts caused by ASSGM is evident [156,218,219], and limited studies are addressing the effects on the health and environmental impacts of artisanal sandstone mining [242] and diamond mining [26,234,243].

The investigations are most frequently related to pollution generated in the soil [244–248], water [249–251], and crops or trees [158,252], which directly influence the health and well-being of humans. Faced with this problem, finding innovative research to eliminate, replace, or reduce environmental pollution in mineral processing is standard. Some examples are the cyanide phytoremediation by water hyacinths (Eichhornia crassipes) in the cyanide effluents treatment in small-scale gold mining [253], hyperaccumulation of zinc by Corydalis davidii in Zn-polluted soils [254], Erato polymnioides as a phytoremediation plant for soils contaminated with Pb, Zn, Cu, and Cd [255], and Heliconia psittacorum in remediating soils and water polluted with heavy metals [256].

Cluster 3 (Green Colour): Mercury Contamination and Its Implication on Health and the Environment

Mercury is a heavy, liquid metal frequently used in artisanal gold mining. This cluster reflects a marked trend of studies focused on the health and environmental effects of mercury or methylmercury contamination in soil, sediments, and water [257–259]. This type of contamination generated significant research on health problems associated with direct or indirect exposure of humans to mercury due to mining activities [167,260–263], as well as studies evaluating the risk posed to human health by ingestion of heavy metals that are present in the water and plants [176,264–266].

Given the implications of mercury on the environment and health, the reason for the emergence of research that highlights the importance of cooperation between government, regional, and local organisations to improve mineral extraction and processing processes through legalisation, financial support, technological innovation, and training [9,212,267,268], as well as studies focused on reducing pollution to ensure human and environmental health [202,269,270], is evident. These include analyses that seek to minimise the use of mercury through price increases [219], laws (agreements) that prohibit its use in mining [269,271–273], promotion of appropriate technology [154,274], and training on improved technologies for gold extraction [275] (e.g., use of cassava to leach gold [276]). Finally, it is essential to highlight how local participation in decision making [277] and indigenous participation due to their ecological knowledge [278] are alternatives to achieving sustainability in ASM mineral processing.

Cluster 4 (Yellow Colour): ASM as Livelihood

In this cluster, the most frequent studies are those related to ASM as a subsistence activity in rural communities with limited resources. Within her research, the women's role in ASM as a means of subsistence due to poverty is emphasised [227,279,280], as well as the need for policies that improve the economic wellbeing of people who depend on ASM regardless of gender [229,281]. On the other hand, considering that several countries chose to ban this type of mining, there is also research related to alternative livelihood strategies for miners who were displaced from their activity [282,283]. Some examples of these strategies are promoting agriculture as an alternative economic source [179,284] or complementary [178], and promoting government support in ASM through regulations that allow regulating activity [194].

3.2.2. Co-Citation Network of Cited Authors

The analysis carried out allowed for the identification of co-cited authors and authors that make up the scientific base of the area studied [285]. This type of analysis proposes that two authors share the same area of research if their documents are cited jointly by one or more documents [286–288]. The author co-citation network (Figure 7), built in the VOSviewer software, groups 512 authors (nodes) into six clusters representing similarities in the topics investigated with more than twenty co-citations.

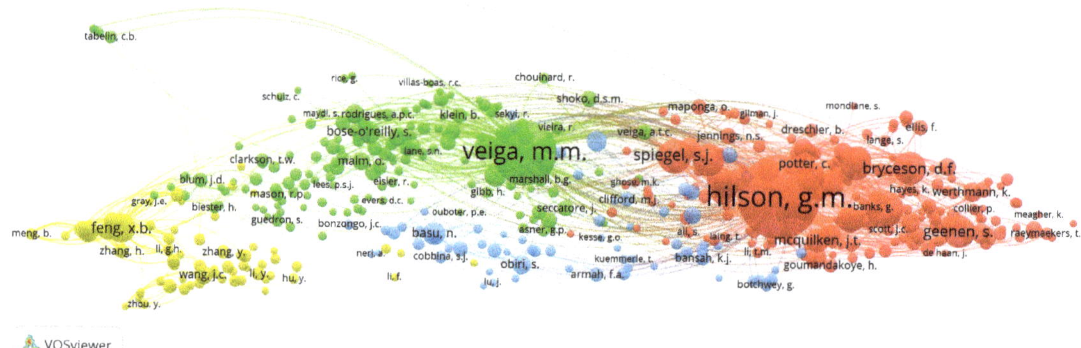

Figure 7. Co-citation network of authors in ASM.

Cluster 1 (red colour), "ASM and implications in society", comprises 206 authors, including Hilson, G.M. (2212), Maconachie, R. (456); Spiegel, S.J. (363); Bryceson, D.F. (353); y Banchirigah, S.M. (337) due to its high number of co-citations. This group of researchers carried out studies within ASM that include: (i) positive and negative effects of artisanal mining formalization [194,197,198,215,289], (ii) ASM and agriculture as a means of subsistence [47,180,284,290,291], and (iii) analysis of alternatives that improve mineral extraction or processing systems [269,292–294].

Within cluster 2 (green colour), "consequences and challenges of Mercury in ASM", the researchers Veiga, M.M.; Beinhoff, C.; Bose-O'reilly, S.; Telmer, K.H.; and y Drasch, G. represent the top five co-cited authors, in a cluster with a total of 166 authors. This research includes studies of mercury contamination in gold mining areas [41,295,296], evaluation of risks to human health due to exposure to mercury by operators, women, and children [167,295,297,298], and strategies to reduce this type of contamination based on the modernization of mineral processing in obtaining gold [148,199,219,299–302].

Cluster 3 (blue colour), "Implications of ASM in health", composed of 73 authors, in which Basu, N.; Pardie, S.; Obiri, S.; Aryee, B.N.A.; and Amankwah, R.K. are the most coveted authors. This cluster mainly includes studies of risk to human health due to exposure to mercury [48,303], environmental impacts of ASM [49], consumption of contaminated food or water [304], or multiple heavy metals [305]. Likewise, the authors expose an interest in providing strategies to reduce pollution produced by ASM, mainly due to the use of mercury [155,216,217,234,242,306].

Finally, cluster 4 (yellow colour) with 67 authors, called "Effects of artisanal mercury extraction", leads to the top five most co-cited authors, represented by Feng, X.B.; Qiu, G.L.; Li, P.; Wang, J.C.; and Wang, S.F. This group of authors dedicate their studies to topics related to Hg contamination in the air [307], water [308], sediments, soil, or crops [309–312] in mercury mining areas, mainly in China. They also analyse the risk posed to miners and people in mining areas when exposed to Hg or methylmercury [313–315].

3.2.3. Journal Co-Citation Network

The analysis considers the similarity between a group of journals based on the citations received when two or more journals are cited jointly by several related documents [316]. The objective of this analysis is based on understanding the structures of the academic areas.

Figure 8 shows the co-citation network of 152 journals (nodes) with more than 20 citations, grouped into four different clusters (differentiated by colours) and their other connections.

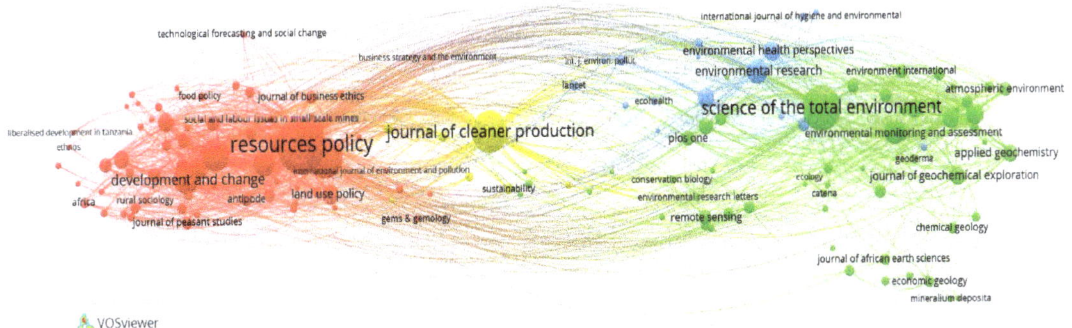

Figure 8. Journal co-citation network in ASM.

Cluster 1 (red colour), "Management, Policy and Development", contains 70 journals representing 8757 citations. In this group, the journals *Resources Policy* (1799 citations, United Kingdom), *Extractive Industries and Society* (939 citations, The Netherlands), *World Development* (578 citations, United Kingdom), *Natural Resources Forum* (526 citations, United Kingdom), and *Development and Change* (502 citations, UK) are shown as the top five of the most talked-about magazines. The studies within this cluster comprise analyses of ASM's political, economic, environmental, and social aspects in different parts of the world.

Cluster 2 (green colour), "Environmental Science and Pollution", with 58 journals and 5675 citations, mainly exposes studies associated with the environmental contamination of ASM and its human implications. In this group are journals such as *Science of the Total Environment* (1170 citations, The Netherlands), *Environmental Science & Technology* (365 citations, United States), *Environmental Pollution* (245 citations, United Kingdom), *Chemosphere* (205 citations, United Kingdom), and *Water, Air and Soil Pollution* (205 citations, The Netherlands), among others.

Cluster 3 (blue), "Environmental Science and Health", has 16 journals and 1327 citations. The journals with the highest citations include *Environmental Research* (322 citations, United States), *Environmental Health Perspectives* (230 citations, United States), *International Journal of Environmental Research and Public Health* (230 citations, Switzerland), *Minerals Engineering* (101 citations, United Kingdom), and *International Journal of Occupational and Environmental Health* (68 citations, UK). Within this cluster, the primary studies focus on evaluating the impact of ASM on human health due to direct or indirect exposure to heavy metals.

Cluster 4 (yellow colour), "Renewable Energy, Sustainability and the Environment", consists of 8 journals with 1305 citations. These journals include research papers focused on mineral extraction and processing sustainability in ASM. The top five most-cited journals are *Journal of Cleaner Production* (1000 citations, UK), *Environmental Science & Policy* (65 citations, The Netherlands), *Ecological Economics* (57 citations, The Netherlands), *Sustainability* (57 citations, Switzerland), and *Journal of Sustainable Mining* (54 citations, Poland).

4. Discussion

The systematic process applied in this study made it possible to identify the intellectual structure of artisanal and small-scale mining (ASM) in the world. Considering the performance analysis carried out, it is apparent that the scientific production of ASM began in 1919, being until 1980 a scarce production (eight articles). Furthermore, the range of years analysed (distributed in three periods) indicates that the research remained relatively constant since 1980 (periods I and II). However, as of 2010 (period III), ASM research increased exponentially worldwide, representing 74.21% of the articles produced (Figure 3).

This marked difference in scientific production could refer to the artisanal mining boom that the world experienced in the last decade, mainly due to the increase in poverty within rural areas. The rise of ASM, characterised by extraction and processing techniques without technical and environmental considerations, clearly represents a risk to humanity and the environment. This is why the increase mentioned above in scientific production focuses its studies on ASM contamination [173,174], mining conflicts [185,186], illegal ASM [187–189], as well as strategies to solve these types of problems [196,197,200,201].

On the other hand, when analyzing scientific production by country, the United States, the United Kingdom, and Canada represent the most significant contributions to research related to ASM (Table 1). Of these countries, the United Kingdom is characterised by its high number of citations (Table 1) and its extensive collaboration (greater than 70%) in studies carried out in African countries (e.g., Ghana and Tanzania). Likewise, this country occupies the number one position with the *Resource Policy* magazine, contributing the highest number of ASM publications (116 articles) (Table 2). On the other hand, the United States and Canada collaborate in studies mainly in South American countries such as Brazil, Peru, and Colombia, and Africa, mainly in Ghana.

Considering the analysis of the intellectual structure through three scientific maps, the study of the co-occurrence of author keywords (Figure 6) made it possible to define, through clusters, four research areas of ASM. Within these areas, "Social conditioning factors of the ASM" and "Mercury contamination and its implication in health and environment" are the most studied topics (e.g., [34,192,209,218,240,252]). On the other hand, it is essential to highlight that cluster 2 ("ASM environmental impacts") and cluster 3 ("Mercury contamination and its implication in health and environment") are strongly related (Figure 6), with studies focused on the impacts of ASM on the environment (e.g., [249,251,258,309]) and health (e.g., [261,263,265]). However, considering a specific orientation and significant scientific production related to mercury, the results reflect the study of mercury as a particular area in this analysis.

Cluster 4 (ASM as livelihood) is an ASM area with relatively less scientific production, strongly related to cluster 1. The objective of ASM as a livelihood area includes research in which ASM is analysed as a means of subsistence and the search for strategies to propose alternative or complementary activities that benefit the living conditions of people who depend economically on this type of activity (e.g., [179,194,280]).

To complement the analysis of the co-occurrence of keywords, the co-citation analysis of authors was carried out, which allowed for the identifying of the relationships between different authors in the references of the research works carried out ion ASM. The results obtained reflect the existence of 512 authors grouped into four clusters, representing the author's areas or lines of research (Figure 7). These areas are very well defined in specific topics; however, they are all within a large area called "Effects of ASM and mitigation measures". Of the clusters obtained, clusters 2 and 3 are firmly related, presenting studies that address similar issues regarding the use and effects of mercury in ASM [216,219,297,303]. On the other hand, it is important to highlight an area aimed at research related to the artisanal extraction of mercury (Cluster 4), in which authors such as Feng, X.B.; Qiu, G.L.; Li, P.; Wang, J.C.; and Wang, S.F. carried out works that include the contamination generated by mercury mines in the soil, water, and air [307–309], as well as the risk it represents for human health [310,313].

Finally, the co-citation analysis of journals was carried out to understand the different academic areas in which ASM studies are published. The results show us four main academic areas (clusters) (Figure 8), defined based on the research topics. For example, in the cluster with the highest number of co-cited journals (cluster 1), the journals *Resources Policy* and *Extractive Industries* stand out with the highest number of co-citations in works oriented to ASM's political, economic, social, and environmental aspects. Likewise, it is essential to highlight that clusters 2 and 3 show related academic areas in which the journals publish research topics on environmental pollution of ASM and its health risks. In these clusters, the journals with the highest number of co-citations correspond to *Science of the Total Environment* (cluster 2) and *Environmental Research* (cluster 3), which add up to 1170 and 322 citations, respectively.

On the other hand, the connection offered by cluster 4 (Renewable Energy, Sustainability, and the Environment) with the other clusters is visible. Being in the centre of the clusters obtained (Figure 8), despite its limited number of journals (eight), its high number of co-citations (1305) highlights the importance of its research topics focused on the socio-environmental aspects of ASM, with the *Journal of Cleaner Production* as the most prominent journal.

Specifically, ASM research exposes excellent studies that identify the causes and effects of the leading social, economic, and environmental problems that compromise environmental and human wellbeing in the short, medium, and long term (e.g., [9,34,73,157,173,218,223,317]). These studies lay the groundwork for issues that must be mitigated and eliminated. The analysed database reflects that, over time, studies developed that focus on solutions to problems generated by ASM (e.g., [238,239,241,253,256,268,278,294,318,319]). However, despite the worldwide importance and impact of research aimed at ASM solutions, it is still scarce (less than 20% of the analysed database). For this reason, the possibility arises that the different authors in ASM strengthen this type of study to the point that in the best of cases, it is considered one of the top research areas in ASM.

The analysis made it possible to evaluate the evolution and trends of research in ASM and propose strengthening innovative studies regarding ASM's environmental, social, legal, and economical solutions. Therefore, this type of research can be included by the representative authors and journals of ASM as a new booming field that represents sustainable solutions for the effects produced by this type of mining activity.

5. Conclusions

The bibliometric analysis allowed us to evaluate the structure of ASM research field within the last four decades. Within the performance analysis, the results obtained show a scientific production with exponential growth in ASM research, with the collaboration of 46 countries, highlighting the United States, United Kingdom, and Canada as the countries with the highest scientific production that address ASM research in mainly Latin American and African countries, respectively. Furthermore, the works are the products of 512 authors published in 468 journals, qualifying ASM as a booming research field.

By analysing the co-occurrence of keywords, four areas of research in ASM were defined: (i) social conditioning factors of ASM, (ii) environmental impacts generated by ASM, (iii) mercury contamination and its implication on health and the environment, and (iv) ASM as a livelihood. Within these areas, a clear trend of studies related to the implications of ASM from the political, social, economic, and environmental points of view is apparent. On the other hand, it is essential to highlight the effects of mercury on the environment and health as topics on the rise, mainly in health risk assessment and strategies that minimise the impact of mercury on ASM. However, studies aimed at finding solutions in ASM to date are scarce and need to be strengthened.

Despite limiting the study to only one database (Scopus) and considering only one type of document (articles) in the English language, the proposed research establishes a global analysis of the ASM study. This analysis can serve as a reference for future researchers in the field for the most researched topics, authors, and outstanding journals; and raise the possibility of forming collaborative networks inside and outside your country.

Author Contributions: Conceptualization, P.C.-M., F.M.-C., N.M.-B. and M.A.-A.; methodology, P.C.-M., F.M.-C., N.M.-B. and M.A.-A.; software, N.M.-B. and M.A.-A.; validation, P.C.-M., F.M.-C. and N.M.-B.; formal analysis, P.C.-M., F.M.-C., N.M.-B. and M.A.-A.; investigation, P.C.-M., F.M.-C., N.M.-B. and M.A.-A.; data curation, N.M.-B. and M.A.-A.; writing—original draft preparation, M.A.-A.; writing—review and editing, P.C.-M., F.M.-C., N.M.-B. and M.A.-A.; supervision, P.C.-M., F.M.-C. and N.M.-B.; project administration, P.C.-M. All authors have read and agreed to the published version of the manuscript.

Funding: This research was funded by the ESPOL Polytechnic University research project called "Registry of geological and mining heritage and its impact on the defense and preservation of geodiversity in Ecuador", CIPAT-01-2018 and "Propuesta de Geoparque Ruta del Oro y su incidencia en el desarrollo territorial", CIPAT-02-2018.

Institutional Review Board Statement: Not applicable.

Informed Consent Statement: Not applicable.

Data Availability Statement: Not applicable.

Acknowledgments: We would like to thank Edgar Berrezueta, (Senior Scientist of the Instituto Geológico y Minero de España (IGME)) for his collaboration in the review process of the structure and content of the manuscript. We would also like to thank the editorial office for the editorial handling and three anonymous reviewers for their constructive comments and corrections.

Conflicts of Interest: The authors declare no conflict of interest.

References

1. Li, Z.; Ma, Z.; van der Kuijp, T.J.; Yuan, Z.; Huang, L. A review of soil heavy metal pollution from mines in China: Pollution and health risk assessment. *Sci. Total Environ.* **2014**, *468–469*, 843–853. [CrossRef] [PubMed]
2. Zhuang, P.; McBride, M.B.; Xia, H.; Li, N.; Li, Z. Health risk from heavy metals via consumption of food crops in the vicinity of Dabaoshan mine, South China. *Sci. Total Environ.* **2009**, *407*, 1551–1561. [CrossRef] [PubMed]
3. UN. *Small-Scale Mining in the Developing Countries*; United Nations: New York, NY, USA, 1972.
4. Vale, E. *Análise Econômica das Pequenas e Médias Empresas de Mineração: Relatório Final*; CPRM: Rio de Janeiro, Brazil, 2002.
5. Vergara, F. Ecuador Registro Oficial, año III, N.517. Available online: http://www.mineriaecuador.com/download/ley_mineriaec.pdf (accessed on 28 June 2022).
6. Veiga, M.M.; Angeloci-Santos, G.; Meech, J.A. Review of barriers to reduce mercury use in artisanal gold mining. *Extr. Ind. Soc.* **2014**, *1*, 351–361. [CrossRef]
7. Seccatore, J.; Veiga, M.; Origliasso, C.; Marin, T.; De Tomi, G. An estimation of the artisanal small-scale production of gold in the world. *Sci. Total Environ.* **2014**, *496*, 662–667. [CrossRef]
8. Marshall, B.G.; Veiga, M.M. Formalization of artisanal miners: Stop the train, we need to get off! *Extr. Ind. Soc.* **2017**, *4*, 300–303. [CrossRef]
9. Hilson, G. Small-scale mining and its socio-economic impact in developing countries. *Nat. Resour. Forum* **2002**, *26*, 3–13. [CrossRef]
10. Veiga, M.M. *Introducing New Technologies for Abatement of Global Mercury Pollution in Latin America*; UNIDO; UBC; CETEM: Rio de Janeiro, Brazil, 1997.
11. Mossa, J.; James, L.A. 13.6 Impacts of Mining on Geomorphic Systems. In *Treatise on Geomorphology*; Elsevier: Amsterdam, The Netherlands, 2013; pp. 74–95.
12. Hentschel, T.; Hruschka, F.; Priester, M. *Global Report on Artisanal & Small-Scale Mining*; IIED: London, UK; WBCSD: Geneva, Switzerland, 2002.
13. D'Souza, K.P.C.J. Artisanal and small-scale mining in Africa: The poor relation. *Geol. Soc. London Spec. Publ.* **2005**, *250*, 95–120. [CrossRef]
14. Buxton, A. *Sustainable Markets, Responding to the Challenges of Artisanal and Small-Scale Mining. How Can Knowledge Networks Help?* IIED: London, UK, 2013.
15. World Bank. *Mining Together Large-Scale Mining Meets ArtisanalMining: A Guide for Action*; Communities and Small-Scale Mining (CASM), The World Bank: Washington, DC, USA, 2013.
16. Maldar, S. *Fairtrade and Fairmined Gold, Empowering Responsible Artisanal and Small-Scale Miners*; Adam Matthew Digital: London, UK, 2011.
17. ICMM. *Working Together: How Large-Scale Mining Can Engage with Artisanal and Small-Scale Miners*; International Council on Mining Metals (ICMM): London, UK, 2010.
18. UNECA. *Minerals and Africa's Development: The International Study Group Report on Africa's Mineral Regimes*; United Nations Economic Commission for Africa (UNECA); Conference Management Section (PCMS): Addis Ababa, Ethiopia, 2011.
19. Hinton, J.J. *Communities and Small-Scale Mining: An Integrated Review for Development Planning*; Communities and Small-Scale Mining (CASM) Initiative Report: Washington, DC, USA, 2005.
20. ILO. *Facts on Small-Scale Mining*; ILO: Geneva, Switzerland, 2003.
21. SDC Swiss Agency for Development Cooperation (SDC). *SDC Experiences with Formalization and Responsible Environmental Practices in Artisanal and Small-Scale Gold Mining in Latin America and Asia (Mongolia)*; Swiss Federal Department of Foreign Affairs (FDFA): Bern, Switzerland, 2011.
22. Hentschel, T.; Hruschka, F.; Priester, M. *Artisanal and Small-Scale Mining: Challenges and Opportunities*; IIED: London, UK, 2003.
23. Labonne, B. Artisanal mining: An economic stepping stone for women. *Nat. Resour. Forum* **1996**, *20*, 117–122. [CrossRef]

24. Schwartz, F.W.; Lee, S.; Darrah, T.H. A Review of the Scope of Artisanal and Small-Scale Mining Worldwide, Poverty, and the Associated Health Impacts. *GeoHealth* **2021**, *5*, e2020GH000325. [CrossRef]
25. Hilson, G.; McQuilken, J. Four decades of support for artisanal and small-scale mining in sub-Saharan Africa: A critical review. *Extr. Ind. Soc.* **2014**, *1*, 104–118. [CrossRef]
26. Maconachie, R. Diamonds, governance and 'local' development in post-conflict Sierra Leone: Lessons for artisanal and small-scale mining in sub-Saharan Africa? *Resour. Policy* **2009**, *34*, 71–79. [CrossRef]
27. Kamlongera, P.J. Making the poor 'poorer' or alleviating poverty? artisanal mining livelihoods in rural malawi. *J. Int. Dev.* **2011**, *23*, 1128–1139. [CrossRef]
28. Bleischwitz, R.; Dittrich, M.; Pierdicca, C. Coltan from Central Africa, international trade and implications for any certification. *Resour. Policy* **2012**, *37*, 19–29. [CrossRef]
29. Hilson, G. Barriers to implementing cleaner technologies and cleaner production (CP) practices in the mining industry: A case study of the Americas. *Miner. Eng.* **2000**, *13*, 699–717. [CrossRef]
30. Carrión-Mero, P.; Aguilar-Aguilar, M.; Morante-Carballo, F.; Domínguez-Cuesta, M.J.; Sánchez-Padilla, C.; Sánchez-Zambrano, A.; Briones-Bitar, J.; Blanco-Torrens, R.; Córdova-Rizo, J.; Berrezueta, E. Surface and Underground Geomechanical Characterization of an Area Affected by Instability Phenomena in Zaruma Mining Zone (Ecuador). *Sustainability* **2021**, *13*, 3272. [CrossRef]
31. Carrión Mero, P.; Blanco Torrens, R.; Borja Bernal, C.; Aguilar Aguilar, M.; Morante Carballo, F.; Briones Bitar, J. Geomechanical characterization and analysis of the effects of rock massif in Zaruma City, Ecuador. In Proceedings of the 17th LACCEI International Multi-Conference for Engineering, Education, and Technology: "Industry, Innovation, and Infrastructure for Sustainable Cities and Communities", Montego Bay, Jamaica, 24–26 July 2019.
32. Carrión-Mero, P.; Loor-Oporto, O.; Andrade-Ríos, H.; Herrera-Franco, G.; Morante-Carballo, F.; Jaya-Montalvo, M.; Aguilar-Aguilar, M.; Torres-Peña, K.; Berrezueta, E. Quantitative and Qualitative Assessment of the "El Sexmo" Tourist Gold Mine (Zaruma, Ecuador) as A Geosite and Mining Site. *Resources* **2020**, *9*, 28. [CrossRef]
33. Mata-Perelló, J.; Carrión, P.; Molina, J.; Villas-Boas, R. Geomining Heritage as a Tool to Promote the Social Development of Rural Communities. In *Geoheritage*; Elsevier: Amsterdam, The Netherlands, 2018; pp. 167–177, ISBN 9780128095423.
34. Siegel, S.; Veiga, M.M. Artisanal and small-scale mining as an extralegal economy: De Soto and the redefinition of "formalization". *Resour. Policy* **2009**, *34*, 51–56. [CrossRef]
35. Berrezueta, E. Caracterización mineralógica y petrográfica de las vetas Vizcaya, Octubrina y Gabi del yacimiento aurífero epitermal Zaruma-Portovelo, Ecuador. *Boletín Geológico Min.* **2021**, *132*, 421–437. [CrossRef]
36. Tonggiroh, A.; Nur, I. Geochemical correlation of gold placer and indication of Au-Cu-Pb-Zn-Ag mineralization at Parigi Moutong, Central Sulawesi, Indonesia. *J. Phys. Conf. Ser.* **2019**, *1341*, 052003. [CrossRef]
37. Ibrahim, E.; Lema, L.; Barnabé, P.; Lacroix, P.; Pirard, E. Small-scale surface mining of gold placers: Detection, mapping, and temporal analysis through the use of free satellite imagery. *Int. J. Appl. Earth Obs. Geoinf.* **2020**, *93*, 102194. [CrossRef]
38. Berrezueta, E.; Castroviejo, R.; Pantoja, F.; Álvarez, R. Mineralogical study and digital image analysis quantification of gold ores from Nariño (Colombia). Application to the improvement of the processing. *Boletín Geol. Min.* **2002**, *113*, 369–379.
39. Iglesias-Martínez, M.; Ordóñez-Casado, B.; Berrezueta, E. Optical image and microchemical analysis of gold grains from a weathered profile of the Minvoul greenstone belt, northern Gabon. *Geol. Mag.* **2020**, *157*, 307–320. [CrossRef]
40. Castroviejo, R.; Berrezueta, E.; Lastra, R. Microscopic digital image analysis of gold ores: A critical test of methodology, comparing reflected light and electron microscopy. *Min. Metall. Explor.* **2002**, *19*, 102–109. [CrossRef]
41. Cordy, P.; Veiga, M.M.; Salih, I.; Al-Saadi, S.; Console, S.; Garcia, O.; Mesa, L.A.; Velásquez-López, P.C.; Roeser, M. Mercury contamination from artisanal gold mining in Antioquia, Colombia: The world's highest per capita mercury pollution. *Sci. Total Environ.* **2011**, *410–411*, 154–160. [CrossRef]
42. Hilson, G. The environmental impact of small-scale gold mining in Ghana: Identifying problems and possible solutions. *Geogr. J.* **2002**, *168*, 57–72. [CrossRef]
43. Salomons, W. Environmental impact of metals derived from mining activities: Processes, predictions, prevention. *J. Geochem. Explor.* **1995**, *52*, 5–23. [CrossRef]
44. Labonne, B. Seminar on Artisanal and Small-Scale Mining in Africa: Identifying Best Practices and Building the Sustainable Livelihoods of Communities. In *The Socioeconomic Impacts of Artisanal and Small-Scale Mining in Developing Countries*; Hilson, G., Ed.; A.A. Balkema: Amsterdam, The Netherlands, 2003; pp. 131–150.
45. Verbrugge, B.; Besmanos, B. Formalizing artisanal and small-scale mining: Whither the workforce? *Resour. Policy* **2016**, *47*, 134–141. [CrossRef]
46. Gamu, J.; Le Billon, P.; Spiegel, S. Extractive industries and poverty: A review of recent findings and linkage mechanisms. *Extr. Ind. Soc.* **2015**, *2*, 162–176. [CrossRef]
47. Hilson, G. Farming, small-scale mining and rural livelihoods in Sub-Saharan Africa: A critical overview. *Extr. Ind. Soc.* **2016**, *3*, 547–563. [CrossRef]
48. Basu, N.; Clarke, E.; Green, A.; Calys-Tagoe, B.; Chan, L.; Dzodzomenyo, M.; Fobil, J.; Long, R.; Neitzel, R.; Obiri, S.; et al. Integrated Assessment of Artisanal and Small-Scale Gold Mining in Ghana—Part 1: Human Health Review. *Int. J. Environ. Res. Public Health* **2015**, *12*, 5143–5176. [CrossRef]

49.   Rajaee, M.; Obiri, S.; Green, A.; Long, R.; Cobbina, S.; Nartey, V.; Buck, D.; Antwi, E.; Basu, N. Integrated Assessment of Artisanal and Small-Scale Gold Mining in Ghana—Part 2: Natural Sciences Review. *Int. J. Environ. Res. Public Health* **2015**, *12*, 8971–9011. [CrossRef] [PubMed]

50.   HIRONS, M. Managing artisanal and small-scale mining in forest areas: Perspectives from a poststructural political ecology. *Geogr. J.* **2011**, *177*, 347–356. [CrossRef]

51.   Cossa, H.; Scheidegger, R.; Leuenberger, A.; Ammann, P.; Munguambe, K.; Utzinger, J.; Macete, E.; Winkler, M.S. Health Studies in the Context of Artisanal and Small-Scale Mining: A Scoping Review. *Int. J. Environ. Res. Public Health* **2021**, *18*, 1555. [CrossRef] [PubMed]

52.   Gibb, H.; O'Leary, K.G. Mercury Exposure and Health Impacts among Individuals in the Artisanal and Small-Scale Gold Mining Community: A Comprehensive Review. *Environ. Health Perspect.* **2014**, *122*, 667–672. [CrossRef] [PubMed]

53.   Zolnikov, T.R.; Ramirez Ortiz, D. A systematic review on the management and treatment of mercury in artisanal gold mining. *Sci. Total Environ.* **2018**, *633*, 816–824. [CrossRef]

54.   Bustamante, N.; Danoucaras, N.; McIntyre, N.; Díaz-Martínez, J.C.; Restrepo-Baena, O.J. Review of improving the water management for the informal gold mining in Colombia. *Rev. Fac. Ing. Univ. Antioq.* **2016**, *79*, 163–172. [CrossRef]

55.   Echchakoui, S. Why and how to merge Scopus and Web of Science during bibliometric analysis: The case of sales force literature from 1912 to 2019. *J. Mark. Anal.* **2020**, *8*, 165–184. [CrossRef]

56.   Muhuri, P.K.; Shukla, A.K.; Abraham, A. Industry 4.0: A bibliometric analysis and detailed overview. *Eng. Appl. Artif. Intell.* **2019**, *78*, 218–235. [CrossRef]

57.   Donthu, N.; Kumar, S.; Pandey, N.; Lim, W.M. Research Constituents, Intellectual Structure, and Collaboration Patterns in Journal of International Marketing: An Analytical Retrospective. *J. Int. Mark.* **2021**, *29*, 1–25. [CrossRef]

58.   Donthu, N.; Kumar, S.; Pandey, N. A retrospective evaluation of Marketing Intelligence and Planning: 1983–2019. *Mark. Intell. Plan.* **2020**, *39*, 48–73. [CrossRef]

59.   Verma, S.; Gustafsson, A. Investigating the emerging COVID-19 research trends in the field of business and management: A bibliometric analysis approach. *J. Bus. Res.* **2020**, *118*, 253–261. [CrossRef] [PubMed]

60.   Gimenez, E.; Salinas, M.; Manzano-Agugliaro, F. Worldwide Research on Plant Defense against Biotic Stresses as Improvement for Sustainable Agriculture. *Sustainability* **2018**, *10*, 391. [CrossRef]

61.   Ellegaard, O.; Wallin, J.A. The bibliometric analysis of scholarly production: How great is the impact? *Scientometrics* **2015**, *105*, 1809–1831. [CrossRef] [PubMed]

62.   Donthu, N.; Kumar, S.; Mukherjee, D.; Pandey, N.; Lim, W.M. How to conduct a bibliometric analysis: An overview and guidelines. *J. Bus. Res.* **2021**, *133*, 285–296. [CrossRef]

63.   Cobo, M.J.; López-Herrera, A.G.; Herrera-Viedma, E.; Herrera, F. An approach for detecting, quantifying, and visualizing the evolution of a research field: A practical application to the Fuzzy Sets Theory field. *J. Informetr.* **2011**, *5*, 146–166. [CrossRef]

64.   Cobo, M.J.; Martínez, M.A.; Gutiérrez-Salcedo, M.; Fujita, H.; Herrera-Viedma, E. 25years at Knowledge-Based Systems: A bibliometric analysis. *Knowl.-Based Syst.* **2015**, *80*, 3–13. [CrossRef]

65.   Van Eck, N.J.; Waltman, L. Software survey: VOSviewer, a computer program for bibliometric mapping. *Scientometrics* **2010**, *84*, 523–538. [CrossRef]

66.   Waltman, L.; van Eck, N.J.; Noyons, E.C.M. A unified approach to mapping and clustering of bibliometric networks. *J. Informetr.* **2010**, *4*, 629–635. [CrossRef]

67.   Kim, J.; McMillan, S.J. Evaluation of Internet Advertising Research: A Bibliometric Analysis of Citations from Key Sources. *J. Advert.* **2008**, *37*, 99–112. [CrossRef]

68.   Carrión-Mero, P.; Montalván-Burbano, N.; Paz-Salas, N.; Morante-Carballo, F. Volcanic Geomorphology: A Review of Worldwide Research. *Geosciences* **2020**, *10*, 347. [CrossRef]

69.   Bellis, N. *Bibliometrics and Citation Analysis*; The Scarecrow Press: Toronto, ON, Canada, 2009.

70.   van Eck, N.J.; Waltman, L. Visualizing Bibliometric Networks. In *Measuring Scholarly Impact*; Springer International Publishing: Cham, Switzerland, 2014; pp. 285–320.

71.   Morante-Carballo, F.; Montalván-Burbano, N.; Carrión-Mero, P.; Espinoza-Santos, N. Cation Exchange of Natural Zeolites: Worldwide Research. *Sustainability* **2021**, *13*, 7751. [CrossRef]

72.   Parish, A.J.; Boyack, K.W.; Ioannidis, J.P.A. Dynamics of co-authorship and productivity across different fields of scientific research. *PLoS ONE* **2018**, *13*, e0189742. [CrossRef] [PubMed]

73.   Hilson, G.; Potter, C. Structural Adjustment and Subsistence Industry: Artisanal Gold Mining in Ghana. *Dev. Change* **2005**, *36*, 103–131. [CrossRef]

74.   Sidorenko, O.; Sairinen, R.; Moore, K. Rethinking the concept of small-scale mining for technologically advanced raw materials production. *Resour. Policy* **2020**, *68*, 101712. [CrossRef]

75.   Lahiri-Dutt, K. *Between the Plough and the Pick: Informal, Artisanal and Small-Scale Mining in the Contemporary World*; ANU Press: Canberra, Australia, 2018.

76.   Andrés, A. *Measuring Academic Research*; Chandos Publishing: Oxford, UK, 2009.

77.   Sánchez, A.D.; de la Cruz Del Río Rama, M.; García, J.Á. Bibliometric analysis of publications on wine tourism in the databases Scopus and WoS. *Eur. Res. Manag. Bus. Econ.* **2017**, *23*, 8–15. [CrossRef]

78. Mongeon, P.; Paul-Hus, A. The journal coverage of Web of Science and Scopus: A comparative analysis. *Scientometrics* **2016**, *106*, 213–228. [CrossRef]

79. Wang, Q.; Waltman, L. Large-scale analysis of the accuracy of the journal classification systems of Web of Science and Scopus. *J. Informetr.* **2016**, *10*, 347–364. [CrossRef]

80. Baas, J.; Schotten, M.; Plume, A.; Côté, G.; Karimi, R. Scopus as a curated, high-quality bibliometric data source for academic research in quantitative science studies. *Quant. Sci. Stud.* **2020**, *1*, 377–386. [CrossRef]

81. Álvarez-García, J.; Durán-Sánchez, A.; del Río-Rama, M.; García-Vélez, D. Active Ageing: Mapping of Scientific Coverage. *Int. J. Environ. Res. Public Health* **2018**, *15*, 2727. [CrossRef]

82. Martín-Martín, A.; Orduna-Malea, E.; Delgado López-Cózar, E. Coverage of highly-cited documents in Google Scholar, Web of Science, and Scopus: A multidisciplinary comparison. *Scientometrics* **2018**, *116*, 2175–2188. [CrossRef]

83. Montalván-Burbano, N.; Velastegui-Montoya, A.; Gurumendi-Noriega, M.; Morante-Carballo, F.; Adami, M. Worldwide Research on Land Use and Land Cover in the Amazon Region. *Sustainability* **2021**, *13*, 6039. [CrossRef]

84. Ruban, D.A. Geotourism—A geographical review of the literature. *Tour. Manag. Perspect.* **2015**, *15*, 1–15. [CrossRef]

85. Doussoulin, J.P.; Mougenot, B. Mapping mining and ecological distribution conflicts in Latin America, a bibliometric analysis. *Resour. Policy* **2022**, *77*, 102650. [CrossRef]

86. Ojeda-Pereira, I.; Campos-Medina, F. International trends in mining tailings publications: A descriptive bibliometric study. *Resour. Policy* **2021**, *74*, 102272. [CrossRef]

87. Martín-Martín, A.; Thelwall, M.; Orduna-Malea, E.; Delgado López-Cózar, E. Google Scholar, Microsoft Academic, Scopus, Dimensions, Web of Science, and OpenCitations' COCI: A multidisciplinary comparison of coverage via citations. *Scientometrics* **2021**, *126*, 871–906. [CrossRef] [PubMed]

88. Montoya, F.G.; Alcayde, A.; Baños, R.; Manzano-Agugliaro, F. A fast method for identifying worldwide scientific collaborations using the Scopus database. *Telemat. Inform.* **2018**, *35*, 168–185. [CrossRef]

89. Parlina, A.; Ramli, K.; Murfi, H. Theme Mapping and Bibliometrics Analysis of One Decade of Big Data Research in the Scopus Database. *Information* **2020**, *11*, 69. [CrossRef]

90. Collazo-Reyes, F. Growth of the number of indexed journals of Latin America and the Caribbean: The effect on the impact of each country. *Scientometrics* **2014**, *98*, 197–209. [CrossRef]

91. Vera-Baceta, M.-A.; Thelwall, M.; Kousha, K. Web of Science and Scopus language coverage. *Scientometrics* **2019**, *121*, 1803–1813. [CrossRef]

92. Martín-Martín, A.; Orduna-Malea, E.; Thelwall, M.; Delgado López-Cózar, E. Google Scholar, Web of Science, and Scopus: A systematic comparison of citations in 252 subject categories. *J. Informetr.* **2018**, *12*, 1160–1177. [CrossRef]

93. Najmi, A.; Rashidi, T.H.; Abbasi, A.; Travis Waller, S. Reviewing the transport domain: An evolutionary bibliometrics and network analysis. *Scientometrics* **2017**, *110*, 843–865. [CrossRef]

94. Chandra, Y. Mapping the evolution of entrepreneurship as a field of research (1990–2013): A scientometric analysis. *PLoS ONE* **2018**, *13*, e0190228.

95. Briones-Bitar, J.; Carrión-Mero, P.; Montalván-Burbano, N.; Morante-Carballo, F. Rockfall Research: A Bibliometric Analysis and Future Trends. *Geosciences* **2020**, *10*, 403. [CrossRef]

96. Aznar-Sánchez, J.A.; Velasco-Muñoz, J.F.; Belmonte-Ureña, L.J.; Manzano-Agugliaro, F. The worldwide research trends on water ecosystem services. *Ecol. Indic.* **2019**, *99*, 310–323. [CrossRef]

97. Pico-Saltos, R.; Carrión-Mero, P.; Montalván-Burbano, N.; Garzás, J.; Redchuk, A. Research Trends in Career Success: A Bibliometric Review. *Sustainability* **2021**, *13*, 4625. [CrossRef]

98. Taşkın, Z.; Aydinoglu, A.U. Collaborative interdisciplinary astrobiology research: A bibliometric study of the NASA Astrobiology Institute. *Scientometrics* **2015**, *103*, 1003–1022. [CrossRef]

99. van Eck, N.J.; Waltman, L.; Noyons, E.C.M.; Buter, R.K. Automatic term identification for bibliometric mapping. *Scientometrics* **2010**, *82*, 581–596. [CrossRef]

100. Morante-Carballo, F.; Montalván-Burbano, N.; Carrión-Mero, P.; Jácome-Francis, K. Worldwide Research Analysis on Natural Zeolites as Environmental Remediation Materials. *Sustainability* **2021**, *13*, 6378. [CrossRef]

101. Ye, C. Bibliometrical Analysis of International Big Data Research: Based on Citespace and VOSviewer. In Proceedings of the 2018 14th International Conference on Natural Computation, Fuzzy Systems and Knowledge Discovery (ICNC-FSKD), Huangshan, China, 28–30 July 2018; pp. 927–932.

102. Li, W.-S.; Yan, Q.; Chen, W.-T.; Li, G.-Y.; Cong, L. Global Research Trends in Robotic Applications in Spinal Medicine: A Systematic Bibliometric Analysis. *World Neurosurg.* **2021**, *155*, e778–e785. [CrossRef]

103. Yu, Y.; Li, Y.; Zhang, Z.; Gu, Z.; Zhong, H.; Zha, Q.; Yang, L.; Zhu, C.; Chen, E. A bibliometric analysis using VOSviewer of publications on COVID-19. *Ann. Transl. Med.* **2020**, *8*, 816. [CrossRef]

104. Wong, C.-S. Science Mapping: A Scientometric Review on Resource Curses, Dutch Diseases, and Conflict Resources during 1993–2020. *Energies* **2021**, *14*, 4573. [CrossRef]

105. Belmonte-Ureña, L.J.; Plaza-Úbeda, J.A.; Vazquez-Brust, D.; Yakovleva, N. Circular economy, degrowth and green growth as pathways for research on sustainable development goals: A global analysis and future agenda. *Ecol. Econ.* **2021**, *185*, 107050. [CrossRef]

106. Cavalieri, A.; Reis, J.; Amorim, M. Circular Economy and Internet of Things: Mapping Science of Case Studies in Manufacturing Industry. *Sustainability* **2021**, *13*, 3299. [CrossRef]
107. Payán-Sánchez, B.; Belmonte-Ureña, L.J.; Plaza-Úbeda, J.A.; Vazquez-Brust, D.; Yakovleva, N.; Pérez-Valls, M. Open Innovation for Sustainability or Not: Literature Reviews of Global Research Trends. *Sustainability* **2021**, *13*, 1136. [CrossRef]
108. Abad-Segura, E.; Batlles-delaFuente, A.; González-Zamar, M.-D.; Belmonte-Ureña, L.J. Implications for Sustainability of the Joint Application of Bioeconomy and Circular Economy: A Worldwide Trend Study. *Sustainability* **2021**, *13*, 7182. [CrossRef]
109. Nobanee, H.; Al Hamadi, F.Y.; Abdulaziz, F.A.; Abukarsh, L.S.; Alqahtani, A.F.; AlSubaey, S.K.; Alqahtani, S.M.; Almansoori, H.A. A Bibliometric Analysis of Sustainability and Risk Management. *Sustainability* **2021**, *13*, 3277. [CrossRef]
110. Flores-Romero, M.B.; Pérez-Romero, M.E.; Álvarez-García, J.; del Río-Rama, M.D.L.C.D. Bibliometric Mapping of Research on Magic Towns of Mexico. *Land* **2021**, *10*, 852. [CrossRef]
111. Cavalcante, W.Q.F.; Coelho, A.; Bairrada, C.M. Sustainability and Tourism Marketing: A Bibliometric Analysis of Publications between 1997 and 2020 Using VOSviewer Software. *Sustainability* **2021**, *13*, 4987. [CrossRef]
112. Mishra, H.G.; Pandita, S.; Bhat, A.A.; Mishra, R.K.; Sharma, S. Tourism and carbon emissions: A bibliometric review of the last three decades: 1990–2021. *Tour. Rev.* **2021**, *77*, 636–658. [CrossRef]
113. Gao, Y.; Xu, Y.; Zhu, Y.; Zhang, J. An analysis of the hotspot and frontier of mine eco-environment restoration based on big data visualization of VOSviewer and CiteSpace. *Geol. Bull. China* **2018**, *37*, 2144–2153.
114. Gizzi, F.T.; Potenza, M.R. The Scientific Landscape of November 23rd, 1980 Irpinia-Basilicata Earthquake: Taking Stock of (Almost) 40 Years of Studies. *Geosciences* **2020**, *10*, 482. [CrossRef]
115. Asli, B.; Eghbali, M.; Ghamami, N.; Abbasabad, H.D.; Rasuli, B.; Rezaie, F. The necessity of developing knowledge map of the world in Earth Sciences and mines field studies based on research activities: A case study of Iran. *Terrae Didat.* **2019**, *15*, e019007. [CrossRef]
116. Herrera-Franco, G.; Montalván-Burbano, N.; Carrión-Mero, P.; Jaya-Montalvo, M.; Gurumendi-Noriega, M. Worldwide Research on Geoparks through Bibliometric Analysis. *Sustainability* **2021**, *13*, 1175. [CrossRef]
117. Noyons, E.C.M.; Moed, H.F.; Van Raan, A.F.J. Integrating research performance analysis and science mapping. *Scientometrics* **1999**, *46*, 591–604. [CrossRef]
118. Alshehhi, A.; Nobanee, H.; Khare, N. The Impact of Sustainability Practices on Corporate Financial Performance: Literature Trends and Future Research Potential. *Sustainability* **2018**, *10*, 494. [CrossRef]
119. Carrión-Mero, P.; Montalván-Burbano, N.; Herrera-Narváez, G.; Morante-Carballo, F. Geodiversity and Mining Towards the Development of Geotourism: A Global Perspective. *Int. J. Des. Nat. Ecodynamics* **2021**, *16*, 191–201. [CrossRef]
120. León-Castro, M.; Rodríguez-Insuasti, H.; Montalván-Burbano, N.; Victor, J.A. Bibliometrics and Science Mapping of Digital Marketing. In *Marketing and Smart Technologies*; Rocha, Á., Reis, J.L., Peter, M.K., Cayolla, R., Loureiro, S., Bogdanović, Z., Eds.; Springer: Singapore, 2021; pp. 95–107.
121. Herrera-Franco, G.; Montalván-Burbano, N.; Carrión-Mero, P.; Bravo-Montero, L. Worldwide Research on Socio-Hydrology: A Bibliometric Analysis. *Water* **2021**, *13*, 1283. [CrossRef]
122. Wormleighton, B.J. Sewerage and Drainage Works in a Small Mining District. *J. R. Sanit. Inst.* **1919**, *40*, 260–263. [CrossRef]
123. Carman, J.S. The Contribution of Small-Scale Mining to World Mineral Production. *Nat. Resour. Forum* **1985**, *9*, 119–124. [CrossRef]
124. Davidson, J. The transformation and successful development of small-scale mining enterprises in developing countries. *Nat. Resour. Forum* **1993**, *17*, 315–326. [CrossRef]
125. Fairbairn, R.A. An Account of a Small Nineteenth-Century Lead Mining Company on Alston Moor. *Ind. Archaeol. Rev.* **1980**, *4*, 245–256. [CrossRef]
126. Godoy, R.A. Small-scale mining and agriculture among the Jukumani Indians, Northern Potosí, Bolivia. *J. Dev. Stud.* **1988**, *24*, 177–196. [CrossRef]
127. Van Vuuren, W.; Hamilton, J. The payoff of developing a small-scale phosphate mine and beneficiating operation in the Mbeya region of Tanzania. *World Dev.* **1992**, *20*, 907–918. [CrossRef]
128. Chachage, C.S.L. New forms of accumulation in Tanzania: The case of gold mining. *Miner. Energy-Raw Mater. Rep.* **1993**, *9*, 2–13. [CrossRef]
129. Wu, Y. Scale, factor intensity and efficiency: An empirical study of the chinese coal industry. *Appl. Econ.* **1993**, *25*, 325–334. [CrossRef]
130. Alpan, S. The Role of Government in Promoting Small-scale Mining. *Nat. Resour. Forum* **1986**, *10*, 95–97. [CrossRef]
131. Kumar, R.; Amaratunga, D. Government policies towards small-scale mining. *Resour. Policy* **1994**, *20*, 15–22. [CrossRef]
132. Tsikata, F.S. The vicissitudes of mineral policy in Ghana. *Resour. Policy* **1997**, *23*, 9–14. [CrossRef]
133. Carma, J.S. Why Small Mining? *Episodes* **1987**, *10*, 159–164. [CrossRef]
134. Amegbey, N.A.; Dankwa, J.B.K.; Al-Hassan, S. Small scale mining in Ghana—Techniques and environmental considerations. *Int. J. Surf. Min. Reclam. Environ.* **1997**, *11*, 135–138. [CrossRef]
135. Traore, P.A. Constraints on small-scale mining in Africa. *Nat. Resour. Forum* **1994**, *18*, 207–212. [CrossRef]
136. Bezerra, O.; Veríssimo, A.; Uhl, C. The regional impacts of small-scale gold mining in Amazonia. *Nat. Resour. Forum* **1996**, *20*, 305–317. [CrossRef]

137. Appleton, J.D.; Williams, T..; Breward, N.; Apostol, A.; Miguel, J.; Miranda, C. Mercury contamination associated with artisanal gold mining on the island of Mindanao, the Philippines. *Sci. Total Environ.* **1999**, *228*, 95–109. [CrossRef]
138. Aldous, P.J.; Smart, P.L.; Black, J.A. Groundwater management problems in abandoned coal-mined aquifers: A case study of the Forest of Dean, England. *Q. J. Eng. Geol. Hydrogeol.* **1986**, *19*, 375–388. [CrossRef]
139. Tingle, T.N.; Borch, R.S.; Hochella, M.F.; Becker, C.H.; Walker, W.J. Characterization of lead on mineral surfaces in soils contaminated by mining and smelting. *Appl. Surf. Sci.* **1993**, *72*, 301–306. [CrossRef]
140. Mighall, T.; Chambers, F.M. The environmental impact of prehistoric mining at Copa Hill, Cwmystwyth, Wales. *Holocene* **1993**, *3*, 260–264. [CrossRef]
141. Tarras-Wahlberg, N.H.; Flachier, A.; Fredriksson, G.; Lane, S.; Lundberg, B.; Sangfors, O. Environmental Impact of Small-scale and Artisanal Gold Mining in Southern Ecuador. *AMBIO J. Hum. Environ.* **2000**, *29*, 484–491. [CrossRef]
142. Wong, I.G.; Humphrey, J.R.; Adams, J.A.; Silva, W.J. Observations of mine seismicity in the eastern Wasatch Plateau, Utah, U.S.A.: A possible case of implosional failure. *Pure Appl. Geophys. PAGEOPH* **1989**, *129*, 369–405. [CrossRef]
143. Šílený, J. The mechanism of small mining tremors from amplitude inversion. *Pure Appl. Geophys. PAGEOPH* **1989**, *129*, 309–324. [CrossRef]
144. Tsuda, T.; Nagira, T.; Yamamoto, M.; Kume, Y. An epidemiological study on cancer in certified arsenic poisoning patients in Toroku. *Ind. Health* **1990**, *28*, 53–62. [CrossRef]
145. Tarras-Wahlberg, N. Environmental management of small-scale and artisanal mining: The Portovelo-Zaruma goldmining area, southern Ecuador. *J. Environ. Manag.* **2002**, *65*, 165–179. [CrossRef] [PubMed]
146. Burke, G. Opportunities for Environmental Management in the Mining Sector in Asia. *J. Environ. Dev.* **2006**, *15*, 224–235. [CrossRef]
147. Roopnarine, L. Small-Scale Gold Mining and Environmental Policy Challenges in Guyana: Protection or Pollution. *Can. J. Lat. Am. Caribb. Stud.* **2006**, *31*, 115–143. [CrossRef]
148. Babut, M.; Sekyi, R.; Rambaud, A.; Potin-Gautier, M.; Tellier, S.; Bannerman, W.; Beinhoff, C. Improving the environmental management of small-scale gold mining in Ghana: A case study of Dumasi. *J. Clean. Prod.* **2003**, *11*, 215–221. [CrossRef]
149. Andrews-Speed, P.; Zamora, A.; Rogers, C.D.; Shen, L.; Cao, S.; Yang, M. A framework for policy formulation for small-scale mines: The case of coal in China. *Nat. Resour. Forum* **2002**, *26*, 45–54. [CrossRef]
150. Hilson, G.; Maponga, O. How has a shortage of census and geological information impeded the regularization of artisanal and small-scale mining? *Nat. Resour. Forum* **2004**, *28*, 22–33. [CrossRef]
151. Hilson, G. Strengthening artisanal mining research and policy through baseline census activities. *Nat. Resour. Forum* **2005**, *29*, 144–153. [CrossRef]
152. Spiegel, S.J. Resource policies and small-scale gold mining in Zimbabwe. *Resour. Policy* **2009**, *34*, 39–44. [CrossRef]
153. Limbong, D.; Kumampung, J.; Rimper, J.; Arai, T.; Miyazaki, N. Emissions and environmental implications of mercury from artisanal gold mining in north Sulawesi, Indonesia. *Sci. Total Environ.* **2003**, *302*, 227–236. [CrossRef]
154. Maponga, O.; Ngorima, C.F. Overcoming environmental problems in the gold panning sector through legislation and education: The Zimbabwean experience. *J. Clean. Prod.* **2003**, *11*, 147–157. [CrossRef]
155. Aryee, B.N.; Ntibery, B.K.; Atorkui, E. Trends in the small-scale mining of precious minerals in Ghana: A perspective on its environmental impact. *J. Clean. Prod.* **2003**, *11*, 131–140. [CrossRef]
156. Tschakert, P.; Singha, K. Contaminated identities: Mercury and marginalization in Ghana's artisanal mining sector. *Geoforum* **2007**, *38*, 1304–1321. [CrossRef]
157. Hylander, L.D.; Goodsite, M.E. Environmental costs of mercury pollution. *Sci. Total Environ.* **2006**, *368*, 352–370. [CrossRef] [PubMed]
158. Feng, X.; Li, G.; Qiu, G. A preliminary study on mercury contamination to the environment from artisanal zinc smelting using indigenous methods in Hezhang County, Guizhou, China: Part 2. Mercury contaminations to soil and crop. *Sci. Total Environ.* **2006**, *368*, 47–55. [CrossRef]
159. Hilson, G.; Potter, C. Why Is Illegal Gold Mining Activity so Ubiquitous in Rural Ghana? *African Dev. Rev.* **2003**, *15*, 237–270. [CrossRef]
160. Lahiri-Dutt, K. Informality in mineral resource management in Asia: Raising questions relating to community economies and sustainable development. *Nat. Resour. Forum* **2004**, *28*, 123–132. [CrossRef]
161. Lahiri-Dutt, K. Informal coal mining in Eastern India: Evidence from the Raniganj Coalbelt. *Nat. Resour. Forum* **2003**, *27*, 68–77. [CrossRef]
162. Banchirigah, S.M. Challenges with eradicating illegal mining in Ghana: A perspective from the grassroots. *Resour. Policy* **2008**, *33*, 29–38. [CrossRef]
163. Hilson, G. Land use competition between small- and large-scale miners: A case study of Ghana. *Land Use Policy* **2002**, *19*, 149–156. [CrossRef]
164. Andrew, J. Potential application of mediation to land use conflicts in small-scale mining. *J. Clean. Prod.* **2003**, *11*, 117–130. [CrossRef]
165. Childs, J. Reforming small-scale mining in sub-Saharan Africa: Political and ideological challenges to a Fair Trade gold initiative. *Resour. Policy* **2008**, *33*, 203–209. [CrossRef]

166. Jønsson, J.B.; Fold, N. Handling uncertainty: Policy and organizational practices in Tanzania's small-scale gold mining sector. *Nat. Resour. Forum* **2009**, *33*, 211–220. [CrossRef]
167. Bose-O'Reilly, S.; Lettmeier, B.; Matteucci Gothe, R.; Beinhoff, C.; Siebert, U.; Drasch, G. Mercury as a serious health hazard for children in gold mining areas. *Environ. Res.* **2008**, *107*, 89–97. [CrossRef] [PubMed]
168. Wickre, J.B.; Folt, C.L.; Sturup, S.; Karagas, M.R. Environmental Exposure and Fingernail Analysis of Arsenic and Mercury in Children and Adults in a Nicaraguan Gold Mining Community. *Arch. Environ. Health Int. J.* **2004**, *59*, 400–409. [CrossRef] [PubMed]
169. Cortes-Maramba, N.; Reyes, J.P.; Francisco-Rivera, A.T.; Akagi, H.; Sunio, R.; Panganiban, L.C. Health and environmental assessment of mercury exposure in a gold mining community in Western Mindanao, Philippines. *J. Environ. Manag.* **2006**, *81*, 126–134. [CrossRef]
170. Heemskerk, M. Self-Employment and Poverty Alleviation: Women's Work in Artisanal Gold Mines. *Hum. Organ.* **2003**, *62*, 62–73. [CrossRef]
171. Ashe, K. Elevated Mercury Concentrations in Humans of Madre de Dios, Peru. *PLoS ONE* **2012**, *7*, e33305. [CrossRef]
172. Castilhos, Z.; Rodrigues-Filho, S.; Cesar, R.; Rodrigues, A.P.; Villas-Bôas, R.; de Jesus, I.; Lima, M.; Faial, K.; Miranda, A.; Brabo, E.; et al. Human exposure and risk assessment associated with mercury contamination in artisanal gold mining areas in the Brazilian Amazon. *Environ. Sci. Pollut. Res.* **2015**, *22*, 11255–11264. [CrossRef]
173. Xiao, R.; Wang, S.; Li, R.; Wang, J.J.; Zhang, Z. Soil heavy metal contamination and health risks associated with artisanal gold mining in Tongguan, Shaanxi, China. *Ecotoxicol. Environ. Saf.* **2017**, *141*, 17–24. [CrossRef]
174. Banza Lubaba Nkulu, C.; Casas, L.; Haufroid, V.; De Putter, T.; Saenen, N.D.; Kayembe-Kitenge, T.; Musa Obadia, P.; Kyanika Wa Mukoma, D.; Lunda Ilunga, J.-M.; Nawrot, T.S.; et al. Sustainability of artisanal mining of cobalt in DR Congo. *Nat. Sustain.* **2018**, *1*, 495–504. [CrossRef]
175. Yard, E.E.; Horton, J.; Schier, J.G.; Caldwell, K.; Sanchez, C.; Lewis, L.; Gastañaga, C. Mercury Exposure Among Artisanal Gold Miners in Madre de Dios, Peru: A Cross-sectional Study. *J. Med. Toxicol.* **2012**, *8*, 441–448. [CrossRef] [PubMed]
176. Plumlee, G.S.; Durant, J.T.; Morman, S.A.; Neri, A.; Wolf, R.E.; Dooyema, C.A.; Hageman, P.L.; Lowers, H.A.; Fernette, G.L.; Meeker, G.P.; et al. Linking Geological and Health Sciences to Assess Childhood Lead Poisoning from Artisanal Gold Mining in Nigeria. *Environ. Health Perspect.* **2013**, *121*, 744–750. [CrossRef] [PubMed]
177. Nakazawa, K.; Nagafuchi, O.; Kawakami, T.; Inoue, T.; Yokota, K.; Serikawa, Y.; Cyio, B.; Elvince, R. Human health risk assessment of mercury vapor around artisanal small-scale gold mining area, Palu city, Central Sulawesi, Indonesia. *Ecotoxicol. Environ. Saf.* **2016**, *124*, 155–162. [CrossRef] [PubMed]
178. Cartier, L.E.; Bürge, M. Agriculture and artisanal gold mining in sierra leone: Alternatives or complements? *J. Int. Dev.* **2011**, *23*, 1080–1099. [CrossRef]
179. Maconachie, R. Re-agrarianising livelihoods in post-conflict sierra leone? Mineral wealth and rural change in artisanal and small-scale mining communities. *J. Int. Dev.* **2011**, *23*, 1054–1067. [CrossRef]
180. Hilson, G.; Garforth, C. 'Agricultural Poverty' and the Expansion of Artisanal Mining in Sub-Saharan Africa: Experiences from Southwest Mali and Southeast Ghana. *Popul. Res. Policy Rev.* **2012**, *31*, 435–464. [CrossRef]
181. Zhao, L.; Anderson, C.W.N.; Qiu, G.; Meng, B.; Wang, D.; Feng, X. Mercury methylation in paddy soil: Source and distribution of mercury species at a Hg mining area, Guizhou Province, China. *Biogeosciences* **2016**, *13*, 2429–2440. [CrossRef]
182. Hilson, G. Poverty traps in small-scale mining communities: The case of sub-Saharan Africa. *Can. J. Dev. Stud. Can. D'études Du Développement* **2012**, *33*, 180–197. [CrossRef]
183. Tonts, M.; Plummer, P.; Lawrie, M. Socio-economic wellbeing in Australian mining towns: A comparative analysis. *J. Rural Stud.* **2012**, *28*, 288–301. [CrossRef]
184. Loayza, N.; Rigolini, J. The Local Impact of Mining on Poverty and Inequality: Evidence from the Commodity Boom in Peru. *World Dev.* **2016**, *84*, 219–234. [CrossRef]
185. Kolk, A.; Lenfant, F. Multinationals, CSR and Partnerships in Central African Conflict Countries. *Corp. Soc. Responsib. Environ. Manag.* **2013**, *20*, 43–54. [CrossRef]
186. Geenen, S. Dispossession, displacement and resistance: Artisanal miners in a gold concession in South-Kivu, Democratic Republic of Congo. *Resour. Policy* **2014**, *40*, 90–99. [CrossRef]
187. Hilson, G.; Hilson, A.; Adu-Darko, E. Chinese participation in Ghana's informal gold mining economy: Drivers, implications and clarifications. *J. Rural Stud.* **2014**, *34*, 292–303. [CrossRef]
188. Van Bockstael, S. The persistence of informality: Perspectives on the future of artisanal mining in Liberia. *Futures* **2014**, *62*, 10–20. [CrossRef]
189. Verbrugge, B. The Economic Logic of Persistent Informality: Artisanal and Small-Scale Mining in the Southern Philippines. *Dev. Change* **2015**, *46*, 1023–1046. [CrossRef]
190. Rakotondrabe, F.; Ndam Ngoupayou, J.R.; Mfonka, Z.; Rasolomanana, E.H.; Nyangono Abolo, A.J.; Ako Ako, A. Water quality assessment in the Bétaré-Oya gold mining area (East-Cameroon): Multivariate Statistical Analysis approach. *Sci. Total Environ.* **2018**, *610–611*, 831–844. [CrossRef]

191. Cobbina, S.; Duwiejuah, A.; Quansah, R.; Obiri, S.; Bakobie, N. Comparative Assessment of Heavy Metals in Drinking Water Sources in Two Small-Scale Mining Communities in Northern Ghana. *Int. J. Environ. Res. Public Health* **2015**, *12*, 10620–10634. [CrossRef]
192. Maconachie, R.; Hilson, G. Safeguarding livelihoods or exacerbating poverty? Artisanal mining and formalization in West Africa. *Nat. Resour. Forum* **2011**, *35*, 293–303. [CrossRef]
193. Geenen, S. A dangerous bet: The challenges of formalizing artisanal mining in the Democratic Republic of Congo. *Resour. Policy* **2012**, *37*, 322–330. [CrossRef]
194. Spiegel, S.J. Governance Institutions, Resource Rights Regimes, and the Informal Mining Sector: Regulatory Complexities in Indonesia. *World Dev.* **2012**, *40*, 189–205. [CrossRef]
195. Teschner, B.A. Small-scale mining in Ghana: The government and the galamsey. *Resour. Policy* **2012**, *37*, 308–314. [CrossRef]
196. Spiegel, S.J. Shifting Formalization Policies and Recentralizing Power: The Case of Zimbabwe's Artisanal Gold Mining Sector. *Soc. Nat. Resour.* **2015**, *28*, 543–558. [CrossRef]
197. Hilson, G.; Hilson, A.; Maconachie, R.; McQuilken, J.; Goumandakoye, H. Artisanal and small-scale mining (ASM) in sub-Saharan Africa: Re-conceptualizing formalization and 'illegal' activity. *Geoforum* **2017**, *83*, 80–90. [CrossRef]
198. Hilson, G.; Maconachie, R. Formalising artisanal and small-scale mining: Insights, contestations and clarifications. *Area* **2017**, *49*, 443–451. [CrossRef]
199. García, O.; Veiga, M.M.; Cordy, P.; Suescún, O.E.; Molina, J.M.; Roeser, M. Artisanal gold mining in Antioquia, Colombia: A successful case of mercury reduction. *J. Clean. Prod.* **2015**, *90*, 244–252. [CrossRef]
200. Marrugo-Negrete, J.; Marrugo-Madrid, S.; Pinedo-Hernández, J.; Durango-Hernández, J.; Díez, S. Screening of native plant species for phytoremediation potential at a Hg-contaminated mining site. *Sci. Total Environ.* **2016**, *542*, 809–816. [CrossRef]
201. Smith, N.M.; Ali, S.; Bofinger, C.; Collins, N. Human health and safety in artisanal and small-scale mining: An integrated approach to risk mitigation. *J. Clean. Prod.* **2016**, *129*, 43–52. [CrossRef]
202. Adler Miserendino, R.; Bergquist, B.A.; Adler, S.E.; Guimarães, J.R.D.; Lees, P.S.J.; Niquen, W.; Velasquez-López, P.C.; Veiga, M.M. Challenges to measuring, monitoring, and addressing the cumulative impacts of artisanal and small-scale gold mining in Ecuador. *Resour. Policy* **2013**, *38*, 713–722. [CrossRef]
203. Gardner, R.M.; Nyland, J.F.; Silva, I.A.; Maria Ventura, A.; Maria de Souza, J.; Silbergeld, E.K. Mercury exposure, serum antinuclear/antinucleolar antibodies, and serum cytokine levels in mining populations in Amazonian Brazil: A cross-sectional study. *Environ. Res.* **2010**, *110*, 345–354. [CrossRef]
204. Sherman, L.S.; Blum, J.D.; Basu, N.; Rajaee, M.; Evers, D.C.; Buck, D.G.; Petrlik, J.; DiGangi, J. Assessment of mercury exposure among small-scale gold miners using mercury stable isotopes. *Environ. Res.* **2015**, *137*, 226–234. [CrossRef]
205. Rajaee, M.; Long, R.; Renne, E.; Basu, N. Mercury Exposure Assessment and Spatial Distribution in A Ghanaian Small-Scale Gold Mining Community. *Int. J. Environ. Res. Public Health* **2015**, *12*, 10755–10782. [CrossRef] [PubMed]
206. Steckling, N.; Boese-O'Reilly, S.; Gradel, C.; Gutschmidt, K.; Shinee, E.; Altangerel, E.; Badrakh, B.; Bonduush, I.; Surenjav, U.; Ferstl, P. Mercury exposure in female artisanal small-scale gold miners (ASGM) in Mongolia: An analysis of human biomonitoring (HBM) data from 2008. *Sci. Total Environ.* **2011**, *409*, 994–1000. [CrossRef] [PubMed]
207. Tschakert, P. Recognizing and nurturing artisanal mining as a viable livelihood. *Resour. Policy* **2009**, *34*, 24–31. [CrossRef]
208. Malpeli, K.C.; Chirico, P.G. The influence of geomorphology on the role of women at artisanal and small-scale mine sites. *Nat. Resour. Forum* **2013**, *37*, 43–54. [CrossRef]
209. Hilson, G.; Zolnikov, T.R.; Ortiz, D.R.; Kumah, C. Formalizing artisanal gold mining under the Minamata convention: Previewing the challenge in Sub-Saharan Africa. *Environ. Sci. Policy* **2018**, *85*, 123–131. [CrossRef]
210. Putzel, L.; Kelly, A.B.; Cerutti, P.O.; Artati, Y. Formalization as Development in Land and Natural Resource Policy. *Soc. Nat. Resour.* **2015**, *28*, 453–472. [CrossRef]
211. Huntington, H.; Marple-Cantrell, K. Customary governance of artisanal and small-scale mining in Guinea: Social and environmental practices and outcomes★. *Land Use Policy* **2021**, *102*, 105229. [CrossRef]
212. Drace, K.; Kiefer, A.M.; Veiga, M.M.; Williams, M.K.; Ascari, B.; Knapper, K.A.; Logan, K.M.; Breslin, V.M.; Skidmore, A.; Bolt, D.A.; et al. Mercury-free, small-scale artisanal gold mining in Mozambique: Utilization of magnets to isolate gold at clean tech mine. *J. Clean. Prod.* **2012**, *32*, 88–95. [CrossRef]
213. Morgan, V.L.; McLamore, E.S.; Correll, M.; Kiker, G.A. Emerging mercury mitigation solutions for artisanal small-scale gold mining communities evaluated through a multicriteria decision analysis approach. *Environ. Syst. Decis.* **2021**, *41*, 413–424. [CrossRef]
214. O'Brien, R.M.; Smits, K.M.; Smith, N.M.; Schwartz, M.R.; Crouse, D.R.; Phelan, T.J. Integrating scientific and local knowledge into pollution remediation planning: An iterative conceptual site model framework. *Environ. Dev.* **2021**, *40*, 100675. [CrossRef]
215. Mohammed Banchirigah, S. How have reforms fuelled the expansion of artisanal mining? Evidence from sub-Saharan Africa. *Resour. Policy* **2006**, *31*, 165–171. [CrossRef]
216. Hilson, G.; Pardie, S. Mercury: An agent of poverty in Ghana's small-scale gold-mining sector? *Resour. Policy* **2006**, *31*, 106–116. [CrossRef]
217. Hilson, G.; Hilson, C.J.; Pardie, S. Improving awareness of mercury pollution in small-scale gold mining communities: Challenges and ways forward in rural Ghana. *Environ. Res.* **2007**, *103*, 275–287. [CrossRef] [PubMed]

218. Veiga, M.M.; Maxson, P.A.; Hylander, L.D. Origin and consumption of mercury in small-scale gold mining. *J. Clean. Prod.* **2006**, *14*, 436–447. [CrossRef]

219. Veiga, M.M.; Hinton, J.J. Abandoned artisanal gold mines in the Brazilian Amazon: A legacy of mercury pollution. *Nat. Resour. Forum* **2002**, *26*, 15–26. [CrossRef]

220. Pestana, M.; Formoso, M. Mercury contamination in Lavras do Sul, south Brazil: A legacy from past and recent gold mining. *Sci. Total Environ.* **2003**, *307*, 125–140. [CrossRef]

221. Hilson, G. Small-scale mining, poverty and economic development in sub-Saharan Africa: An overview. *Resour. Policy* **2009**, *34*, 1–5. [CrossRef]

222. Culnan, M.J.; O'Reilly III, C.A.; Chatman, J.A. Intellectual structure of research in organizational behavior, 1972–1984: A cocitation analysis. *J. Am. Soc. Inf. Sci.* **1990**, *41*, 453–458. [CrossRef]

223. Bebbington, A.; Humphreys Bebbington, D.; Bury, J.; Lingan, J.; Muñoz, J.P.; Scurrah, M. Mining and Social Movements: Struggles Over Livelihood and Rural Territorial Development in the Andes. *World Dev.* **2008**, *36*, 2888–2905. [CrossRef]

224. Fisher, E. Occupying the Margins: Labour Integration and Social Exclusion in Artisanal Mining in Tanzania. *Dev. Change* **2007**, *38*, 735–760. [CrossRef]

225. Zupic, I.; Čater, T. Bibliometric Methods in Management and Organization. *Organ. Res. Methods* **2015**, *18*, 429–472. [CrossRef]

226. Hilson, G. Family Hardship and Cultural Values: Child Labor in Malian Small-Scale Gold Mining Communities. *World Dev.* **2012**, *40*, 1663–1674. [CrossRef]

227. Brottem, L.V.; Ba, L. Gendered livelihoods and land tenure: The case of artisanal gold miners in Mali, West Africa. *Geoforum* **2019**, *105*, 54–62. [CrossRef]

228. Okoh, G.; Hilson, G. Poverty and livelihood diversification: Exploring the linkages between smallholder farming and artisanal mining in rural ghana. *J. Int. Dev.* **2011**, *23*, 1100–1114. [CrossRef]

229. Stewart, J.; Kibombo, R.; Rankin, L.P. Gendered livelihoods in the artisanal mining sector in the Great Lakes Region. *Can. J. African Stud./Rev. Can. Des Études Afr.* **2020**, *54*, 37–56. [CrossRef]

230. Hilson, G. Shootings and burning excavators: Some rapid reflections on the Government of Ghana's handling of the informal Galamsey mining 'menace. *Resour. Policy* **2017**, *54*, 109–116. [CrossRef]

231. Byemba, G.K. Formalization of artisanal and small-scale mining in eastern Democratic Republic of the Congo: An opportunity for women in the new tin, tantalum, tungsten and gold (3TG) supply chain? *Extr. Ind. Soc.* **2020**, *7*, 420–427. [CrossRef]

232. Siwale, A.; Siwale, T. Has the promise of formalizing artisanal and small-scale mining (ASM) failed? The case of Zambia. *Extr. Ind. Soc.* **2017**, *4*, 191–201. [CrossRef]

233. Dery Tuokuu, F.X.; Idemudia, U.; Bawelle, E.B.G.; Baguri Sumani, J.B. Criminalization of "galamsey" and livelihoods in Ghana: Limits and consequences. *Nat. Resour. Forum* **2020**, *44*, 52–65. [CrossRef]

234. Amankwah, R.K.; Anim-Sackey, C. Strategies for sustainable development of the small-scale gold and diamond mining industry of Ghana. *Resour. Policy* **2003**, *29*, 131–138. [CrossRef]

235. Zvarivadza, T.; Nhleko, A.S. Resolving artisanal and small-scale mining challenges: Moving from conflict to cooperation for sustainability in mine planning. *Resour. Policy* **2018**, *56*, 78–86. [CrossRef]

236. Zvarivadza, T. Artisanal and Small-Scale Mining as a challenge and possible contributor to Sustainable Development. *Resour. Policy* **2018**, *56*, 49–58. [CrossRef]

237. Muduli, K.; Barve, A. Establishment of a sustainable development framework in small scale mining supply chains in India. *Int. J. Intell. Enterp.* **2013**, *2*, 84. [CrossRef]

238. Sinan Erzurumlu, S.; Erzurumlu, Y.O. Sustainable mining development with community using design thinking and multi-criteria decision analysis. *Resour. Policy* **2015**, *46*, 6–14. [CrossRef]

239. Mtegha, H.D.; Cawood, F.T.; Minnitt, R.C.A. National minerals policies and stakeholder participation for broad-based development in the southern African development community (SADC). *Resour. Policy* **2006**, *31*, 231–238. [CrossRef]

240. Maconachie, R.; Conteh, F. Artisanal mining policy reforms, informality and challenges to the Sustainable Development Goals in Sierra Leone. *Environ. Sci. Policy* **2021**, *116*, 38–46. [CrossRef]

241. Masuku, S. An indigenous knowledge-based approach to environmental conservation in Zimbabwe. *Afr. Renaiss.* **2019**, *16*, 165–183. [CrossRef]

242. Armah, F.A.; Boamah, S.A.; Quansah, R.; Obiri, S.; Luginaah, I. Working conditions of male and female artisanal and small-scale goldminers in Ghana: Examining existing disparities. *Extr. Ind. Soc.* **2016**, *3*, 464–474. [CrossRef]

243. Yelpaala, K.; Ali, S.H. Multiple scales of diamond mining in Akwatia, Ghana: Addressing environmental and human development impact. *Resour. Policy* **2005**, *30*, 145–155. [CrossRef]

244. Odumo, B.O.; Carbonell, G.; Angeyo, H.K.; Patel, J.P.; Torrijos, M.; Rodríguez Martín, J.A. Impact of gold mining associated with mercury contamination in soil, biota sediments and tailings in Kenya. *Environ. Sci. Pollut. Res.* **2014**, *21*, 12426–12435. [CrossRef]

245. Podolský, F.; Ettler, V.; Šebek, O.; Ježek, J.; Mihaljevič, M.; Kříbek, B.; Sracek, O.; Vaněk, A.; Penížek, V.; Majer, V.; et al. Mercury in soil profiles from metal mining and smelting areas in Namibia and Zambia: Distribution and potential sources. *J. Soils Sediments* **2015**, *15*, 648–658. [CrossRef]

246. Sun, Z.; Xie, X.; Wang, P.; Hu, Y.; Cheng, H. Heavy metal pollution caused by small-scale metal ore mining activities: A case study from a polymetallic mine in South China. *Sci. Total Environ.* **2018**, *639*, 217–227. [CrossRef] [PubMed]

247. Mandeng, E.P.B.; Bidjeck, L.M.B.; Bessa, A.Z.E.; Ntomb, Y.D.; Wadjou, J.W.; Doumo, E.P.E.; Dieudonné, L.B. Contamination and risk assessment of heavy metals, and uranium of sediments in two watersheds in Abiete-Toko gold district, Southern Cameroon. *Heliyon* **2019**, *5*, e02591. [CrossRef]
248. Yevugah, L.L.; Darko, G.; Bak, J. Does mercury emission from small-scale gold mining cause widespread soil pollution in Ghana? *Environ. Pollut.* **2021**, *284*, 116945. [CrossRef] [PubMed]
249. Tomiyasu, T.; Hamada, Y.K.; Kodamatani, H.; Hidayati, N.; Rahajoe, J.S. Transport of mercury species by river from artisanal and small-scale gold mining in West Java, Indonesia. *Environ. Sci. Pollut. Res.* **2019**, *26*, 25262–25274. [CrossRef]
250. Niane, B.; Moritz, R.; Guédron, S.; Ngom, P.M.; Pfeifer, H.R.; Mall, I.; Poté, J. Effect of recent artisanal small-scale gold mining on the contamination of surface river sediment: Case of Gambia River, Kedougou region, southeastern Senegal. *J. Geochemical Explor.* **2014**, *144*, 517–527. [CrossRef]
251. Goix, S.; Maurice, L.; Laffont, L.; Rinaldo, R.; Lagane, C.; Chmeleff, J.; Menges, J.; Heimbürger, L.-E.; Maury-Brachet, R.; Sonke, J.E. Quantifying the impacts of artisanal gold mining on a tropical river system using mercury isotopes. *Chemosphere* **2019**, *219*, 684–694. [CrossRef]
252. Wang, X.; Yuan, W.; Lin, C.-J.; Wu, F.; Feng, X. Stable mercury isotopes stored in Masson Pinus tree rings as atmospheric mercury archives. *J. Hazard. Mater.* **2021**, *415*, 125678. [CrossRef]
253. Ebel, M.; Evangelou, M.W.H.; Schaeffer, A. Cyanide phytoremediation by water hyacinths (Eichhornia crassipes). *Chemosphere* **2007**, *66*, 816–823. [CrossRef]
254. Lin, W.; Xiao, T.; Wu, Y.; Ao, Z.; Ning, Z. Hyperaccumulation of zinc by Corydalis davidii in Zn-polluted soils. *Chemosphere* **2012**, *86*, 837–842. [CrossRef]
255. Chamba, I.; Gazquez, M.J.; Selvaraj, T.; Calva, J.; Toledo, J.J.; Armijos, C. Selection of a suitable plant for phytoremediation in mining artisanal zones. *Int. J. Phytoremediation* **2016**, *18*, 853–860. [CrossRef] [PubMed]
256. Samuel, W.; Richard, B.; Nyantakyi, J.A. Phytoremediation of heavy metals contaminated water and soils from artisanal mining enclave using Heliconia psittacorum. *Model. Earth Syst. Environ.* **2021**, *8*, 591–600. [CrossRef]
257. Gerson, J.R.; Driscoll, C.T.; Hsu-Kim, H.; Bernhardt, E.S. Senegalese artisanal gold mining leads to elevated total mercury and methylmercury concentrations in soils, sediments, and rivers. *Elem. Sci. Anthr.* **2018**, *6*, 11. [CrossRef]
258. Niane, B.; Guédron, S.; Feder, F.; Legros, S.; Ngom, P.M.; Moritz, R. Impact of recent artisanal small-scale gold mining in Senegal: Mercury and methylmercury contamination of terrestrial and aquatic ecosystems. *Sci. Total Environ.* **2019**, *669*, 185–193. [CrossRef] [PubMed]
259. Pinedo-Hernández, J.; Marrugo-Negrete, J.; Díez, S. Speciation and bioavailability of mercury in sediments impacted by gold mining in Colombia. *Chemosphere* **2015**, *119*, 1289–1295. [CrossRef]
260. Bose-O'Reilly, S.; Schierl, R.; Nowak, D.; Siebert, U.; William, J.F.; Owi, F.T.; Ir, Y.I. A preliminary study on health effects in villagers exposed to mercury in a small-scale artisanal gold mining area in Indonesia. *Environ. Res.* **2016**, *149*, 274–281. [CrossRef]
261. Gyamfi, O.; Sorenson, P.B.; Darko, G.; Ansah, E.; Bak, J.L. Human health risk assessment of exposure to indoor mercury vapour in a Ghanaian artisanal small-scale gold mining community. *Chemosphere* **2020**, *241*, 125014. [CrossRef]
262. Calao-Ramos, C.; Bravo, A.G.; Paternina-Uribe, R.; Marrugo-Negrete, J.; Díez, S. Occupational human exposure to mercury in artisanal small-scale gold mining communities of Colombia. *Environ. Int.* **2021**, *146*, 106216. [CrossRef]
263. Camacho-delaCruz, A.A.; Espinosa-Reyes, G.; Rebolloso-Hernández, C.A.; Carrizales-Yáñez, L.; Ilizaliturri-Hernández, C.A.; Reyes-Arreguín, L.E.; Díaz-Barriga, F. Holistic health risk assessment in an artisanal mercury mining region in Mexico. *Environ. Monit. Assess.* **2021**, *193*, 541. [CrossRef]
264. de Souza, E.S.; Texeira, R.A.; da Costa, H.S.C.; Oliveira, F.J.; Melo, L.C.A.; do Carmo Freitas Faial, K.; Fernandes, A.R. Assessment of risk to human health from simultaneous exposure to multiple contaminants in an artisanal gold mine in Serra Pelada, Pará, Brazil. *Sci. Total Environ.* **2017**, *576*, 683–695. [CrossRef]
265. Tirima, S.; Bartrem, C.; von Lindern, I.; von Braun, M.; Lind, D.; Anka, S.M.; Abdullahi, A. Food contamination as a pathway for lead exposure in children during the 2010–2013 lead poisoning epidemic in Zamfara, Nigeria. *J. Environ. Sci.* **2018**, *67*, 260–272. [CrossRef] [PubMed]
266. Veiga, M.M.; Fadina, O. A review of the failed attempts to curb mercury use at artisanal gold mines and a proposed solution. *Extr. Ind. Soc.* **2020**, *7*, 1135–1146. [CrossRef]
267. Silvestre, B.S.; Silva Neto, R.E. Are cleaner production innovations the solution for small mining operations in poor regions? The case of Padua in Brazil. *J. Clean. Prod.* **2014**, *84*, 809–817. [CrossRef]
268. Shandro, J.A.; Veiga, M.M.; Chouinard, R. Reducing mercury pollution from artisanal gold mining in Munhena, Mozambique. *J. Clean. Prod.* **2009**, *17*, 525–532. [CrossRef]
269. Hilson, G. Abatement of mercury pollution in the small-scale gold mining industry: Restructuring the policy and research agendas. *Sci. Total Environ.* **2006**, *362*, 1–14. [CrossRef]
270. Rodriguez, L.A.; Pfaff, A.; Velez, M.A. Graduated stringency within collective incentives for group environmental compliance: Building coordination in field-lab experiments with artisanal gold miners in Colombia. *J. Environ. Econ. Manag.* **2019**, *98*, 102276. [CrossRef]

271. Spiegel, S.; Keane, S.; Metcalf, S.; Veiga, M. Implications of the Minamata Convention on Mercury for informal gold mining in Sub-Saharan Africa: From global policy debates to grassroots implementation? *Environ. Dev. Sustain.* **2015**, *17*, 765–785. [CrossRef]

272. Clifford, M.J. Future strategies for tackling mercury pollution in the artisanal gold mining sector: Making the Minamata Convention work. *Futures* **2014**, *62*, 106–112. [CrossRef]

273. Spiegel, S.J. Labour challenges and mercury management at gold mills in Zimbabwe: Examining production processes and proposals for change. *Nat. Resour. Forum* **2009**, *33*, 221–232. [CrossRef]

274. Vieira, R. Mercury-free gold mining technologies: Possibilities for adoption in the Guianas. *J. Clean. Prod.* **2006**, *14*, 448–454. [CrossRef]

275. Spiegel, S.J. Socioeconomic dimensions of mercury pollution abatement: Engaging artisanal mining communities in Sub-Saharan Africa. *Ecol. Econ.* **2009**, *68*, 3072–3083. [CrossRef]

276. Torkaman, P.; Veiga, M.M.; de Andrade Lima, L.R.P.; Oliveira, L.A.; Motta, J.S.; Jesus, J.L.; Lavkulich, L.M. Leaching gold with cassava: An option to eliminate mercury use in artisanal gold mining. *J. Clean. Prod.* **2021**, *311*, 127531. [CrossRef]

277. Owusu, O.; Bansah, K.J.; Mensah, A.K. "Small in size, but big in impact": Socio-environmental reforms for sustainable artisanal and small-scale mining. *J. Sustain. Min.* **2019**, *18*, 38–44. [CrossRef]

278. Hennessy, L. Where There Is No Company: Indigenous Peoples, Sustainability, and the Challenges of Mid-Stream Mining Reforms in Guyana's Small-Scale Gold Sector. *New Polit. Econ.* **2015**, *20*, 126–153. [CrossRef]

279. Kumah, C.; Hilson, G.; Quaicoe, I. Poverty, adaptation and vulnerability: An assessment of women's work in Ghana's artisanal gold mining sector. *Area* **2020**, *52*, 617–625. [CrossRef]

280. Velásquez-López, P.C.; Páez-Varas, C.; Benavides-Zúñiga, X.; Gallegos, F.; Fallon, G. Women mine-rock waste collectors in artisanal and small-scale mining in Ecuador: Challenges and opportunities. *Extr. Ind. Soc.* **2020**, *7*, 1579–1586. [CrossRef]

281. Bashwira, M.-R.; Cuvelier, J.; Hilhorst, D.; van der Haar, G. Not only a man's world: Women's involvement in artisanal mining in eastern DRC. *Resour. Policy* **2014**, *40*, 109–116. [CrossRef]

282. Adonteng-Kissi, O.; Adonteng-Kissi, B. Precarious work or sustainable livelihoods? Aligning Prestea's Programme with the development dialogue on artisanal and small-scale mining. *Nat. Resour. Forum* **2018**, *42*, 123–137. [CrossRef]

283. Mabe, F.N.; Owusu-Sekyere, E.; Adeosun, O.T. Livelihood coping strategies among displaced small scale miners in Ghana. *Resour. Policy* **2021**, *74*, 102291. [CrossRef]

284. Banchirigah, S.M.; Hilson, G. De-agrarianization, re-agrarianization and local economic development: Re-orientating livelihoods in African artisanal mining communities. *Policy Sci.* **2010**, *43*, 157–180. [CrossRef]

285. Carrión-Mero, P.; Montalván-Burbano, N.; Morante-Carballo, F.; Quesada-Román, A.; Apolo-Masache, B. Worldwide Research Trends in Landslide Science. *Int. J. Environ. Res. Public Health* **2021**, *18*, 9445. [CrossRef] [PubMed]

286. Díez-Martín, F.; Blanco-González, A.; Prado-Román, C. The intellectual structure of organizational legitimacy research: A co-citation analysis in business journals. *Rev. Manag. Sci.* **2021**, *15*, 1007–1043. [CrossRef]

287. Kim, H.J.; Jeong, Y.K.; Song, M. Content- and proximity-based author co-citation analysis using citation sentences. *J. Informetr.* **2016**, *10*, 954–966. [CrossRef]

288. White, H.D.; Griffith, B.C. Author cocitation: A literature measure of intellectual structure. *J. Am. Soc. Inf. Sci.* **1981**, *32*, 163–171. [CrossRef]

289. Hilson, G.; Maconachie, R. Artisanal and small-scale mining and the Sustainable Development Goals: Opportunities and new directions for sub-Saharan Africa. *Geoforum* **2020**, *111*, 125–141. [CrossRef]

290. Maconachie, R.; Binns, T. 'Farming miners' or 'mining farmers'?: Diamond mining and rural development in post-conflict Sierra Leone. *J. Rural Stud.* **2007**, *23*, 367–380. [CrossRef]

291. Bryceson, D.F. The Scramble in Africa: Reorienting Rural Livelihoods. *World Dev.* **2002**, *30*, 725–739. [CrossRef]

292. Hilson, G.; Ackah-Baidoo, A. Can Microcredit Services Alleviate Hardship in African Small-scale Mining Communities? *World Dev.* **2011**, *39*, 1191–1203. [CrossRef]

293. Spiegel, S.J.; Savornin, O.; Shoko, D.; Veiga, M.M. Mercury Reduction in Munhena, Mozambique: Homemade Solutions and the Social Context for Change. *Int. J. Occup. Environ. Health* **2006**, *12*, 215–221. [CrossRef]

294. Spiegel, S.J.; Veiga, M.M. Building Capacity in Small-Scale Mining Communities: Health, Ecosystem Sustainability, and the Global Mercury Project. *Ecohealth* **2005**, *2*, 361–369. [CrossRef]

295. Castilhos, Z.C.; Rodrigues-Filho, S.; Rodrigues, A.P.C.; Villas-Bôas, R.C.; Siegel, S.; Veiga, M.M.; Beinhoff, C. Mercury contamination in fish from gold mining areas in Indonesia and human health risk assessment. *Sci. Total Environ.* **2006**, *368*, 320–325. [CrossRef] [PubMed]

296. Telmer, K.H.; Veiga, M.M. World Emissions of Mercury from Artisanal and Small Scale Gold Mining. In *Mercury Fate and Transport in the Global Atmosphere*; Springer: Boston, MA, USA, 2009; pp. 131–172.

297. Bose-O'Reilly, S.; Drasch, G.; Beinhoff, C.; Rodrigues-Filho, S.; Roider, G.; Lettmeier, B.; Maydl, A.; Maydl, S.; Siebert, U. Health assessment of artisanal gold miners in Indonesia. *Sci. Total Environ.* **2010**, *408*, 713–725. [CrossRef] [PubMed]

298. Drasch, G.; Böse-O'Reilly, S.; Beinhoff, C.; Roider, G.; Maydl, S. The Mt. Diwata study on the Philippines 1999—Assessing mercury intoxication of the population by small scale gold mining. *Sci. Total Environ.* **2001**, *267*, 151–168. [CrossRef]

299. Hinton, J.J.; Veiga, M.M.; Veiga, A.T.C. Clean artisanal gold mining: A utopian approach? *J. Clean. Prod.* **2003**, *11*, 99–115. [CrossRef]
300. Spiegel, S.J.; Veiga, M.M. International guidelines on mercury management in small-scale gold mining. *J. Clean. Prod.* **2010**, *18*, 375–385. [CrossRef]
301. Velásquez-López, P.C.; Veiga, M.M.; Hall, K. Mercury balance in amalgamation in artisanal and small-scale gold mining: Identifying strategies for reducing environmental pollution in Portovelo-Zaruma, Ecuador. *J. Clean. Prod.* **2010**, *18*, 226–232. [CrossRef]
302. Sousa, R.N.; Veiga, M.M.; Klein, B.; Telmer, K.; Gunson, A.J.; Bernaudat, L. Strategies for reducing the environmental impact of reprocessing mercury-contaminated tailings in the artisanal and small-scale gold mining sector: Insights from Tapajos River Basin, Brazil. *J. Clean. Prod.* **2010**, *18*, 1757–1766. [CrossRef]
303. Basu, N.; Horvat, M.; Evers, D.C.; Zastenskaya, I.; Weihe, P.; Tempowski, J. A State-of-the-Science Review of Mercury Biomarkers in Human Populations Worldwide between 2000 and 2018. *Environ. Health Perspect.* **2018**, *126*, 106001. [CrossRef]
304. Obiri, S. Determination of Heavy Metals in Water from Boreholes in Dumasi in the Wassa West District of Western Region of Republic of Ghana. *Environ. Monit. Assess.* **2007**, *130*, 455–463. [CrossRef]
305. Basu, N.; Nam, D.-H.; Kwansaa-Ansah, E.; Renne, E.P.; Nriagu, J.O. Multiple metals exposure in a small-scale artisanal gold mining community. *Environ. Res.* **2011**, *111*, 463–467. [CrossRef]
306. Styles, M.T.; Amankwah, R.K.; Al Hassan, S.; Nartey, R.S. The identification and testing of a method for mercury-free gold processing for artisanal and small-scale gold miners in Ghana. *Int. J. Environ. Pollut.* **2010**, *41*, 289. [CrossRef]
307. Li, P.; Feng, X.; Qiu, G.; Shang, L.; Wang, S.; Meng, B. Atmospheric mercury emission from artisanal mercury mining in Guizhou Province, Southwestern China. *Atmos. Environ.* **2009**, *43*, 2247–2251. [CrossRef]
308. Qiu, G.; Feng, X.; Wang, S.; Fu, X.; Shang, L. Mercury distribution and speciation in water and fish from abandoned Hg mines in Wanshan, Guizhou province, China. *Sci. Total Environ.* **2009**, *407*, 5162–5168. [CrossRef] [PubMed]
309. Feng, X.; Dai, Q.; Qiu, G.; Li, G.; He, L.; Wang, D. Gold mining related mercury contamination in Tongguan, Shaanxi Province, PR China. *Appl. Geochem.* **2006**, *21*, 1955–1968. [CrossRef]
310. Ping, L.; Feng, X.; Shang, L.; Qiu, G.; Meng, B.; Liang, P.; Zhang, H. Mercury pollution from artisanal mercury mining in Tongren, Guizhou, China. *Appl. Geochem.* **2008**, *23*, 2055–2064. [CrossRef]
311. Meng, B.; Feng, X.; Qiu, G.; Cai, Y.; Wang, D.; Li, P.; Shang, L.; Sommar, J. Distribution Patterns of Inorganic Mercury and Methylmercury in Tissues of Rice (Oryza sativa L.) Plants and Possible Bioaccumulation Pathways. *J. Agric. Food Chem.* **2010**, *58*, 4951–4958. [CrossRef]
312. Qiu, G.; Feng, X.; Wang, S.; Shang, L. Mercury and methylmercury in riparian soil, sediments, mine-waste calcines, and moss from abandoned Hg mines in east Guizhou province, southwestern China. *Appl. Geochem.* **2005**, *20*, 627–638. [CrossRef]
313. Feng, X.; Li, P.; Qiu, G.; Wang, S.; Li, G.; Shang, L.; Meng, B.; Jiang, H.; Bai, W.; Li, Z.; et al. Human Exposure To Methylmercury through Rice Intake in Mercury Mining Areas, Guizhou Province, China. *Environ. Sci. Technol.* **2008**, *42*, 326–332. [CrossRef]
314. Iwata, T.; Sakamoto, M.; Feng, X.; Yoshida, M.; Liu, X.-J.; Dakeishi, M.; Li, P.; Qiu, G.; Jiang, H.; Nakamura, M.; et al. Effects of mercury vapor exposure on neuromotor function in Chinese miners and smelters. *Int. Arch. Occup. Environ. Health* **2007**, *80*, 381–387. [CrossRef]
315. Li, P.; Feng, X.; Qiu, G.; Li, Z.; Fu, X.; Sakamoto, M.; Liu, X.; Wang, D. Mercury exposures and symptoms in smelting workers of artisanal mercury mines in Wuchuan, Guizhou, China. *Environ. Res.* **2008**, *107*, 108–114. [CrossRef]
316. Moya-Anegón, F.; Herrero-Solana, V.; Jiménez-Contreras, E. A connectionist and multivariate approach to science maps: The SOM, clustering and MDS applied to library and information science research. *J. Inf. Sci.* **2006**, *32*, 63–77. [CrossRef]
317. Veiga, M.M.; Angeloci, G.; Hitch, M.; Colon Velasquez-Lopez, P. Processing centres in artisanal gold mining. *J. Clean. Prod.* **2014**, *64*, 535–544. [CrossRef]
318. Gunson, A.J.; Klein, B.; Veiga, M.; Dunbar, S. Reducing mine water network energy requirements. *J. Clean. Prod.* **2010**, *18*, 1328–1338. [CrossRef]
319. Veiga, M.M.; Angeloci, G.; Ñiquen, W.; Seccatore, J. Reducing mercury pollution by training Peruvian artisanal gold miners. *J. Clean. Prod.* **2015**, *94*, 268–277. [CrossRef]

International Journal of
Environmental Research and Public Health

MDPI

Review

Mercury Exposure and Its Health Effects in Workers in the Artisanal and Small-Scale Gold Mining (ASGM) Sector—A Systematic Review

Kira Taux, Thomas Kraus and Andrea Kaifie *

Institute for Occupational, Social and Environmental Medicine, Medical Faculty, RWTH Aachen University, Pauwelsstraße 30, 52074 Aachen, Germany; kira.taux@rwth-aachen.de (K.T.); tkraus@ukaachen.de (T.K.)
* Correspondence: akaifie@ukaachen.de

Abstract: Gold is one of the most valuable materials but is frequently extracted under circumstances that are hazardous to artisanal and small-scale gold miners' health. A common gold extraction method uses liquid mercury, leading to a high exposure in workers. Therefore, a systematic review according to the PRISMA criteria was conducted in order to examine the health effects of occupational mercury exposure. Researching the databases PubMed®, EMBASE® and Web of Science™ yielded in a total of 10,589 results, which were screened by two independent reviewers. We included 19 studies in this review. According to the quantitative assessment, occupational mercury exposure may cause a great variety of signs and symptoms, in particular in the field of neuro-psychological disorders, such as ataxia, tremor or memory problems. However, many reported symptoms were largely unspecific, such as hair loss or pain. Most of the included studies had a low methodological quality with an overall high risk of bias rating. The results demonstrate that occupational mercury exposure seriously affects miners' health and well-being.

Keywords: work; health; disease; intoxication; heavy metal; neuro-psychological disorders

Citation: Taux, K.; Kraus, T.; Kaifie, A. Mercury Exposure and Its Health Effects in Workers in the Artisanal and Small-Scale Gold Mining (ASGM) Sector—A Systematic Review. *Int. J. Environ. Res. Public Health* **2022**, *19*, 2081. https://doi.org/10.3390/ijerph19042081

Academic Editors: Masayuki Sakakibara, Win Thiri Kyaw and José Luis Rivera Parra

Received: 20 January 2022
Accepted: 9 February 2022
Published: 13 February 2022

Publisher's Note: MDPI stays neutral with regard to jurisdictional claims in published maps and institutional affiliations.

1. Introduction

1.1. Gold Mining

Since ancient times, gold has been one of the most desired and noble elements in the world, with an outstanding variety of application areas. One of the most impressive examples for its use in art is the world-famous death mask of the Egyptian pharaoh Tutankhamun. The jewellery industry processes gold for all conceivable kinds of products. Furthermore, gold is also of considerable relevance as a financial reserve for national banks. Germany, for instance, possessed about 3400 t of gold ingots in 2020 [1]. Hence, the gold price has shown an overall upward trend over the last five decades and amounted to approximately 1800 USD/fine troy ounce in 2020 [2,3], indicating it to remain a promising market.

The total annual gold production is, according to the U.S. Geological Survey, about 3300 t in 2019 [4], of which artisanal and small-scale gold mining (ASGM) is estimated to account for 380–450 t per year [5]. As defined by the United Nations (U.N.), artisanal and small-scale gold miners (ASG miners) are persons who engage in mining as "individual miners or [in] small enterprises with limited capital investment and production" [6]. Nevertheless, official figures on the exact number of labourers are, to our knowledge, not obtainable, but approximate values suggest that about 16 million people work as ASG miners [5]. This form of gold mining is predominantly practised in countries of the global south, for example, in Ghana, Ecuador and Indonesia [5].

Gold mining, and ASGM in particular, is of great importance for low- and middle-income countries where it reveals both positive and negative aspects. On the one hand, a striking positive aspect is its economic weight, for which Ghana is presented as an example:

the gold mining industry alone amounted to 7.1% of Ghana's national gross domestic product (GDP) in 2019 (provisional data) [7] with a general increase in revenues [8]. In addition, the companies that are members of the Ghana Chamber of Mines also support society by, for example, financially sponsoring projects in the fields of health or education [8]. On the other hand, negative effects concern the environment [9–12], with parts of the tropical rainforest in South America being destroyed in the context of gold mining [9]. Mining waste including chemicals such as cyanide or mercury can pollute the environment, such as water, sediments and soil, finally affecting the human food chain, as well [10–12]. In addition, the workers' health is harmed by their work itself. Nakua et al. observed that ASG miners have a high risk of being injured at work, while the safety precautions are at a low standard [13]. Mercury exposure is known to cause a considerable burden of disease in miners, being responsible for up to more than 2 million DALYs per year, especially in countries of the global south [14].

In general, the mining process can be carried out using various methods, depending on available materials, equipment and knowledge [15,16]. Among other methods, the mercury amalgamation is still used for extracting gold [15,16]. Here, gold-containing rocks are ground and afterwards pulverized into small pieces, and the material thus obtained is mixed with elemental mercury to form an amalgam out of gold and mercury [15,16]. After gathering, this gold amalgam is further processed by smelting, so that the mercury vaporizes and the gold remains behind [15,16].

1.2. Mercury

Mercury (also known as Hg or quicksilver) is a chemical element with a silver–grey colour. It is the only metal that exists in a liquid aggregate state under standard conditions and evaporates on contact with the ambient air. According to an official European Union directive, mercury is categorized as a threat to aquatic ecosystems, as toxic through inhalation and as hazardous to human's health [17].

Natural processes, such as volcanism, cause atmospheric mercury emissions. However, anthropogenic sources, such as coal combustion or ASGM contribute to at least three-fold higher mercury masses in the atmosphere [18–20]. Currently, the ASGM sector is the largest contributor to anthropogenic—man-made, non-natural—mercury emissions [20]. In 2015, its airborne emissions amounted up to 838 t, which represents approximately 38% of the total mercury emissions worldwide [20].

The consequences of mercury exposure are highly dependent on its chemical form (organic or inorganic compounds) as well as its dose and exposure pathway. Mercury can be absorbed via different pathways. ASG miners heat the amalgam to extract gold, leading to an evaporation of metallic (inorganic) mercury [15,16]. Therefore, this paper puts the emphasis on the metallic form. The main absorption route of elemental mercury is via inhalation. In an experimental setting, exposed individuals absorbed 67–87% of the entire inhaled mercury vapour [21]), while absorption of vapour through skin contact and gastrointestinal absorption of ingested metallic mercury played only a subordinate role [22,23]. The exhalative half-life of mercury amounts approximately 2 days [24,25], while the half-life for the urinary excretion is 63 days [25]. However, mercury accumulation occurs in organ tissues, but its exact distribution depends on its chemical form [26,27]. After inhalation of inorganic mercury, the distribution takes place via blood as it crosses most cell membranes, such as the blood–brain barrier or the placenta. The oxidation of elemental mercury in the erythrocytes influences the uptake in the brain with its corresponding typical mercury-related signs and symptoms. In contrast, occupational exposure to organic mercury compounds can be found exemplarily in the production of mercury fulminate. Organic mercury is highly lipophilic and is mainly absorbed via skin and inhalation. The main target organ of organic mercury compounds is the brain.

Internal human mercury burden can be detected in certain biological materials such as blood or urine, but due to the varying specificity for the different forms of Hg, other materials are suitable for the measurements, such as hair [28].

Acute cases of mercury poisoning can occur after contact with its metallic form [29–31]. Due to the inhalation of toxic fumes, the resulting symptoms affect, in particular, the pulmonary system with cough and dyspnoea up to acute respiratory distress syndrome (ARDS) [29–31]. Additionally, mercury exposure can lead to other unspecific symptoms, such as nausea, diarrhoea, fever or lymphadenopathy [30,31]. The course of mercury poisoning can end lethal, depending on its severity [31]. Likewise, chronic exposure to elemental Hg vapour can trigger a multitude of symptoms, mainly neuro-psychological disorders, such as tremor or erethism [32,33], which may persist in a reduced intensity even after the end of the exposure [32,34]. Further typical symptoms include a dark discolouration of the gum, gingivitis and renal damage [32,33].

In order to curb anthropogenic mercury pollution and to avoid the resulting health and environmental impacts in the future, the U.N. adopted the Minamata Convention on Mercury in 2013 [6]. This convention provided and established regulations to diminish mercury-using practices [6]. Although most of the participating countries have already signed and ratified this convention [35], mercury pollution in the artisanal and small-scale gold mining sector remains a major challenge. The aim of this systematic review was to examine the situation of occupational mercury exposure and mercury-related health effects among ASG miners in middle- and low-income countries in order to give a comprehensive overview on affected workers and their symptoms and diseases.

2. Materials and Methods

2.1. Conceptualization and Literature Research

The foundation of this systematic review was the "The Preferred Reporting Items for Systematic reviews and Meta-Analyses (PRISMA)" checklist [36]. A systematic study protocol was submitted to and accepted by the International prospective register of systematic reviews (PROSPERO), in order to ensure the methodological quality of this review. The review protocol is available online (PROSPERO-ID: CRD42021235289).

The main research question was developed according to the PECO (Population, Exposure, Comparison, Outcome) criteria [37]: How does a direct ongoing occupational mercury exposure (E) among gold miners (P) influence their health (such as mercury-related symptoms, diseases and intoxication) (O) in comparison to individuals without an occupational mercury exposure (C) (Table 1)?

Exclusively, studies with a peer-reviewed full-text article in English or German were included. These articles had to be published between 1 January 1980 and 31 December 2020. The study design was restricted to the inclusion of prospective studies, observational studies, cross-sectional studies, case-control studies, systematic reviews, non-systematic reviews and meta-analyses. All other kinds of studies were intentionally excluded. Furthermore, only studies that use the amalgam method (or another mercury-related method of gold mining) as well as studies in low- and middle-income countries were included.

To perform a systematic literature research, a search string was created considering three categories: population, exposure and outcome. The selected keywords were combined with the Boolean operators AND (to combine the topics) and OR (to combine the words in a category) to ensure the creation of senseful and complete results. Several words were also ended with *, so that all possible endings were included in the search (for example, symptom* can mean symptom as well as symptoms or symptomatic etc.). For every database (PubMed®, EMBASE® and Web of Science™), a specific search string was individually established according to the given guidelines.

The research was implemented on the 16 February 2021 by one reviewer (Kira Taux). In addition, the complete reference lists of certain key reviews [38–42] were also screened for eligible literature.

Table 1. Inclusion and Exclusion Criteria.

PECO Scheme	Inclusion Criteria	Exclusion Criteria
Population	Workers (adults as well as children and adolescents under the age of 18 years) with an ongoing occupation as gold miner in middle- and low-income countries	Adults as well as children and adolescents under the age of 18 years with no activity in gold mining; gold miners who interrupted their gold mining activities; former gold miners; residents; workers from high income countries
Exposure	Ongoing direct occupational-related mercury exposure; use of the amalgamation method for gold extraction	No ongoing direct occupational-related mercury exposure; use of other methods for gold extraction (e.g., cyanide)
Comparison (if available)	Adults as well as children and adolescents under the age of 18 years with no direct gold mining activities; residents	Workers (children and adolescents under the age of 18 years and adults) with direct relation to gold mining activities; former gold miners
Outcome	Primary outcome: all health outcomes must be a direct consequence of mercury exposure; mercury-related diseases; symptoms of acute and chronic mercury intoxication; long-time health	Studies that do not match the inclusion criteria

2.2. Screening Process

All duplicates were removed from the list of results; then the selection of literature started according to the a priori designed review protocol. The entire screening process was conducted by the same two independent reviewers (Kira Taux and Andrea Kaifie).

A screening of the articles' titles, then abstracts and finally full texts was performed by these reviewers to determine whether the aforementioned inclusion criteria were met or not. After suitable studies were selected by each reviewer, the results were compared, and disagreement was solved by discussion until consensus was achieved. Each exclusion of a study was individually documented, including the specific reason for exclusion.

2.3. Data Extraction (Quantitative Assessment)

A table containing the following categories was created to outline all relevant data: author, year, study design, setting, time, participants, exposure, measurements, outcome, effect parameters.

Depending on the given statistics, the statistical mean was preferred to report any kind of socio-demographic data (e.g., age or working time), while median values were reported as outcome (for example, laboratory parameters). If no effect parameters were given, odds ratios (OR) were calculated by one author (Andrea Kaifie) if the underlying data were available [43–48].

All data were extracted and summarized by one reviewer (Kira Taux), while the second reviewer (Andrea Kaifie) controlled the precision and completeness of the data extraction. Although considered, the creation of a meta-analysis out of the selected studies was not possible because the given data were too heterogeneous.

2.4. Bias Assessment (Qualitative Assessment)

The methodological quality evaluation of each included study was based on the assessment of the following items: selection bias, performance bias, detection bias, attrition bias, reporting bias and other source(s) of bias. Considering this classification, a table for the assessment was established according to the following references.

To evaluate the selection bias, two subsidiary categories (bias and confounder) were considered according to questions 14–26 of the checklist first published by Downs and Black [49]. Answer options were yes, no or unable to determine. If all questions were answered with no, this category was rated as low risk; if at least one question was answered with yes, the category was rated as high risk. If at least one question was answered with unable to determine, the category was rated as unable to determine. The checklist had to be adapted to the design of the included studies, which was mainly cross-sectional. Therefore, questions about participants' blinding and their compliance concerning the intervention (numbers 14 and 19) were considered as not applicable since the studies did not deal with any intervention. The questions concerning the follow-up (numbers 17 and 26) could not be answered adequately because none of the studies had a follow-up. Ultimately, the questions dealing with random sequence generation and allocation concealment (numbers 23 and 24) were omitted since none of the studies was randomized.

The remaining bias categories (performance, detection, attrition, reporting and other source(s) of bias) were evaluated according to the Cochrane Collaboration [50]. Each topic was rated in analogy to the aforementioned categories: the rating was low risk when all questions of an item were judged as low risk, the rating was high risk when at least one question was assessed as high risk and the rating was unclear risk when at least one question was answered as unable to determine.

A conclusive judgement over all categories was done after all items were evaluated and controlled: studies with an overall low risk of bias had all items assessed as low risk, studies with an overall high risk of bias had at least one category rated as high risk and studies with an unclear risk of bias had at least one category judged as unable to determine.

A blank protocol for this bias assessment is attached as Supplemental Material (Table S1).

3. Results

3.1. Literature Research and Screening Process

The literature research yielded in a total of 10,589 results. After the exclusion of duplicates, 6562 publications were finally taken into consideration. Although the reviewed publications included studies from all over the world, a major portion had to be excluded due to failure to meet the aforementioned inclusion criteria. Title and abstract screening led to the further exclusion of 6464 studies, while the literature review of selected articles [38–42] yielded 5 additional studies to be included in the full-text screening. The subsequent evaluation of the remaining 103 full-texts resulted in the exclusion of 84 articles due to their failure to fulfil the following eligibility criteria: study population ($n = 55$), outcome ($n = 16$), exposure ($n = 6$), study design ($n = 4$), language ($n = 2$) or context ($n = 1$). Eventually, 19 publications were included in this systematic review [43–48,51–63] (Figure 1).

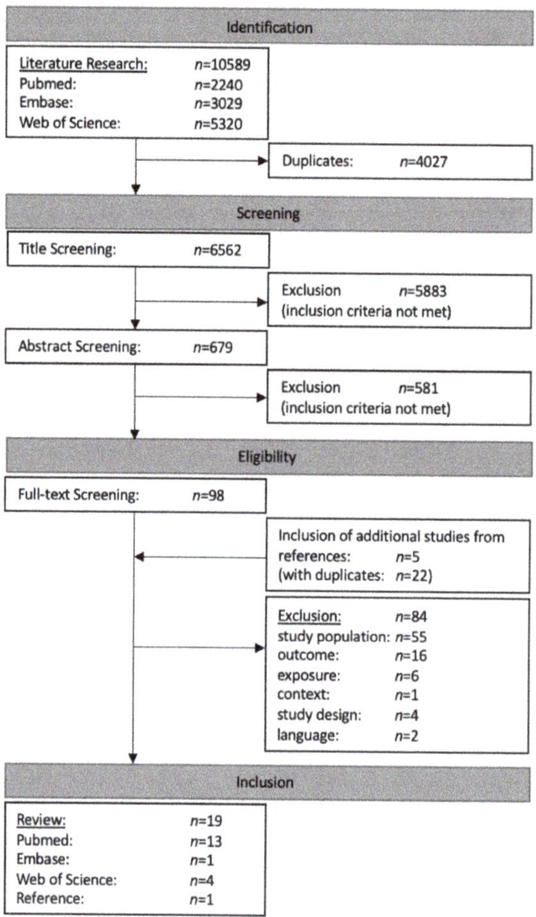

Figure 1. Flow chart of literature research and screening process.

3.2. Data Extraction (Quantitative Assessment)

The selected articles included 18 cross-sectional studies [43–47,51–63] and 1 case series [48], whose publication years ranged from 1993 to 2020 [43–48,51–63]. The following geographical regions were covered:

1. Africa: Ghana [43,44,51,52], Tanzania [45,53], Zimbabwe [46,54], Sudan [55], Burkina Faso [56], Uganda [57].
2. Asia: Indonesia [46,47,58], Pakistan [59,60].
3. South America: Brazil [48,61], Ecuador [62,63] (all Table 2).

Table 2. Data extraction.

Author, Year	Study Design, Setting, Time, Participants, Exposure	Measurements, Examinations	Outcome	Effect Parameters (Bold Indicates Statistically Significance)
Afrifa et al. (2017) [43]	cross-sectional study Bibiani/Western Region, Ghana year not reported gold miners (occupational (direct) Hg exposure): n = 110 (male, various inclusion criteria, e.g., $\geq$1 year occupational Hg exposure, no kidney-affecting disease) • exposed (B–Hg $\geq$ 5.0 µg/L): n = 61 (mean age 35.8 years, mean work duration 14.7 years) • non-exposed (B–Hg < 5.0 µg/L): n = 49 (mean age 34 years, mean work duration 10.8 years)	laboratory: spot urine sample: proteinuria blood sample: Hg, serum creatinine, eGFR examination: questionnaire: socio-demographics, anamnesis, occupation occupational assessment interview	laboratory (only exposed): significantly elevated levels: • B–Hg (mean 18.4 µg/L) • urine protein (mean 41.7 mg/dL) • serum creatinine (mean 2.2 µmol/L) significantly reduced levels: eGFR (mean 57 mL/min/1.73 m^2) examination: symptoms associated with Hg exposure: • non-significant more frequent in exposed: skin rash, cough, fever, itchy eyes, fatigue, headache, muscle ache, numbness, hair loss • non-significant less frequent in exposed: metallic taste	correlation: B–Hg: • significantly positive: proteinuria (**r = 0.7**) • significantly negative: eGFR (**r = −0.8**) odds ratio OR (95% CI): Hg exposure (age-adjusted): • urine protein $\geq$ 10 mg/dL: OR = **50.3 (11–230.5)** • serum creatinine > 106 µmol/L: OR = **101.1 (25.2–404.9)** • eGFR $\leq$ 90 mL/min/1.73 m^2: OR = **263.2 (48.8–1420)** Hg exposure and symptoms (OR (95% CI)) (calculated by the authors of this systematic review from available data in the original publication): • neuro-psychological symptoms: fatigue (OR = 1.1 (0.5–2.6)), headache (1.3 (0.5–3)), numbness (OR = 1.9 (0.9–4.1)) • other symptoms: skin rash (OR = 1.4 (0.7–3)), cough (OR = 1.7 (0.8–3.6)), fever (OR = 1.1 (0.5–2.4)), itchy eyes (OR = 1.2 (0.5–2.9)), muscle ache (OR = 1.5 (0.7–3.2)), hair loss (OR = 1.2 (0.2–7.6)), metallic taste (OR = 0.7 (0.3–1.5))

Table 2. *Cont.*

Author, Year	Study Design, Setting, Time, Participants, Exposure	Measurements, Examinations	Outcome	Effect Parameters (Bold Indicates Statistically Significance)

Afrifa et al. (2018) [51]	cross-sectional study
	Bibiani/Western Region, Ghana
	January 2017–March 2018
	gold miners (occupational (direct) Hg exposure):
	n = 137 (male, various inclusion criteria, e.g., ≥1 year occupational Hg exposure, no diseases affecting liver or thyroid)
	• exposed (B–Hg ≥ 5.0 µg/L): n = 80 (median age 26 years, median work duration 8 years)
	• non-exposed (B–Hg < 5.0 µg/L): n = 57 (median age 30 years, median work duration 4 years)
		laboratory:
		blood sample:
		• thyroid hormones: T4, T3, TSH
		• Hg
		questionnaire:
		socio-demographics, anamnesis, occupation
			laboratory:
			exposed:
			• significantly elevated levels: B–Hg (median 8 µg/L)
			• significantly reduced levels:
			  – T4 (mean 5.4 µg/dl)
			  – T3 (mean 1.5 nmol/L)
			• non-significantly elevated levels: TSH (mean 1.7 mIU/L)
			both groups: normal range for thyroid parameters
				correlation:
				B–Hg:
				• significantly negative: T4 **(r = −0.7)**, T3 **(r = −0.3)**
				• non-significantly positive: TSH (r = 0.1)

Rajaee et al. (2015) [44]	cross-sectional study Upper East Region, Ghana • Kejetia (miners + residents) • Gorogo (controls) May–July 2011 gold miners (occupational (direct) Hg exposure): n = 70 (n = 42 male, n = 28 female, mean age 30.6 years) residents (environmental (indirect) Hg exposure): n = 26 (n = 7 male, n = 19 female, mean age 33.8 years) controls (no known (background) Hg exposure): n = 75 (n = 34 male, n = 41 female, mean age 51.5 years)	laboratory: spot urine sample (n = 91): Hg hair sample (n = 69): MeHg examination: interview: socio-demographics, anamnesis, occupation, lifestyle medical parameters: • pulse • blood pressure: SBP, DBP, MAP, PP	laboratory (only miners' median values reported): U–Hg: significantly elevated levels in miners (4.2 µg/L, 5.2 µg/L (SG-adjusted)) (trend: miners > residents > controls) H–Hg: significantly elevated levels in miners (0.9 µg/g) (trend: miners > residents > controls) examination (only miners): mean blood pressure: 122.6/75.2 mmHg hypertension: n = 11	correlations (only Kejetia): significantly negative: U–Hg (SG-adjusted) + pulse **(ρ = −0.2)** associations (only miners): H–Hg: • non-significantly positive: PP (ß = 1.9), SBP (ß = 1.4), MAP (ß = 0.1) • non-significantly negative: DBP (ß = −0.5), pulse (ß = −0.3) U–Hg (SG-adjusted): • non-significantly positive: PP (ß = 0.2), SBP (ß = 0.1) • non-significantly negative: DBP (ß = −0.1), pulse (ß = −0.02), MAP (ß = −0.04) odds ratio (OR (95% CI)) (calculated by the authors of this systematic review from available data in the original publication) (Hg exposure, miners vs. residents): hypertension: (OR = 0.8 (0.3–2))

Table 2. *Cont.*

Author, Year	Study Design, Setting, Time, Participants, Exposure	Measurements, Examinations	Outcome	Effect Parameters (Bold Indicates Statistically Significance)

Mensah et al. (2016) [52]	cross-sectional study
	Prestea/Western Region, Ghana
	2012
	gold miners (occupational (direct) Hg exposure):
	n = 343 (n = 323 male, n = 20 female, age 15–70 years, mean age 29.5 years, work duration 1–38 years, mean work duration 7.2 years)
	• exposure:
	  – exposed (U–Hg $\geq$ 5.0 µg/L): n = 160
	  – non-exposed (U–Hg < 5.0 µg/L): n = 183
	• workplace (previous occupation in another mine): n = 25
		laboratory:
		morning urine sample: Hg
		examination:
		questionnaire:
		socio-demographics, anamnesis, occupation
		interview + observation:
		occupational assessment
			laboratory:
			exposed: U–Hg 5–50.5 µg/L (mean 14.8 µg/L)
				associations (Hg exposure):
				non-significant in all miners (n = 343):
				• neuro-psychological symptoms: headache (χ^2 = 0.7), fatigue (χ^2 = 0.01), insomnia (χ^2 = 1), numbness (χ^2 = 0.00)
				• other symptoms red eyes (χ^2 = 0.3), skin rash (χ^2 = 3.5), cough (χ^2 = 0.2), fever (χ^2 = 0.1), metallic taste (χ^2 = 0.4), muscle ache (χ^2 = 0.1), sinusitis (χ^2 = 0.4), hair loss (χ^2 = 0.04)
				significant in miners with a previous occupation in another mine (n = 25): **numbness (χ^2 = 5)**

Bose-O'Reilly et al. (2010a) [45]	cross-sectional study
	Tanzania:
	• Rwamagasa/Geita District
	• Katoro
	October–November 2003
	gold miners (occupational (direct) Hg exposure):
	n = 138 (from Rwamagasa)
	• non-smelters: n = 34 (n = 20 male, n = 14 female, age 14–50 years, mean age 26.6 years)
	• smelters: n = 104 (n = 87 male, n = 17 female, age 14–57 years, mean age 33.8 years)
	residents (environmental (indirect) Hg exposure):
	n = 52 (from Rwamagasa, n = 21 male, n = 31 female, age 11–57 years, mean age 32.3 years)
	controls (no known (background) Hg exposure):
	n = 31 (from Katoro, n = 12 male, n = 19 female, age 15–51 years, mean age 32.4 years)
	→ exposed: smelters, non-smelters, residents
		laboratory:
		blood sample: Hg
		urine sample (n = 218): Hg
		hair sample (n = 188): T–Hg, I–Hg, O–Hg
		examination:
		questionnaire: anamnesis, Hg exposure
		clinical examination with special focus on neurology
		neuro-psychological tests:
		memory test, matchbox test, pencil tapping test, Frostig score
		diagnosis of intoxication:
		algorithm including symptoms + laboratory parameters
			laboratory (only miners' median values):
			significantly elevated levels in exposed groups: smelters >> non-smelters > residents > controls
			• U–Hg: non-smelters 1 µg/L (0.8 µg/gCr), smelters 5.9 µg/L (3.6 µg/gCr)
			• B–Hg: non-smelters 1.6 µg/L, smelters 2.5 µg/L
			• hair:
			  – T–Hg: non-smelters 0.5 µg/g, smelters 0.8 µg/g
			  – I–Hg: non-smelters 0.1 µg/g, smelters 0.4 µg/g
			examination:
			significantly more frequent in exposed
				odds ratio (OR (95% CI)) (calculated by the authors of this systematic review from available data in the original publication) (Hg exposure, gold miners vs. residents):
				• neuro-psychological symptoms: sleep problems (OR = 1.2 (0.6–2.5)), tremor (OR = 3.1 (0.9–10.9)), nervousness (OR = 1.2 (0.5–2.9)), sadness (OR = 1.9 (0.9–3.7)), headache (OR = 1.3 (0.7–2.4)), numbness (OR = 1.1 (0.6–2.1))
				• other symptoms: loss of appetite (OR = 1.7 (0.8–3.6)), hair loss (OR = 2.7 (0.1–53.4)), metallic taste (OR = 1.1 (0.2–5.8)), salivation (OR = 2.5 (1–6.4)), palpitations (OR = 1.4 (0.7–2.7)), nausea (OR = 1.1 (0.5–2.4))

Table 2. *Cont.*

Author, Year	Study Design, Setting, Time, Participants, Exposure	Measurements, Examinations	Outcome	Effect Parameters (Bold Indicates Statistically Significance)
			<ul><li>worsened health</li><li>neuro-psychological symptoms: memory problems, tiredness, tremor finger-to-nose + eyelid, sadness, problems to find words, ataxia of gait, sensory disturbance, abnormal ASR + BSR</li><li>other symptoms: less appetite, salivation, discoloured gum</li></ul>no significant difference:<ul><li>neuro-psychological symptoms: sleep problems, concentration problems, thinking problems, nervousness, headache, numbness, problems with impetus, dysdiadochokinesia, intentional tremor heel-to-shin, ataxia heel-to-shin, bradykinesia, hypomimia, abnormal PSR</li><li>other symptoms: loss of hair, metallic taste, less muscle strength, weakness, eyestrain problems, palpitations, nausea, stomatitis, gingivitis</li></ul>neuro-psychological tests:<ul><li>significantly worse results in exposed: pencil tapping test, matchbox test</li><li>no significant differences: memory test, Frostig score</li></ul>intoxication: smelters (n = 104) n = 25, residents (n = 31) n = 1	

Table 2. *Cont.*

Author, Year	Study Design, Setting, Time, Participants, Exposure	Measurements, Examinations	Outcome	Effect Parameters (Bold Indicates Statistically Significance)
Harada et al. (1999) [53]	cross-sectional study Tanzania: • seven gold mines near Lake Victoria • one gold mine in the inland • three fishing villages around Lake Victoria • one fishing village in the inland • Mwanza/State of Mwanza 1996–97 (three spot investigations) gold miners (occupational (direct) Hg exposure): n = 150 (age 7–70 years, n = 136 male, n = 14 female, work duration 3–10 years) fishermen (environmental (indirect) Hg exposure): n = 103 (fishermen + families, age 6–70 years, n = 87 male, n = 16 female) controls (no known (background) Hg exposure): n = 19 (inhabitants of Mwanza, age 0.5-46 years, n = 11 male, n = 8 female)	laboratory: hair: T–Hg, MeHg (n = 9) fish (Lake Victoria): T–Hg examination (n = 225): questionnaire clinical examination with focus on neurology diagnosis of intoxication: elevated T–Hg + symptoms	laboratory: gold miners: • mean T–Hg 1–81.9 ppm • mean T–Hg 1–4 ppm (exclusion of six cases >50 ppm) MeHg/T–Hg ratio (n = 3 from each group, marker for direct non-dietary Hg exposure): higher in gold miners (1–20.5%) fish: tilapia 63 ppb, sardine 12 ppb, catfish 8.9 ppb examination of gold miners (n = 118): • neuro-psychological symptoms (trembling 21.2%, headache 11.9%, numbness 9.3%, sensory disturbance 8.5%, taste problems 8.5%, tremor 8.5%, abnormal reflex 5.1%/2.5%, memory problems 7.6%, smell problems 5.9%, insomnia 5.1%, vertigo/dizziness 5.1%, neurasthenia 3.4%, night blindness 3.4%) • pain (chest pain 8.5%, limb pain 5.9%) • respiratory problems (dyspnoea 7.6%, cough 6.8%) • other symptoms: palpitation (5.9%), gingivitis (13.6%) • mild I–Hg intoxication: n = 14	—

Table 2. *Cont.*

Author, Year	Study Design, Setting, Time, Participants, Exposure	Measurements, Examinations	Outcome	Effect Parameters (Bold Indicates Statistically Significance)

Bose-O'Reilly et al. (2008) [46]	cross-sectional study
	Sulawesi/Indonesia, Kadoma/Zimbabwe
	2003–04
	gold miners (occupational (direct) Hg exposure): n = 80 (median age 12 years, n = 60 male, n = 20 female)
	residents (environmental (indirect) Hg exposure): n = 36 (living in gold mining areas, median age 11 years, n = 11 male, n = 25 female)
	controls (no known (background) Hg exposure): n = 50 (median age 12 years, n = 24 male, n = 26 female, living in areas without gold mining)
		laboratory:
		spot urine sample: Hg
		blood sample: Hg
		hair sample (n = 150): T–Hg, I–Hg, O–Hg
		examination:
		questionnaire: anamnesis, Hg exposure, confounders
		clinical examination with special focus on neurology
		neuro-psychological tests: memory test, matchbox test, pencil tapping test, Frostig score
		diagnosis of intoxication:
		algorithm including symptoms + laboratory parameters
			laboratory (only miners' values):
			significantly elevated levels in miners + residents (highest burden in miners):
			• B–Hg: median 7.8 µg/L
			• U–Hg: median 10.1 µg/L, 7.1 µg/gCr
			• hair:
			  – T–Hg: median 2.3 µg/g
			  – I–Hg: median 0.9 µg/g
			elevated in miners > residents: factor U–Hg/B–Hg (U–Hg > B–Hg, miners: 3.8)
			examination:
			significantly more common in miners > residents
			• neuro-psychological symptoms (ataxia, dysdiadochokinesia, abnormal reflexes)
			• other symptoms: salivation, metallic taste, blue coloration of gums
			no significant difference:
			• neuro-psychological symptoms (headache, memory problems, nausea, numbness/prickling feet, concentration problems, sleeping problems, tremor, hypomimia, mento-labial reflex)
			• other symptoms: gingivitis, stomatitis, proteinuria
			neuro-psychological tests:
			• significantly worse performance in miners + residents: matchbox test, pencil tapping test
			• no significant difference: memory test, Frostig score
			intoxication: miners: n = 20, residents: n = 4
				odds ratio (OR (95% CI)) (calculated by the authors of this systematic review from available data in the original publication) (Hg exposure, gold miners vs. controls):
				neuro-psychological symptoms: headache (OR = 0.7 (0.3–1.9)), memory problems (OR = 0.9 (0.2–5.8)), numbness/prickling feet (OR = 2 (0.4–10)), sleeping problems (OR = 1.9 (0.2–18.9)), concentration problems (OR = 4 (0.5–34)), dysdiadochokinesia **(OR = 4.6 (1.5–14.4))**, hypomimia (OR = 3.2 (0.2–68.4))
				• ataxia: heel-to-shin **(OR = 3.3 (1.1–10.5))**, gait **(OR = 5.2 (1.5–18.6))**
				• abnormal reflex: mento-labial **(OR = 0.6 (0.3–1.6))**, ankle jerk **(OR = 5.5 (1.8–17))**, biceps brachii **(OR = 8 (1.8–35.9))**, quadriceps **(OR = 10.4 (1.3–81.7))**
				• tremor (OR = 2 (0.2–19.3)): finger-to-nose (OR = 4.6 (0.2–90.2)), heel-to-shin (OR = 1.9 (0.1–47.7)), eyelid (OR = 1.5 (0.7–3))
				other symptoms: salivation (OR = 15 (0.9–262.7)), metallic taste (OR = 8.8 (0.5–159.9)), nausea (OR = 0.9 (0.2–5.8)), blue coloration of gums (OR = 8.8 (0.5–159.9)), gingivitis (OR = 0.6 (0–32.1)), stomatitis (OR = 0.6 (0–32.1)), proteinuria (OR = 0.1 (0–1.7))

Table 2. *Cont.*

Author, Year	Study Design, Setting, Time, Participants, Exposure	Measurements, Examinations	Outcome	Effect Parameters (Bold Indicates Statistically Significance)
Steckling et al. (2014) [54]	cross-sectional study Zimbabwe • Kadoma (gold miners) • Chikwaka (controls) 2004, (2006) gold miners (occupational (direct) Hg exposure): n = 181 (age 9–75 years, mean age 27 years, n = 122 male, n = 59 female, n = 33 non-smelters, n = 148 smelters, occupational Hg exposure 1–23 years, mean occupational Hg exposure 4 years) controls (no known (background) Hg exposure): n = 91 (age 11–59 years, mean age 24 years, n = 24 male, n = 67 female)	definition of chronic Hg intoxication: algorithm including symptoms + laboratory parameters diagnosis of chronic Hg intoxication disease prevalence in Zimbabwe: socio-demographic data DALY = YLD (years lived with disability) + YLL (years of life lost)	laboratory (miners' median values): higher values in miners 80–90% miners + few controls exceed limit values U–Hg: 26.1 µg/L, 25.8 µg/gCr B–Hg (n = 152): 11.4 µg/L H–Hg (n = 158): 3.3 µg/L symptoms (miners): medical score sum: significant higher significantly more frequent: • neuro-psychological symptoms: tremor at work + finger-to-nose, ataxia of gait, dysdiadochokinesia • other symptoms: metallic taste, blue discoloured gum • worse results in neuro-psychological tests: Frostig test, pencil tapping test non-significantly more frequent: • worsened health • neuro-psychological symptoms: heel-to-knee ataxia, heel-to-knee-tremor, abnormal reflex, sleeping problems • other symptoms: salivation, proteinuria • worse results in neuro-psychological tests: memory test, matchbox test	–

Table 2. *Cont.*

Author, Year	Study Design, Setting, Time, Participants, Exposure	Measurements, Examinations	Outcome	Effect Parameters (Bold Indicates Statistically Significance)
			intoxication: n = 131 miners gold miners: total 350,000 adults (≥15 years): 85% (male 70%, female 30%) children (9–14 years): 15% (male 73%, female 27%) chronic Hg intoxication: population: 2% (3% occupational Hg exposure) miners: 72% (adults 72% (male 90%, female 40%), children 76%) 95,400 DALYs (8 DALYs/1000 persons) • most affected: male (78,400 DALYs, 13 DALYs/1000 persons) • female (17,000 DALYs, 3 DALYs/1000 persons)	
Tayrab (2017) [55]	cross-sectional study Sudan: • Abuhamad gold mining area/River Nile State (gold miners) • Khartoum State (controls) August 2012–November 2014 gold miners (occupational (direct) Hg exposure): n = 83 (male, age 18–55 years, mean age 30.5 years, occupation > 6 months: 44.6% wells, 32.5% mills, 16.9% washing, 6% moulding) controls (no known (background) Hg exposure): n = 50 (male, mean age 28.1 years)	laboratory: blood sample: TSH, TT3, TT4, FT3, FT4 clinical examination	significantly elevated levels in miners: • TSH (mean 5.1 µIU/mL) • TT4 (mean 86.3 pmol/L) significantly reduced levels in miners: • TT3 (mean 1.2 ng/dL) • FT3 (mean 1.3 pg/mL) • FT4 (mean 6.4 ng/dL) elevated in miners: FT4/FT3 ratio (mean 4.8) laboratory results (TSH, FT3, TT3) compatible with hypothyroidism	–

Table 2. *Cont.*

Author, Year	Study Design, Setting, Time, Participants, Exposure	Measurements, Examinations	Outcome	Effect Parameters (Bold Indicates Statistically Significance)
Tomicic et al. (2011) [56]	cross-sectional study Burkina Faso (eight gold mining areas in six regions) • Bagassi • Bouda • Fandjora II + III • Mossobadougou • Pousghin (Macara) • Safané • Zinigma year not reported participants: n = 1090 • gold miners (occupational (direct) Hg exposure): n = 779 – most susceptible to Hg: n = 93 (n = 82 male, n = 11 female, age 17–56 years, mean age 31.7 years, occupational Hg exposure 1–12 years, mean occupational Hg exposure 4 years) – gold dealers: n = 146 (susceptible: n = 52) – ore washers: n = 151 (susceptible: n = 33) – others (susceptible: n = 8) • non-miners (environmental (indirect) Hg exposure): e.g., miners' families	laboratory: spot urine sample: Hg (n = 93), albumin, creatinine examination: questionnaire: socio-demographics, anamnesis, occupation, Hg exposure, consumption medical parameters	laboratory: U–Hg: 3–3493 µg/L, 4.3–1707 µg/gCr • 69% >35 µg/gCr, 16% >350 µg/gCr (reference < 3 µg/gCr) • statistically significant trend: dealers > ore washers > others albuminuria: 40.9% (most susceptible miners) examination: more frequent in gold miners: • walking problems (8.2%) • rhinitis (9.5%) more frequent in susceptible miners: • neuro-psychological symptoms (headache 53.3%, sleep problems 25.3%, dizziness 53.8%, tiredness 33%, trembling 31.9%, sensory problems on hands/feet 23.1%, visual problems 30.8%) • other symptoms: mouth irritations/wounds (22%), cough (27.5%), thoracic pain (34.1%)	association: significantly positive: • ore washing + trembling: χ^2 **6.6** • packaging Hg + rhinitis: χ^2 **5.6** • U–Hg: – thoracic pain: χ^2 **10.4** – grabbing problems: χ^2 **6.8** non-significantly positive: • gold dealer + trembling: χ^2 3.7 • heating Hg + thoracic pain: χ^2 3.5

Table 2. *Cont.*

Author, Year	Study Design, Setting, Time, Participants, Exposure	Measurements, Examinations	Outcome	Effect Parameters (Bold Indicates Statistically Significance)

Wanyana et al. (2020) [57]	cross-sectional study
	Uganda (two mining areas each):
	• Amudat/Karamoja Region
	• Busia/Eastern Region
	• Ibanda/Western Region
	• Mubende/Central Region
	
	June–July 2018
	gold miners (occupational (direct) and environmental (indirect) Hg exposure): n = 183 (n = 133 male, n = 50 female, age 15–65 years, occupation $\geq$ 1/2 year, mean Hg exposure 5.4 years), including
	• extractors
	• processors
	• burners/buyers
		laboratory:
		blood sample (n = 31): Hg
		urine sample (n = 31): Hg
		environmental samples (n = 26, water + topsoil): Hg
		examination:
		questionnaire:
		socio-demographics, Hg exposure
		clinical examination: anamnesis, neurological examination
			laboratory:
			blood: B–Hg 26.3–205 µg/L (median 67.5 µg/L, all samples exceed HBM-II limit value)
			• significantly higher: Mubende, OHS knowledge
			urine: U–Hg 37.5–296 µg/L (median 70.8 µg/L, all samples exceed HBM-II limit value)
			• significantly higher: Mubende, female, panners, OHS knowledge
			environment:
			• drinking water: mean 23.8 µg/L (in all samples above WHO limit (6 µg/L))
			• soil: mean 0.2 µg/L
			symptoms (examination):
			• least in Ibanda
			• statistically significant more frequent:
			  – female: swollen legs, psychiatric problems, stomachache, memory problems, diarrhoea, respiratory problems
			  – male: headache
			• no significant differences between male + female: injuries, numbness, shaking hands, pain (back, chest, joint, feet), eye problems, general malaise, dizziness, fatigue + stress
				odds ratio (OR (95% CI)) (Hg exposure):
				odds ratio (adjustment for confounders)/odds ratio
				neuro-psychological symptoms:
				• shaking of hands + head: **(OR = 24.1 (1.7–338.7)/OR = 7.8 (2.7–22))**
				• eye problems: **(OR = 11 (2–62.5)/OR = 9.2 (3.7–23.2))**
				• numbness: **(OR = 8.5 (2.1–34.4)/OR = 7.9 (7.9–3.6))**
				• fatigue + stress: **(OR = 5.4 (1.9–14.9)/OR = 6.5 (3.5–12.3))**
				• headache: **(OR = 4.7 (1.9–11.3)/OR = 6.4 (3.4–12))**
				• dizziness: **(OR = 3.8 (1.5–9.7)/OR = 6 (2.9–12.2))**
				pain:
				• chest: **(OR = 9 (3.3–24.6)/OR = 8 (4–16))**
				• back: **(OR = 6.2 (2.2–17.5)/OR = 6.1 (3.3–11.2))**
				• joint: **(OR = 3.2 (1.3–8.3)/OR = 6.1 (2.9–12.9))**
				respiratory problems: **(OR = 3.2 (1–10.1)/OR = 6.8 (2.7–17.4))**

Table 2. *Cont.*

Author, Year	Study Design, Setting, Time, Participants, Exposure	Measurements, Examinations	Outcome	Effect Parameters (**Bold** Indicates Statistically Significance)

Bose-O'Reilly et al. (2010b) [47]	cross-sectional study
	Indonesia:
	• Tatelu in North Sulawesi
	• Kerang Pangi in Central Kalimantan
	August–September 2003
	gold miners (occupational (direct) Hg exposure):
	• non-smelters: n = 47 (n = 31 female, n = 16 male, age 19–59 years, mean age 33.7 years for Kalimantan/36 years for Sulawesi)
	• smelters: n = 129 (n = 28 female, n = 101 male, age 19–58 years, mean age 31.9 years for Kalimantan/35.2 years for Sulawesi)
	residents (environmental (indirect) Hg exposure):
	n = 84 (n = 78 female, n = 6 male, age 19–57, mean age 31.6 years for Kalimantan/39.1 years for Sulawesi)
	controls (no known (background) Hg exposure):
	n = 21 (only from North Sulawesi, n = 4 female, n = 17 male, age 21–46 years, mean age 27.1 years)
	r → exposed: smelters, non-smelters, residents
		laboratory:
		blood sample: Hg
		urine sample: Hg
		hair sample: T–Hg, I–Hg, O–Hg
		examination:
		questionnaire: anamnesis, Hg exposure
		clinical examination with special focus on neurology
		neuro-psychological tests: memory test, matchbox test, pencil tapping test, Frostig score
		diagnosis of intoxication:
		algorithm including symptoms + laboratory parameters
			laboratory (only miners' median values):
			all parameters significantly elevated in exposed (trend: smelter > non-smelter > resident > control):
			• U–Hg: non-smelter 5.3–7.8 µg/L (3.3–3.7 µg/gCr), smelter 10.2–22.4 µg/L (5.3–10.2 µg/gCr)
			• B–Hg: non-smelter 9.2–9.5 µg/L, smelter 10.6–13.3 µg/L
			• hair:
			  – T–Hg: non-smelter 3–3.8 µg/g, smelter 3.9–4.9 µg/g
			  – I–Hg: non-smelter 1.1–1.2 µg/g, smelter 1.3–2 µg/g
			examination:
			more symptoms in exposed significantly more frequent in exposed groups:
			• neuro-psychological symptoms (ataxia of gait + heel-to-shin, eyelid tremor, hypomimia)
			more frequent in different locations:
			• Central Kalimantan: hair loss, salivation, numbness, fatigue
			• North Sulawesi: sleeping problems, fatigue (for smelters)
			no significant difference: finger-to-nose tremor, dysdiadochokinesia
			neuro-psychological tests: worse results in all tests for exposed
			• significantly worse results: memory test, pencil tapping test, matchbox test
				odds ratio (OR (95% CI)) (calculated by the authors of this systematic review from available data in the original publication) (Hg exposure): intoxication (gold miners vs. residents):
				• Kalimantan: (**OR = 2.8 (1.5–5.4)**)
				• Sulawesi: (**OR = 4.6 (1.2–17.3)**)

Table 2. *Cont.*

Author, Year	Study Design, Setting, Time, Participants, Exposure	Measurements, Examinations	Outcome	Effect Parameters (Bold Indicates Statistically Significance)
			• non-significantly worse results: Frostig score intoxication: significantly more cases in exposed (all from Kalimantan, smelters from Sulawesi) • smelters: n = 33 (Sulawesi), n = 43 (Kalimantan) • non-smelters: n = 4 (Sulawesi), n = 13 (Kalimantan) • residents: n = 3 (Sulawesi), n = 21 (Kalimantan)	
Ekawanti and Krisnayanti (2015) [58]	cross-sectional study Sekotong ASGM/West Nusa Tenggara Province, Indonesia year not reported gold miners (occupational (direct) Hg exposure): n = 71 (male, occupational Hg exposure for ≥5 years, detailed work duration not reported, smokers) non-miners (environmental (indirect) Hg exposure): n = 29 (n = 25 female, n = 4 male, miners' families as wives + children, living in the mining area for ≥1 year)	laboratory: blood sample: haemoglobin, haematocrit urine sample: Hg, proteinuria hair sample: Hg questionnaire: Hg exposure, duration + handling	laboratory: urine: • significantly elevated levels in miners: — urine protein (mean 1.7 g/L) — U–Hg (mean 69.4 µg/L) • more frequent in miners: proteinuria (92.6%) blood: • significantly reduced in miners: — haemoglobin (mean 12.7 g/dL) — haematocrit (mean 38.2%) • more frequent in miners: anaemia (57.7% for standard limits, 67.6% for smokers' limits) H–Hg: non-significantly elevated in miners: (mean 2.8 µg/L)	—

Table 2. *Cont.*

Author, Year	Study Design, Setting, Time, Participants, Exposure	Measurements, Examinations	Outcome	Effect Parameters (Bold Indicates Statistically Significance)
Khan et al. (2012) [59]	cross-sectional study Gilgit-Baltistan Province, Pakistan • Astor • Dainyor • Gulmiti • Gupis • Haramosh • Ishkumen • Jaglot • Jalalabad • Jutal • Khari • Shimshal year not reported gold miners (occupational (direct) Hg exposure): n = unknown adults: n = unknown (age 18–50 years, male + female) children: n = unknown (age 8–15 years, male + female) control group (environmental (indirect) Hg exposure): n = unknown (male, female, children)	laboratory: blood sample: • RBC • plasma urine sample: Hg questionnaire: socio-demographics, anamnesis, Hg exposure, consumption significance level: <0.01	laboratory (only mean values for miners): urine: significantly elevated levels of U–Hg • adults: male 57.1 µg/L, female 68.5 µg/L (98% exceed WHO limit (50 µg/L)) • children: male 24.5 µg/L, female 13.6 µg/L blood (no detailed numbers reported): significantly elevated levels of T–Hg, I–Hg examination (only miners): higher percentage of symptoms symptoms (male, female): kidney disease (56%, 30%), skin rash (38%, 48%), neuro-psychological symptoms (tiredness + headache 67%, 23%; cognitive problems 45%, 20%; sensory problems 31%, 28%; tremor 16%, 9%; memory problems 12%, 7%; abnormal reflexes 6%, 4%; smell + taste problems 16%, 12%; night blindness 9%, 2%; neurasthenia 7%, 8%), pain (chest 53%, 67%; limb 10%, 8%), children: slow growth (67%, 71%), cough (31%, 22%), palpitation (21%, 28%)	–

Table 2. *Cont.*

Author, Year	Study Design, Setting, Time, Participants, Exposure	Measurements, Examinations	Outcome	Effect Parameters (Bold Indicates Statistically Significance)
Riaz et al. (2016) [60]	cross-sectional study Gilgit-Baltistan Province, Pakistan • Chalt • Chirmish • Goharabad • Jalalabad Khari • Minor Khari • Nomal • Yashokaldas year not reported gold miners (occupational (direct) Hg exposure): n = 45 (male + female, detailed distribution of age, sex + work duration not reported)	laboratory: blood sample: • plasma: T–Hg, I–Hg, O–Hg • RBC: T–Hg, I–Hg, O–Hg urine sample: T–Hg, I–Hg, O–Hg hair sample: Hg nail sample: Hg questionnaire: socio-demographics, anamnesis, occupation, consumption significance level: ≤0.01	laboratory (only mean values for urine reported): RBC: results exceed limits of WHO + USEPA plasma: results exceed limits of USEPA (5.8 µg/L) urine: results exceed limits of WHO (50 ng/mL) • T–Hg: male 61.4 µg/L, female 51.7 µg/L • I–Hg: male 40.5 µg/L, female 36.5 µg/L H–Hg: male 2.7 g/kg, female 1.8 µg/kg N–Hg: male 2.2 µg/kg, female 2.8 µg/kg examination: more health problems in occupational context health problems (male, female): kidney disease (67%, 54%), gastrointestinal problems (stomach problems 89%, 75%; teeth problem 58%, 46%; belly pain 68%, 46%), neck pain (45%, 58%), joints problem (35%, 22%), hernia (25%, 18%), skin burn (75%, 98%), stunted growth in children (56%, 34%), heart problem (45%, 35%), inhalation problem (76%, 86%)	–

Table 2. *Cont.*

Author, Year	Study Design, Setting, Time, Participants, Exposure	Measurements, Examinations	Outcome	Effect Parameters (Bold Indicates Statistically Significance)
Lacerda et al. (2020) [61]	cross-sectional study Pará State, Brazil Itaituba (riverines) Serra Pelada (gold miners) year not reported gold miners (occupational (direct) Hg exposure): n = 34 (male, mean age 45.9 years) riverines (environmental (indirect) Hg exposure) n = 10 (male, mean age 40.9 years) controls (only for hue ordering test, no known (background) Hg exposure): n = 41 (male, age-matched, urban inhabitants, no disease affecting the visual outcome, visual acuity 20/20)	laboratory: hair sample: Hg examination: questionnaire: socio-demographics, anamnesis, occupation, consumption, smoking visual test: participants with visual acuity ≥20/40 on both eyes (better one tested) visual perimetry hue ordering test	laboratory: significantly reduced levels in miners: H–Hg examination: perimetry: • riverines: all below reference, significantly smaller perimetric area than miners • miners: 61.8% below reference hue ordering test: • miners: significantly more errors than controls, no significant difference to riverines	associations (only miners): H–Hg: • visual perimetry: partial r = 0.1 • colour vision: partial r = 0.5 correlations (only miners): non-significant for visual outcome: r = −0.2
Branches et al. (1993) [48]	case series Brazil • Tapajós/Pará region • Tapajós/Amazonas region 1986–91 participants: n = 55 (hospital patients with suspicion or anamnesis of Hg exposure, age 8–75 years) • exposed group (occupational (direct) Hg exposure): – gold shop workers (n = 11, male, mean age 37 years, mean occupational Hg exposure 5.3 years) – gold miners (n = 22, male, mean age 43 years, mean occupational Hg exposure 16.3 years) • residents (environmental (non-occupational) Hg exposure): n = 22 (urban inhabitants, n = 11 male, n = 10 female, mean age 37 years)	laboratory: blood sample: Hg spot urine sample: Hg examination questionnaire: anamnesis, occupation clinical examination	laboratory (gold miners): B–Hg: mean 2.2 µg/dL, median 2.3 µg/dL U–Hg (n = 6): mean 35.4 µg/L, median 25 µg/L examination (gold miners): • neuro-psychological symptoms (dizziness 45%, headache 45%, tremor 45%/9%, insomnia 41%, numbness 41%, visual problems 41%/18%, memory problems 32%, nervousness 23%, balance problems 18%/14%, coordination problems 18%, fatigue 18%, seagull sign 18%, cramps 14%, fear 14%, ataxia 5%, deep sensibility problems 5%, tactile problems 5%) • pain (chest pain 27%, abdominal pain 14%)	odds ratio (OR (95% CI) (calculated by the authors of this systematic review from available data in the original publication) (Hg exposure): • neuro-psychological symptoms: dizziness (OR = 0.8 (0.3–2.7)), headache (OR = 1.2 (0.5–2.9)), tremor (OR = ₂.8 (0.5–6.1)), numbness (OR = 1.5 (0.5–4.7)), insomnia (OR = 1.9 (0.5–6.6)), nervousness (OR = 0.3 (0.1–1.1)), visual problems (OR = 4.4 (1–19.4)), memory problems (OR = 2.1 (0.5–8.6)), cramps (OR = 0.7 (0.1–3.6)), fatigue (OR = 1.4 (0.3–7.2)), fear (OR = 0.7 (0.1–3.6)) • pain: chest (OR = 3.6 (0.7–21.2)), abdomen (OR = 1 (0.2–5.6))

Table 2. *Cont.*

Author, Year	Study Design, Setting, Time, Participants, Exposure	Measurements, Examinations	Outcome	Effect Parameters (Bold Indicates Statistically Significance)
			• other symptoms: palpitations (36%), dyspnoea (27%), oedema (27%/18%), hair loss (27%/5%), hepatomegaly (27%) + splenomegaly (23%) (positive anamnesis for malaria), loss of appetite (27%), impotence (23%), weakness (23%), weight loss (23%), tonsillar hypotrophy (18%), no physical pathologies (14%), premature aging (14%), pruritus (5%), gingivitis (5%)	• other symptoms: palpitations (OR = 0.5 (0.1–1.6)), dyspnoea (OR = 1 (0.3–3.8)), loss of appetite (OR = 1 (0.3–3.8)), weakness (OR = 0.5 (0.1–1.9)), hair loss (OR = 2.4 (0.5–11.1)), pruritus (OR = 0.5 (0–5.7)), impotence (OR = 14.1 (0.7–273.4)), weight loss (OR = 2.9 (0.5–17.1)), oedema (OR = 7.9 (0.9–72.1))

Harari et al. (2012) [62]	cross-sectional study three mining sites, Ecuador year not reported gold miners (occupational (direct) Hg exposure): n = 200 (male, mean age 37 years, occupational exposure 0–36 years, mean occupational exposure 9 years, intermittent amalgam burning: 146 burned in last 200 days) gold merchants (occupational (direct) Hg exposure): n = 37 (male, mean age 31 years, occupational exposure 1–14 years, mean occupational exposure 9 years, regular amalgam burning: daily) controls (environmental (non-occupational) Hg exposure): n = 72 (male, mean age 38 years)	laboratory: blood sample: Hg spot urine sample: Hg, creatinine examination: questionnaire: socio-demographics, lifestyle, occupation, Hg exposure neurological examination: postural tremor, hand coordination, reaction time, postural stability	laboratory: significantly elevated levels in miners in comparison to controls: U–Hg + P–Hg (no difference in B–Hg) B–Hg: merchants >>> miners > controls
			• 11% miners + 71.4% merchants exceed BEI limit (>15 µg/L)
			• miners: 0.7–100 µg/L (mean 5.3 µg/L)
			P–Hg (no detailed numbers reported): merchants >>> miners > controls U–Hg: merchants >>> miners > controls (miners: depends very much on time since last burning amalgam)
			• 5.1% miners + 61.1% merchants exceed BEI limit (>35 µg/L)
			• miners: 0.3–170 µg/gCr (mean 3.3 µg/gCr)
				associations: significantly positive:
				• B–Hg:
				  – tremor (centre frequency left) **(r$_s$ = 0.1)**
				  – reaction time **(r$_s$ = 0.2)**
				• U–Hg:
				  – tremor (centre frequency) **(r$_s$ = 0.1, r$_s$ = 0.1)**
				  – reaction time **(r$_s$ = 0.2)**
				significantly negative:
				• B–Hg: postural sway (velocity) **(r$_s$ = −0.2)**
				• U–Hg: postural sway (velocity) **(r$_s$ = −0.2)**

Table 2. *Cont.*

Author, Year	Study Design, Setting, Time, Participants, Exposure	Measurements, Examinations	Outcome	Effect Parameters (Bold Indicates Statistically Significance)
Schutzmeier et al. (2016) [63]	cross-sectional study Ecuador • Portovelo • Zaruma August 2015 gold miners (occupational (direct) Hg exposure): n = 865 (age 18–65 years) • participants for a pharmaceutical study: n = 44 (male, age 19–59 years, mean age 38.6 years, work duration 1/2-40 years, mean work duration 11.2 years, U–Hg $\geq$ 15 µg/L)	laboratory: spot urine sample: Hg examination: clinical examination drug screening (alcohol, other drugs) diagnosis of intoxication: examination + laboratory parameters	U–Hg (laboratory): gold miners (n = 865): • <0.5–163 µg/L (median 1.8 µg/L) • 78.3% < HMB-I (7 µg/L), 15.8% 7–25 µg/L, 5.9% >HBM-II (25 µg/L), 3.4% >BAT (35 µg/L) participants (n = 44): 15–163 µg/L (median 35 µg/L) examination (n = 44): medical score sum (most common symptoms (n = 36)): • neuro-psychological symptoms (sleeping problems 91.7%, dysdiadochokinesia 88.9%, ataxia of gait 72.2%, tremor 13.9%) • problems in neuro-psychological tests (matchbox test 91.7%, heel-to-shin test 88.9%, pencil tapping test 75%) • other symptoms: discoloured gum 33.3%, salivation 33.3%, proteinuria 2.8% most frequent other symptoms (n = 24): physical + mental fatigue 62.5%, social problems (54.2%), irritability 54.2%	correlations: significantly positive (weak/moderate): U–Hg + medical score sum ($r_s = 0.4$)

3.2.1. Africa

Ghana

The first study by Afrifa et al. focused on mercury's impact on kidney function and discovered that Hg exposure among gold miners was significantly positively associated with elevated levels of urine protein and serum creatinine, while its association with eGFR was significantly negative [43]. None of the surveyed symptoms showed any statistically significant association with mercury exposure [43].

Concerning the thyroid function of gold miners, Afrifa and co-authors found a significantly negative association between mercury blood levels with T3 as well as T4 concentrations; in contrast, its association with TSH was non-significantly positive [51]. However, all measured thyroid values were still in a physiological range [51].

The connection of mercury with blood pressure was illustrated by the study from Rajaee et al. [44]. A significant association between mercury exposure and hypertension could not be demonstrated for miners versus residents [44].

Mensah et al. observed in their study that mercury exposure did not show significant associations with any medical symptoms for all miners, while the association with numbness was statistically significant for the sub-group with previous work experience in another mine [52].

Tanzania

The first study described a variety of symptoms to be more frequent in exposed than in controls, in particular neuro-psychological symptoms, increased salivation or discoloured gums [45]. However, the association between the surveyed symptoms and Hg exposure lacked significance for miners compared to residents [45].

The second study detected a variety of symptoms, such as neuro-psychological disorders (e.g., trembling or numbness), gingivitis or several respiratory symptoms in gold miners [53]. The combination of certain observed symptoms with elevated mercury levels in hair led to the diagnosis of mercury intoxication in approximately 10% of the examined miners [53].

Zimbabwe

Bose-O'Reilly et al. investigated the medical consequences of occupational mercury exposure in children in Zimbabwe and Indonesia [46]. Certain symptoms, such as neuro-psychological disorders (e.g., ataxia or dysdiadochokinesia), increased salivation or discoloured gums, were more frequent in exposed children [46]. In addition, mercury exposure was significantly associated with certain neuro-psychological disorders, for example, ataxia or dysdiadochokinesia [46].

The situation in Zimbabwe was also analysed regarding the disease burden presented in DALYs (Disability-Adjusted Life Years) [54]. The examinations also revealed that a number of symptoms were significantly more common in miners, for example, certain neuro-psychological symptoms [54]. Approximately 3% of the Zimbabwean population was occupationally exposed to Hg, while a total of 2% (which was equivalent to be around 72% of all miners) was considered to be intoxicated, which corresponded to 95,400 DALYs triggered by Hg exposure in the context of ASGM [54].

Sudan

Concerning gold miners in Sudan, one study determined variations in thyroid hormones and demonstrated that both TSH and TT4 were significantly elevated in miners in comparison to a control group, while FT3, FT4 and TT3 were significantly reduced [55]. These results were considered generally compatible to the laboratory parameters of hypothyroidism [55].

Burkina Faso

Highly exposed workers showed certain medical symptoms, such as neuro-psychological disorders (for example, headache or trembling), thoracic pain or cough [56]. A statistically significant positive association was observed between the mercury values in urine with problems to grab as well as with thoracic pain [56].

Uganda

Wanyana et al. described that observed symptoms showed a significant different distribution according to sex: males reported more frequently about headache, while females reported more frequently about psychiatric disorders or memory problems [57]. In addition, mercury exposure was statistically significantly associated with neuro-psychological disorders (such as headache, numbness, dizziness), certain kinds of pain and respiratory symptoms [57].

3.2.2. Asia

Indonesia

Following Bose-O'Reilly et al., the examination revealed that certain symptoms, such as certain neuro-psychological disorders, were significantly more frequent in exposed persons [47]. Moreover, cases of mercury intoxication were significantly more frequent in exposed participants [47].

Ekawanti and Krisnayanti focused on changes in haematological and renal parameters and highlighted that miners' haemoglobin and haematocrit was significantly reduced in comparison to a control group, which led to a higher frequency of anaemia [58]. In addition, urine protein was significantly elevated as well, leading to a frequent proteinuria among miners [58].

Pakistan

Khan et al. demonstrated that miners complained about various symptoms, such as neuro-psychological disorders, kidney diseases or different kinds of pain [59].

The second study by Riaz et al. also observed that gold miners showed a wide variety of symptoms, for example, gastrointestinal disorders, kidney diseases or respiratory symptoms [60].

3.2.3. South America

Brazil

Lacerda et al. examined the visual performance of gold miners and observed a reduced perimetric area in gold miners. In addition, the colour vision was also significantly worse compared to controls [61]. However, none of the calculated associations or correlations were statistically significant for miners [61].

The study by Branches et al. detected various symptoms in gold miners, for example, neuro-psychological disorders, certain kinds of pain or gingivitis [48]. Nevertheless, the associations between mercury exposure and diagnosed symptoms were not statistically significant [48].

Ecuador

The first study presented a significantly positive association between mercury in blood and urine with both tremor and reaction time, while their association with postural sway showed a significantly inverse association [62].

The second study highlighted that a fluctuating proportion of the clinically examined gold miners manifested particular symptoms, such as neuro-psychological disorders, discolouration of the gums or social problems [63]. A statistically significantly positive correlation between the urinary mercury content and the number of medical symptoms could be detected [63].

3.3. Risk of Bias Assessment (Qualitative Assessment)

The detailed results of the methodological quality assessment can be found in Table 3. The overall risk of bias rating is attached in the Supplemental Material (Table S2). Four studies were evaluated with an overall low risk of bias [43,45–47], while fifteen studies were assessed with an overall high risk of bias [44,48,51–63]. This rating as high-risk was based on a substantial potential for internal validity bias [44,48,52,53,55,56,58–62], internal validity confounder [48,52–56,58–60,62], detection bias [44,48,53,54,56–59,61,63] and/or attrition bias [44,48,51,53,55,56,58–61]. Other remarks that could not be assigned to the fixed categories but could lead to an increased risk of bias were identified for nine studies [43,48,53,54,57–61].

Table 3. Bias assessment. (XS: cross sectional study).

Study	Study Design	Internal Validity—Bias	Internal Validity—Confounder (Selection Bias)	Perform-ance Bias	Detection Bias	Attrition Bias	Reporting Bias	Others
Afrifa et al. (2017) [43]	XS	low risk	low risk	low risk	low risk	low risk	low risk	contradictory data (The text (page 6) reports a different odds ratio for serum creatinine than the corresponding table (page 7).)
Afrifa et al. (2018) [51]	XS	low risk	low risk	low risk	low risk	high risk	low risk	none
Rajaee et al. (2015) [44]	XS	high risk	low risk	low risk	high risk	high risk	low risk	none
Mensah et al. (2016) [52]	XS	high risk	high risk	low risk	low risk	low risk	low risk	none
Bose-O'Reilly et al. (2010a) [45]	XS	low risk	low risk	low risk	low risk	low risk	low risk	none
Harada et al. (1999) [53]	XS	high risk	high risk	low risk	high risk	high risk	low risk	(1) H–Hg as general marker for Hg exposure (2) MeHg values: n = 9
Bose-O'Reilly et al. (2008) [46]	XS	low risk	low risk	low risk	low risk	low risk	low risk	none
Steckling et al. (2014) [54]	XS	low risk	high risk	low risk	high risk	low risk	low risk	only the study in Zimbabwe could be taken into account
Tayrab (2017) [55]	XS	high risk	high risk	low risk	low risk	high risk	low risk	none
Tomicic et al. (2011) [56]	XS	high risk	high risk	low risk	high risk	high risk	low risk	none
Wanyana et al. (2020) [57]	XS	low risk	low risk	low risk	high risk	low risk	low risk	contradictory data (The text says that 75.8% of the miners use PPE (page 5), while the table says that 75.8% of the miners do not use PPE (Table 2).)
Bose-O'Reilly et al. (2010b) [47]	XS	low risk	low risk	low risk	low risk	low risk	low risk	none

Table 3. *Cont.*

Study	Study Design	Internal Validity—Bias	Internal Validity—Confounder (Selection Bias)	Perform-ance Bias	Detection Bias	Attrition Bias	Reporting Bias	Others
Ekawanti and Krisnayanti (2015) [58]	XS	high risk	high risk	low risk	high risk	high risk	low risk	no detailed information about child labour in the control group
Khan et al. (2012) [59]	XS	high risk	high risk	low risk	high risk	high risk	low risk	(1) contradictory data (The text (page 2) gives the concentrations of T–Hg in RBC and plasma for children, who work as miners, in another order than the table (Table 1).) (2) missing data (There is no number of participants reported (n=unknown).)
Riaz et al. (2016) [60]	XS	high risk	high risk	low risk	low risk	high risk	low risk	contradictory data (The text says that the values of T–Hg in hair are higher for female miners, although this is—according to the reported data—not the case.)
Lacerda et al. (2020) [61]	XS	high risk	low risk	low risk	high risk	high risk	low risk	(1) H–Hg as general marker for Hg exposure (2) contradictory data (The text (page 2) reports different ages for the gold miners and the riverines than Table 1.)
Branches et al. (1993) [48]	Case series	high risk	high risk	low risk	high risk	high risk	low risk	contradictory data (The number of urban inhabitants differs in the text (n = 21 on page 4, n = 22 on page 8).)
Harari et al. (2012) [62]	XS	high risk	high risk	low risk	low risk	low risk	low risk	none
Schutzmeier et al. (2016) [63]	XS	low risk	low risk	low risk	high risk	low risk	low risk	none

4. Discussion

4.1. Data Extraction (Quantitative Assessment)

Although a systematic literature research was conducted, the eligibility criteria only applied to 19 studies, which indicates that there is only limited literature available on the specific topic of mercury-related health effects in ASG miners.

In summary, mercury exposure in the ASGM sector was reported to cause an extraordinarily wide variety of symptoms in diverse organ systems [43–48,51–63]. Interestingly, many symptoms such as pain, hair loss or cough were largely unspecific [43–48,51–63] and not easy to relate to mercury exposure. However, a substantial part of the reported symptoms could be classified to mercury-related neuro-psychological disorders (e.g., tremor, ataxia, memory problems) [43,45–48,52–54,56,57,59,62,63], although the extent varied considerably among the studies. It also has to be considered that working in the ASGM sector in low- or middle-income countries is related to low safety standards and a high probability to suffer from work-related injuries [13]. Challenging working conditions can lead to a high psychological burden, which may cause unspecific somatic symptoms. In a variety of countries with ASGM, medical care availability is often restricted, in particular in terms of occupational health. Therefore, a medical undersupply of work- and non-work-related diseases has to be assumed. Nevertheless, it must be underlined that mercury-related symptoms mean a severe burden of disease in affected persons, in particular for gold miners, who show a high risk of occupational-related mercury exposure.

The organ specific toxicity of mercury exposure has been described in animal studies before [64–66]. Akgül and colleagues showed that exposure to mercury vapour caused histological renal damage in rats [64]. Renal damage caused by mercury exposure also has been described in humans, where significantly elevated levels of urine protein or serum creatinine could be observed [43,58]. However, two included studies in this review could not demonstrate significant changes in proteinuria [46,54]. Regarding neuro-psychological abnormalities, Altunkaynak and colleagues observed histological damages in rats' cerebellum after exposure to mercury vapour [65]. In addition, a further toxicological study in mice detected that post-natal exposure to mercury vapour affects the neuro-behavioral function, such as locomotive activity [66]. These observations support our findings that mercury exposure is in particular connected to neuro-psychological abnormalities [43,45–48,52–54,56,57,59,62,63].

Occupational mercury exposure is not limited to ASGM; it can also occur in other industrial sectors [67–69], such as in fluorescent lamp production companies [67]. Mercury-exposed labourers from the fluorescent lamp industry showed statistically significant more frequent psychological symptoms and a worse performance in neuro-psychological tests in comparison to non-exposed controls [67]. These findings confirmed our results, where mercury-exposed gold miners suffered significantly more often from neuro-psychological disorders [45–47,54] and showed somewhat significant worse results in neuro-psychological tests [45–47,54].

Another sector with an occupational exposure to mercury is the e-waste industry [68]. Decharat et al. examined e-waste workers and highlighted that neuro-psychological disorders, such as headache, were statistically significantly more frequent in exposed workers in comparison to office staff from the same company [68]. These results only partly agree with the findings of this review. For example, we could not observe statistically significant differences for headache between exposed and non-exposed groups [43,45,46]. Only one study could observe a statistically significant association between mercury exposure and headache [57]. However, headache must be considered as an unspecific symptom, which can be caused by a wide range of circumstances and diseases, such as migraine, meningitis or intra-cranial tumours. Consequently, this symptom could have also been caused by other circumstances or pre-existing conditions.

In addition, chlor-alkali workers are occupationally exposed to mercury [69]. These workers showed, according to Neghab et al., a statistically significant association between mercury exposure and neuro-psychological disorders (e.g., memory problems), while other associations with symptoms, such as tremor, showed no statistical significance [69]. However, only two studies in this review found significant differences regarding memory problems [45,47], but in contrast, we detected three studies with non-significant differences [45,46,54]. The reviewed studies used different methods to detect memory problems, from self-reported outcomes to objective neuro-psychological tests [45–47,54]. In particular, self-reported memory problems [45,46] are difficult to compare because of a missing generally applicable definition of memory problems, which makes them prone to a recall bias. Similar differing findings have been observed for tremor in miners or mercury-exposed participants [45–47,54], where the results differed both in subjective symptoms and clinical examinations [45–47,54]. In has to be considered that health consequences of mercury exposure are highly dependent of its chemical form, exposure pathway and dose. In particular the inhalation of inorganic mercury during the amalgamation process leads to a high absorption of and therefore high body burden of toxic metallic mercury. Although ASGM activities are mainly related to occupational inorganic mercury exposure, a burden with organic mercury compounds may be attributed by nutritional habits, such as fish or the ingestion of plant production products. These variables could at least partly explain the differing results.

4.2. General Methodology

Since all included studies were cross-sectional studies or case series, none of them was able to randomize their participants into groups [43–48,51–63]. In addition, a follow-up was not available [43–48,51–63]. This would have been helpful in order to understand the long-term effects of mercury and the clinical course in an exposed population.

Considering the study design, the availability of a control group was handled differently among the studies. The majority of the studies had a control group [44–48,53–56,58,59,61,62]; only six studies lacked a comparable group [43,51,52,57,60,63]. In addition, the composition of control groups was very heterogeneous among the studies. Some control groups consisted exclusively of indirectly exposed participants (e.g., residents from the same area) [48,56,58,59,62] or non-exposed controls (e.g., participants with no known mercury exposure) [54,55]. In contrast, six studies included more than one control group, respectively [44–47,53,61].

The included studies also differed in terms of data collection and diagnostic procedures [43–48,51–63]. The data were mainly collected through measurements of defined parameters [43–48,51–63], surveys [43–48,51–53,56–62] and/or clinical examinations [44–48,53–57,61–63]. Since no generally applicable definition of the signs and symptoms of a mercury intoxication existed, the diagnosis standards varied between studies, as well [45–47,53,54,63].

4.3. Specific Methodology

Six studies reported contradictory data in their manuscripts (see Table 3) [43,48,57,59–61]. In addition, one of these studies did not provide all the required information, such as the total number of included participants [59], which made it difficult to assess the results.

Another methodological aspect was the inhomogeneous inclusion of different job categories in the miners' study groups or in the control group. Exemplarily, three studies also included other occupations, such as gold traders, in the miners' exposure group [56,57,63]. Therefore, a comparability between the individual studies in general, and in particular their specific results, was limited. Moreover, the study conducted by Ekawanti and Krisnayanti also included child gold miners in the control group [58]. Unfortunately, this point was only mentioned in the Section 4, and no further information was made available in this regard [58].

4.4. Strengths and Limitations of This Review

The key strength of this review is the application of a systematic methodology according to the PRISMA scheme [36]. The underlying review protocol was drafted in advance and submitted to PROSPERO (PROSPERO-ID: CRD42021235289) to secure standard methodological claims and transparency. A further strength was the risk of bias assessment, which was created on the basis of relevant literature in order to classify the different findings of the included studies [49,50]. This review provided an overview of the extensive health-related problems caused by ongoing inorganic mercury exposure.

In contrast, the major limitation of this review was the failed attempt to carry out a meta-analysis due to the unsuitable data for this procedure. Another restriction was the limited publication period, ending on 31 December 2020. Therefore, articles that were published later and could have also fulfilled the eligibility criteria could not be considered.

5. Conclusions

This systematic review underlines the substantial adverse health effects in ASG miners in low- and middle-income countries. Research should therefore continue to focus on the situation of workers in the ASGM sector. More high-quality studies are urgently necessary, as most of the included studies only had a low methodological quality, resulting in a high risk of bias. These should include a defined control group, a clear definition of mercury-related diseases and a diagnostic standard to detect mercury intoxication. A comprehensive assessment of confounding factors of reported symptoms and diseases is necessary, as well, since a variety of symptoms observed in the included studies were quite unspecific. Due to the cross-sectional design, a causal relation was difficult to derive. Prospective studies that detect a clear causation between mercury exposure and mercury-related outcome are urgently required in order to increase the pressure for change. Mercury remains a significant health threat; this has been described in several toxicological animal studies before and should be underlined by epidemiological analyses in humans in the ASGM sector [64–66].

In the past, a first attempt was made to reduce the mercury burden with the Minamata Convention that initiates National Action Plans in the affected countries [6,35]. As the presented results indicate, these efforts were not as sufficient as intended. Mercury still causes serious health problems in ASG miners, and the topic demands intensified attention. Awareness must be raised in miners and their environment. Mercury-related symptoms can persist for a long time, even after termination of the corresponding exposure [32,34]. Hence, governments should take action to improve the working conditions for miners. Miners need to be convinced of replacing mercury-containing practices in gold extraction with non-hazardous techniques, as have already been attempted in model projects [70,71]. Finally, mercury-intoxicated miners must also be given the opportunity to receive sufficient medical care.

Supplementary Materials: The following are available online at https://www.mdpi.com/article/10.3390/ijerph19042081/s1, Table S1: Methodological Evaluation (Blank Protocol), Table S2: Overall Risk of Bias Rating.

Author Contributions: Elaboration of the research question: K.T. and A.K.; Study protocol—original draft: K.T.; Study protocol—review and editing: K.T. and A.K.; Literature research: K.T.; Screening of the results: K.T. and A.K.; Data extraction and bias assessment—original draft: K.T.; Data extraction and bias assessment—review and editing: K.T. and A.K.; Calculation of missing effect parameters: A.K.; Manuscript—original draft: K.T.; Manuscript—review and editing: K.T., A.K. and T.K. All authors have read and agreed to the published version of the manuscript.

Funding: This research received no external funding.

Institutional Review Board Statement: Not applicable.

Informed Consent Statement: Not applicable.

Data Availability Statement: Data can be requested by the authors of this review.

Conflicts of Interest: The authors declare no conflict of interest.

Abbreviations

ASG miner	artisanal and small-scale gold miner
ASGM	artisanal and small-scale gold mining
B–Hg	mercury concentration in blood
H–Hg	mercury concentration in hair
HBM	human biological monitoring
Hg	mercury
I–Hg	inorganic mercury concentration
MeHg	methylmercury
N–Hg	mercury concentration in nails
O–Hg	organic mercury concentration
P–Hg	mercury concentration in plasma
T–Hg	total mercury concentration
U–Hg	mercury concentration in urine
UBA	German Umweltbundesamt, German Environment Agency
µg/gCR	microgram per gram creatinine

References

1. Deutsche Bundesbank. Goldbestand der Deutschen Bundesbank. Available online: https://www.bundesbank.de/resource/blob/743058/c70ab6299bc0a4edae91b321e784e081/mL/goldbarrenliste-data.pdf (accessed on 22 February 2021).
2. World Bank Group. *Commodity Markets Outlook: Causes and Consequences of Metal Price Shocks, April 2021*; World Bank: Washington, DC, USA, 2021; License: Creative Commons Attribution CC BY 3.0 IGO.
3. LBMA. LBMA Precious Metal Prices. Available online: https://www.lbma.org.uk/prices-and-data/precious-metal-prices (accessed on 30 September 2021).
4. U.S. Geological Survey. *Mineral Commodity Summaries 2021: U.S. Geological Survey*; US Geological Survey: Reston, VA, USA, 2021; p. 200. [CrossRef]
5. Seccatore, J.; Veiga, M.; Origliasso, C.; Marin, T.; De Tomi, G. An Estimation of the Artisanal Small-Scale Production of Gold in the World. *Sci. Total Environ.* **2014**, *496*, 662–667. [CrossRef] [PubMed]
6. United Nations. Minamata Convention on Mercury. Available online: https://treaties.un.org/doc/Treaties/2013/10/20131010%2011-16%20AM/CTC-XXVII-17.pdf (accessed on 23 September 2021).
7. Ghana Statistical Service (GSS). Rebased 2013-2019 Annual Gross Domestic Product. Available online: https://statsghana.gov.gh/gssmain/storage/img/marqueeupdater/Annual_2013_2019_GDP.pdf (accessed on 22 September 2021).
8. The Ghana Chamber of Mines. 2020 Mining Industry Statistics and Data. Available online: http://ghanachamberofmines.org/wp-content/uploads/2021/09/2020-Mining-Industry-Statistics-and-Data.pdf (accessed on 22 September 2021).
9. Alvarez-Berríos, N.L.; Aide, T.M. Global Demand for Gold Is Another Threat for Tropical Forests. *Environ. Res. Lett.* **2015**, *10*, 014006. [CrossRef]
10. Marshall, B.G.; Veiga, M.M.; da Silva, H.A.M.; Guimarães, J.R.D. Cyanide Contamination of the Puyango-Tumbes River Caused by Artisanal Gold Mining in Portovelo-Zaruma, Ecuador. *Curr. Envir. Health Rpt.* **2020**, *7*, 303–310. [CrossRef] [PubMed]
11. Appleton, J.D.; Weeks, J.M.; Calvez, J.P.S.; Beinhoff, C. Impacts of Mercury Contaminated Mining Waste on Soil Quality, Crops, Bivalves, and Fish in the Naboc River Area, Mindanao, Philippines. *Sci. Total Environ.* **2006**, *354*, 198–211. [CrossRef]
12. Marshall, B.G.; Veiga, M.M.; Kaplan, R.J.; Adler Miserendino, R.; Schudel, G.; Bergquist, B.A.; Guimarães, J.R.D.; Sobral, L.G.S.; Gonzalez-Mueller, C. Evidence of Transboundary Mercury and Other Pollutants in the Puyango-Tumbes River Basin, Ecuador–Peru. *Environ. Sci. Processes Impacts* **2018**, *20*, 632–641. [CrossRef]
13. Nakua, E.K.; Owusu-Dabo, E.; Newton, S.; Koranteng, A.; Otupiri, E.; Donkor, P.; Mock, C. Injury Rate and Risk Factors among Small-Scale Gold Miners in Ghana. *BMC Public Health* **2019**, *19*, 1368. [CrossRef]
14. Steckling, N.; Tobollik, M.; Plass, D.; Hornberg, C.; Ericson, B.; Fuller, R.; Bose-O'Reilly, S. Global Burden of Disease of Mercury Used in Artisanal Small-Scale Gold Mining. *Ann. Glob. Health* **2017**, *83*, 234–247. [CrossRef]
15. Stoffersen, B.; Køster-Rasmussen, R.; Cardeño, J.I.C.; Appel, P.W.U.; Smidth, M.; Na-Oy, L.D.; Lardizabal, D.L.; Onos, R.W. Comparison of Gold Yield with Traditional Amalgamation and Direct Smelting in Artisanal Small-Scale Gold Mining in Uganda. *J. Health Pollut.* **2019**, *9*. [CrossRef]
16. Hylander, L.D.; Plath, D.; Miranda, C.R.; Lücke, S.; Öhlander, J.; Rivera, A.T.F. Comparison of Different Gold Recovery Methods with Regard to Pollution Control and Efficiency. *Clean Soil Air Water* **2007**, *35*, 52–61. [CrossRef]
17. European Union. Regulation (EC) No 1272/2008 of the European Parliament and of the Council of 16 December 2008 on Classification, Labelling and Packaging of Substances and Mixtures, Amending and Repealing Directives 67/548/EEC and 1999/45/EC, and Amending Regulation (EC) No 1907/2006. Available online: https://eur-lex.europa.eu/legal-content/EN/TXT/PDF/?uri=CELEX:32008R1272&from=en (accessed on 23 September 2021).

18. Bagnato, E.; Allard, P.; Parello, F.; Aiuppa, A.; Calabrese, S.; Hammouya, G. Mercury Gas Emissions from La Soufrière Volcano, Guadeloupe Island (Lesser Antilles). *Chem. Geol.* **2009**, *266*, 267–273. [CrossRef]
19. Cui, Z.; Li, Z.; Zhang, Y.; Wang, X.; Li, Q.; Zhang, L.; Feng, X.; Li, X.; Shang, L.; Yao, Z. Atmospheric Mercury Emissions from Residential Coal Combustion in Guizhou Province, Southwest China. *Energy Fuels* **2019**, *33*, 1937–1943. [CrossRef]
20. AMAP/UN Environment. Technical Background Report for the Global Mercury Assessment 2018. In *Arctic Monitoring and Assessment Programme, Oslo, Norway/UN Environment Programme*; Viii + 426 Pp Including E-Annexes; Chemicals and Health Branch: Geneva, Switzerland, 2019.
21. Kudsk, F.N. Absorption of Mercury Vapour from the Respiratory Tract in Man. *Acta Pharmacol. Et Toxicol.* **1965**, *23*, 250–262. [CrossRef] [PubMed]
22. Hursh, J.B.; Clarkson, T.W.; Miles, E.F.; Goldsmith, L.A. Percutaneous Absorption of Mercury Vapor by Man. *Arch. Environ. Health Int. J.* **1989**, *44*, 120–127. [CrossRef] [PubMed]
23. Geijersstam, E.; Sandborgh-Englund, G.; Jonsson, F.; Ekstrand, J. Mercury Uptake and Kinetics after Ingestion of Dental Amalgam. *J. Dent. Res.* **2001**, *80*, 1793–1796. [CrossRef]
24. Sandborgh-Englund, G.; Elinder, C.-G.; Johanson, G.; Lind, B.; Skare, I.; Ekstrand, J. The Absorption, Blood Levels, and Excretion of Mercury after a Single Dose of Mercury Vapor in Humans. *Toxicol. Appl. Pharmacol.* **1998**, *150*, 146–153. [CrossRef]
25. Jonsson, F.; Sandborgh-Englund, G.; Johanson, G. A Compartmental Model for the Kinetics of Mercury Vapor in Humans. *Toxicol. Appl. Pharmacol.* **1999**, *155*, 161–168. [CrossRef]
26. Falnoga, I.; Tušek-Žnidarič, M.; Horvat, M.; Stegnar, P. Mercury, Selenium, and Cadmium in Human Autopsy Samples from Idrija Residents and Mercury Mine Workers. *Environ. Res.* **2000**, *84*, 211–218. [CrossRef]
27. Matsuo, N.; Suzuki, T.; Akagi, H. Mercury Concentration in Organs of Contemporary Japanese. *Arch. Environ. Health Int. J.* **1989**, *44*, 298–303. [CrossRef]
28. Institut für Wasser-, Boden- und Lufthygiene des Umweltbundesamtes. Stoffmonographie Quecksilber – Referenz- Und Human-Biomonitoring-(HBM)-Werte. *Bundesgesundheitsblatt Gesundh. Gesundh.* **1999**, *42*, 522–532. [CrossRef]
29. Cortes, J.; Peralta, J.; Díaz-Navarro, R. Acute Respiratory Syndrome Following Accidental Inhalation of Mercury Vapor. *Clin. Case Rep.* **2018**, *6*, 1535–1537. [CrossRef]
30. Cicek-Senturk, G.; Altay, F.A.; Ulu-Kilic, A.; Gurbuz, Y.; Tutuncu, E.; Sencan, I. Acute Mercury Poisoning Presenting as Fever of Unknown Origin in an Adult Woman: A Case Report. *J. Med. Case Rep.* **2014**, *8*, 266. [CrossRef] [PubMed]
31. Rowens, B.; Guerrero-Betancourt, D.; Gottlieb, C.A.; Boyes, R.J.; Eichenhorn, M.S. Respiratory Failure and Death Following Acute Inhalation of Mercury Vapor. *Chest* **1991**, *99*, 185–190. [CrossRef] [PubMed]
32. Kishi, R.; Doi, R.; Fukuchi, Y.; Satoh, H.; Satoh, T.; Ono, A.; Moriwaka, F.; Tashiro, K.; Takahata, N. Subjective Symptoms and Neurobehavioral Performances of Ex-Mercury Miners at an Average of 18 Years after the Cessation of Chronic Exposure to Mercury Vapor. *Environ. Res.* **1993**, *62*, 289–302. [CrossRef] [PubMed]
33. Li, P.; Feng, X.; Qiu, G.; Li, Z.; Fu, X.; Sakamoto, M.; Liu, X.; Wang, D. Mercury Exposures and Symptoms in Smelting Workers of Artisanal Mercury Mines in Wuchuan, Guizhou, China. *Environ. Res.* **2008**, *107*, 108–114. [CrossRef] [PubMed]
34. Zachi, E.C.; Taub, A.; Faria, M.D.A.M.; Ventura, D.F. Neuropsychological Alterations in Mercury Intoxication Persist Several Years after Exposure. *Dement. Neuropsychol.* **2008**, *2*, 91–95. [CrossRef]
35. Minamata Convention on Mercury. Parties and Signatories. Available online: https://www.mercuryconvention.org/en/parties (accessed on 23 September 2021).
36. Page, M.J.; McKenzie, J.E.; Bossuyt, P.M.; Boutron, I.; Hoffmann, T.C.; Mulrow, C.D.; Shamseer, L.; Tetzlaff, J.M.; Akl, E.A.; Brennan, S.E.; et al. The PRISMA 2020 Statement: An Updated Guideline for Reporting Systematic Reviews. *BMJ* **2021**, *88*, 105906. [CrossRef]
37. Morgan, R.L.; Whaley, P.; Thayer, K.A.; Schünemann, H.J. Identifying the PECO: A Framework for Formulating Good Questions to Explore the Association of Environmental and Other Exposures with Health Outcomes. *Environ. Int.* **2018**, *121*, 1027–1031. [CrossRef]
38. Afrifa, J.; Opoku, Y.K.; Gyamerah, E.O.; Ashiagbor, G.; Sorkpor, R.D. The Clinical Importance of the Mercury Problem in Artisanal Small-Scale Gold Mining. *Front. Public Health* **2019**, *7*, 131. [CrossRef]
39. Gibb, H.; O'Leary, K.G. Mercury Exposure and Health Impacts among Individuals in the Artisanal and Small-Scale Gold Mining Community: A Comprehensive Review. *Environ. Health Perspect.* **2014**, *122*, 667–672. [CrossRef]
40. Berzas Nevado, J.J.; Rodríguez Martín-Doimeadios, R.C.; Guzmán Bernardo, F.J.; Jiménez Moreno, M.; Herculano, A.M.; do Nascimento, J.L.M.; Crespo-López, M.E. Mercury in the Tapajós River Basin, Brazilian Amazon: A Review. *Environ. Int.* **2010**, *36*, 593–608. [CrossRef]
41. Vianna, A.D.S.; Matos, E.P.D.; Jesus, I.M.D.; Asmus, C.I.R.F.; Câmara, V.D.M. Human Exposure to Mercury and Its Hematological Effects: A Systematic Review. *Cad. Saude Publica* **2019**, *35*, e00091618. [CrossRef] [PubMed]
42. Chirico, F.; Scoditti, E.; Viora, C.; Magnavita, N. How Occupational Mercury Neurotoxicity Is Affected by Genetic Factors. A Systematic Review. *Appl. Sci.* **2020**, *10*, 7706. [CrossRef]
43. Afrifa, J.; Essien-Baidoo, S.; Ephraim, R.K.D.; Nkrumah, D.; Dankyira, D.O. Reduced Egfr, Elevated Urine Protein and Low Level of Personal Protective Equipment Compliance among Artisanal Small Scale Gold Miners at Bibiani-Ghana: A Cross-Sectional Study. *BMC Public Health* **2017**, *17*, 601. [CrossRef] [PubMed]

44. Rajaee, M.; Sánchez, B.; Renne, E.; Basu, N. An Investigation of Organic and Inorganic Mercury Exposure and Blood Pressure in a Small-Scale Gold Mining Community in Ghana. *Int. J. Environ. Res. Public Health* **2015**, *12*, 10020–10038. [CrossRef]
45. Bose-O'Reilly, S.; Drasch, G.; Beinhoff, C.; Tesha, A.; Drasch, K.; Roider, G.; Taylor, H.; Appleton, D.; Siebert, U. Health Assessment of Artisanal Gold Miners in Tanzania. *Sci. Total Environ.* **2010**, *408*, 796–805. [CrossRef]
46. Bose-O'Reilly, S.; Lettmeier, B.; Gothe, R.M.; Beinhoff, C.; Siebert, U.; Drasch, G. Mercury as a Serious Health Hazard for Children in Gold Mining Areas. *Environ. Res.* **2008**, *107*, 89–97. [CrossRef]
47. Bose-O'Reilly, S.; Drasch, G.; Beinhoff, C.; Rodrigues-Filho, S.; Roider, G.; Lettmeier, B.; Maydl, A.; Maydl, S.; Siebert, U. Health Assessment of Artisanal Gold Miners in Indonesia. *Sci. Total Environ.* **2010**, *408*, 713–725. [CrossRef]
48. Branches, F.J.; Erickson, T.B.; Aks, S.E.; Hryhorczuk, D.O. The Price of Gold: Mercury Exposure in the Amazonian Rain Forest. *J. Toxicol. Clin. Toxicol.* **1993**, *31*, 295–306. [CrossRef]
49. Downs, S.H.; Black, N. The Feasibility of Creating a Checklist for the Assessment of the Methodological Quality Both of Randomised and Non-Randomised Studies of Health Care Interventions. *J. Epidemiol. Community Health* **1998**, *52*, 377–384. [CrossRef]
50. Higgins, J.; Altman, D.; Sterne, J. Chapter 8: Assessing Risk of Bias in Included Studies. In *Cochrane Handbook for Systematic Reviews of Interventions Version 5.1.0 (Updated March 2011)*; Higgins, J.P.T., Green, S., Eds.; The Cochrane Collaboration: London, UK, 2011; Available online: www.handbook.cochrane.org (accessed on 23 September 2021).
51. Afrifa, J.; Ogbordjor, W.D.; Duku-Takyi, R. Variation in Thyroid Hormone Levels Is Associated with Elevated Blood Mercury Levels among Artisanal Small-Scale Miners in Ghana. *PLoS ONE* **2018**, *13*, e0203335. [CrossRef]
52. Mensah, E.K.; Afari, E.; Wurapa, F.; Sackey, S.; Quainoo, A.; Kenu, E.; Nyarko, K.M. Exposure of Small-Scale Gold Miners in Prestea to Mercury, Ghana, 2012. *Pan Afr. Med. J.* **2016**, *25*, 6. [CrossRef] [PubMed]
53. Harada, M.; Nakachi, S.; Cheu, T.; Hamada, H.; Ono, Y.; Tsuda, T.; Yanagida, K.; Kizaki, T.; Ohno, H. Monitoring of Mercury Pollution in Tanzania: Relation between Head Hair Mercury and Health. *Sci. Total Environ.* **1999**, *227*, 249–256. [CrossRef]
54. Steckling, N.; Bose-O'Reilly, S.; Pinheiro, P.; Plass, D.; Shoko, D.; Drasch, G.; Bernaudat, L.; Siebert, U.; Hornberg, C. The Burden of Chronic Mercury Intoxication in Artisanal Small-Scale Gold Mining in Zimbabwe: Data Availability and Preliminary Estimates. *Environ. Health* **2014**, *13*, 111. [CrossRef] [PubMed]
55. Tayrab, E. Thyroid Function in Sudanese Gold Miners with Chronic Mercury Exposure. *Eur. J. Pharm. Med. Res.* **2017**, *4*, 177–180.
56. Tomicic, C.; Vernez, D.; Belem, T.; Berode, M. Human Mercury Exposure Associated with Small-Scale Gold Mining in Burkina Faso. *Int. Arch. Occup. Environ. Health* **2011**, *84*, 539–546. [CrossRef] [PubMed]
57. Wanyana, M.W.; Agaba, F.E.; Sekimpi, D.K.; Mukasa, V.N.; Kamese, G.N.; Douglas, N.; Ssempebwa, J.C. Mercury Exposure Among Artisanal and Small-Scale Gold Miners in Four Regions in Uganda. *J. Health Pollut.* **2020**, *10*, 200613. [CrossRef]
58. Ekawanti, A.; Krisnayanti, B.D. Effect of Mercury Exposure on Renal Function and Hematological Parameters among Artisanal and Small-Scale Gold Miners at Sekotong, West Lombok, Indonesia. *J. Health Pollut.* **2015**, *5*, 25–32. [CrossRef]
59. Khan, S.; Shah, M.T.; Din, I.U.; Rehman, S. Mercury Exposure of Workers and Health Problems Related with Small-Scale Gold Panning and Extraction. *J. Chem. Soc. Pak.* **2012**, *34*, 870–876.
60. Riaz, A.; Khan, S.; Shah, M.T.; Li, G.; Gul, N.; Shamshad, I. Mercury Contamination in the Blood, Urine, Hair and Nails of the Gold Washers and Its Human Health Risk during Extraction of Placer Gold along Gilgit, Hunza and Indus Rivers in Gilgit-Baltistan, Pakistan. *Environ. Technol. Innov.* **2016**, *5*, 22–29. [CrossRef]
61. Lacerda, E.M.D.C.B.; Souza, G.D.S.; Cortes, M.I.T.; Rodrigues, A.R.; Pinheiro, M.C.N.; Silveira, L.C.D.L.; Ventura, D.F. Comparison of Visual Functions of Two Amazonian Populations: Possible Consequences of Different Mercury Exposure. *Front. Neurosci.* **2020**, *13*, 1428. [CrossRef]
62. Harari, R.; Harari, F.; Gerhardsson, L.; Lundh, T.; Skerfving, S.; Strömberg, U.; Broberg, K. Exposure and Toxic Effects of Elemental Mercury in Gold-Mining Activities in Ecuador. *Toxicol. Lett.* **2012**, *213*, 75–82. [CrossRef] [PubMed]
63. Schutzmeier, P.; Berger, U.; Bose-O'Reilly, S. Gold Mining in Ecuador: A Cross-Sectional Assessment of Mercury in Urine and Medical Symptoms in Miners from Portovelo/Zaruma. *Int. J. Environ. Res. Public Health* **2016**, *14*, 34. [CrossRef] [PubMed]
64. Akgül, N.; Altunkaynak, B.Z.; Altunkaynak, M.E.; Deniz, Ö.G.; Ünal, D.; Akgül, H.M. Inhalation of Mercury Vapor Can Cause the Toxic Effects on Rat Kidney. *Ren. Fail.* **2016**, *38*, 465–473. [CrossRef] [PubMed]
65. Altunkaynak, B.Z.; Akgül, N.; Yahyazedeh, A.; Makaracı, E.; Akgül, H.M. A Stereological Study of the Effects of Mercury Inhalation on the Cerebellum. *Biotech. Histochem.* **2019**, *94*, 42–47. [CrossRef]
66. Yoshida, M.; Lee, J.-Y.; Satoh, M.; Watanabe, C. Neurobehavioral Effects of Postnatal Exposure to Low-Level Mercury Vapor and/or Methylmercury in Mice. *J. Toxicol. Sci.* **2018**, *43*, 11–17. [CrossRef]
67. Al-Batanony, M.A.; Abdel-Rasul, G.M.; Abu-Salem, M.A.; Al-Dalatony, M.M.; Allam, H.K. Occupational Exposure to Mercury among Workers in a Fluorescent Lamp Factory, Quisna Industrial Zone, Egypt. *Int. J. Occup. Environ. Med.* **2013**, *4*, 149–156.
68. Decharat, S. Urinary Mercury Levels Among Workers in E-Waste Shops in Nakhon Si Thammarat Province, Thailand. *J. Prev. Med. Public Health* **2018**, *51*, 196–204. [CrossRef]
69. Neghab, M.; Amin Norouzi, M.; Choobineh, A.; Reza Kardaniyan, M.; Hassan Zadeh, J. Health Effects Associated With Long-Term Occupational Exposure of Employees of a Chlor-Alkali Plant to Mercury. *Int. J. Occup. Saf. Ergon.* **2012**, *18*, 97–106. [CrossRef]

70. Appel, P.W.U.; Andersen, A.; Na-Oy, L.D.; Onos, R. Introduction of Mercury-Free Gold Extraction Methods to Medium-Scale Miners and Education of Health Care Providers to Reduce the Use of Mercury in Sorata, Bolivia. *J. Health Pollut.* **2015**, *5*, 12–17. [CrossRef]
71. Køster-Rasmussen, R.; Westergaard, M.L.; Brasholt, M.; Gutierrez, R.; Jørs, E.; Thomsen, J.F. Mercury Pollution from Small-Scale Gold Mining Can Be Stopped by Implementing the Gravity-Borax Method–A Two-Year Follow-Up Study from Two Mining Communities in the Philippines. *New Solut.* **2016**, *25*, 567–587. [CrossRef]

MDPI
St. Alban-Anlage 66
4052 Basel
Switzerland
Tel. +41 61 683 77 34
Fax +41 61 302 89 18
www.mdpi.com

International Journal of Environmental Research and Public Health Editorial Office
E-mail: ijerph@mdpi.com
www.mdpi.com/journal/ijerph